CHILTON'S
TIMING BELT
SERVICE MANUAL
1980-00

Covers Cars, Trucks, Vans and SUVs

C.E.O.
Rick Van Dalen

President
Dean F. Morgantini, S.A.E.

Vice President—Finance
Barry L. Beck

Vice President—Sales
Glenn D. Potere

Executive Editor
Kevin M. G. Maher, A.S.E.

Manager—Consumer Automotive
Richard Schwartz, A.S.E.

Manager—Professional Automotive
Richard J. Rivele

Manager—Marine/Recreation
James R. Marotta, A.S.E.

Electronic Fulfillment Manager
Will Kessler, A.S.E., S.A.E.

Production Specialists
Brian Hollingsworth, Melinda Possinger

Project Managers
Thomas A. Mellon, A.S.E., S.A.E., Christine Sheeky, S.A.E.,
Todd W. Stidham, A.S.E., Ron Webb

Schematics Editor
Christopher G. Ritchie

Editors
Tim Crain, A.S.E., Robert E. Doughten, Scott A. Freeman, Robert McAnally, John McCabe,
Eric Michael Mihalyi, A.S.E., S.A.E., S.T.S., Norman D. Norville, A.S.E., S.A.E., S.T.S.,
Richard T. Smith, Joseph D'Orazio, A.S.E., Paul D'Santo, A.S.E.

CHILTON AUTOMOTIVE INFORMATION
PUBLISHED BY W. G. NICHOLS, INC.

Manufactured in USA, © 1999 W. G. Nichols, 1020 Andrew Drive, West Chester, PA 19380
ISBN 0-8019-9305-9
Library of Congress Catalog Card No. 99-76173

44.95

Table of Contents

Table of Contents

HOW TO USE THIS MANUAL

Locating Information

This book is divided into three sections. Section One covers domestic cars; Section Two covers imported cars; Section Three covers trucks, vans and SUVs.

The Table of Contents, located at the front of the book, lists manufacturer and engine in the order in which they appear.

To find a particular timing belt procedure, you need only look in the Table of Contents. Once you have found the proper listing, simply turn to the indicated page to find the applicable information.

Safety Notice

Proper service and repair procedures are vital to the safe, reliable operation of all motor vehicles, as well as the personal safety of those performing the repairs. This manual outlines procedures for servicing and repairing vehicles using safe effective methods. The procedures contain many NOTES, WARNINGS and CAUTIONS which should be followed along with standard safety procedures to eliminate the possibility of personal injury or improper service which could damage the vehicle or compromise its safety.

It is important to note that repair procedures and techniques, tools and parts for servicing vehicles, as well as the skill and experience of the individual performing the work vary widely. It is not possible to anticipate all of the conceivable ways or conditions under which vehicles may be serviced, or to provide cautions as to all of the possible hazards that may result. Standard and accepted safety precautions and equipment should be used when handling toxic or flammable fluids, and safety goggles or other protection should be used during cutting, grinding, chiseling, prying, or any other process that can cause material removal or projectiles.

Some procedures require the use of tools specially designed for a specific purpose. Before substituting another tool or procedure, you must be completely satisfied that neither your personal safety, nor the performance of the vehicle will be endangered.

Although information in this manual is based on industry sources and is as complete as possible at the time of publication, the possibility exists that some vehicle manufacturers made later changes which could not be included here. Information on very late models may not be available in some circumstances. While striving for total accuracy, Nichols Publishing cannot assume responsibility for any errors, changes, or omissions that may occur in the compilation of this data.

Part Numbers

Part numbers listed in this book are not recommendations by Nichols Publishing for any product by brand name. They are references that can be used with interchanges manuals and aftermarket supplier catalogs to locate each brand supplier's discrete part number.

Special Tools

Special tools are recommended by the vehicle manufacturer to perform their specific job. Use has been kept to a minimum, but where absolutely necessary, they are referred to in the text by the part number of the tool manufacturer. These tools may be purchased, under the appropriate part number, from your local dealer or regional distributor, or an equivalent tool can be purchased locally from a tool supplier or parts outlet. Before substituting any tool for the one recommended, read the previous Safety Notice.

Acknowledgments

This publication contains material that is reproduced and distributed under a license from Ford Motor Company. No further reproduction or distribution of the Ford Motor Company material is allowed without the expressed written permission from Ford Motor Company.

Portions of the material contained herein have been reprinted with permission of General Motors Corporation, Service Technology Group.

Nichols Publishing would like to express thanks to all of the fine companies who participate in the production of our books. Hand tools supplied by Craftsman are used during all phases of our vehicle teardown and photography. Many of the fine specialty tools used in our procedures were provided courtesy of Lisle Corporation. Lincoln Automotive Products (1 Lincoln Way, St. Louis, MO 63120) has provided their industrial shop equipment, including jacks (engine, transmission and floor), engine stands, fluid and lubrication tools, as well as shop presses. Rotary Lifts, the largest automobile lift manufacturer in the world, offering the biggest variety of surface and in-ground lifts available (1-800-640-5438 or www.Rotary-Lift.com), have fulfilled our shop's lift needs. Much of our shop's electronic testing equipment was supplied by Universal Enterprises Inc. (UEI).

Copyright Notice

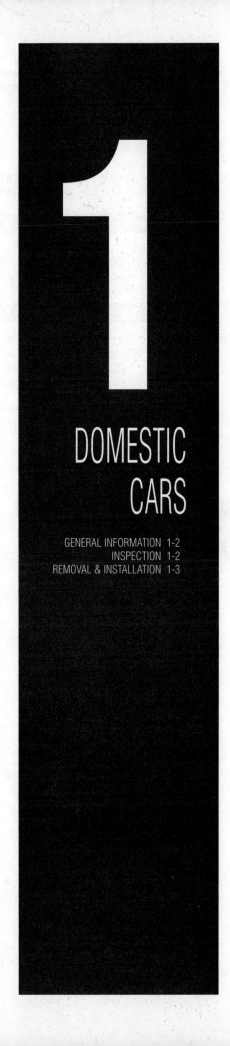

1

DOMESTIC CARS

GENERAL INFORMATION

Whenever a vehicle with an unknown service history comes into your repair facility or is recently purchased, here are some points that should be asked to help prevent costly engine damage:

- Does the owner know if or when the belt was replaced?
- If the vehicle purchased is used or the condition and mileage of the last timing belt replacement are unknown, it is recommended to inspect, replace or at least inform the owner that the vehicle is equipped with a timing belt.
- Note the mileage of the vehicle. The average replacement interval for a timing belt is approximately 60,000 miles (96,000 km).

INSPECTION

The average replacement interval for a timing belt is approximately 60,000 miles (96,000km). If, however, the timing belt is inspected earlier or more frequently than suggested, and shows signs of wear or defects, the belt should be replaced at that time.

※※ WARNING

Never allow antifreeze, oil or solvents to come into with a timing belt. If this occurs immediately wash the solution from the timing belt. Also, never excessively bend or twist the timing belt; this can damage the belt so that its lifetime is severely shortened.

Inspect both sides of the timing belt. Replace the belt with a new one if any of the following conditions exist:
- Hardening of the rubber—back side is glossy without resilience and leaves no indentation when pressed with a fingernail
- Cracks on the rubber backing
- Cracks or peeling of the canvas backing
- Cracks on rib root
- Cracks on belt sides
- Missing teeth or chunks of teeth
- Abnormal wear of belt sides—the sides are normal if they are sharp, as if cut by a knife.

If none of these conditions exist, the belt does not need replacement unless it is at the recommended interval. The belt MUST be replaced at the recommended interval.

※※ WARNING

On interference engines, it is very important to replace the timing belt at the recommended intervals, otherwise expensive engine damage will likely result if the belt fails.

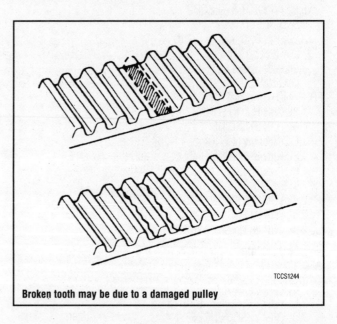

Broken tooth may be due to a damaged pulley

Never bend or twist a timing belt excessively, and do not allow solvents, antifreeze, gasoline, acid or oil to come into contact with the belt

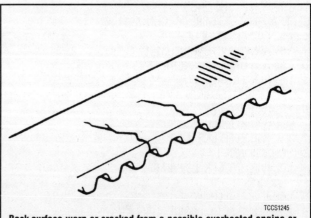

Back surface worn or cracked from a possible overheated engine or interference with the belt cover

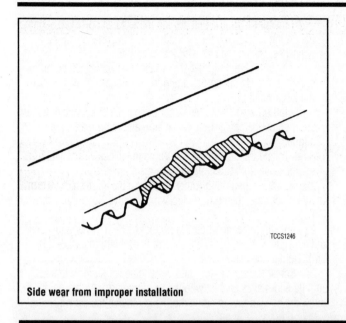

Side wear from improper installation

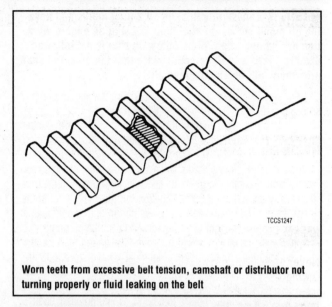

Worn teeth from excessive belt tension, camshaft or distributor not turning properly or fluid leaking on the belt

REMOVAL & INSTALLATION

Chrysler Corporation

1.6L (VIN Y) & 2.0L (VIN F, R AND U) ENGINES

1. Disconnect the negative battery cable.
2. Remove the timing belt upper and lower covers.
3. Rotate the crankshaft clockwise and align the timing marks so No. 1 piston will be at Top Dead Center (TDC) of the compression stroke. At this time the timing marks on the camshaft sprocket and the upper surface of the cylinder head should coincide, and the dowel pin of the camshaft sprocket should be at the upper side.

➡**Always rotate the crankshaft in a clockwise direction. Make a mark on the back of the timing belt indicating the direction of rotation so it may be reassembled in the same direction if it is to be reused.**

4. Remove the auto tensioner and remove the outermost timing belt.
5. Remove the timing belt tensioner pulley, tensioner arm, idler pulley.
6. Locate the access plug on the side of block. Remove the plug and install a Phillips screwdriver. Remove the oil pump sprocket nut, oil pump sprocket, special washer, flange and spacer.
7. Remove the silent shaft (inner) belt tensioner and remove the belt.
   To install:
8. Align the timing marks on the crankshaft sprocket and the silent shaft sprocket.
9. Fit the inner timing belt over the crankshaft and silent shaft sprocket. Ensure that there is no slack in the belt.
10. While holding the inner timing belt tensioner with your fingers, adjust the timing belt tension by applying a force towards the center of the belt, until the tension side of the belt is taut. Tighten the tensioner bolt.

➡**When tightening the bolt of the tensioner, ensure that the tensioner pulley shaft does not rotate with the bolt. Allowing it to rotate with the bolt can cause excessive tension on the belt.**

11. Check belt for proper tension by depressing the belt on long side with your finger and noting the belt deflection. The desired reading is 0.20–0.28 in. (5–7mm). If tension is not correct, readjust and check belt deflection.
12. Install the flange, crankshaft and washer to the crankshaft. The flange on the crankshaft sprocket must be installed towards the inner timing belt sprocket. Tighten bolt to 80–94 ft. lbs. (110–130 Nm).

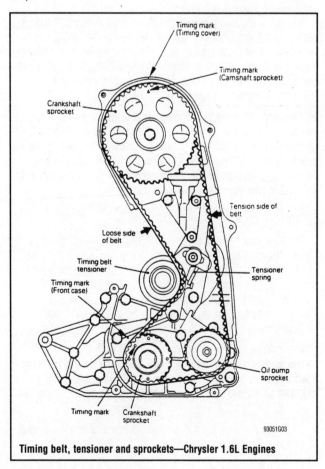

Timing belt, tensioner and sprockets—Chrysler 1.6L Engines

➡ **There is a possibility to align all timing marks and have the oil pump sprocket out of time, causing an engine vibration during operation. If the following step is not followed exactly, there is a 50 percent chance that the oil pump shaft alignment will be 180 degrees off.**

13. Before installing the timing belt, ensure that the oil pump sprocket shaft is in the correct position as follows:

 a. Remove the plug from the rear side of the block and insert a Phillips screwdriver with shaft diameter of 0.31 in. (8mm) into the hole.

 b. With the timing marks still aligned, the shaft of the screwdriver must be able to go in at least 2.36 in. (60mm). If the tool can only go in 0.79–0.98 in. (20–25mm), the shaft is not in the correct orientation and will cause a vibration during engine operation. Remove the tool from the hole and turn the oil pump sprocket 1 complete revolution. Realign the timing marks and insert the tool. The shaft of the tool must go in at least 2.36 in. (60mm).

 c. Recheck and realign the timing marks.

 d. Leave the tool in place to hold the oil pump silent shaft while continuing.

14. To install the oil pump sprocket and tighten the nut to 36–43 ft. lbs. (50–60 Nm).

15. Position the auto-tensioner into a vise with soft jaws. The plug at the rear of tensioner protrudes, be sure to use a washer as a spacer to protect the plug from contacting vise jaws.

16. Slowly push the rod into the tensioner until the set hole in rod is aligned with set hole in the auto-tensioner.

17. Insert a 0.055 in. (1.4mm) wire into the aligned set holes. Unclamp the tensioner from vise and install to vehicle. Tighten tensioner to 17 ft. lbs. (24 Nm).

18. When installing the timing belt, turn the 2 camshaft sprockets so their dowel pins are located on top. Align the timing marks facing each other with the top surface of the cylinder head. When you let go of the exhaust camshaft sprocket, it will rotate 1 tooth in the counterclockwise direction. This should be taken into account when installing the timing belts on the sprocket.

➡ **The same sprockets are used for the intake and exhaust camshafts and are provided with 2 timing marks. When the sprocket is mounted on the exhaust camshaft, use the timing mark on the right with the dowel pin hole on top. For the intake camshaft sprocket, the timing mark is on the left with the dowel pin hole on top.**

19. Align the crankshaft sprocket and oil pump sprocket timing marks.

20. After alignment of the oil pump sprocket timing marks, remove the plug on the cylinder block and insert a Phillips screwdriver with a shaft diameter of 0.31 in. (8mm) through the hole. If the shaft can be inserted 2.4 in. deep, the silent shaft is in the correct position. If the shaft of the tool can only be inserted 0.8–1.0 in. (20–25mm) deep, turn the oil pump sprocket 1 turn and realign the marks. Reinsert the tool making sure it is inserted 2.4 in. deep. Keep the tool inserted in hole for the remainder of this procedure.

➡ **The above step assures that the oil pump socket is in correct orientation to the silent shafts. This step must not be skipped or a vibration may develop during engine operation.**

21. Install the timing belt as follows:

 a. Install the timing belt around the intake camshaft sprocket and retain it with 2 spring clips or binder clips.

 b. Install the timing belt around the exhaust sprocket, aligning the timing marks with the cylinder head top surface using 2 wrenches. Retain the belt with 2 spring clips.

 c. Install the timing belt around the idler pulley, oil pump sprocket, crankshaft sprocket and the tensioner pulley. Remove the 2 spring clips.

 d. Lift upward on the tensioner pulley in a clockwise direction and tighten the center bolt. Make sure all timing marks are aligned.

 e. Rotate the crankshaft ¼ turn counterclockwise. Then, turn in clockwise until the timing marks are aligned again.

22. To adjust the timing (outer) belt, turn the crankshaft ¼ turn counterclockwise, then turn it clockwise to move No. 1 cylinder to TDC.

23. Loosen the center bolt. Using tool MD998738 or equivalent and a torque wrench, apply a torque of 22–24inch lbs. (2.6–2.8 Nm). Tighten the center bolt.

24. Screw the special tool into the engine left support bracket until its end makes contact with the tensioner arm. At this point, screw the special tool in some more and remove the set wire

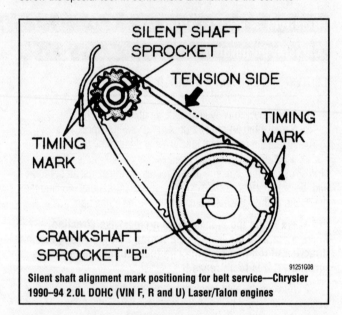

Silent shaft alignment mark positioning for belt service—Chrysler 1990–94 2.0L DOHC (VIN F, R and U) Laser/Talon engines

Proper camshaft timing mark alignment—Chrysler 1990 1.6L (VIN Y) and 1990–94 2.0L DOHC (VIN F, R and U) Laser/Talon engines

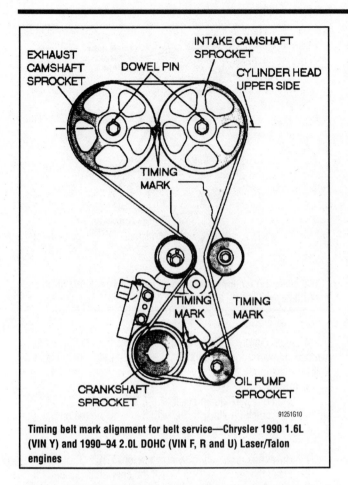

Timing belt mark alignment for belt service—Chrysler 1990 1.6L (VIN Y) and 1990–94 2.0L DOHC (VIN F, R and U) Laser/Talon engines

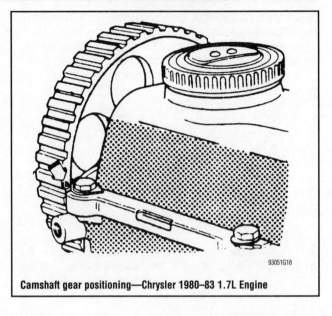

Camshaft gear positioning—Chrysler 1980–83 1.7L Engine

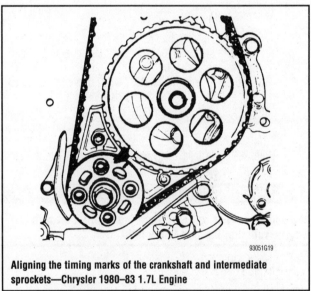

Aligning the timing marks of the crankshaft and intermediate sprockets—Chrysler 1980–83 1.7L Engine

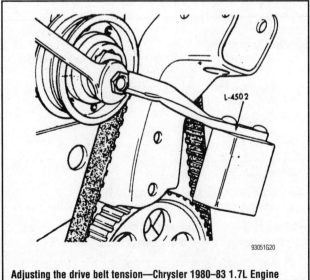

Adjusting the drive belt tension—Chrysler 1980–83 1.7L Engine

attached to the auto tensioner, if the wire was not previously removed. Then remove the special tool.

25. Rotate the crankshaft 2 complete turns clockwise and let it sit for approximately 15 minutes. Then, measure the auto tensioner protrusion (the distance between the tensioner arm and auto tensioner body) to ensure that it is within 0.15–0.18 in. (3.8–4.5mm). If out of specification, repeat belt adjustment procedure until the specified value is obtained.

26. If the timing belt tension adjustment is being performed with the engine mounted in the vehicle, and clearance between the tensioner arm and the auto tensioner body cannot be measured, the following alternative method can be used:

    a. Screw in special tool MD998738 or equivalent, until its end makes contact with the tensioner arm.

    b. After the special tool makes contact with the arm, screw it in some more to retract the auto tensioner pushrod while counting the number of turns the tool makes until the tensioner arm is brought into contact with the auto tensioner body. Make sure the number of turns the special tool makes conforms with the standard value of 2.5–3 turns.

    c. Install the rubber plug to the timing belt rear cover.

27. Install the timing belt covers and all related items.
28. Connect the negative battery cable.

1.7L ENGINE

1980–83

1. Disconnect the negative battery cable.
2. Remove the No. 1 spark plug.

3. Rotate the crankshaft to position the No. 1 piston on the Top Dead Center (TDC) of its compression stroke.

4. Remove the air conditioning compressor, the alternator and the power steering pump drive belts and set the equipment aside.

5. Raise and safely support the front of the vehicle and remove the fender splash shield.

6. Remove the idler pulley assembly, the crankshaft pulley and the upper/lower front cover.

7. Lower the vehicle and place a jack under the engine with a board to cushion the engine.

8. Remove the right side engine mounting bolt and raise the engine slightly.

9. If reusing the timing belt, mark the direction of rotation on the belt.

10. Loosen the timing belt tensioner and remove the timing belt.

To install:

11. Turn the crankshaft and the intermediate sprockets until both markings on the sprockets are aligned facing each other.

12. Turn the camshaft sprocket until the mark on the sprocket is aligned with the cylinder head cover.

13. Install the timing belt.

14. To adjust the timing belt tension, perform the following procedure:

a. Place a Belt Tension tool No. L-4502 or equivalent, horizontally on the large hex of the timing belt tensioner pulley and loosen the tensioner locknut.

b. Reset the Belt Tension tool No. L-4502 or equivalent, index, if necessary, to have the axis within 15 degrees of horizontal.

c. Rotate the crankshaft clockwise from TDC, 2 revolutions, to TDC.

d. Tighten the locknut to 32 ft. lbs. (44 Nm).

15. Complete the installation by reversing the removal procedures.

16. Connect the negative battery cable.

✳✳ WARNING

If a whirring noise is heard from the timing belt with the engine running, the belt is too tight.

2.0L (VIN C) SOHC ENGINE

1. On Cirrus, Stratus, Breeze and Sebring Convertible models, disconnect the negative battery cable from the left strut tower. The ground cable is equipped with an insulator grommet which should be placed on the stud to prevent the negative battery cable from accidentally grounding.

2. On Neon models, disconnect the negative battery cable.

3. Remove the drive belts and accessories.

4. Raise and safely support the vehicle.

5. Remove the right inner splash-shield.

6. Remove the crankshaft damper.

7. Remove the right engine mount

8. Place a support under the engine.

9. Remove the engine mount bracket

10. Remove the timing belt cover.

11. Loosen the timing belt tensioner bolts.

12. Remove the timing belt and the tensioner.

➡**When tensioner is removed from the engine it is necessary to compress the plunger into the tensioner body.**

13. Place the tensioner into a soft-jawed vise to compress the tensioner.

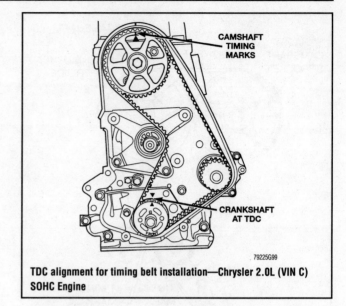

TDC alignment for timing belt installation—Chrysler 2.0L (VIN C) SOHC Engine

14. After compressing the tensioner place a pin (⁵⁄₆4 in. Allen wrench will work) into the plunger side hole to retain the plunger until installation.

To install:

15. Set the crankshaft sprocket to Top Dead Center (TDC) by aligning the notch on the sprocket with the arrow on the oil pump housing, then back off the sprocket 3 notches before TDC.

16. Set the camshaft to align the timing marks.

17. Move the crankshaft to ½ notch before TDC.

18. Install the timing belt starting at the crankshaft, around the water pump, then around the camshaft last.

19. Move the crankshaft to TDC to take up the belt slack.

20. Reinstall the tensioner to the block but do not tighten it.

21. Using a torque wrench apply 21 ft. lbs. (28 Nm) of torque to the tensioner pulley.

22. With torque being applied to the tensioner pulley, move the tensioner up against the tensioner bracket and tighten the fasteners to 23 ft. lbs. (31 Nm).

23. Remove the tensioner plunger pin, the tension is correct when the plunger pin can be removed and replaced easily.

24. Rotate the crankshaft 2 revolutions and recheck the timing marks.

25. Reinstall the timing belt cover.

26. Reinstall the engine mount bracket.

27. Reinstall the right engine mount.

28. Remove the engine support.

29. Reinstall the crankshaft damper and tighten to 105 ft. lbs. (142 Nm).

30. Reinstall the drive belts and accessories.

31. Reinstall the right inner splash-shield.

32. Perform the crankshaft and camshaft relearn alignment procedure using the DRB scan tool or equivalent.

2.0L (VIN F) ENGINE

1. Disconnect the negative battery cable.

2. Remove the engine undercover.

3. Remove the engine mount bracket.

4. Remove the drive belts.

5. Remove the belt tensioner pulley.

6. Remove the water pump pulleys.

7. Remove the crankshaft pulley.

8. Remove the stud bolt from the engine support bracket and remove the timing belt covers.

9. Rotate the crankshaft clockwise to align the camshaft timing marks. Always turn the crankshaft in the forward direction only.

10. Loosen the tension pulley center bolt.

➡ **If the timing belt is to be reused, mark the direction of rotation on the flat side of the belt with an arrow.**

11. Move the tension pulley towards the water pump and remove the timing belt.

12. Remove the crankshaft sprocket center bolt using special tool MB990767 to hold the crankshaft sprocket while removing the

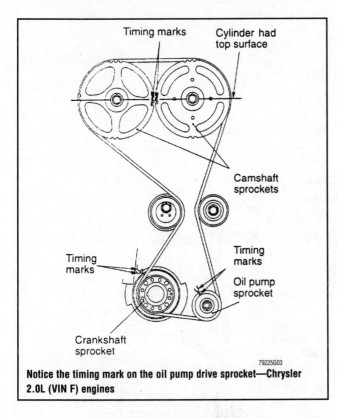

Notice the timing mark on the oil pump drive sprocket—Chrysler 2.0L (VIN F) engines

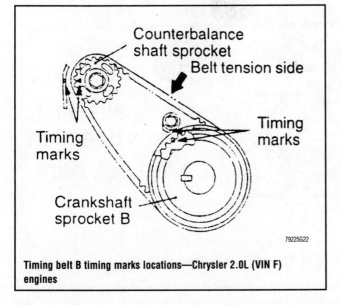

Timing belt B timing marks locations—Chrysler 2.0L (VIN F) engines

center bolt. Then, use MB998778 or equivalent puller to remove the sprocket.

13. Mark the direction of rotation on the timing belt **B** with an arrow.

14. Loosen the center bolt on the tensioner and remove the belt.

15. To remove the camshaft sprocket, remove the cylinder head cover. Use a wrench to hold the hexagonal part of the camshaft and remove the sprocket mounting bolt.

✳✳ WARNING

Do not rotate the camshafts or the crankshaft while the timing belt is removed.

To install:

16. Use a wrench to hold the camshaft, and install the sprocket and mounting bolt. Tighten the bolt(s) to 65 ft. lbs. (88 Nm).

17. Install the cylinder head cover.

18. Place the crankshaft sprocket on the crankshaft. Use tool MB990767, or equivalent, to hold the crankshaft sprocket while tightening the center bolt. Tighten the center bolt to 80–94 ft. lbs. (108–127 Nm).

19. Align the timing marks on the crankshaft sprocket **B** and the balance shaft.

20. Install timing belt **B** on the sprockets. Position the center of the tensioner pulley to the left and above the center of the mounting bolt.

21. Push the pulley clockwise toward the crankshaft to apply tension to the belt and tighten the mounting bolt to 14 ft. lbs. (19 Nm). Do not let the pulley turn when tightening the bolt because it will cause excessive tension on the belt. The belt should deflect 0.20–0.28 in. (5–7mm) when finger pressure is applied between the pulleys.

22. Install the crankshaft sensing blade and the crankshaft sprocket. Apply engine oil to the mounting bolt and tighten the bolt to 80–94 ft. lbs. (108–127 Nm).

23. Use a press or vise to compress the auto tensioner pushrod. Insert a set pin when the holed are aligned.

✳✳ WARNING

Do not compress the pushrod too quickly, damage to the pushrod can occur.

24. Install the auto tensioner on the engine.

25. Align the timing marks on the camshaft sprocket, crankshaft sprocket and the oil pump sprocket.

26. After aligning the mark on the oil pump sprocket, remove the cylinder block plug and insert a Phillips screwdriver in the hole to check the position of the counter balance shaft. The screwdriver should go in at least 2.36 in. (60mm) or more. If not, rotate the oil pump sprocket once and realign the timing mark so the screwdriver goes in. Do not remove the screwdriver until the timing belt is installed.

27. Install the timing belt on the intake camshaft and secure it with a clip.

28. Install the timing belt on the exhaust camshaft. Align the timing marks with the cylinder head top surface using 2 wrenches. Secure the belt with another clip.

29. Install the belt around the idler pulley, oil pump sprocket, crankshaft sprocket and the tensioner pulley.

30. Turn the tension pulley so the pinholes are at the bottom. Press the pulley lightly against the timing belt.

31. Screw the special tool into the left engine support bracket until it contacts the tensioner arm, then screw the tool in a little more and remove the pushrod pin from the auto tensioner. Remove the special tool and tighten the center bolt to 35 ft. lbs. (48 Nm).

32. Turn the crankshaft ¼ turn counterclockwise, then clockwise until the timing marks are aligned.

33. Loosen the center bolt. Install special tool MD998767 or equivalent, on the tension pulley. Turn the tension pulley counterclockwise with a torque of 31 inch lbs. (3.5 Nm) and tighten the center bolt to 35 ft. lbs. (48 Nm). Do not let the tension pulley turn with the bolt.

34. Turn the crankshaft 2 revolutions to the right and align the timing marks. After 15 minutes, measure the protrusion of the pushrod on the auto tensioner. The standard measurement is 0.150–0.177 in. (3.8–4.5mm). If the protrusion is out of specification, loosen the tension pulley, apply the proper torque to the belt and retighten the center bolt.

35. Install the crankshaft pulley. Tighten the mounting bolts to 18 ft. lbs. (25 Nm).

36. Install the water pump. Tighten the mounting bolts to 78 inch lbs. (8.8 Nm).

37. Install and adjust the drive belts.

38. Install the engine mount bracket.

39. Install the engine undercover.

40. Connect the negative battery cable.

2.0L (VIN Y) ENGINE

Valve timing is critical to engine operation. Use care when servicing the timing belt. There are a number of timing marks that must be properly aligned or engine damage will result. If the timing belt has not broken, or jumped teeth, it is recommended that the crankshaft be turned by hand (clockwise) to Top Dead Center (TDC) of the No. 1 cylinder compression stroke (firing position) before beginning work. This should align all the timing marks and serve as a reference for later work. Some technicians will apply a small amount of white paint to all timing marks. This helps make them more visible under the low-light conditions found underhood.

1. Disconnect the negative battery cable.

2. Remove the accessory drive belts.

3. Using Crankshaft Holding tool C-3281 or equivalent, remove the crankshaft pulley center retaining bolt.

4. Using puller tools 1026 and 6827 or equivalent, remove the crankshaft pulley.

5. Remove the power steering pump from the bracket and position it aside. Do not disconnect the hoses.

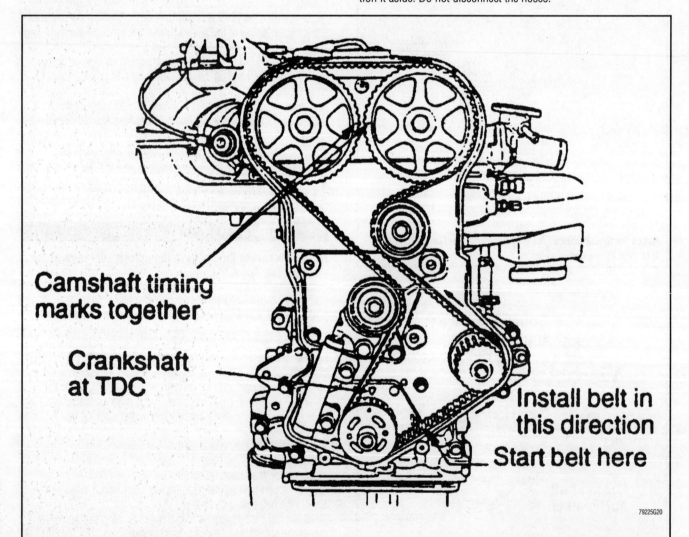

Camshaft timing marks together

Crankshaft at TDC

Install belt in this direction

Start belt here

79225G20

Install the timing belt by starting at the crankshaft sprocket and working around the other pulleys in a counterclockwise direction—Chrysler 2.0L (VIN Y) engine

6. Remove the power steering pump bracket from the engine.

7. Use a floor jack with a piece of wood on it and raise the engine to take the weight off of the engine mount.

8. Remove the engine mount and bracket.

9. Remove the front timing belt cover.

10. If not done so previously, align the timing marks. Loosen the timing belt tensioner and remove the belt.

✳✳ WARNING

Do not rotate the crankshaft or camshafts after removing the timing belt or valvetrain components may be damaged. Always align the timing marks before removing the timing belt.

11. For 1995–97 models, perform the following procedures:

a. If the timing belt tensioner is to be replaced, remove the retaining bolts and remove the timing belt tensioner. When the timing belt tensioner is removed from the engine it is necessary to compress the plunger into the tensioner body.

b. Place the tensioner in a vise and slowly compress the plunger.

➡**Position the tensioner in the vise the same way it will be installed on the engine. This is to ensure proper pin orientation for when the tensioner is installed on the engine.**

c. When the plunger is compressed into the tensioner body, install a pin through the body and plunger to hold the plunger in place until the tensioner is installed.

12. For 1998–00 models, place an 8mm Allen wrench into the belt tensioner, then using the long end of a 1/8 in. (3mm) Allen wrench, rotate the tensioner counterclockwise until it slides into the locking hole.

13. Remove the timing belt.

To install:

14. Check that all timing marks are still aligned. Bring the crankshaft sprocket to 1/2 a notch before TDC.

15. Install the timing belt. Starting at the crankshaft, route the belt around the water pump sprocket, idler pulley, camshaft sprockets, then around the tensioner pulley.

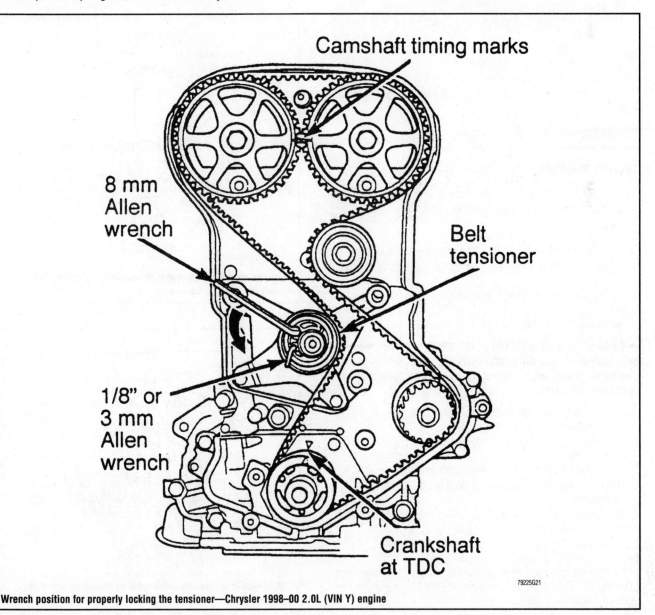

Camshaft timing marks

8 mm Allen wrench

Belt tensioner

1/8" or 3 mm Allen wrench

Crankshaft at TDC

79225G21

Wrench position for properly locking the tensioner—Chrysler 1998–00 2.0L (VIN Y) engine

16. Move the crankshaft to TDC to take up the slack in the belt. Install the tensioner to the block but do not tighten the retaining bolts.

17. For 1995–97 models:

    a. Using a torque wrench on the tensioner pulley, apply 21 ft. lbs. (28 Nm) of torque to the pulley.

    b. With the torque being applied to the tensioner pulley, move the tensioner up against the tensioner pulley bracket and tighten the retaining bolts to 23 ft. lbs. (31 Nm).

    c. Remove the tensioner plunger pin. Pretension is correct when the pin can be removed and installed.

18. For 1998–00 models, remove the wrenches from the tensioner.

19. Rotate the crankshaft 2 revolutions and check the timing marks. If the timing marks are not properly aligned, remove the belt and reinstall it as described.

20. Install the front timing belt cover.

21. Lower the engine enough to install the engine mount bracket.

22. Install the bracket and remove the floor jack.

23. Install the power steering pump bracket and pump.

24. Install the crankshaft pulley using C-4685-C or an equivalent pulley installer.

25. Tighten the mounting bolt to 105 ft. lbs. (142 Nm).

26. Install the accessory drive belts.

27. Connect the negative battery cable.

28. Start the engine. Check for leaks and proper engine operation.

2.2L ENGINE

1988–89 Medallion

1. Disconnect the negative battery cable.

2. Remove the drive belts, vibration damper, pulley and Woodruff® key.

3. Remove the timing belt cover.

4. Make a mark on the back of the timing belt indicating the direction of rotation so it may be reassembled in the same direction, if to be reused.

5. Loosen the timing belt tensioner pivot bolt and locking bolt.

6. Remove the timing belt.

➡ If coolant or engine oil comes in contact with the timing belt, they will drastically shorten its life. Also, do not allow engine oil or coolant to contact the timing belt sprockets or tensioner assembly.

7. Inspect all parts for damage and wear. If any of the following is found, replacement is necessary:

    a. Timing belt—cracks on back surface, sides, bottom and separated canvas.

    b. Tensioner pulleys—turn the pulleys and check for binding, excessive play, unusual noise or if there is a grease leak.

To install:

8. Position the camshaft sprocket timing index aligned with the static timing mark.

9. Position the crankshaft so that the No. 1 piston is at Top Dead Center (TDC) on the compression stroke.

10. Remove the access hole plug in the cylinder block and insert tool Mot. 861 (TDC Rod) or equivalent, into the TDC slot in the crankshaft counterweight.

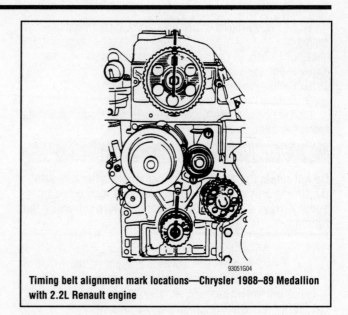

Timing belt alignment mark locations—Chrysler 1988–89 Medallion with 2.2L Renault engine

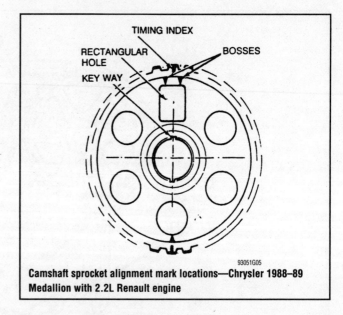

Camshaft sprocket alignment mark locations—Chrysler 1988–89 Medallion with 2.2L Renault engine

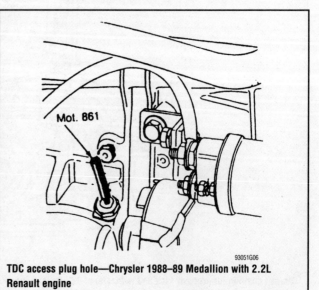

TDC access plug hole—Chrysler 1988–89 Medallion with 2.2L Renault engine

11. Loosen the timing belt tensioner bolts. Push the tensioner pulley towards the water pump to compress the tensioner spring. Tighten the tensioner bolts. This allows for easier installation of the timing belt.

12. Install the timing belt on the sprockets. If the original timing belt is being reused, install the timing belt with the arrow previously made, pointing in the proper direction of rotation.

13. Loosen the tensioner bolts and allow the spring loaded tensioner to contact the belt. This will automatically tension the belt. Then, tighten the tensioner retaining bolts.

14. Position the timing belt cover over the sprockets and check the position of the camshaft sprocket timing mark with the index on the cover.

15. Remove tool Mot. 861 or equivalent, and install the cylinder block plug.

16. Check the timing belt tension adjustment by performing the following procedure:

   a. Rotate the crankshaft clockwise 2 complete revolutions.

   b. Loosen the tensioner bolts ¼ turn.

   c. The spring loaded timing belt tensioner will automatically adjust to the correct position.

   d. Tighten the bottom tensioner bolt first. Then, tighten the upper bolt. Torque both bolts to 18 ft. lbs. (25 Nm).

   e. Check the timing belt tension with a tension gauge tool, Ele. 346, or equivalent. The deflection should be 0.216–0.276 in. (5.5–7.0mm).

17. Complete the installation by reversing the removal procedures.

2.2L SOHC & 2.5L ENGINES

1981–94

1. Disconnect the negative battery cable.
2. Position the engine so the No. 1 piston is at Top Dead Center (TDC) of the compression stroke.
3. Remove the nuts and bolts that attach the upper cover to the valve cover, block or cylinder head.
4. Remove the bolt that attaches the upper cover to the lower cover.
5. Remove the upper cover.

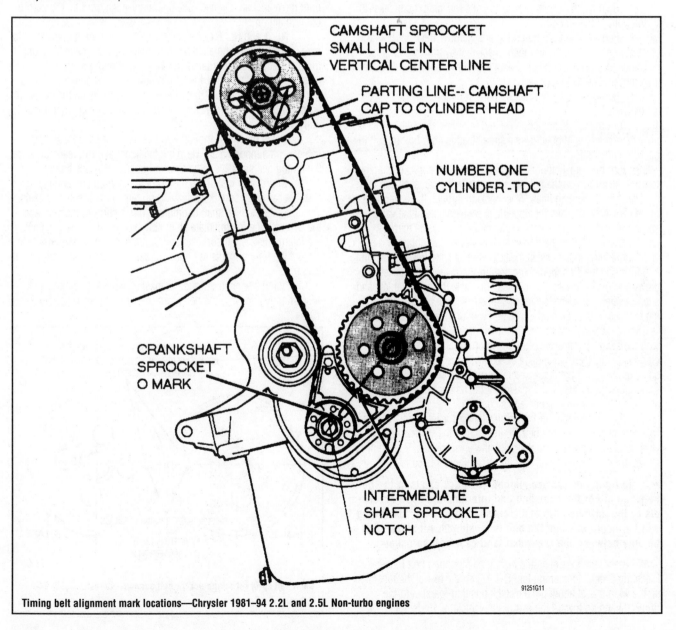

Timing belt alignment mark locations—Chrysler 1981–94 2.2L and 2.5L Non-turbo engines

6. Remove the right tire and wheel assembly. Remove the right-side inner splash shield.

7. Remove the crankshaft pulley, water pump pulley and the accessory drive belt(s).

8. Remove the lower cover attaching bolts and the cover from the engine.

9. Remove the timing belt tensioner and allow the belt to hang free.

10. Place a floor jack under the engine and separate the right motor mount.

11. Remove the air conditioning compressor belt idler pulley, if equipped, and remove the mounting stud. Remove the compressor/alternator bracket as follows:

   a. Remove the alternator pivot bolt and remove the alternator from the bracket. Turn the alternator so the wire connections are facing up and disconnect the harness connectors from the rear of the alternator.

   b. Remove the air conditioning compressor belt idler.

   c. Remove the right engine mount yoke screw securing engine mount support strut to the engine.

   d. Remove the 5 side mounting bolts retaining the bracket to the front of the engine.

   e. Remove the front mounting nut. Remove the front bolt and strut and rotate the solid mount bracket away from the engine. Slide the bracket on the stud until free of the mounting studs and remove from the engine.

   f. Remove the timing belt from the vehicle.

To install:

12. Turn the crankshaft sprocket and intermediate shaft sprocket until the marks are aligned. Use a straight-edge from bolt to bolt to confirm alignment.

13. Turn the camshaft until the small hole in the sprocket is at the top and the arrows on the hub are aligned with the camshaft cap to cylinder head mounting lines. When looking through the hole on top of the camshaft sprocket, the uppermost center nipple of the valve cover end seal should be at the center of the hole. Use a mirror to check the alignment of the arrows so it is viewed straight on and not at an angle from above. Install the belt but let it hang free at this point.

14. Install the air conditioning compressor/alternator bracket, idler pulley and motor mount. Remove the floor jack. Raise the vehicle and support safely. Have the tensioner at an arm's reach because the timing belt will have to be held in position with one hand.

15. To properly install the timing belt, reach up and engage it with the camshaft sprocket. Turn the intermediate shaft counter-clockwise slightly, then engage the belt with the intermediate shaft sprocket. Hold the belt against the intermediate shaft sprocket and turn clockwise to take up all tension; if the timing marks are out of alignment, repeat until alignment is correct.

16. Using a wrench, turn the crankshaft sprocket counterclockwise slightly and wrap the belt around it. Turn the sprocket clockwise so there is no slack in the belt between sprockets; if the timing marks are out of alignment, repeat until alignment is correct.

➡**If the timing marks are aligned but slack exists in the belt between either the camshaft and intermediate shaft sprockets or the intermediate and crankshaft sprockets, the timing will be incorrect when the belt is tensioned. All slack must be only between the crankshaft and camshaft sprockets.**

17. Install the tensioner and install the mounting bolt loosely. Place the special tensioning tool C-4703 on the hex of the tensioner so the weight is at about the 9 o'clock position (parallel to the ground, hanging toward the rear of the vehicle) ± 15 degrees.

18. Hold the tool in position and tighten the bolt to 45 ft. lbs. (61 Nm). Do not pull the tool past the 9 o'clock position; this will make the belt too tight and will cause it to howl or possibly break.

19. Lower the vehicle and recheck the camshaft sprocket positioning. If it is correct install the timing belt covers and all related parts.

20. Connect the negative battery cable and road test the vehicle.

2.2L DOHC ENGINE

1. Position the engine so the No. 1 piston is at Top Dead Center (TDC) of the compression stroke. Disconnect the negative battery cable.

2. Remove the timing belt covers.

3. Remove the timing belt tensioner and allow the belt to hang free.

4. Place a floor jack under the engine and separate the right motor mount.

5. Remove the timing belt from the vehicle.

To install:

6. Turn the crankshaft sprocket and intermediate shaft sprocket until the marks are aligned. Use a straight-edge from bolt to bolt to confirm alignment.

7. No. 1 and No. 6 camshaft journals have aligning pin holes to index with the blind holes in the camshaft. Turn the camshafts until the pin holes in the journals align with the aligning holes in the corresponding bearing caps. Install pin punches to secure this timing position. At this position, the sprocket timing holes on the camshaft sprockets should both be centered at the cylinder head mounting surface line.

8. Install the motor mount. Remove the floor jack. Raise the vehicle and support safely. Have the tensioner at arm's reach because the timing belt will have to be held in position with one hand.

9. To properly install the timing belt, reach up and engage it with the camshaft sprockets, leaving no tension between sprockets. Turn the intermediate shaft counterclockwise slightly, then engage the belt with the intermediate shaft sprocket. Hold the belt against the intermediate shaft sprocket and turn clockwise to take up all tension; if the timing marks are out of alignment, repeat until alignment is correct.

10. Using a wrench, turn the crankshaft sprocket counterclockwise slightly and wrap the belt around it. Turn the sprocket clock-

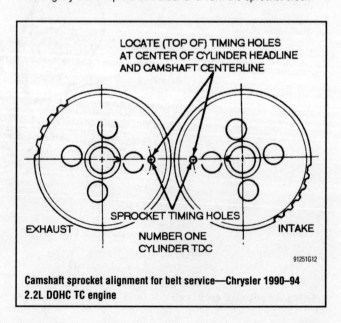

Camshaft sprocket alignment for belt service—Chrysler 1990–94 2.2L DOHC TC engine

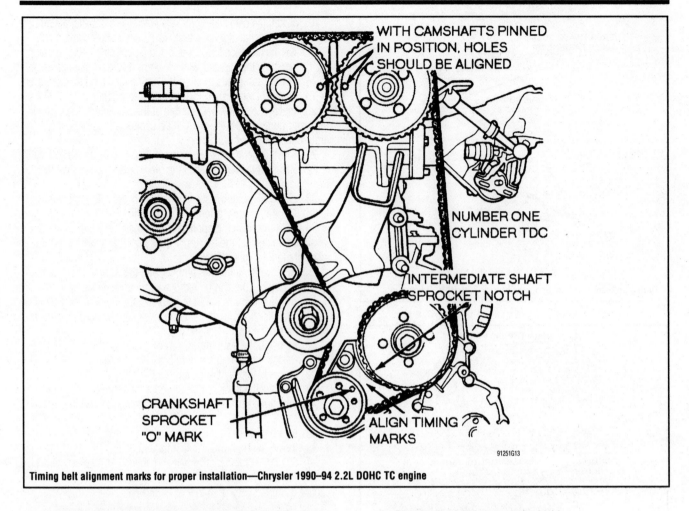

WITH CAMSHAFTS PINNED
IN POSITION, HOLES
SHOULD BE ALIGNED

NUMBER ONE
CYLINDER TDC

INTERMEDIATE SHAFT
SPROCKET NOTCH

CRANKSHAFT
SPROCKET
"O" MARK

ALIGN TIMING
MARKS

91251G13

Timing belt alignment marks for proper installation—Chrysler 1990–94 2.2L DOHC TC engine

wise so there is no slack in the belt between sprockets; if the timing marks are out of alignment, repeat until alignment is correct.

➡**If the timing marks are aligned but slack exists in the belt anywhere except on the tensioner side, the timing will be incorrect when the belt is tensioned. All slack must be only between the crankshaft and exhaust camshaft sprockets.**

11. Install the tensioner and install the mounting bolt loosely. Remove the pin punches from the camshafts. Place the special tensioning tool C-4703 on the hex of the tensioner so the weight is at about the 9 o'clock position (parallel to the ground, hanging toward the rear of the vehicle) ± 15 degrees.

12. Hold the tool in position and tighten the bolt to 45 ft. lbs. (61 Nm). Do not pull the tool past the 9 o'clock position; this will make the belt too tight and will cause it to howl or possibly break.

13. Rotate the crankshaft 2 full revolutions. With the No. 1 cylinder at TDC, all timing marks must be aligned. Repeat the procedure if the timing is not correct.

14. Install the timing belt covers and all related parts.

15. Connect the negative battery cable and road test the vehicle.

2.4L (VIN X) ENGINE

1. Disconnect the negative battery cable from the left strut tower. The ground cable is equipped with a insulator grommet which should be placed on the stud to prevent the negative battery cable from accidentally grounding.

2. Remove the right inner splash-shield.

3. Remove the accessory drive belts.

4. Remove the crankshaft damper.

5. Remove the right engine mount.

6. Place a suitable floor jack under the vehicle to support the engine.

7. Remove the engine mount bracket

8. Remove the timing belt cover.

➡**Do not rotate the crankshaft or the camshafts after the timing belt has been removed. Damage to the valve components may occur. Before removing the timing belt, always align the timing marks.**

9. Align the timing marks of the timing belt sprockets to the timing marks on the rear timing belt cover and oil pump cover. Loosen the timing belt tensioner bolts.

10. Remove the timing belt and the tensioner.

11. Remove the camshaft timing belt sprockets.

12. Remove the crankshaft timing belt sprocket using special removal tool No. 6793 or equivalent.

13. Place the tensioner into a soft-jawed vise to compress the tensioner.

14. After compressing the tensioner, place a pin (5⁄64 in. Allen wrench will work) into the plunger side hole to retain the plunger until installation.

To install:

15. Using special tool No. 6792 or its equivalent, install the crankshaft timing belt sprocket onto the crankshaft.

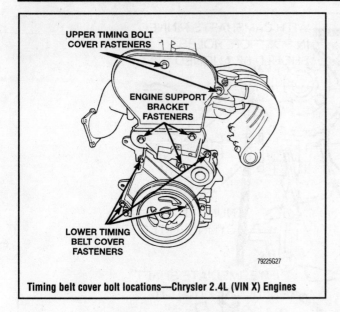

Timing belt cover bolt locations—Chrysler 2.4L (VIN X) Engines

79225G27

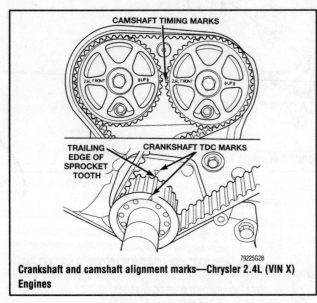

Crankshaft and camshaft alignment marks—Chrysler 2.4L (VIN X) Engines

79225G28

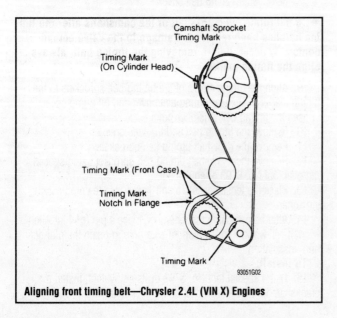

Aligning front timing belt—Chrysler 2.4L (VIN X) Engines

93051G02

16. Install the camshaft sprockets onto the camshafts. Install and tighten the camshaft sprocket bolts to 75 ft. lbs. (101 Nm).

17. Set the crankshaft sprocket to Top Dead Center (TDC) by aligning the notch on the sprocket with the arrow on the oil pump housing.

18. Set the camshafts to align the timing marks on the sprockets.

19. Move the crankshaft to ½ notch before TDC.

20. Install the timing belt starting at the crankshaft, then around the water pump sprocket, idler pulley, camshaft sprockets, then around the tensioner pulley.

21. Move the crankshaft sprocket to TDC to take up the belt slack.

22. Reinstall the tensioner to the block, but do not tighten it at this time.

23. Using a torque wrench on the tensioner pulley, apply 21 ft. lbs. (28 Nm) of torque to the tensioner pulley.

24. With torque being applied to the tensioner pulley, move the tensioner up against the tensioner pulley bracket and tighten the fasteners to 23 ft. lbs. (31 Nm).

25. Remove the tensioner plunger pin, the tension is correct when the plunger pin can be removed and replaced easily.

26. Rotate the crankshaft 2 revolutions and recheck the timing marks. Wait several minutes, then recheck that the plunger pin can easily be removed and installed.

27. Reinstall the front timing belt cover.

28. Reinstall the engine mount bracket.

29. Reinstall the right engine mount.

30. Remove the floor jack from under the vehicle.

31. Install the crankshaft damper and tighten to 105 ft. lbs. (142 Nm).

32. Install and adjust the accessory drive belts.

33. Install the right inner splash-shield.

34. Reconnect the negative battery cable.

35. Perform the crankshaft and camshaft relearn alignment procedure using the DRB scan tool or equivalent.

2.5L (VIN H) ENGINE

1. Disconnect the negative battery cable from the left strut tower. The ground cable is equipped with a insulator grommet which should be placed on the stud to prevent the negative battery cable from accidentally grounding.

2. Raise and safely support the vehicle. Remove the right inner splash-shield.

3. Remove the accessory drive belts.

4. Remove the crankshaft damper.

5. Remove the right engine mount

6. Place a suitable floor jack under the vehicle to support the engine.

7. Remove the right engine mount bracket

8. Remove the timing belt upper left cover, upper right cover and lower cover.

9. Loosen the timing belt tensioner bolts.

➡**Before removing timing belt, be sure to align the sprocket timing marks to the timing marks on the rear timing belt cover.**

10. If the present timing belt is going to be reused, mark the running direction of the timing belt for installation. Remove the timing belt and the tensioner.

11. Remove the camshaft timing belt sprockets from the camshaft, if necessary.

12. Remove the crankshaft timing belt sprocket and key.

13. Place the tensioner into a soft-jawed vise to compress the tensioner.

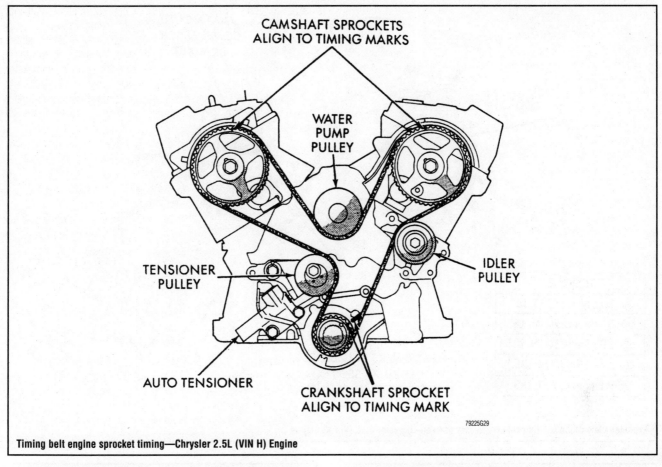

CAMSHAFT SPROCKETS
ALIGN TO TIMING MARKS

WATER
PUMP
PULLEY

IDLER
PULLEY

TENSIONER
PULLEY

AUTO TENSIONER

CRANKSHAFT SPROCKET
ALIGN TO TIMING MARK

79225G29

Timing belt engine sprocket timing—Chrysler 2.5L (VIN H) Engine

14. After compressing the tensioner place a pin into the plunger side hole to retain the plunger until installation.

To install:

15. If removed, reinstall the camshaft sprockets onto the camshaft. Install the camshaft sprocket bolt and tighten to 65 ft. lbs. (88 Nm).

16. If removed, reinstall the crankshaft timing belt sprocket and key onto the crankshaft.

17. Set the crankshaft sprocket to Top Dead Center (TDC) by aligning the notch on the sprocket with the arrow on the oil pump housing, then back off the sprocket 3 notches before TDC.

18. Set the camshafts to align the timing marks on the sprockets with the marks on the rear timing belt cover.

19. Install the belt on the rear camshaft sprocket first.

20. Install a binder clip on the belt to the sprocket so it won't slip out of position.

21. Keeping the belt taut, install it under the water pump pulley and around the front camshaft sprocket.

22. Install a binder on the front sprocket and belt.

23. Rotate the crankshaft to TDC.

24. Continue routing the belt by the idler pulley and around the crankshaft sprocket to the tensioner pulley.

25. Move the crankshaft sprocket clockwise to TDC to take up the belt slack. Check that all timing marks are in alignment.

26. Reinstall the tensioner to the block but do not tighten it at this time.

27. Using special tool No. MD998767 or equivalent, and a torque wrench on the tensioner pulley, apply 39 inch lbs. (4.4 Nm) of torque to the tensioner. Tighten the tensioner pulley bolt to 35 ft. lbs. (48 Nm).

28. With torque being applied to the tensioner pulley, move the tensioner up against the tensioner bracket and tighten the fasteners to 17 ft. lbs. (23 Nm).

29. Remove the tensioner plunger pin, the tension is correct when the plunger pin can be removed and replaced easily.

30. Rotate the crankshaft 2 revolutions clockwise and recheck the timing marks. Check to be sure the tensioner plunger pin can be easily installed and removed. If the pin does not remove and install easily, perform the procedure again.

31. Reinstall the timing belt cover.

32. Reinstall the engine mount bracket.

33. Reinstall the right engine mount.

34. Remove the engine support.

35. Install the crankshaft damper and tighten to 134 ft. lbs. (182 Nm).

36. Reinstall the accessory drive belts and adjust them.

37. Reinstall the right inner splash-shield.

38. Perform the crankshaft and camshaft relearn alignment procedure using the DRB scan tool or equivalent.

2.5L (VIN K) ENGINE

1986–95

Working on any engine (especially overhead camshaft engines) requires much care be given to valve timing. It is good practice to set the engine up to Top Dead Center (TDC) of the compression stroke of the No. 1 cylinder firing position. Verify that all timing marks on the crankshaft and camshaft sprockets are properly aligned before removing the timing belt. This serves as a point of

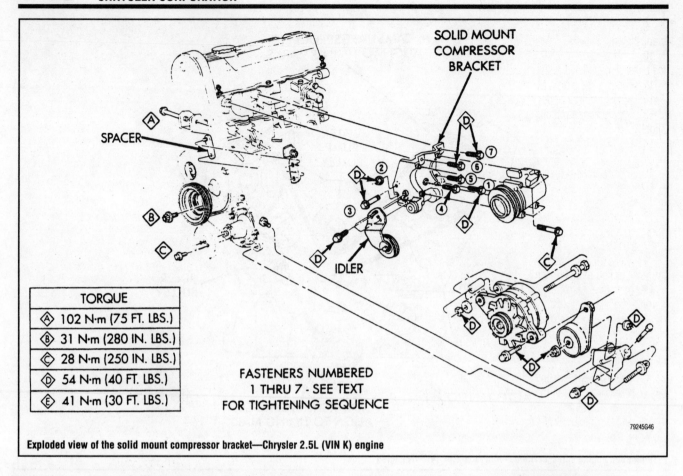

TORQUE	
Ⓐ	102 N·m (75 FT. LBS.)
Ⓑ	31 N·m (280 IN. LBS.)
Ⓒ	28 N·m (250 IN. LBS.)
Ⓓ	54 N·m (40 FT. LBS.)
Ⓔ	41 N·m (30 FT. LBS.)

FASTENERS NUMBERED
1 THRU 7 - SEE TEXT
FOR TIGHTENING SEQUENCE

Exploded view of the solid mount compressor bracket—Chrysler 2.5L (VIN K) engine

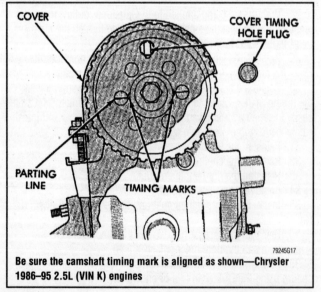

Be sure the camshaft timing mark is aligned as shown—Chrysler 1986–95 2.5L (VIN K) engines

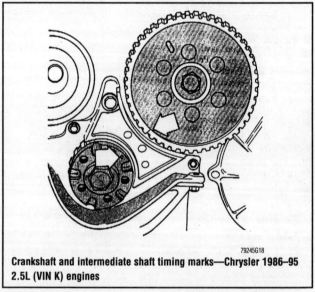

Crankshaft and intermediate shaft timing marks—Chrysler 1986–95 2.5L (VIN K) engines

reference for all work that follows. Valve timing is most important and engine damage will result if the work is incorrect.

➡This is an interference engine. Do not rotate the crankshaft or the camshafts after the timing belt has been removed. Damage to the valve components may occur. Before removing the timing belt, always align the timing marks.

1. Disconnect the negative battery cable.
2. Remove the accessory drive belts.
3. Remove the right engine mount yoke screw.

4. Remove the air conditioning compressor and set it aside. Remove the solid mount compressor bracket mounting bolts.
5. Turn the solid mount bracket away from the engine and slide it on the No. 2 stud until it is free. The front bolt and spacer will be removed with the bracket.
6. Remove the alternator and the drive belt idler.
7. Raise and safely support the vehicle. Remove the right inner fender splash shield.
8. Loosen and remove the 3 water pump pulley mounting bolts and remove the pulley.

9. Remove the 4 crankshaft pulley retaining bolts and the crankshaft pulley.

10. Remove the nuts at the upper portion of the timing cover and the bolts from the lower portion, then remove both halves of the cover.

11. Remove the timing belt covers.

12. Position a jack under the engine.

13. Separate the right engine mount and raise the engine slightly.

14. Loosen the timing belt tensioner bolt, rotate the hex nut, and remove timing belt.

15. Remove the timing belt tensioner, if necessary.

16. Remove the crankshaft sprocket with a suitable puller tool and a bolt approximately 6 in. (15 cm) long.

17. Remove the camshaft sprocket and intermediate shaft sprocket, if necessary.

To install:

18. Clean all parts well. A small amount of white paint on the sprocket timing marks may make alignment easier.

19. Install the crankshaft sprocket. Tighten the crankshaft sprocket bolt to 85 ft. lbs. (115 Nm).

20. If necessary, turn the crankshaft and intermediate shaft until markings on both sprockets are aligned.

21. Rotate the camshaft so the arrows on the hub are aligned with the No. 1 camshaft cap-to-cylinder headline. The small hole in the cam sprocket should be centered in the vertical centerline.

22. If removed, install the timing belt tensioner.

23. Install the timing belt over the drive sprockets and adjust.

24. Tighten the tensioner by turning the tensioner hex to the right. Tension should be correct when the belt can be twisted 90 degrees with the thumb and forefinger, midway between the camshaft and intermediate sprocket.

❈❈ WARNING

If any binding is felt when adjusting the timing belt tension by turning the crankshaft, STOP turning the engine, because the pistons may be hitting the valves.

25. Turn the engine clockwise from TDC, 2 complete revolutions with the crankshaft bolt. Check the timing marks for correct alignment.

❈❈ WARNING

Do not use the camshaft or intermediate shaft to rotate the engine. Do not allow oil or solvent to contact the timing belt as they will deteriorate the belt and cause slipping.

26. Tighten the locknut on the tensioner, while holding the weighted wrench tool C-4503 or equivalent, in position, to 45 ft. lbs. (61 Nm).

27. Lower the engine onto the right engine mount and install the fasteners. Remove the support from the engine.

28. Some engines use a foam stuffer block inside the timing belt housing. Inspect the foam block's condition and position. The stuffer block should be intact and secure within the engine bracket tunnel.

29. Install the timing belt cover. Secure the upper section to the cylinder head with nuts and the lower section to the cylinder block with screws. Tighten all of the timing belt cover fasteners to 40 inch lbs. (4 Nm).

30. Check valve timing again. With the timing belt cover installed,

and with No. 1 cylinder at TDC, the small hole in the sprocket must be centered in the timing belt cover hole. If the hole is not aligned correctly, perform the timing belt installation procedure again.

31. Install the water pump pulley and the crankshaft pulley. Tighten the water pump pulley bolts to 21 ft. lbs. (28 Nm). Tighten the crankshaft pulley bolts to 24 ft. lbs. (31 Nm).

32. Install the inner fender splash shield. Lower the vehicle.

33. Install the solid mount compressor bracket. The bracket mounting fasteners must be tightened to 40 ft. lbs. (54 Nm).

34. Install the alternator and drive belt idler. Tighten mounting bolts to 40 ft. lbs. (54 Nm).

35. Install the right engine mount yoke bolt and tighten to 100 ft. lbs. (133 Nm).

36. Install the accessory drive belts and adjust them to the proper tension.

➡**With the timing belt cover installed and the piston in the No. 1 cylinder at TDC, the small hole in the cam sprocket should be centered in timing belt cover hole.**

37. Reconnect the negative battery cable. Road test the vehicle.

2.5L (VIN N) ENGINE

1. Disconnect the negative battery cable.

2. Remove the accessory drive belts.

3. Using Crankshaft Holding tools MB990767 and MB998754, or their equivalents, remove the crankshaft bolt and remove the pulley.

4. Remove the heated oxygen sensor connection.

5. Remove the power steering pump with the hose attached and position it aside.

6. Remove the power steering pump bracket.

7. Place a floor jack under the engine oil pan, with a block of wood in between, and raise the engine so that the weight of the engine is no longer being applied to the engine support bracket.

8. Remove the upper engine mount. Spraying lubricant, slowly remove the reamer (alignment) bolt and remaining bolts and remove the engine support bracket.

➡**The reamer bolt is sometimes heat-seized on the engine support bracket**

9. Remove the front timing belt covers.

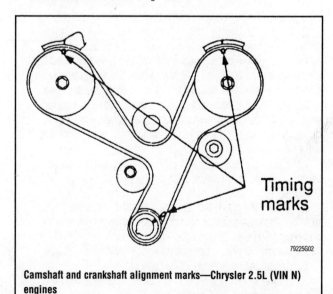

Camshaft and crankshaft alignment marks—Chrysler 2.5L (VIN N) engines

79225G02

10. If the timing belt is to be reused, draw an arrow indicating the direction of rotation on the back of the belt for reinstallation.

11. Align the timing marks by turning the crankshaft with MD998769 Crankshaft Turning tool or its equivalent. Loosen the center bolt on the timing belt tensioner pulley and remove the belt.

✳✳ WARNING

Do not rotate the crankshaft or camshaft after removing the timing belt, or valvetrain components may be damaged. Always align the timing marks before removing the timing belt.

12. Check the belt tensioner for leaks and check the pushrod for cracks.

13. If the timing belt tensioner is to be replaced, remove the retaining bolts and remove the timing belt tensioner. When the timing belt tensioner is removed from the engine it is necessary to compress the plunger into the tensioner body.

14. Place the tensioner in a vise and slowly compress the plunger. Take care not to damage the pushrod.

➡**Position the tensioner in the vise the same way it will be installed on the engine. This is to ensure proper pin orientation for when the tensioner is installed on the engine.**

15. When the plunger is compressed into the tensioner body, install a pin through the body and plunger to hold the plunger in place until the tensioner is installed.

To install:

16. Install the timing belt tensioner and tighten the retaining bolts to 17 ft. lbs. (24 Nm), but do not remove the pin at this time.

17. Check that all timing marks are still aligned.

18. Use bulldog clips (large paper binder clips) or other suitable tool to secure the timing belt and to prevent it from slacking. Install the timing belt. Starting at the crankshaft, go around the idler pulley, then the front camshaft sprocket, the water pump pulley, the rear camshaft sprocket, then around the tensioner pulley.

19. Be sure the belt is tight between the crankshaft and front camshaft sprocket, between the camshaft sprockets and the water pump. Gently raise the tensioner pulley, so that the belt does not sag, and temporarily tighten the center bolt.

20. Move the crankshaft ¼ turn counterclockwise, then turn it clockwise to the position where the timing marks are aligned.

21. Loosen the center bolt of the tensioner pulley. Using MD998767 or equivalent tensioner tool, and a torque wrench apply 40 inch lbs. (4.4 Nm) tensional torque to the timing belt and tighten the center bolt to 35 ft. lbs. (48 Nm). When tightening the bolt, be sure that the tensioner pulley shaft does not rotate with the bolt.

22. Remove the tensioner plunger pin. Pretension is correct when the pin can be removed and installed easily. If the pin cannot be easily removed and installed it is still satisfactory as long as it is within its standard value.

23. Check that the tensioner pushrod is within the standard value. When the tensioner is engaged the pushrod should measure 0.149–0.177 in. (3.8–4.5mm).

24. Rotate the crankshaft 2 revolutions and check the timing marks. If the timing marks are not properly aligned, remove the belt and repeat Steps 17 through 23.

25. Install the timing belt covers.

26. Install the engine mounting bracket.

27. Lower the engine enough to install the engine mount onto bracket and remove the floor jack.

28. Install the power steering pump bracket and pump.

29. Install the crankshaft pulley and tighten the retaining bolt to 13 ft. lbs. (18 Nm).

30. Install the accessory drive belts.

31. Properly fill the cooling system.

32. Connect the negative battery cable.

33. Check for leaks and proper engine and cooling system operation.

3.0L (VIN 3, H & S) SOHC ENGINES

1986–94

1. Disconnect the negative battery cable.

✳✳ CAUTION

Wait at least 90 seconds after the negative battery cable is disconnected to prevent possible deployment of the air bag.

2. Position the engine so the No. 1 cylinder is at Top Dead Center (TDC) of the compression stroke. Disconnect the negative battery cable.

3. Remove the engine undercover.

4. Remove the cruise control actuator.

5. Remove the accessory drive belts.

6. Remove the air conditioner compressor tension pulley assembly.

7. Remove the tension pulley bracket.

8. Using the proper equipment, slightly raise the engine to take the weight off the side engine mount. Remove the engine mounting bracket.

9. Disconnect the power steering pump pressure switch connector. Remove the power steering pump with hoses attached and wire aside.

10. Remove the engine support bracket.

11. Remove the crankshaft pulley.

12. Remove the timing belt cover cap.

13. Remove the timing belt upper and lower covers.

14. If the same timing belt will be reused, mark the direction of the timing belt's rotation for installation in the same direction. Make sure the engine is positioned so the No. 1 cylinder is at the Top Dead Center (TDC) of its compression stroke and the timing marks are aligned with the engine's timing mark indicators.

15. Loosen the timing belt tensioner bolt and remove the belt. If the tensioner is not being removed, position it as far away from the center of the engine as possible and tighten the bolt.

16. If the tensioner is being removed, paint the outside of the spring to ensure that it is not installed backwards. Unbolt the tensioner and remove it along with the spring.

✳✳ WARNING

Do not rotate the camshafts when the timing belt is removed from the engine. Turning the camshaft when the timing belt is removed could cause the valves to interfere with the pistons thus causing severe internal engine damage.

To install:

17. Install the tensioner, if removed, and hook the upper end of the spring to the water pump pin and the lower end to the tensioner in exactly the same position as originally installed. If not already done, position both camshafts so the marks align with those on the

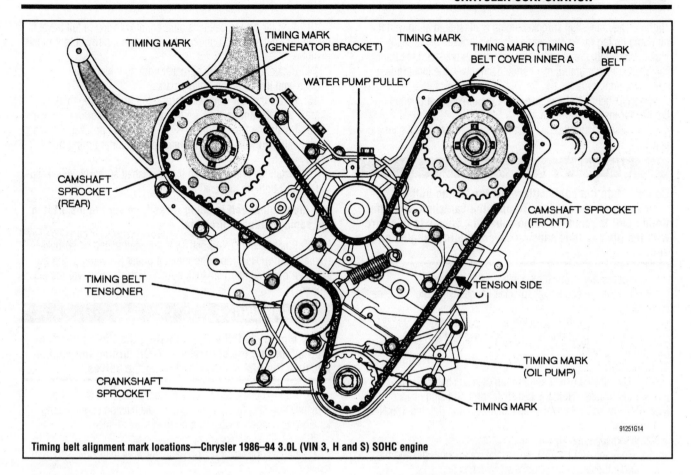

Timing belt alignment mark locations—Chrysler 1986–94 3.0L (VIN 3, H and S) SOHC engine

rear. Rotate the crankshaft so the timing mark aligns with the mark on the front cover.

18. Install the timing belt on the crankshaft sprocket and while keeping the belt tight on the tension side, install the belt on the front camshaft sprocket.

19. Install the belt on the water pump pulley, then the rear camshaft sprocket and the tensioner.

20. Rotate the front camshaft counterclockwise to tension the belt between the front camshaft and the crankshaft. If the timing marks became misaligned, repeat the procedure.

21. Install the crankshaft sprocket flange.

22. Loosen the tensioner bolt and allow the spring to apply tension to the belt.

23. Turn the crankshaft 2 full turns in the clockwise direction until the timing marks align again. Now that the belt is properly tensioned, torque the tensioner lock bolt to 21 ft. lbs. (29 Nm). Measure the belt tension between the rear camshaft sprocket and the crankshaft with belt tension gauge. The specification is 46–68 lbs. (210–310 N).

24. Install the timing covers. Make sure all pieces of packing are positioned in the inner grooves of the covers when installing.

25. Install the crankshaft pulley. Tighten the bolt to 108–116 ft. lbs. (150–160 Nm).

26. Install the engine support bracket.

27. Install the power steering pump and reconnect wire harness at the power steering pump pressure switch.

28. Install the engine mounting bracket and remove the engine support fixture.

29. Install the tension pulleys and drive belts.

30. Install the cruise control actuator.

31. Install the engine undercover.

32. Connect the negative battery cable and road test the vehicle.

1995–96

The timing belt can be inspected by removing the upper front outer timing belt cover.

Working on any engine (especially overhead camshaft engines) requires much care be given to valve timing. It is good practice to set the engine up at Top Dead Center (TDC) of the compression stroke of the No. 1 cylinder firing position before beginning work. Verify that all timing marks on the crankshaft and camshaft sprockets are properly aligned before removing the timing belt and starting camshaft service. This serves as a point of reference for all work that follows. Valve timing is very important and engine damage will result if the work is incorrect.

1. Disconnect the negative battery cable.

✳ CAUTION

Wait at least 90 seconds after the negative battery cable is disconnected to prevent possible deployment of the air bag.

2. Remove the accessory drive belts. Remove the engine mount insulator from the engine support bracket.

3. Remove the engine support bracket. Remove the crankshaft pulleys and torsional damper. Remove the timing belt covers.

4. Rotate the crankshaft until the sprocket timing marks are

aligned. The crankshaft sprocket timing mark should align with the oil pump timing mark. The rear camshaft sprocket timing mark should align with the generator bracket timing mark and the front camshaft sprocket timing mark should align with the inner timing belt cover timing mark.

5. If the belt is to be reused, mark the direction of rotation on the belt for installation reference.

6. Loosen the timing belt tensioner bolt and remove the timing belt.

✳✳ WARNING

Do not rotate the camshafts when the timing belt is removed from the engine. Turning the camshaft when the timing belt is removed could cause the valves to interfere with the pistons thus causing severe internal engine damage.

7. If necessary, remove the timing belt tensioner.

8. Remove the crankshaft sprocket flange shield and crankshaft sprocket.

9. Hold the camshaft sprocket using spanner tool MB990775 or equivalent, and remove the camshaft sprocket bolt and washer. Remove the camshaft sprocket.

To install:

10. Install the camshaft sprocket on the camshaft with the retaining bolt and washer. Hold the camshaft sprocket using spanner tool MB990775 or equivalent, and tighten the bolt to 70 ft. lbs. (95 Nm).

11. Install the crankshaft sprocket.

12. If removed, install the timing belt tensioner and tensioner spring. Hook the spring upper end to the water pump pin and the lower end to the tensioner bracket with the hook out.

13. Turn the timing belt tensioner counterclockwise full travel in the adjustment slot and tighten the bolt to temporarily hold it in this position.

14. Rotate the crankshaft sprocket until its timing mark is aligned with the oil pump timing mark.

15. Rotate the rear camshaft sprocket until its timing mark is aligned with the timing mark on the generator bracket.

16. Rotate the front (radiator side) camshaft sprocket until its mark is aligned with the timing mark on the inner timing belt cover.

17. Install the timing belt on the crankshaft sprocket while keeping the belt tight on the tension side.

➡ **If the original belt is being reused, be sure to install it in the same rotational direction.**

18. Position the timing belt over the front camshaft sprocket (radiator side). Next, position the belt under the water pump pulley, then over the rear camshaft sprocket and finally over the tensioner.

✳✳ WARNING

If any binding is felt when adjusting the timing belt tension by turning the crankshaft, STOP turning the engine, because the pistons may be hitting the valves.

19. Apply rotating force in the opposite direction to the front camshaft sprocket (radiator side) to create tension on the timing belt tension side. Check that all timing marks are aligned.

20. Install the crankshaft sprocket flange.

21. Loosen the tensioner bolt and allow the tensioner spring to tension the belt.

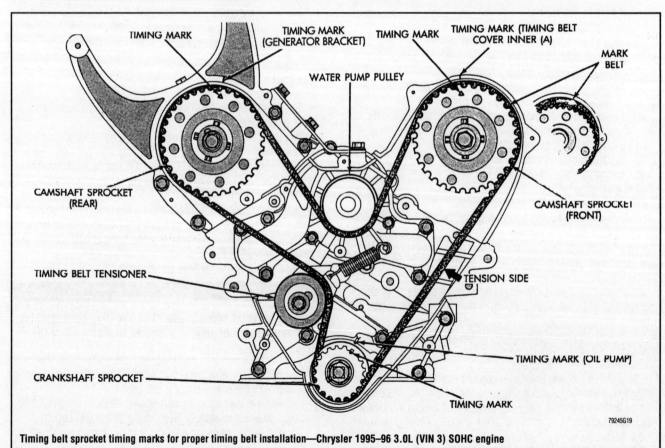

Timing belt sprocket timing marks for proper timing belt installation—Chrysler 1995–96 3.0L (VIN 3) SOHC engine

22. Rotate the crankshaft 2 full turns in a clockwise direction. Turn the crankshaft smoothly and in a clockwise direction only.

23. Again align the timing marks. If all marks are aligned, tighten the tensioner bolt to 21 ft. lbs. (28 Nm). Otherwise repeat the installation procedure.

24. Install the timing belt covers. Install the engine support bracket. Tighten the support bracket mounting bolts to 35 ft. lbs. (47 Nm).

25. Install the engine mount insulator, torsional damper and crankshaft pulleys. Tighten the crankshaft pulley bolt to 112 ft. lbs. (151 Nm).

26. Install the accessory drive belts and adjust them to the proper tension.

27. Reconnect the negative battery cable.

28. Run the engine and check for proper operation. Road test the vehicle.

3.0L (VIN B, C, K & J) DOHC ENGINES

1. Disconnect the negative battery cable.

✖ CAUTION

Wait at least 90 seconds after the negative battery cable is disconnected to prevent possible deployment of the air bag.

2. Position the engine so the No. 1 cylinder is at Top Dead Center (TDC) of the compression stroke. Disconnect the negative battery cable.

3. Remove the engine undercover.

4. Remove the cruise control actuator.

5. Remove the alternator. Remove the air hose and pipe.

6. Remove the belt tensioner assembly and the power steering belt.

7. Remove the crankshaft pulley.

8. Disconnect the brake fluid level sensor.

9. Remove the timing belt upper cover.

10. Using the proper equipment, slightly raise the engine to take the weight off the side engine mount. Remove the engine mount bracket.

11. Remove the alternator/air conditioner idler pulley.

12. Remove the engine support bracket. The mounting bolts are different lengths; mark them for proper installation.

13. Remove the timing belt lower cover. Timing bolt cover mounting bolts are different in length, note their position during removal.

14. If the same timing belt will be reused, mark the direction of the timing belt's rotation for installation in the same direction. Make sure the engine is positioned so the No. 1 cylinder is at the TDC of its compression stroke and the timing marks are aligned with the engine's timing mark indicators on the valve covers or head.

15. Loosen the center bolt of tensioner pulley and unbolt auto-tensioner assembly. The auto-tensioner assembly must be reset to correctly adjust belt tension. Remove the timing belt.

✖ WARNING

Do not rotate the camshafts when the timing belt is removed from the engine. Turning the camshaft when the timing belt is removed could cause the valves to interfere with the pistons thus causing severe internal engine damage.

16. Remove and position the auto-tensioner into a vise with soft jaws. The plug at the rear of tensioner protrudes, be sure to use a washer as a spacer to protect the plug from contacting vise jaws.

17. Slowly push the rod into the tensioner until the set hole in rod is aligned with set hole in the auto-tensioner.

18. Insert a 0.055 in. (1.4mm) wire into the aligned set holes. Unclamp the tensioner from vise and install to vehicle. Tighten tensioner to 17 ft. lbs. (24 Nm).

19. On 1991 DOHC 3.0L engines, clean and inspect both auto tensioner mounting bolts. Coat the threads of the old bolts with Mopar thread sealer 4318034 or equivalent. If new bolts are installed, inspect the heads of the new bolts. If there is white paint on the bolt head, no sealer is required. If there is no paint on the head of the bolt, apply a coat of thread sealer to the bolt. Install both bolts and tighten to 17 ft. lbs. (24 Nm).

To install:

✖ WARNING

Turning the camshaft sprocket when the timing belt is removed could cause the valves to interfere with the pistons.

20. Align the mark on the crankshaft sprocket with the mark on the front case. Then move the sprocket 3 teeth clockwise to lower the piston so the valves do not touch the piston if the camshafts are being moved.

21. Turn each camshaft sprocket 1 at a time to align the timing marks with the mark on the valve cover or head. If the intake and exhaust valves of the same cylinder are opened simultaneously, they could interfere with each other. Therefore, if any resistance is felt, turn the other camshaft to move the valve.

22. Using paper clips to secure the timing belt to sprockets, install the timing belt in the following order. Be sure camshafts to cylinder heads and crankshaft to front cover timing marks are aligned.

  a. Exhaust camshaft sprocket (front bank).
  b. Intake camshaft sprocket (front bank).
  c. Water pump pulley.
  d. Intake camshaft sprocket (rear bank).
  e. Exhaust camshaft sprocket (rear bank).
  f. Idler pulley.
  g. Crankshaft pulley.
  h. Tensioner pulley.

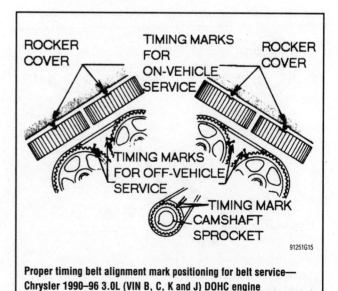

Proper timing belt alignment mark positioning for belt service—
Chrysler 1990–96 3.0L (VIN B, C, K and J) DOHC engine

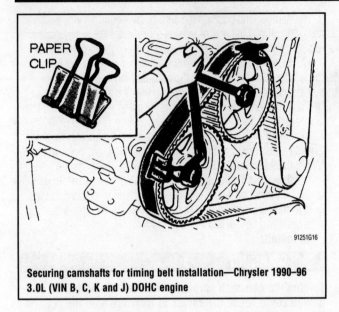

Securing camshafts for timing belt installation—Chrysler 1990–96 3.0L (VIN B, C, K and J) DOHC engine

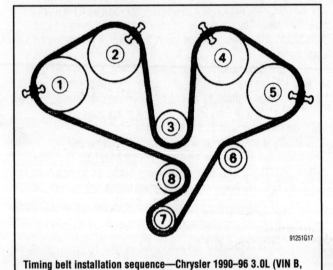

Timing belt installation sequence—Chrysler 1990–96 3.0L (VIN B, C, K and J) DOHC engine

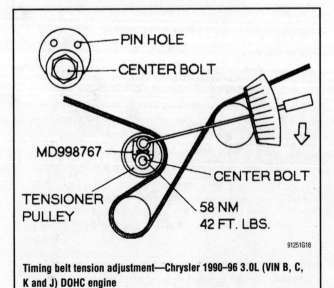

Timing belt tension adjustment—Chrysler 1990–96 3.0L (VIN B, C, K and J) DOHC engine

➡ **Since the camshaft sprockets turn easily, secure them with box wrenches to install timing belt.**

23. Align all timing marks and raise tensioner pulley against belt to remove slack, snug tensioner bolt.

24. Loosen the center bolt on the tensioner pulley. Using tool MD998767 or equivalent and a torque wrench, apply a torque of 86 inch lbs. (10 Nm). Tighten the tensioner bolt to 42 ft. lbs. (58 Nm) and make sure the tensioner does not rotate with the bolt.

25. Remove the set wire attached to the auto tensioner.

26. Rotate the crankshaft 2 complete turns clockwise and let it sit for approximately 5 minutes. Then, check that the set pin can easily be inserted and removed from the hole in the auto tensioner.

➡ **Even if the set pin cannot be easily inserted, the auto tensioner is normal if its rod protrusion is within specification.**

27. Measure the auto tensioner protrusion (the distance between the tensioner arm and auto tensioner body) to ensure that it is within 0.15–0.18 in. (3.8–4.5mm). If out of specification, repeat adjustment procedure until the specified value is obtained.

28. Check again that the timing marks on all sprockets are in proper alignment.

29. Make sure all pieces of packing are positioned in the inner grooves of the lower cover, position cover on engine and install mounting bolts in their original location.

30. Install the engine support bracket and secure using mounting bolts in their original location. Lubricate the reaming area of the reamer bolt and tighten slowly.

31. Install the idler pulley.

32. Install the engine mount bracket. Remove the engine support fixture.

33. Make sure all pieces of packing are positioned in the inner grooves of the upper cover and install.

34. Connect the brake fluid level sensor.

35. Install the crankshaft pulley. Tighten the bolt to 130–137 ft. lbs. (180–190 Nm).

36. Install the belt tensioner assembly and the power steering belt.

37. Install the air hose and pipe.

38. Install the alternator.

39. Install the cruise control actuator.

40. Install the engine undercover.

41. Connect the negative battery cable.

3.2L (VIN J) & 3.5L (VIN F, G) ENGINES

Use care when servicing a timing belt. Valve timing is absolutely critical to engine performance. If the valve timing marks on all drive sprockets are not properly aligned, engine damage will result. If only the belt and tensioner are being serviced, do not loosen the camshaft drive sprockets unless they are to be replaced. The sprockets have oversized openings and can be rotated several degrees in each direction on their shafts. This means the sprockets must be retimed, requiring some special tools.

✸✸ CAUTION

Fuel injection systems remain under pressure, even after the engine has been turned off. The fuel system pressure must be relieved before disconnecting any fuel lines. Failure to do so may result in fire and/or personal injury.

1. Disconnect the negative battery cable.
2. Rotate the engine to Top Dead Center (TDC) on the compression stroke for cylinder No. 1.
3. Release the fuel system pressure using the recommended procedure.
4. Place a pan under the radiator and drain the coolant.
5. Remove the radiator and cooling fan assemblies.
6. Remove the accessory drive belts.
7. Remove the upper radiator hose.
8. Remove the crankshaft damper with a quality puller tool gripping the inside of the pulley.
9. Remove the stamped steel cover. Do not remove the sealer on the cover; it may be reusable.
10. Remove the left-side cast cover. If necessary, remove the lower belt cover, located behind the crankshaft damper.
11. If the timing belt is to be reused, mark the timing belt with the running direction for installation.
12. Align the camshaft sprockets with the marks on the rear covers.
13. Remove the timing belt and tensioner.
14. If it is necessary to service the camshaft sprockets, use the following procedure:

    a. Hold the camshaft sprocket with a 36mm box end wrench, loosen and remove the sprocket retaining bolt and washer.

➡**To remove the camshaft sprocket retainer bolt while the engine is in the vehicle, it may be necessary to raise that side of the engine due to the length of the retainer bolt. The right bolt is 8⅜ in. (213mm) long, while the left bolt is 10 in. (254mm) long. These bolts are not interchangeable and their original location during removal should be noted.**

    b. Remove the camshaft sprocket from the camshaft. The camshaft sprockets are not interchangeable from side-to-side.
    c. Remove the crankshaft sprocket using Puller L-4407A or equivalent.

To install:

15. If it was necessary to remove the camshaft sprockets, use the following procedure:

➡**This procedure can only be used when the camshaft sprockets have been loosened or removed from the camshafts. Each sprocket has a D-shaped hole that allows it to be rotated several degrees in each direction on its shaft. This design must be timed with the engine to ensure proper performance.**

    a. Install the crankshaft sprocket, using tool C-4685-C1, thrust bearing, washer and 12mm bolt.
    b. When the camshaft sprockets are loosened or removed, the camshafts must be timed to the engine. Install the Camshaft Alignment tools 6642-A or their exact equivalents, to the rear of the cylinder heads. These tools lock the camshafts in the proper position.
16. Preload the belt tensioner as follows:
    a. Place the tensioner in a vise the same way it is mounted on the engine.
    b. Slowly compress the plunger into the tensioner body.
    c. When the plunger is compressed into the tensioner body, install a pin through the body and plunger to retain the plunger in place until the tensioner is installed.
17. Install both camshaft sprockets to the appropriate shafts. The left camshaft sprocket has the DIS pick-up as part of the sprocket.

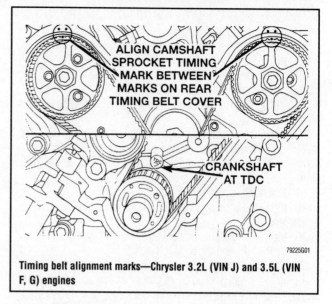

Timing belt alignment marks—Chrysler 3.2L (VIN J) and 3.5L (VIN F, G) engines

➡**The right bolt is 8⅜ in. (213mm) long, while the left bolt is 10 in. (254mm) long. These bolts are not interchangeable.**

18. Apply Loctite®271 or equivalent, to the threads of the camshaft sprocket retainer bolts and install to the appropriate shafts. Do not tighten the bolts at this time.
19. Align the camshaft sprockets between the marks on the rear belt covers.
20. Align the crankshaft sprocket with the TDC mark on the oil pump cover.
21. Install the timing belt, starting at the crankshaft sprocket and going in a counterclockwise direction. After the belt is installed on the right sprocket, keep tension on the belt until it is past the tensioner pulley.
22. Holding the tensioner pulley against the belt, install the timing belt tensioner into the housing and tighten to 21 ft. lbs. (28 Nm).
23. When the tensioner is in place, pull the retainer pin out to allow tensioner to extend to the pulley bracket.

➡**Be sure that the timing marks on the cam sprockets are still between the marks on the rear cover.**

24. Remove the spark plug in the No. 1 cylinder and install a dial indicator to check for Top Dead Center (TDC) of the piston. Rotate the crankshaft until the piston is exactly at TDC.
25. Hold the camshaft sprocket hex with a 36mm wrench and tighten the right camshaft sprocket bolt to 75 ft. lbs. (102 Nm) plus an additional 90 degree turn. Tighten the left camshaft sprocket bolt to 85 ft. lbs. (115 Nm) plus an additional 90 degree turn.
26. Remove the dial indicator. Install the spark plug and tighten to 20 ft. lbs. (28 Nm).
27. Remove the camshaft alignment tools from the back of the cylinder heads and install the cam covers with new O-rings.
28. Tighten the fasteners to 20 ft. lbs. (27 Nm). Repeat this procedure on the other camshaft.
29. Rotate the crankshaft sprocket 2 revolutions and check for proper alignment of the timing marks on the camshaft and the crankshaft. If the timing marks do not align, repeat the procedure.
30. Before installing, inspect the sealer on the stamped steel

cover. If some sealer is missing, use MOPAR Silicone Rubber Adhesive sealant or equivalent to replace the missing sealer.

31. Install the lower belt cover behind the crankshaft damper, if necessary.

32. Install the stamped steel cover and the left-side cast cover. Tighten the 6mm bolts to 105 inch lbs. (12 Nm), the 8mm bolts to 21 ft. lbs. (28 Nm) and the 10mm bolts to 40 ft. lbs. (54 Nm).

33. Install the crankshaft damper using special tool L-4524, a 6 in. (15cm) long bolt, thrust bearing and washer or equivalent damper installation tools. Tighten the center bolt to 85 ft. lbs. (115 Nm).

34. Install the upper radiator hose.

35. Install the accessory drive belts and adjust them to the proper tension.

36. Install the radiator and cooling fan assemblies.

37. Refill and bleed the cooling system.

38. Connect the negative battery cable.

39. With the radiator cap off so coolant can be added, run the engine. Watch for leaks and listen for unusual engine noises.

Ford Motor Company

1.3L (VIN H & K) ENGINE

1988–94

1. Disconnect the negative battery cable. Remove the drive belts.

2. Remove the 3 water pump pulley attaching bolts and remove the water pump pulley.

3. Raise and safely support the vehicle.

4. Remove the right front wheel and tire assembly and the right inner fender panel.

5. Remove the 4 attaching bolts and the screws from the crankshaft pulley. Remove the spacer and outer pulley, if equipped. Remove the inner spacer, inner pulley and the baffle or guide plates, as required.

6. Remove the attaching bolts and the upper and lower covers.

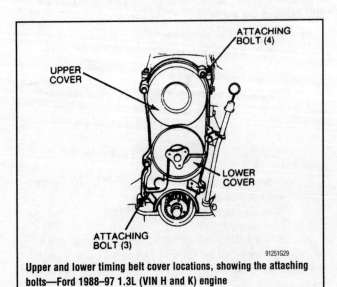

Upper and lower timing belt cover locations, showing the attaching bolts—Ford 1988–97 1.3L (VIN H and K) engine

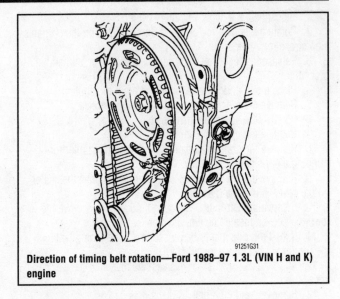

Direction of timing belt rotation—Ford 1988–97 1.3L (VIN H and K) engine

7. Mark the direction of rotation of the timing belt, if the belt is to be reused.

8. Remove the timing belt tensioner spring and retaining bolt. Remove the timing belt.

To install:

9. Align the camshaft and crankshaft timing marks with the marks located on the cylinder head and oil pump housing.

10. If reusing the original timing belt, install the timing belt with the mark made indicating the direction of rotation.

11. Install the timing belt tensioner spring and cover on the pulley. Position the tensioner and spring assembly on the engine and install the attaching bolt. Do not tighten the bolt at this time.

12. Reconnect the free end of the spring to the spring anchor. Torque the tensioner bolt to 14–19 ft. lbs. (19–26 Nm).

13. Install the upper and lower covers. Install the attaching bolts and tighten to 69–95 inch lbs. (8–11 Nm).

14. Install the crankshaft pulley baffle with the curved lip facing outward or install the large guide plate and then the small guide plate, as required.

15. Install the inner pulley with the deep recess facing outward. Install the spacer and then the outer pulley, spacer and screws. Install the pulley bolts and tighten to 109–152 inch lbs. (12–17 Nm).

16. Install the inner fender panel and the wheel and tire assembly. Lower the vehicle.

17. Install the water pump pulley and tighten the bolts to 36–45 ft. lbs. (49–61 Nm).

18. Install the drive belts. Connect the negative battery cable.

1995–97

1. Disconnect the negative battery cable.

2. Remove the accessory drive belts.

3. Remove the 3 water pump pulley attaching bolts and remove the water pump pulley.

4. Raise and safely support the vehicle on jackstands.

5. Remove the right front wheel and tire assembly and the right inner fender panel.

6. Remove the 4 attaching bolts and the screws from the crankshaft pulley. Remove the spacer and outer pulley, if equipped. Remove the inner spacer, inner pulley and the baffle or guide plates, as required.

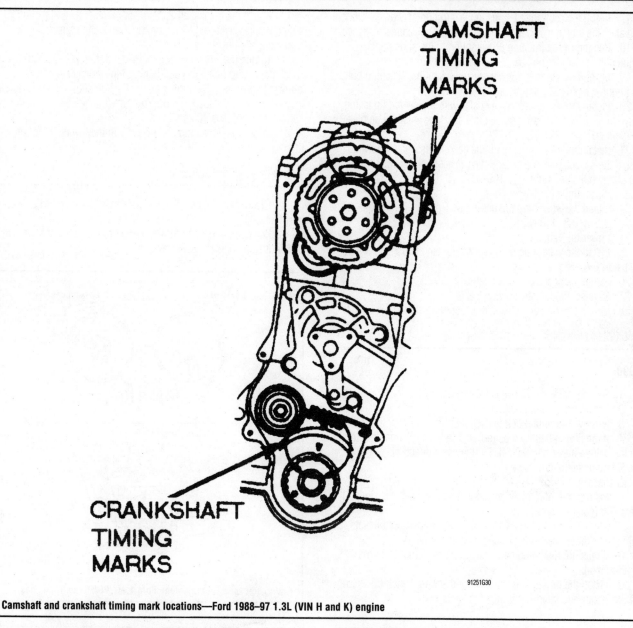

CAMSHAFT
TIMING
MARKS

CRANKSHAFT
TIMING
MARKS

91251G30

Camshaft and crankshaft timing mark locations—Ford 1988–97 1.3L (VIN H and K) engine

7. Remove the attaching bolts and the upper and lower covers.

8. Rotate the crankshaft until the sprocket timing marks are aligned.

9. Remove the timing belt tensioner spring, spring cover and timing belt tensioner bolt. Remove the timing belt.

➡**If the timing belt is to be reused, mark the direction of rotation on the belt, using a crayon, so the belt can be reinstalled in the same direction.**

10. If the camshaft sprocket requires removal, proceed as follows:
    a. Hold the camshaft stationary with an open end wrench and remove the camshaft sprocket retaining bolt.
    b. Pull the camshaft sprocket with the dowel pin off of the camshaft. Use care not to drop the dowel pin.

11. If the crankshaft sprocket requires removal, proceed as follows:
    a. Remove the crankshaft pulley retaining bolt.
    b. Pull the crankshaft pulley hub, sprocket and key from the crankshaft. Be sure not to drop the crankshaft key.

To install:

12. If removed, install the crankshaft sprocket as follows:
    a. Install the sprocket with the key onto the crankshaft.
    b. Install the crankshaft pulley hub.
    c. Clean the threads of the crankshaft pulley bolt and coat with a non-hardening sealer.
    d. Install the bolt and tighten to 80–85 ft. lbs. (108–118 Nm).

13. If removed, install the camshaft sprocket as follows:
    a. Position the sprocket and dowel pin to the camshaft and install the retaining bolt.
    b. Hold the camshaft stationary with an open end wrench and tighten the retaining bolt to 36–45 ft. lbs. (49–61 Nm).

14. Align the camshaft and crankshaft timing marks with the marks located on the cylinder head and oil pump housing.

15. If reusing the original timing belt, install the timing belt with the mark made indicating the direction of rotation.

16. Install the timing belt tensioner spring and cover on the tensioner. Position the tensioner and spring assembly on the engine and install the attaching bolt. Do not tighten the bolt at this time.

17. Rotate the crankshaft 2 turns in the direction of normal rotation and align the timing marks. Ensure all marks are still correctly aligned.

18. Reconnect the free end of the spring to the spring anchor. Tighten the tensioner bolt to 14–19 ft. lbs. (19–26 Nm).

19. Install the upper and lower covers. Install the attaching bolts and tighten to 71–97 inch lbs. (8–11 Nm).

20. Install the crankshaft pulley baffle with the curved lip facing outward, or install the large guide plate, then the small guide plate, as required.

21. Install the inner pulley with the deep recess facing outward. Install the spacer, then the outer pulley, spacer and screws. Install the pulley bolts and tighten to 109–152 inch lbs. (12–17 Nm).

22. Install the inner fender panel.

23. Install the wheel and tire assembly. Tighten the lug bolts to 65–87 ft. lbs. (88–118 Nm).

24. Lower the vehicle.

25. Install the water pump pulley and tighten the bolts to 36–45 ft. lbs. (49–61 Nm).

26. Install the accessory drive belts.

27. Connect the negative battery cable.

28. Run the engine and check for proper operation.

1.6L (VIN 5) ENGINE

1990

1. Disconnect the negative battery cable.
2. Remove the timing belt covers.
3. Remove the timing belt tensioner spring and retaining bolt.
4. If the timing belt is to be reused, mark the rotation direction on the belt so it can be reinstalled in the same direction.
5. Remove the timing belt.

To install:

6. Inspect the timing belt, tensioner and sprockets for signs of wear and replace, as necessary.

7. Align the marks on the camshaft and crankshaft sprockets with the cylinder head and oil pump alignment marks.

8. If reusing the timing belt, install it in the direction of the rotation mark.

9. Install the timing belt tensioner and spring. Install the spring on its anchor and hand-tighten the tensioner bolt.

10. Rotate the crankshaft 2 complete revolutions and realign the timing marks. Reaffirm that the timing marks are aligned; if not, repeat the alignment procedure.

11. Torque the tensioner bolt to 14–19 ft. lbs. (19–25 Nm) and check the timing belt deflection between the crankshaft and camshaft sprockets. The timing belt deflection should be 0.35–0.39 in. (9–10mm) at 22 lbs. (10 kg). Pressure. If the deflection is not correct, repeat Steps 10 and 11.

12. Install the remaining components in the reverse order of their removal.

1.6L (VIN Z & 6) ENGINES

1. Disconnect the negative battery cable. Raise and safely support the vehicle.

2. Remove the right front wheel and tire assembly and remove the right splash guard. Lower the vehicle.

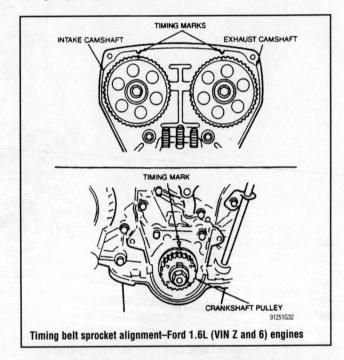

Timing belt sprocket alignment—Ford 1.6L (VIN Z and 6) engines

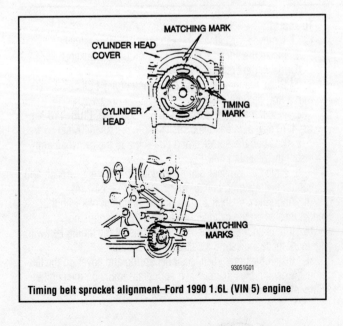

Timing belt sprocket alignment—Ford 1990 1.6L (VIN 5) engine

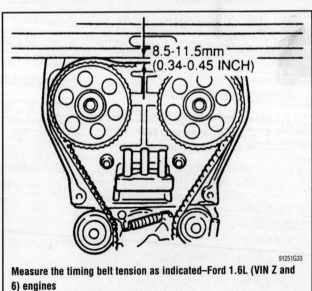

Measure the timing belt tension as indicated—Ford 1.6L (VIN Z and 6) engines

3. Remove the spark plugs. Set the engine position to Top Dead Center (TDC) of the compression stroke on the No. 1 cylinder.

4. Remove the alternator and power steering belts. Remove the oil dipstick and the water pump pulley.

5. Remove the crankshaft pulley, damper and baffle plate.

6. Remove the upper, center and lower timing belt covers.

7. Remove the timing belt tension spring and loosen the timing belt tension pulley.

8. Support the engine with a floor jack and remove the right engine mount.

9. Mark the timing belt rotation direction and remove the timing belt.

10. Inspect the timing belt and timing sprockets for wear and/or damage and replace as necessary.

11. Check the free length of the timing belt tension spring. It should be 2.315 in. (58.8mm). Replace if out of specification.

To install:

Make sure the timing marks are properly positioned on the camshafts and crankshaft. The intake camshaft should have the letter **I** aligned with the arrow on the belt cover. The exhaust camshaft should have the letter **E** aligned with the arrow on the belt covers

12. The crankshaft key should align with the arrow.

13. Tighten the tension pulley with the tension spring fully extended.

14. Install the timing belt. Keep tension on the opposite side of the tensioner as tight as possible. Make sure the rotation mark on the belt is correct.

15. Turn the crankshaft 2 full turns. Check the alignment marks. If any mark is not aligned, remove the timing belt and reset the timing.

16. Loosen the tension pulley retaining bolt to allow the tension spring to tighten the belt.

17. Tighten the tension pulley retaining bolt to 27–38 ft. lbs. (37–52 Nm). Rotate the engine 2 full turns. Make sure the timing marks are aligned.

18. Measure the timing belt tension between the camshaft pulleys. Belt deflection should be 0.33–0.45 in. (8.5–11.5mm). If incorrect, loosen the tension pulley and repeat the procedure. If proper tension cannot be achieved, replace the tension spring.

19. Install the lower, center and upper timing belt covers. Tighten the retaining bolts to 71–97 inch lbs. (8–11 Nm).

20. Install the right engine mount and lower the engine. Tighten the retaining nuts to 44–63 ft. lbs. (60–85 Nm).

21. Install the crankshaft pulley, damper and baffle. Tighten the baffle and damper retaining screws to 109–152 inch lbs. (12–17 Nm). Tighten the pulley retaining bolts to 109–152 inch lbs. (12–17 Nm).

22. Install the water pump pulley. Tighten the retaining bolts to 69–95 inch lbs. (8–11 Nm).

23. Install the alternator and power steering belts. Install the dipstick. Raise and safely support the vehicle.

24. Install the splash guard and the right front wheel and tire assembly.

25. Lower the vehicle and install the spark plugs. Start the engine and check for proper operation.

1.8L (VIN 8) ENGINE

1. Disconnect the negative battery cable.
2. Remove the timing belt upper cover and gasket.
3. Remove the accessory drive belts.
4. Remove the water pump pulley bolts and remove the pulley.

5. Raise and safely support the vehicle on jackstands.

6. Remove the right front wheel and tire assembly.

7. Remove the right upper and lower splash-shields.

8. Remove the timing belt middle and lower covers along with the gaskets.

9. Remove the crankshaft pulley hub bolt and hub.

10. Rotate the crankshaft and align the timing marks located on the camshaft sprockets and seal plate.

11. Check that the crankshaft sprocket and the oil pump are aligned.

➡**If the timing belt is to be reused, mark an arrow on the belt to indicate it's rotational direction for installation reference.**

12. Loosen the timing belt tensioner bolt.

13. Turn the timing belt tensioner counterclockwise and hand-tighten the tensioner bolt to relieve the tension on the timing belt.

14. Remove the timing belt.

15. If the camshaft sprockets are to be removed, continue as follows:

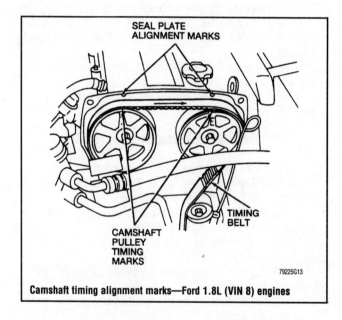

Camshaft timing alignment marks—Ford 1.8L (VIN 8) engines

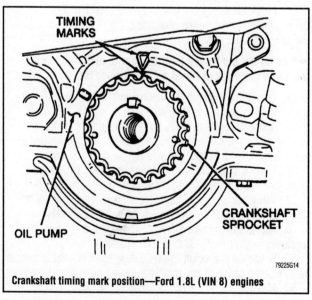

Crankshaft timing mark position—Ford 1.8L (VIN 8) engines

a. Disconnect and tag the ignition wires and vacuum lines blocking the removal of the cylinder head cover.

b. Remove the cylinder head cover retaining bolts and remove the cover and gasket.

c. While holding the camshaft with a wrench, remove the camshaft sprocket retaining bolt.

d. Remove the camshaft sprocket.

e. If removing both camshaft sprockets, tag the sprockets for identification at reassembly.

16. If removing the crankshaft sprocket, remove the crankshaft pulley bolt and hub, if not already done. Slide the crankshaft sprocket off the crankshaft.

17. Inspect the timing belt tensioner and spring, replace if necessary.

To install:

18. If the crankshaft sprocket was removed, install the crankshaft key with the tapered end facing the oil pump. Install the crankshaft sprocket onto the crankshaft while making sure to match the alignment grooves.

19. If removed, install the camshaft sprockets as follows

a. Turn the camshaft until the dowel pins face straight up.

b. Install the camshaft sprocket with the **I** mark straight up for the intake camshaft or with the **E** mark straight up for the exhaust camshaft.

c. Align the camshaft sprockets with the timing marks on the seal plate.

d. While holding each camshaft with a wrench, install the camshaft sprocket retaining bolts. Tighten the bolts to 36–45 ft. lbs. (49–61 Nm).

e. Install a new cylinder head cover gasket onto the cylinder head.

f. Place the cylinder head cover into its mounting position and install the retaining bolts. Tighten the cylinder head cover bolts to 43–78 inch lbs. (4.9–8.8 Nm).

g. Install the ignition wires to the spark plugs and connect the vacuum hoses to the cylinder head cover.

20. Temporarily secure the timing belt tensioner in the far left position.

21. Verify that the timing marks on the crankshaft sprocket and the oil pump are aligned.

22. Verify that the timing marks on the camshaft sprockets and the seal plate are aligned.

23. Install the timing belt in a counterclockwise motion. Be sure there is no looseness on the idler side of the timing belt or between the camshaft sprockets.

➡ **If using the old timing belt, be sure to install the belt in the same direction of travel as it was removed.**

24. Loosen the timing belt tensioner bolt. Allow the tensioner spring to apply tension to the timing belt.

25. Rotate the crankshaft 1⅚ turns clockwise and align the timing belt pulley mark with the tension set mark which is located at approximately the 10 o'clock position.

26. Turn the crankshaft 2 turns clockwise and align the crankshaft sprocket with the tension set mark on the oil pump.

27. Verify that all timing marks are aligned. If not, remove the timing belt and repeat the installation procedures.

28. Apply tension to the timing belt tensioner and tighten the tensioner lockbolt to 27–38 ft. lbs. (37–52 Nm).

29. Rotate the crankshaft 2⅙ (780 degrees) turns clockwise and verify that the camshaft and crankshaft timing marks are aligned.

30. Measure the timing belt deflection by applying 22 lbs. (10

kg) of pressure on the timing belt between the camshaft sprockets. The timing belt deflection should be 0.35–0.45 in. (9–11.5mm). If necessary to adjust the timing belt deflection, rotate the crankshaft 2 turns clockwise and ensure that the timing marks are still aligned. If the timing marks are not aligned, repeat the installation procedure.

31. Install the crankshaft pulley hub and tighten the retaining bolt to 80–87 ft. lbs. (108–118 Nm).

32. Install the crankshaft pulley and washer. Install the retaining bolts and tighten to 109–152 inch lbs. (12–17 Nm).

33. Install the timing belt middle and lower covers with the gaskets. Tighten the middle and lower timing belt cover retaining bolts to 65–95 inch lbs. (7.8–11 Nm).

34. Install the power steering drive belt.

35. Install the water pump pulley and retaining bolts. Tighten the bolts to 69–95 inch lbs. (7.8–11.0 Nm).

36. Install the alternator/water pump drive belt.

37. Install the splash-shields. Tighten the bolts to 69–95 inch lbs. (7.8–11.0 Nm).

38. Install the right wheel and tire assembly. Tighten the lug nuts to 65–87 ft. lbs. (88–118 Nm).

39. Lower the vehicle.

40. Install the timing belt upper cover and gasket. Tighten the bolts to 69–95 inch lbs. (7.8–11.0 Nm).

41. Connect the negative battery cable.

42. Run the engine and check for leaks.

43. Road test the vehicle and check for proper engine operation.

1.6L (VIN 2, 4, 5, 7, 8) & 1.9L (VIN J, 9) ENGINES

1981–90

1. Disconnect the negative battery cable.

2. Remove the timing belt cover.

3. Align the timing mark on the camshaft sprocket with the timing mark on the cylinder head.

4. Install the timing belt cover and confirm that the timing mark on the crankshaft damper aligns with the Top Dead Center (TDC) of the compression stroke on the front cover.

5. Remove the timing belt cover.

6. Loosen both timing belt tensioner attaching bolts.

7. Pry the belt tensioner away from the belt as far as possible and tighten 1 of the tensioner attaching bolts.

8. Remove crankshaft damper as follows:

a. Properly support the engine and remove the right-side engine mount bolt.

b. Lower the engine at the right-side until the crankshaft damper bolt clears the frame rail and remove the damper bolt.

c. Raise the engine and remove the damper.

9. Remove the timing belt. If the belt is to be reused, mark the direction of rotation on the belt so it can be reinstalled in the same direction.

➡ **With the timing belt removed and the No. 1 piston at TDC, Do not rotate the camshaft. If the camshaft must be rotated, align the crankshaft damper 90 degrees BTDC.**

To install:

10. Install the timing belt over the sprockets in a counterclockwise direction starting at the crankshaft. Keep the belt span from the

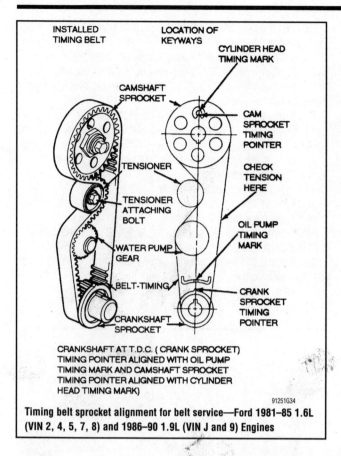

Timing belt sprocket alignment for belt service—Ford 1981–85 1.6L (VIN 2, 4, 5, 7, 8) and 1986–90 1.9L (VIN J and 9) Engines

crankshaft to the camshaft tight as the belt is installed over the remaining sprocket.

11. Loosen belt tensioner attaching bolts and allow the tensioner to snap against the belt.

12. Tighten 1 of the tensioner attaching bolts.

13. Install the crankshaft damper, driveplate and damper attaching bolt. Hold the crankshaft damper stationary and tighten the attaching bolt to 81–96 ft. lbs. (110–130 Nm).

14. To seat the belt on the sprocket teeth, proceed as follows:

    a. Connect the negative battery terminal.

    b. Crank engine several revolutions.

    c. Disconnect the negative battery terminal.

    d. Turn camshaft, as necessary, to align the timing pointer on the cam sprocket with the timing mark on the cylinder head.

➡**Do not turn the engine counterclockwise to align the timing marks.**

    e. Position the timing belt cover on the engine and check to see that the timing mark on the crankshaft aligns with the TDC pointer on the cover. If the timing marks do not align, remove the belt, align the timing marks and return to Step 10.

15. Loosen the belt tensioner attaching bolt tightened in Step 12. The tensioner spring will apply the proper load on the belt. Tighten the belt tensioner bolt.

➡**The engine must be at room temperature. Do not set belt tension on a hot engine.**

16. Install timing belt cover.

17. Install accessory drive belts.

18. Connect negative battery cable.

1991–96

1. Disconnect the negative battery cable.

2. Remove the accessory drive belt automatic tensioner and the accessory drive belt.

3. Remove the timing belt cover.

4. Align the timing mark on the camshaft sprocket with the timing mark on the cylinder head.

5. Confirm that the timing mark on the crankshaft sprocket is aligned with the timing mark on the oil pump housing.

6. Loosen the belt tensioner attaching bolt, pry the tensioner away from the timing belt and retighten the bolt.

7. Remove the spark plugs. Remove the right engine mount.

8. Raise and safely support the vehicle on jackstands.

9. Remove the right-side splash-shield.

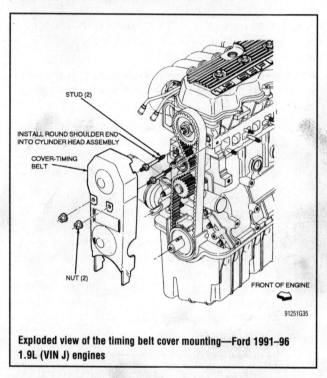

Exploded view of the timing belt cover mounting—Ford 1991–96 1.9L (VIN J) engines

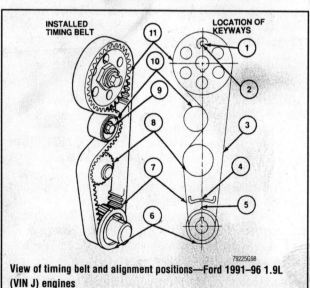

View of timing belt and alignment positions—Ford 1991–96 1.9L (VIN J) engines

10. Remove the flywheel inspection shield.

11. Use a suitable tool to hold the flywheel in place.

12. Remove the crankshaft damper bolt and washer and remove the bolt.

13. Remove the timing belt.

➡ **With the timing belt removed and the No. 1 piston at TDC, do not rotate the camshaft. If the camshaft must be rotated, align the crankshaft damper 90 degrees BTDC.**

To install:

14. Install the timing belt over the sprockets in a counterclockwise direction starting at the crankshaft. Keep the belt span from the crankshaft to the camshaft tight while installing over the remaining sprocket.

15. Loosen the belt tensioner attaching bolt, allowing the tensioner to snap against the belt.

16. Rotate the crankshaft clockwise 2 complete revolutions, stopping at TDC. This will allow the tensioner spring to load the timing belt.

➡ **Do not turn the engine counterclockwise to align the timing marks. Do not rotate the crankshaft with the spark plugs installed.**

17. Recheck the camshaft and crankshaft timing marks for alignment, to be sure the timing belt has not skipped a tooth during rotation. Repeat the procedure if the timing marks are not aligned.

18. Tighten the tensioner attaching bolt to 17–22 ft. lbs. (23–30 Nm).

19. Install the crankshaft damper and the bolt and washer. Tighten the bolt to 81–96 ft. lbs. (110–130 Nm).

20. Install the flywheel inspection shield.

21. Install the splash-shield and lower the vehicle.

22. Install the right engine mount. Install the spark plugs.

23. Install the timing belt cover.

24. Install the accessory drive belt automatic tensioner and the accessory drive belt.

25. Connect the negative battery cable.

2.0L (VIN 3) ENGINE

When installing a timing belt, tensioner spring (6L277) and retaining bolt (W700001-S309) must be purchased and properly installed on the engine. First check to see if these parts are already installed. The tensioner spring will adjust the timing belts tension and should not require further adjustments.

1. Disconnect the negative battery cable.

2. Remove the engine air intake resonators.

3. Label and remove the ignition wires from the spark plugs. Move the ignition wires aside.

4. Remove the spark plugs.

5. Manually rotate the crankshaft to Top Dead Center (TDC) for the No. 1 piston on its compression stroke. Be sure to align the timing marks.

6. Disconnect the retaining bracket for the power steering pressure hose from the engine lifting eye.

7. Install the Three Bar Engine Support D88L-6000-A or equivalent, onto the engine lifting eyes and slightly raise the engine.

8. Remove the upper camshaft timing belt cover retaining bolts and the cover from the engine.

➡ **Mark the location of the upper front engine support bracket before removing it from the engine support bracket.**

9. Remove the upper front engine support bracket retainer nuts, the bracket and the upper front engine support insulator.

10. If equipped, remove the wiring harness connector from the low coolant level sensor at the radiator coolant recovery reservoir.

11. Remove the radiator coolant recovery reservoir retainers and move the reservoir aside.

12. Remove the upper front engine support insulator.

13. Set the coolant recovery reservoir back into position temporarily.

14. Loosen the water pump pulley retaining bolts. Do not remove the bolts completely.

15. Remove the accessory drive belt.

16. Remove the drive belt idler pulley retaining bolt and pulley from the alternator mounting bracket.

17. Finish removing the water pump retaining bolts and remove the water pump pulley.

18. Remove the center camshaft timing belt cover retaining bolts and the cover from the engine.

19. Raise and safely support the vehicle on jack stands.

20. Remove the crankshaft pulley.

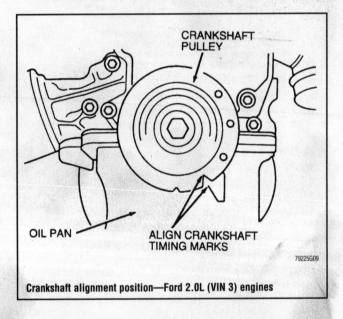

Crankshaft alignment position—Ford 2.0L (VIN 3) engines

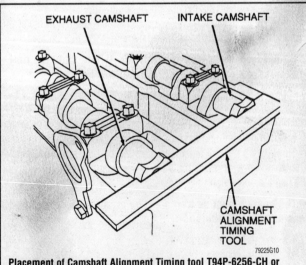

Placement of Camshaft Alignment Timing tool T94P-6256-CH or equivalent—Ford 2.0L (VIN 3) engines

21. Remove the lower camshaft timing belt cover bolts and the cover from the engine.

22. Remove the valve cover as follows:

 a. Disconnect the crankcase ventilation tube from the valve cover.

 b. Remove the retaining bolt and nut for the power steering pressure hose retaining bracket and move the hose aside.

 c. Remove the valve cover retaining bolts in a standard removal sequence starting from the outside of the valve cover and working toward the inside of the valve cover.

 d. Remove the valve cover and gasket from the engine.

23. Place Camshaft Alignment Timing tool T94P-6256-CH or equivalent, into the slots of both camshafts at the rear of the cylinder head to lock the camshafts into position.

24. Loosen the camshaft timing belt tensioner pulley retaining bolt and move the tensioner pulley to relieve the tension on the timing belt.

25. Temporarily tighten the tensioner in this position.

➡ **If the timing belt is to be reused, mark the belt for the direction of rotation before removing to prevent premature wear or failure.**

26. Remove the timing belt.

27. If required, remove the sprockets as follows:

 a. Hold the camshaft with the Camshaft Sprocket Holding tool T74P-6256-B or equivalent.

 b. Loosen and remove the camshaft sprocket retaining bolt.

 c. Remove the sprocket from the camshaft.

 d. Repeat the procedure for the 2nd camshaft sprocket.

 e. Remove the crankshaft sprocket.

To install:

28. Slide the crankshaft sprocket onto the crankshaft aligning the keyway.

29. Align the camshafts using the Camshaft Alignment Timing tool T94P-6256-CH.

30. Reinstall the sprockets onto the camshafts and loosely install the camshaft retaining bolts.

31. Tighten the camshaft sprocket retaining bolts to 47–53 ft. lbs. (64–72 Nm).

32. Loosely install the crankshaft pulley to verify that the engine is at TDC. Realign the marks if they have moved.

33. Verify that the camshafts are aligned.

➡ **It is recommended to purchase a tensioner spring and retaining bolt through the dealer parts to apply the proper tension for used or new belt installations. The spring is bolted to the tensioner assembly and becomes a part of the engine. Ignore this notice if the tensioner spring is already installed.**

34. Reinstall the retaining bolt (W700001-S309) into the hole provided in the cylinder block and place the tensioner spring (6L277) between the bolt and the camshaft timing belt tensioner pulley.

35. Tighten the retainer bolt to 71–97 inch lbs. (8–11 Nm).

36. Remove the crankshaft pulley and install the timing belt onto the crankshaft sprocket, then onto the camshaft sprockets working in a counterclockwise direction.

37. Tighten the camshaft sprocket retaining bolts to 47–53 ft. lbs. (64–72 Nm).

38. Be sure that the span of the camshaft timing belt between the crankshaft sprocket and the exhaust camshaft sprocket is not loose.

39. Be sure that the camshaft timing belt is securely aligned on all sprockets.

40. Reinstall the lower timing belt cover and tighten the retaining bolts to 53–71 inch lbs. (6–8 Nm).

41. Apply silicone sealer to the keyway of the crankshaft pulley and install. Tighten the retaining bolt to 81–89 ft. lbs. (110–120 Nm).

42. Inspect the timing mark on the crankshaft pulley to verify that the engine is still at TDC.

43. Loosen the camshaft timing belt tensioner pulley retaining bolt and allow the tensioner spring attached to the pulley to draw the tensioner pulley against the camshaft timing belt.

44. Remove the camshaft alignment timing tool from the camshafts at the rear of the engine.

45. Turn the crankshaft 2 revolutions in a clockwise direction.

46. Tighten the camshaft timing belt tensioner pulley retaining bolt to 26–30 ft. lbs. (35–40 Nm).

47. Recheck that the crankshaft timing mark is at TDC for the No. 1 piston, and that both camshafts are in alignment using the camshaft alignment timing tool.

➡ **A slight adjustment of the camshafts to allow the insertion of the camshaft alignment timing tool is permissible as long as the crankshaft stays at the TDC location.**

48. Camshaft Sprocket Holding tool T74P-6256-b or equivalent, can be used to move the camshaft sprocket(s) if a slight adjustment is required.

49. If a camshaft is not properly aligned, perform the following procedure:

 a. Loosen the retaining bolt securing the sprocket to the camshaft while holding the camshaft sprocket from turning with the sprocket holding tool.

 b. Turn the camshaft until the camshaft alignment timing tool can be installed.

 c. Verify that the crankshaft timing mark is at TDC for the No. 1 cylinder.

 d. While holding the camshaft sprocket with the camshaft sprocket holding tool, tighten the retaining bolt to 47–53 ft. lbs. (64–72 Nm).

 e. Remove the tool and rotate the crankshaft 2 revolutions (clockwise).

 f. Verify that the camshafts are aligned and that the crankshaft is at TDC for the No. 1 cylinder.

50. Reinstall the valve cover as follows:

 a. Clean the gasket sealing surfaces.

 b. Inspect the valve cover gasket and O-rings; replace as required.

 c. Reinstall the valve cover retaining bolts and tighten them in a standard sequence starting from the center and working towards the outside of the valve cover to 53–71 inch lbs. (6–8 Nm).

 d. Reinstall the power steering hose retaining bracket and the power steering hose.

 e. Reinstall the crankcase ventilation tube to the valve cover.

51. Position the center camshaft timing belt cover.

52. Reinstall the center camshaft timing belt cover retaining bolts and tighten them to 53–71 inch lbs. (6–8 Nm).

53. Reinstall the water pump pulley and the retaining bolts. Reinstall the bolts, finger-tight.

54. Reinstall the drive belt idler pulley.

55. Reinstall the drive belt idler pulley retaining bolt and tighten it to 35 ft. lbs. (48 Nm).

56. Reinstall the accessory drive belt.

57. Tighten the water pump pulley retaining bolts to 89–124 inch lbs. (10–14 Nm).

58. Move the radiator coolant recovery reservoir aside.

59. Reinstall the upper front engine support insulator.

60. Position the radiator coolant recovery reservoir and install the retainers.

61. If equipped, install the wiring harness to the low coolant level sensor on the coolant recovery reservoir.

62. Reinstall the upper front engine support bracket to the engine and the upper front engine support insulator using the mark made during the removal procedure for reference.

63. Install the upper camshaft timing belt cover.

64. Reinstall the upper camshaft timing belt cover retaining bolts and tighten to 27–44 inch lbs. (3–5 Nm).

65. Remove the engine support.

66. Reinstall the retaining bracket for the power steering pressure hose to the engine lifting eye.

67. Reinstall the spark plugs and the ignition wires.

68. Reinstall the engine air intake resonators.

69. Replace the engine oil.

70. Reconnect the negative battery cable.

71. Run the engine and check for leaks and proper operation.

2.0L (VIN A) ENGINE

1. Disconnect the negative battery cable.

2. Label and disconnect the spark plug wires and clips from the cylinder head cover. Remove the ignition distributor with wiring and set it aside.

3. Remove the power steering hose brackets from the cylinder head cover. If necessary disconnect the crankshaft position sensor.

4. Disconnect the breather tube and PCV valve from the cylinder head cover.

5. Loosen the cylinder head cover bolts in 2–3 steps. Remove the cylinder head cover.

6. Remove the power steering belt shield. Loosen the power steering adjusting bolt, lockbolt and through-bolt and remove the power steering belt.

7. Loosen the alternator adjusting bolt and upper mounting bolt. Remove the alternator belt.

8. Support the engine with Engine Support tool 014-00750 or equivalent. Raise the engine slightly with a jack and remove the right-side engine support insulator (mount).

9. Remove the oil level indicator bolt and 4 upper timing belt cover bolts and remove the upper timing belt cover.

10. Raise and safely support the vehicle.

11. Remove the splash-shields. Using Holder tool T92C-6316-AH, or equivalent, hold the crankshaft pulley and remove the pulley bolt. Use a suitable puller to remove the pulley, then remove the guide plate.

12. Remove the 4 lower timing belt cover bolts and remove the lower timing belt cover.

13. Temporarily install the crankshaft pulley bolt.

14. Turn the crankshaft until the timing mark on the crankshaft sprocket aligns with the timing mark on the oil pump and the camshaft sprocket timing marks, E and I, align on the camshaft sprockets.

15. Lower the vehicle.

16. Insert camshaft sprocket holding tool T92C-6256-AH or equivalent, between the camshaft sprockets.

17. Turn the timing belt tensioner with an Allen wrench and remove the tensioner spring from the tensioner spring pin.

18. If the timing belt is to be reused, mark the direction of rotation on the timing belt. Remove the timing belt.

19. If it is necessary to remove the sprockets, remove the camshaft sprocket holding tool. Hold the camshaft by placing a suitable wrench on the hexagon which is cast into the camshaft. Place another wrench onto the camshaft sprocket retaining bolt and loosen the bolt.

➡ **Before removing the camshaft sprocket(s), be sure that the camshafts are still in alignment and tag each sprocket to the camshaft from which it was removed.**

20. Remove the camshaft sprocket bolt and the camshaft sprocket from the camshaft.

21. Repeat the camshaft sprocket removal procedure for the opposite camshaft if required.

22. Raise and safely support the vehicle.

23. Slide off the crankshaft sprocket and remove the crankshaft key.

To install:

24. Install the crankshaft key and slide the crankshaft sprocket into position.

25. Lower the vehicle.

26. Install the camshaft sprocket onto the proper camshaft, making sure to align the dowel pin.

27. Be sure that the I and E are in alignment.

28. Install the camshaft sprocket bolt. Hold the hexagon on the camshaft with a suitable wrench and tighten the sprocket bolt to 35–48 ft. lbs. (47–65 Nm). Be sure the camshaft sprockets are still properly aligned and reinstall the sprocket holding tool.

29. Be sure the timing marks on the camshaft and crankshaft sprockets are still aligned.

30. Install the timing belt. If reusing the original timing belt, be sure it is installed in the same direction of rotation.

31. Turn the tensioner clockwise with an Allen wrench and install the tensioner spring. Remove the holding tool from between the camshaft sprockets.

32. Rotate the crankshaft clockwise 2 turns and align the timing marks. Be sure all marks are still correctly aligned.

➡ **The timing chain tensioner automatically adjusts the tension on the timing belt.**

33. Raise and safely support vehicle on jackstands.

34. Install the timing belt lower cover and tighten the 4 bolts to 71–88 inch lbs. (8–10 Nm).

35. Install the guide plate, crankshaft pulley and pulley bolt. Secure the pulley with the holder tool and tighten the bolt to 116–123 ft. lbs. (157–167 Nm).

36. Install the splash-shields and lower the vehicle.

37. Raise the engine slightly with the jack and install the right-side engine mount. Tighten the mount through-bolt to 63–86 ft. lbs. (86–116 Nm) and the mount attaching nuts to 54–75 ft. lbs. (74–103 Nm). Remove the engine support tool.

38. Install the upper timing belt cover and tighten the bolts to 71–88 inch lbs. (8–10 Nm).

39. Clean the cylinder head and valve cover mating surfaces thoroughly.

40. Apply silicone sealant to the cylinder head surface in the area adjacent to the front camshaft bearing caps. Apply sealant to a new gasket and install it on the cylinder head cover.

41. Install the cylinder head cover and tighten the bolts.

42. Install the power steering hose brackets and tighten the bolts to 71–88 inch lbs. (8–10 Nm). Connect the spark plug wires and

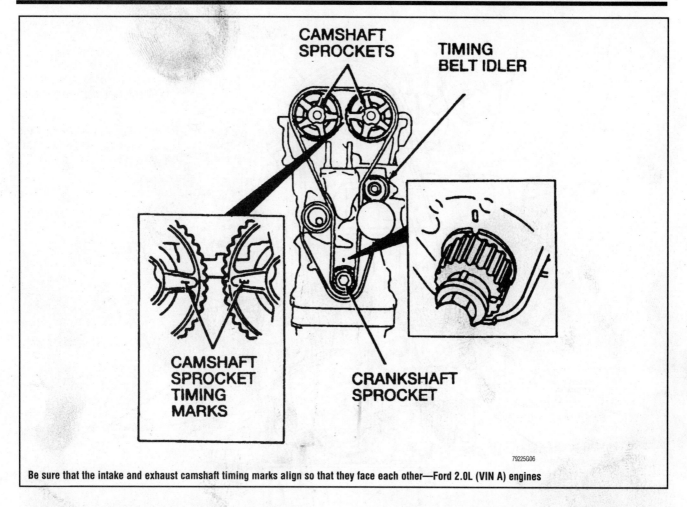

CAMSHAFT SPROCKETS

TIMING BELT IDLER

CAMSHAFT SPROCKET TIMING MARKS

CRANKSHAFT SPROCKET

79225G06

Be sure that the intake and exhaust camshaft timing marks align so that they face each other—Ford 2.0L (VIN A) engines

wire clips. Connect the breather tube and PCV valve. If necessary, connect the crankshaft position sensor.

43. Install the alternator belt and adjust the tension. Tighten the upper mounting bolt to 14–18 ft. lbs. (19–25 Nm) and the lower through-bolt to 27–38 ft. lbs. (37–52 Nm).

44. Install the power steering belt and adjust the tension. Tighten the through-bolt to 32–45 ft. lbs. (43–61 Nm) and the lockbolt to 23–34 ft. lbs. (31–46 Nm). Install the power steering belt shield and tighten the bolts to 61–86 inch lbs. (7–9 Nm).

45. Connect the negative battery cable.

46. Run the engine and check for leaks and proper engine operation.

2.0L (VIN H) DIESEL ENGINE

1983–87

➡The following procedures require that the engine be removed from the vehicle in order to replace the timing belt.

FRONT BELT

1. Disconnect the negative battery cable and drain the cooling system.

2. Remove the engine from the vehicle and place it on an engine stand.

3. Remove the drive belt(s) and the necessary components to allow for the front timing belt cover removal.

4. Rotate the crankshaft to position the No. 1 piston on the Top Dead Center (TDC) of its compression stroke.

5. Remove the upper timing belt cover.

6. Install a Flywheel Holding tool No. T84P6375A, or equivalent.

7. Remove the 6 crankshaft pulley-to-crankshaft sprocket bolts.

8. Install a Crankshaft Pulley Remover tool No. T58P-6316-D, or equivalent, using Adapter tool No. T74P-6700-B or equivalent, and remove the pulley.

9. Remove the front timing belt lower cover.

10. If reusing the timing belt, mark the rotational direction on it.

11. Loosen the timing belt tensioning pulley and remove the timing belt.

To install:

12. Align the camshaft sprocket with the timing mark.

➡**Check the crankshaft sprocket to see that the timing marks are aligned.**

13. Remove the tensioner spring from the pocket in the front timing belt upper cover and install it in the slot in the tensioner lever and over the stud in the crankcase.

14. Push the tensioner lever toward the water pump as far as it will travel and tighten the lockbolt snug.

15. Noting the rotational direction, install the timing belt.

16. Install the tensioner spring in the belt tensioner lever and over the stud mounted on the front of the crankcase.

17. Adjust the timing belt tension by performing the following procedure:

    a. Loosen the tensioner pulley lockbolt.

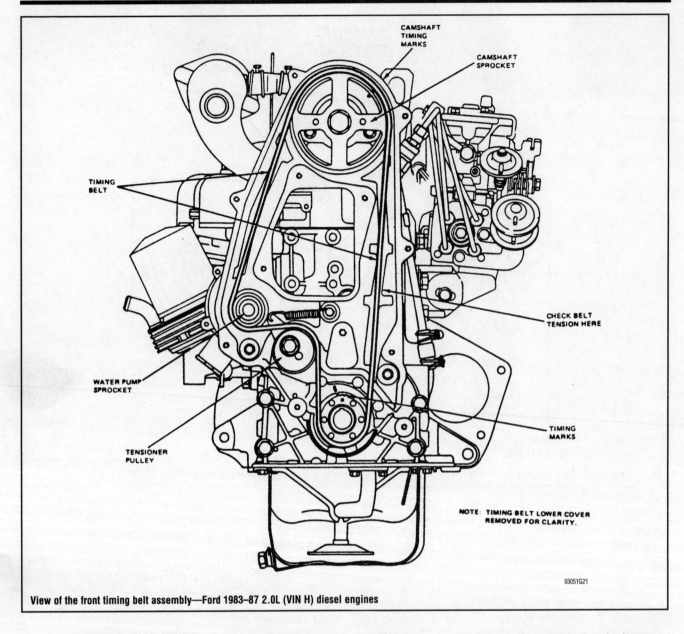

CAMSHAFT
TIMING
MARKS

CAMSHAFT
SPROCKET

TIMING
BELT

CHECK BELT
TENSION HERE

WATER PUMP
SPROCKET

TIMING
MARKS

TENSIONER
PULLEY

NOTE: TIMING BELT LOWER COVER
REMOVED FOR CLARITY.

93051G21

View of the front timing belt assembly—Ford 1983–87 2.0L (VIN H) diesel engines

b. Rotate the crankshaft pulley 2 complete clockwise revolutions until the flywheel TDC timing mark aligns with the pointer on the rear cover plate.

c. Check the front camshaft sprocket to see that it is aligned with its timing mark.

d. Tighten the tensioner lockbolt to 23–34 ft. lbs. (31–46 Nm).

e. Using a Belt Tension Gauge model 21-0028 or equivalent, check the timing belt tension; it should be 33–44 lbs. (15–20 kg).

18. Install the front timing belt lower cover and torque the bolts to 60–84 inch lbs. (6.8–9.5 Nm).

19. Install the crankshaft pulley and torque the bolts to 17–24 ft. lbs. (23–33 Nm).

20. Install the front timing belt upper cover and torque the bolts to 60–84 inch lbs. (6.8–9.5 Nm).

21. Install any components that were removed and install the drive belt(s).

22. Install the engine into the vehicle.

23. Refill the cooling system and connect the negative battery cable.

REAR BELT

1. Disconnect the negative battery cable and drain the cooling system.

2. Remove the engine from the vehicle and place it on an engine stand.

3. Rotate the crankshaft to position the No. 1 piston on the Top Dead Center (TDC) of its compression stroke.

4. Remove the rear timing belt cover.

5. Remove the flywheel timing mark cover from the clutch housing.

6. Check that the injection pump and camshaft sprocket timing marks are aligned.

7. Loosen the tensioner locknut. With a suitable tool inserted in the slot provided, rotate the tensioner clockwise to relieve the belt tension. Tighten the locknut snug.

8. If reusing the timing belt, mark the rotational direction on it.

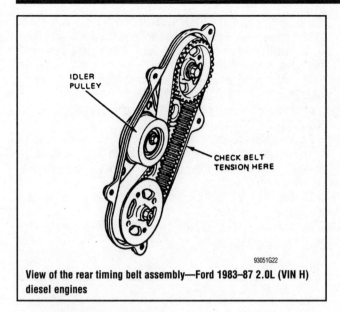

View of the rear timing belt assembly—Ford 1983–87 2.0L (VIN H) diesel engines

9. Remove the rear timing belt.
To install:
10. Noting the rotational direction, install the rear timing belt.
11. Adjust the timing belt tension by performing the following procedure:

    a. Loosen the tensioner pulley lockbolt.

    b. Rotate the crankshaft pulley 2 complete clockwise revolutions until the flywheel TDC timing mark aligns with the pointer on the rear cover plate.

    c. Check the front camshaft sprocket and the injection pump sprocket are aligned with there timing marks.

    d. Tighten the tensioner lockbolt to 15–20 ft. lbs. (20–27 Nm).

    e. Using a Belt Tension Gauge model 21-0028 or equivalent, check the timing belt tension; it should be 22–33 lbs. (10–15 Nm).

12. Install the rear timing belt cover and torque the 6mm bolts to 60–84 inch lbs. (6.8–9.5 Nm) and the 8mm bolt to 12–16 ft. lbs. (16–22 Nm).

13. Install the flywheel timing mark cover.

14. Install any components that were removed and install the drive belt(s).

15. Install the engine into the vehicle.

16. Refill the cooling system and connect the negative battery cable.

2.2L (VIN C & L) ENGINES

1989–92

1. Bring the No. 1 cylinder piston to Top Dead Center (TDC) on the compression stroke. The notch on the crankshaft damper should align with the TDC mark on the front cover.

2. Disconnect the negative battery cable.

3. Loosen the air conditioning compressor and alternator adjusting and pivot bolts, rotate the compressor and alternator toward the engine and remove the drive belts.

4. Raise and safely support the vehicle.

5. Remove the right front wheel and tire assembly and the right inner fender panel. Remove the 6 bolts, the crankshaft pulley and baffle plate.

6. Lower the vehicle.

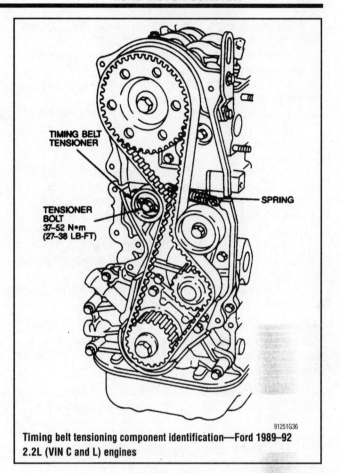

Timing belt tensioning component identification—Ford 1989–92 2.2L (VIN C and L) engines

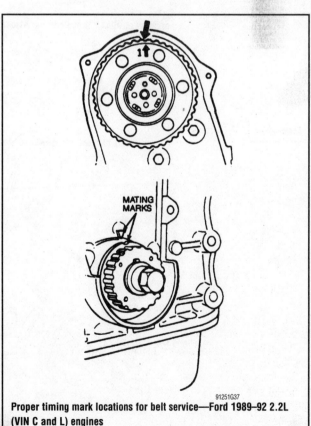

Proper timing mark locations for belt service—Ford 1989–92 2.2L (VIN C and L) engines

7. Support the engine with a floor jack. Remove the 2 nuts and dowels from the right engine mount and remove the mount.

8. Remove the 7 bolts that retain the timing belt covers and remove the covers.

9. Remove the timing belt tensioner spring and retaining bolt. Remove the idler pulley retaining bolt.

10. If the timing belt is to be reused, mark the direction of rotation so it can be reinstalled in the same direction.

11. Remove the timing belt.

To install:

12. Align the camshaft and crankshaft sprockets with the marks on the cylinder head front housing and the oil pump housing.

13. Install the timing belt. If reusing the old belt, observe the direction of rotation mark made during the removal procedure.

14. Place the timing belt tensioner and spring in position. Temporarily secure the tensioner with the spring fully extended. Make sure the timing belt is installed so there is no looseness at the water pump pulley at the idler side.

15. Loosen the idler bolt. Turn the crankshaft twice in the direction of rotation; align the timing marks.

➡**Always turn the crankshaft in the correct direction of rotation only. If the crankshaft is turned in the opposite direction, the timing belt may lose tension and correct belt timing may be lost.**

16. Check to see that the timing marks are correctly aligned. If they are not aligned, remove the timing belt and align the timing marks, then repeat Steps 8–11.

17. Tighten the tensioner bolt to 27–38 ft. lbs. (37–52 Nm).

18. Measure the belt deflection between the crankshaft and camshaft pulleys. The correct deflection should be 0.30–0.33

(7.5–8.5mm) at 22 ft. lbs. (98 Nm) of pressure. If the deflection is not correct, loosen the tensioner bolt and repeat Steps 10 and 11.

19. Install the lower cover gasket and the lower cover. Tighten the bolts to 61–87 inch lbs. (7–10 Nm).

20. Install the upper cover gasket and the upper cover. Tighten the bolts to 61–87 inch lbs. (7–10 Nm).

21. Position the engine mount on the engine and install the 2 nuts and dowels. Remove the floor jack.

22. Install the crankshaft sprocket baffle with the curved outer lip facing outward. Install the crankshaft pulley with the deep recess facing out and install the 6 bolts. Tighten the bolts to 109–152 inch lbs. (12–17 Nm).

23. Install the drive belts. Adjust the belt tension and tighten the adjusting and pivot bolts.

24. Install the right inner fender panel and wheel and tire assembly. Connect the negative battery cable.

2.3L ENGINES

1980–93

1. Disconnect the negative battery cable and drain the cooling system. Remove the 4 water pump pulley bolts.

2. Remove the automatic belt tensioner and accessory drive belt. Remove the upper radiator hose.

3. Remove the crankshaft pulley bolt and pulley. Remove the thermostat housing and gasket.

4. Remove the timing belt outer cover retaining bolt(s). Release the cover interlocking tabs, if equipped, and remove the cover.

5. Loosen the belt tensioner adjustment screw, position belt tensioner tool T74P-6254-A or equivalent, on the tension spring roll

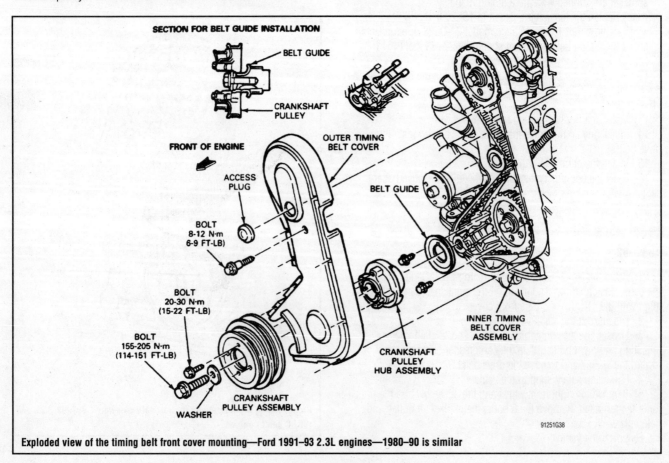

Exploded view of the timing belt front cover mounting—Ford 1991–93 2.3L engines—1980–90 is similar

91251G38

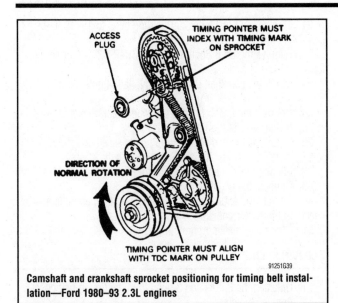

ACCESS PLUG

TIMING POINTER MUST INDEX WITH TIMING MARK ON SPROCKET

DIRECTION OF NORMAL ROTATION

TIMING POINTER MUST ALIGN WITH TDC MARK ON PULLEY

91251G39

Camshaft and crankshaft sprocket positioning for timing belt installation—Ford 1980–93 2.3L engines

pin and release the belt tensioner. Tighten the adjustment screw to hold the tensioner in the released position.

6. On 1991–93 vehicles, remove the bolts holding the timing sensor in place and pull the sensor assembly free of the dowel pin.

7. Remove the crankshaft pulley, hub and belt guide. Remove the timing belt. If the belt is to be reused, mark the direction of rotation so it may be reinstalled in the same direction.

To install:

8. Position the crankshaft sprocket to align with the TDC mark and the camshaft sprocket to align with the camshaft timing pointer. On 1980–90 vehicles, remove the distributor cap and set the rotor to the No. 1 firing position by turning the auxiliary shaft.

9. Install the timing belt over the crankshaft sprocket and then counterclockwise over the auxiliary and camshaft sprockets. Align the belt fore-and-aft on the sprockets.

10. Loosen the tensioner adjustment bolt to allow the tensioner to move against the belt. If the spring does not have enough tension to move the roller against the belt, it may be necessary to manually push the roller against the belt and tighten the bolt.

11. To make sure the belt does not jump time during rotation in Step 10, remove a spark plug from each cylinder.

12. Rotate the crankshaft 2 complete turns in the direction of normal rotation to remove the slack from the belt. Tighten the tensioner adjustment to 29–40 ft. lbs. (40–55 Nm) and pivot bolts to 14–22 ft. lbs. (20–30 Nm). Check the alignment of the timing marks.

13. Install the crankshaft belt guide.

14. On 1980–90 vehicles, install the crankshaft pulley and tighten the retaining bolt to 103–133 ft. lbs. (140–180 Nm). On 1991–93 vehicles, proceed as follows:

a. Install the timing sensor onto the dowel pin and tighten the 2 longer bolts to 14–22 ft. lbs. (20–30 Nm).

b. Rotate the crankshaft 45 degrees counterclockwise and install the crankshaft pulley and hub assembly. Tighten the bolt to 114–151 ft. lbs. (155–205 Nm).

c. Rotate the crankshaft 90 degrees clockwise so the vane of the crankshaft pulley engages with timing sensor positioner tool T89P-6316-A or equivalent. Tighten the 2 shorter sensor bolts to 14–22 ft. lbs. (20–30 Nm).

d. Rotate the crankshaft 90 degrees counterclockwise and remove the sensor positioner tool.

e. Rotate the crankshaft 90 degrees clockwise and measure the outer vane to sensor air gap. The air gap must be 0.018–0.039 in. (0.458–0.996mm).

15. Position the timing belt front cover. Snap the interlocking tabs into place, if necessary. Install the timing belt outer cover retaining bolt(s) and tighten to 71–106 inch lbs. (8–12 Nm).

16. Install the thermostat housing and a new gasket. Install the upper radiator hose.

17. Install the crankshaft pulley and retaining bolt. Tighten to 103–133 ft. lbs. (140–180 Nm) on 1980–90 vehicles or 114–151 ft. lbs. (155–205 Nm) on 1991–94 vehicles.

18. Install the water pump pulley and the automatic belt tensioner. Install the accessory drive belt.

19. Install the spark plugs and remaining components.

20. Connect the negative battery cable, start the engine and check the ignition timing.

2.4L (VIN L) DIESEL ENGINE

1984–85

1. Disconnect the negative battery cable.

2. Drain the cooling system.

3. Remove the accessory drive belts.

4. Remove the fan assembly and the water pump assembly.

5. Remove the vibration damper and pulley.

6. Disconnect the heater hose from the thermostat housing.

7. Remove the 4 bolts attaching the camshaft drive belt cover to the crankcase and remove the cover.

8. Remove the rocker arm cover.

9. Rotate the crankshaft until the No. 1 cylinder is at Top Dead Center (TDC) on its compression stroke (intake and exhaust valves on base circle).

10. Install TDC Aligning Pin T84P-6256-A or equivalent.

➡**Flat side of nut or cam position tool should be facing down.**

11. Using a piece of chalk or similar marker, mark the direction of engine rotation on the drive belt, unless a new belt is to be installed.

12. Loosen both belt tensioner bolts.

13. Remove the camshaft drive belt.

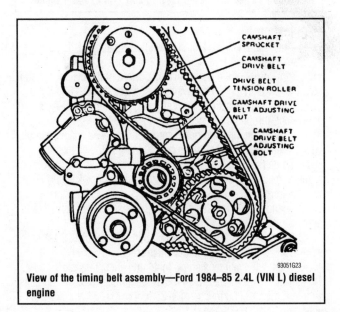

CAMSHAFT SPROCKET

CAMSHAFT DRIVE BELT

DRIVE BELT TENSION ROLLER

CAMSHAFT DRIVE BELT ADJUSTING NUT

CAMSHAFT DRIVE BELT ADJUSTING BOLT

93051G23

View of the timing belt assembly—Ford 1984–85 2.4L (VIN L) diesel engine

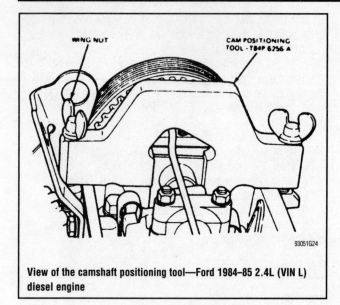

View of the camshaft positioning tool—Ford 1984–85 2.4L (VIN L) diesel engine

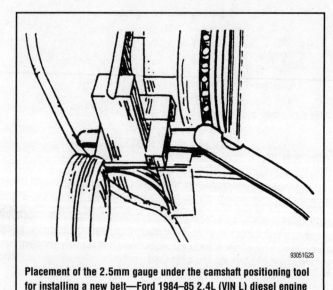

Placement of the 2.5mm gauge under the camshaft positioning tool for installing a new belt—Ford 1984–85 2.4L (VIN L) diesel engine

To install:

14. Insert a 0.098 in. (2.5mm) thick feeler gauge blade between the Cam Positioning tool No. T84P-6256-A or equivalent, at the right front corner of the gasket mating surface of the cylinder head if using a new drive belt or a drive belt used less than 10,000 miles.

15. Install the Injection Pump Aligning Pin T84P-9000-A or equivalent, through the injection pump sprocket.

16. Rotate the injection pump sprocket clockwise against the pin.

17. Install the camshaft drive belt. Starting at the crankshaft, route the belt around the intermediate shaft sprocket, injection pump sprocket, camshaft sprocket and tension roller, keeping slack to a minimum.

18. Hand tighten the belt with the belt tensioner until all slack is gone.

19. Remove the Injection Pump Aligning Pin T84P-9000-A or equivalent, from the injection pump sprocket.

20. Adjust the belt tension by tightening the belt tensioner. Tighten the belt tensioner to 34–36 ft. lbs. (15.5–16.3 Nm) on belts with less than 10,000 miles or 23–25 ft. lbs. (31–34 Nm) for belts with more than 10,000 miles.

21. Tighten both belt tensioner holding bolts to 15–18 ft. lbs. (20–24.5 Nm).

22. Remove the Cam Positioning tool T84P-6265-A or equivalent.

23. Install the camshaft drive belt cover and tighten bolts to 72–84 inch lbs. (8–9.5 Nm).

24. Connect the heater hose to the thermostat housing.

25. Install the vibration damper.

26. Install the fan and water pump pulley assembly.

27. Install and adjust the accessory drive belts.

28. Fill and bleed the cooling system.

29. Connect the negative battery cable.

30. Operate the engine to normal operating temperatures and check for oil and/or coolant leaks.

31. Check the injection pump timing.

2.0L (VIN P) ENGINE

1. Unfasten the 3 nuts and 3 bolts from the timing belt cover.

2. Remove the cover.

3. Raise and safely support the vehicle.

4. Remove the right-hand splash shield and the crankshaft pulley.

5. Align the timing marks as illustrated in the accompanying illustration.

6. Refer to the accompanying illustration and remove the timing belt as follows:

   a. Loosen the timing belt tensioner bolt (1).

   b. Use an 8mm Allen wrench, and turn the tensioner (2) counterclockwise ¼ turn.

   c. Insert a ⅛ inch drill bit in the hole (3) to lock the belt tensioner in place.

   d. Remove the timing belt (4).

   e. Inspect the belt for damage and signs of oil leakage.

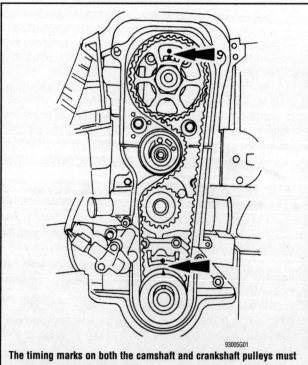

The timing marks on both the camshaft and crankshaft pulleys must be aligned like this before removing or installing the timing belt—Ford 2.0L (VIN P) SOHC engines

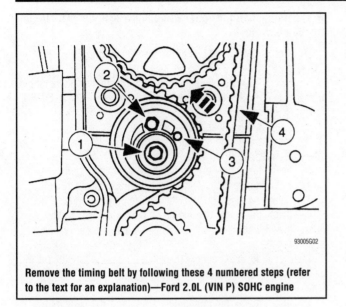

Remove the timing belt by following these 4 numbered steps (refer to the text for an explanation)—Ford 2.0L (VIN P) SOHC engine

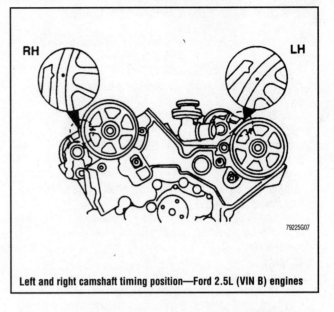

Left and right camshaft timing position—Ford 2.5L (VIN B) engines

To install:

➡Install the timing belt over the sprocket in a counterclockwise direction starting at the crankshaft. Keep the belt span between the crankshaft and camshaft tight when installing the belt over the camshaft.

7. Install the timing belt and remove the drill bit.
8. Tighten the tensioner bolt to 15–22 ft. lbs. (20–30 Nm).
9. Rotate the engine 2 complete revolutions and make sure the timing marks are aligned.
10. Install the timing belt cover and tighten the nuts and bolts to 71–97 inch lbs. (8–11 Nm).
11. Install the pulley and the bolt. Tighten the bolt to 81–98 ft. lbs. (110–120 Nm).
12. Install the splash shield(s).
13. Lower the vehicle and install the drive belt.
14. Connect the negative battery cable.
15. Start the vehicle and check for proper operation.

2.5L (VIN B) ENGINE

1. Disconnect the negative battery cable.
2. Label and disengage the electrical connectors from the coolant elbow. Label and remove the electrical connectors from the knock sensor (if required) and crankshaft position sensor.
3. Loosen the drive belt tensioner locknuts and adjusting bolts. Remove the accessory drive belts.
4. Raise and safely support the vehicle on jackstands.
5. Remove the lower bolt from the air conditioning and alternator tensioner bracket.
6. Remove the right wheel and splash-shields.
7. Hold the crankshaft pulley (damper) with Holder tool T92C-6316-AH or equivalent, and remove the crankshaft pulley bolt. Remove the crankshaft pulley, using a puller if needed.
8. Remove the 5 front timing belt cover bolts.
9. Hold the water pump pulley with Holder tool T92C-6312-AH or equivalent, remove the 4 bolts and the water pump pulley.
10. Hold the power steering pump pulley with Strap Wrench D85L-6000-A or equivalent and remove the power steering pump pulley nut and pulley.
11. Lower the vehicle.

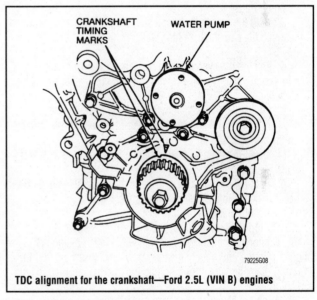

TDC alignment for the crankshaft—Ford 2.5L (VIN B) engines

12. Remove the upper bolt from the belt idler bracket and remove the bracket.
13. Remove the engine oil dipstick tube retaining bolt and the tube.
14. Remove the 8 rear timing belt cover retaining bolts and remove the timing belt covers.
15. Temporarily reinstall the crankshaft pulley bolt.
16. Remove the 3 nuts and through-bolt from the right-hand engine support insulator and remove the support insulator. Remove the support insulator bracket.
17. Raise and safely support the vehicle on jack stand.
18. Turn the crankshaft to TDC No. 1 cylinder in the direction of normal rotation. Be sure that the timing mark on the crankshaft sprocket aligns with the timing mark on the oil pump.
19. Remove the 2 bolts from the timing belt tensioner arm, removing the lower bolt first.
20. Remove the timing belt tensioner arm.
21. If the timing belt is to be reused, mark the direction of rotation on the timing belt.

22. Loosen the Allen bolt on the timing belt tensioner.
23. Remove the timing belt.
24. If the timing belt sprockets are to be removed, proceed as follows:

a. Remove the intake manifold.

b. Label and disconnect the necessary hoses from the cylinder head covers.

c. Label and disconnect the spark plug wires from the spark plugs.

d. Remove the cylinder head cover retaining bolts and remove the cylinder head covers.

e. Hold the camshaft using a suitable wrench on the hexagon cast into the camshaft. Remove the camshaft sprocket bolts and the camshaft sprockets.

f. Use Crankshaft Damper Puller T74P-6316-A or equivalent, to remove the crankshaft sprocket. Remove the crankshaft sprocket key.

To install:

25. If the timing belt sprockets were removed, proceed as follows:

a. Install the crankshaft sprocket key and crankshaft sprocket.

b. Install the camshaft sprockets on the camshafts with the retaining bolts.

c. Hold the camshaft using a suitable wrench on the hexagon cast into the camshaft. Tighten the camshaft sprocket bolts to 90–103 ft. lbs. (123–140 Nm).

d. Be sure the cylinder head cover and cylinder head contact surfaces are clean and free of dirt, oil and old sealant and gasket material.

e. Apply silicone sealant to the cylinder heads in the area adjacent to the front and rear camshaft caps. Install new gaskets on the cylinder heads.

f. Install the cylinder head covers and tighten the retaining bolts.

g. Connect the spark plug wires to the spark plugs and connect the hoses to the cylinder head covers.

h. Install the intake manifold.

26. Position the timing belt tensioner arm in a suitable press.
27. Compress the tensioner until the hole in the piston is aligned with the 2nd hole in the tensioner case. Insert a 0.060 in. (1.6mm) diameter wire or pin through the 2nd hole to keep the piston compressed.
28. Align the camshaft sprockets to TDC.
29. Turn the crankshaft counterclockwise until the crankshaft sprocket is offset from TDC by 1 tooth.
30. Install the timing belt.
31. If the original belt is being reused, be sure it is installed in the same direction of rotation.
32. Turn the crankshaft in the direction of normal engine rotation until the crankshaft sprocket timing mark is at TDC. This should place all of the belt slack in the timing belt tensioner portion of the timing belt.
33. Install the timing belt tensioner arm and 2 bolts. Tighten the bolts to 14–18 ft. lbs. (19–25 Nm).
34. Remove the wire or pin from the tensioner.

➡**When properly timed, the crankshaft timing marks will align and the crankshaft sprocket timing mark will no longer be 1 tooth off.**

35. Rotate the crankshaft 2 complete turns in the direction of normal rotation and align the timing marks. Be sure all marks are still correctly aligned. This will also set the timing belt tension.

➡**The timing belt tensioner will automatically adjust the timing belt tension.**

36. Tighten the timing belt tensioner Allen bolt to 28–32 ft. lbs. (35–51 Nm).
37. Install the right-hand engine support insulator. Tighten the 3 nuts to 54–76 ft. lbs. (74–103 Nm) and the through-bolt to 50–68 ft. lbs. (67–93 Nm).
38. Remove the crankshaft damper bolt.
39. Install the timing belt covers with the rear 8 bolts. Tighten to 71–88 inch lbs. (8–10 Nm).
40. Install the engine oil dipstick tube and retaining nut.
41. Install the belt idler bracket and the upper retaining bolt.
42. Raise and safely support the vehicle on jackstands.
43. Install the power steering pump pulley and nut. Tighten the nut to 36–43 ft. lbs. (49–59 Nm) while holding the pulley with a strap wrench.
44. Install the water pump pulley and 4 bolts. Secure the pulley with the holder tool and tighten the bolts to 71–88 inch lbs. (8–10 Nm).
45. Install the 5 front timing belt cover bolts and tighten to 71–88 inch lbs. (8–10 Nm).
46. Install the crankshaft pulley (damper) with the bolt. Hold the crankshaft pulley with the holding tool and tighten to 116–122 ft. lbs. (157–166 Nm).
47. Install the splash-shields.
48. Install the wheel and tighten the lug nuts to 65–87 ft. lbs. (88–118 Nm).
49. Install the lower bolt into the air conditioning and alternator tensioner bracket.
50. Lower the vehicle.
51. Install the accessory drive belts and adjust the tension.
52. Engage the electrical connectors to the sensors at the coolant elbow and the knock sensor and crankshaft position sensor.
53. Connect the negative battery cable.
54. Run the engine and check for leaks and proper engine operation.

3.0L (VIN Y) SHO ENGINE

1989–95

1. Disconnect both battery cables, negative cable first. Remove the battery.
2. Remove the right engine roll damper. Disconnect the wiring to the ignition module.
3. Remove the intake manifold crossover tube bolts. Loosen the intake manifold tube hose clamps. Remove the intake manifold crossover tube.
4. Loosen the alternator/air conditioning belt tensioner pulley and remove the drive belt.
5. Loosen the water pump/power steering belt tensioner pulley and remove the drive belt.
6. Remove the alternator/air conditioning belt tensioner pulley and bracket assembly.
7. Remove the water pump/power steering belt tensioner pulley only. Remove the upper timing belt cover.
8. Unplug the Crankshaft Position (CKP) sensor electrical connector. Place the transaxle gear selector in **N** (Neutral).
9. Rotate the crankshaft until the piston for the No. 1 cylinder is at Top Dead Center (TDC) of the compression stroke. Be sure the white mark on the crankshaft damper aligns with the **0** degree index mark on

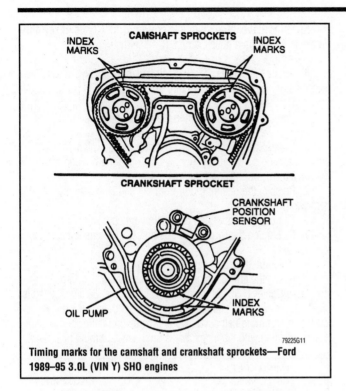

Timing marks for the camshaft and crankshaft sprockets—Ford 1989–95 3.0L (VIN Y) SHO engines

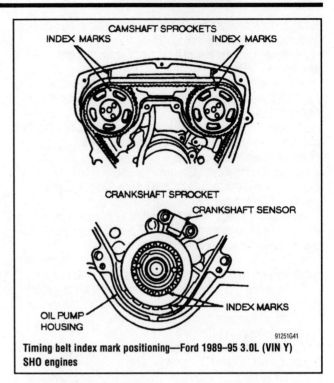

Timing belt index mark positioning—Ford 1989–95 3.0L (VIN Y) SHO engines

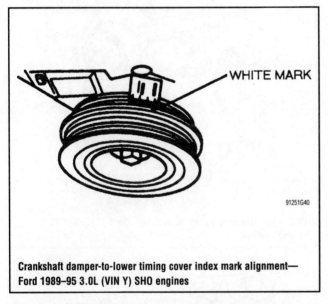

Crankshaft damper-to-lower timing cover index mark alignment—Ford 1989–95 3.0L (VIN Y) SHO engines

the lower timing belt cover and the marks on the intake camshaft sprockets align with the index marks on the metal timing belt cover.

10. Raise and safely support the vehicle on jackstands. Remove the right front wheel and tire assembly. Loosen the fender splash-shield and place it aside.

11. Remove the crankshaft damper and pulley retaining bolt. Using Puller T67L-3600-A or equivalent, remove the crankshaft damper and pulley.

12. Remove the lower timing belt cover.

13. Remove the center timing belt cover and disconnect the CKP sensor wire and grommet from the slot in the cover and the stud on the water pump.

14. Loosen the timing belt tensioner idler pulley. Rotate the idler pulley 180 degrees clockwise and tighten the tensioner nut to hold the pulley in an unloaded position.

15. Lower the vehicle.

16. Remove the timing belt. If the belt is to be reused, use crayon to mark an arrow on the belt to indicate the direction of rotation, for installation reference.

17. If removing one or more camshaft sprockets, remove 2 retaining bolts securing each camshaft sprocket and remove the sprocket, noting the location of the dowel pin.

18. If removing the crankshaft sprocket, install Puller T67L-3600-A or equivalent and pull the crankshaft sprocket off the crankshaft using care not to damage the pulse wheel.

To install:

19. If the crankshaft sprocket was removed, install the crankshaft sprocket by aligning the keyway and pushing the sprocket on by hand.

20. If one or more camshaft sprockets were removed, install each camshaft sprocket by aligning the timing marks on the sprockets with the camshaft using the dowel pin as a guide. Install 2 retaining bolts and tighten to 10–13 ft. lbs. (14–18 Nm).

➡**Before installing the timing belt, inspect it for cracks, wear or other damage and replace, if necessary. Do not allow the timing belt to come into contact with gasoline, oil or coolant. Do not twist or turn the belt inside out.**

21. Be sure the engine is at TDC for the No. 1 cylinder. Check that the camshaft sprocket marks align with the index marks on the upper steel belt cover and that the crankshaft sprocket aligns with the index mark on the oil pump housing.

➡**The timing belt has 3 yellow lines. Each line aligns with the index marks.**

22. Install the timing belt over the crankshaft and camshaft sprockets. The lettering on the belt **KOA** should be readable from the rear of the engine (top of the lettering to the front of the engine). Be sure the yellow lines are aligned with the index marks on the sprockets.

23. Release the timing belt tensioner idler pulley locknut. Leave the nut loose. Raise and safely support the vehicle on jackstands.

24. Install the center timing belt cover. Be sure the CKP sensor

wiring and grommet are installed and routed properly. Tighten the mounting bolts to 60–90 inch lbs. (7–11 Nm).

25. Install the lower timing belt cover. Tighten the bolts to 60–90 inch lbs. (7–11 Nm).

26. Install the crankshaft damper and pulley using Installer T88T-6701-A or equivalent. Install the retaining bolt and tighten to 113–126 ft. lbs. (152–172 Nm).

27. Rotate the crankshaft 2 revolutions in the clockwise direction until the yellow mark on the damper aligns with the **0** degree mark on the lower timing belt cover.

28. Remove the plastic door in the lower timing belt cover. Tighten the tensioner locknut to 25–37 ft. lbs. (33–51 Nm) and install the plastic door.

29. Rotate the crankshaft 60 degrees more in the clockwise direction until the white mark on the damper aligns with the **0** degree mark on the lower timing belt cover.

30. Lower the vehicle. Be sure the index marks on the camshaft sprockets align with the marks on the rear metal timing belt cover.

31. Route the CKP sensor wiring and connect it to the engine wiring harness.

32. Install the upper timing belt cover. Tighten the bolts to 60–90 inch lbs. (7–11 Nm).

33. Install the water pump/power steering tensioner pulley. Tighten the nut to 11–17 ft. lbs. (15–23 Nm).

34. Install the alternator/air conditioning tensioner pulley and bracket assembly. Tighten the bolts to 11–17 ft. lbs. (15–23 Nm).

35. Install the water pump/power steering and alternator/air conditioning drive belts, and set the tension. Tighten the idler pulley nut to 25–36 ft. lbs. (34–50 Nm).

36. Install the intake manifold crossover tube. Tighten the bolts to 11–17 ft. lbs. (15–23 Nm).

37. Install the engine roll damper. Install the battery. Connect the wiring to the ignition module.

38. Connect both battery cables, negative cable last. Raise and safely support the vehicle on jack stands.

39. Install the splash-shield. Install the right front wheel and tire assembly.

40. Lower the vehicle. Run the engine and check for proper operation.

3.2L (VIN P) SHO ENGINE

1990–95

1. Disconnect the battery cables, negative cable first.
2. Remove the battery.
3. Remove the right engine roll damper.
4. Disconnect the wiring to the ignition module.

5. Remove the intake manifold crossover tube bolts. Loosen the intake manifold tube hose clamps. Remove the intake manifold crossover tube.

6. Rotate the accessory drive belt tensioner clockwise to relieve belt tension. Remove the accessory drive belt.

7. Disconnect the surge tank fitting.

8. Remove the bolts retaining the upper and lower idler pulleys to the engine and remove the pulleys.

9. Using Strap Wrench D85L-6000-A, or equivalent, hold the power steering pump pulley. Remove the retaining nut and washer. Remove the power steering pulley.

10. Remove the retaining bolt from the belt tensioner and remove the tensioner.

11. Remove the upper and center timing belt covers.

12. Disengage the Crankshaft Position (CKP) sensor electrical connector.

13. Place the transaxle selector in **N** (Neutral).

14. Rotate the crankshaft until the piston for No. 1 cylinder is at Top Dead Center (TDC) of the compression stroke. Be sure the white mark on the crankshaft damper aligns with the **0** degree index mark on the lower timing belt cover and the marks on the intake camshaft sprockets align with the index marks on the metal timing belt cover.

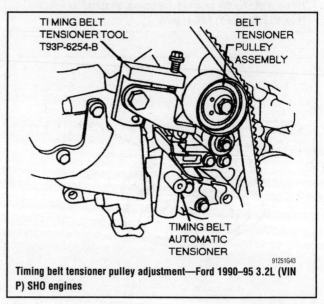

Timing belt tensioner pulley adjustment—Ford 1990–95 3.2L (VIN P) SHO engines

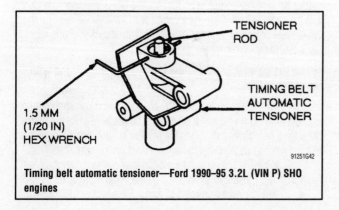

Timing belt automatic tensioner—Ford 1990–95 3.2L (VIN P) SHO engines

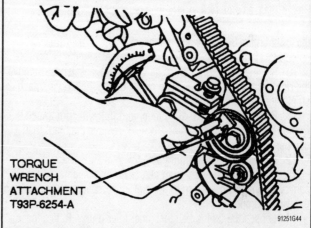

Timing belt tensioner pulley torque adjustment—Ford 1990–95 3.2L (VIN P) SHO engines

15. Raise and safely support the vehicle on jackstands.
16. Remove the right front wheel and tire assembly.
17. Loosen the fender splash-shield and place it aside.
18. Remove the crankshaft pulley and damper using Puller T67L-3600-A with the appropriate adapters or equivalent.
19. Remove the lower timing belt cover and belt guide.
20. Remove the upper timing belt tensioner bolt.
21. Slowly loosen the lower timing belt tension bolt and remove the tensioner.
22. Lower the vehicle.
23. Remove the timing belt. If the belt is to be reused, use crayon to mark an arrow on the belt to indicate the direction of rotation, for installation reference.
24. If removing one or more camshaft sprockets, remove 2 retaining bolts securing each camshaft sprocket and remove the sprocket noting the location of the dowel pin.
25. If removing the crankshaft sprocket, install Puller T67L-3600-A or equivalent and pull the crankshaft sprocket off the crankshaft using care not to damage the pulse wheel.

To install:
26. If the crankshaft sprocket was removed, install the crankshaft sprocket by aligning the keyway and pushing the sprocket on by hand.
27. If one or more camshaft sprockets were removed, install each camshaft sprocket by aligning the timing marks on the camshaft sprocket with the camshaft using the dowel pin as a guide. Install 2 retaining bolts and tighten to 10–13 ft. lbs. (14–18 Nm).

➡**Before installing the timing belt, inspect it for cracks, wear or other damage and replace as necessary. Do not allow the timing belt to come into contact with gasoline, oil or coolant. Do not twist or turn the belt inside out.**

28. Slowly compress the timing belt tensioner in a soft-jawed vise until the hole in the tensioner housing aligns with the hole in the tensioner rod.

❋❋ CAUTION

Use care when compressing the timing belt tensioner in the vise to insure that the tensioner does not slip from the vise.

29. Insert a ¹⁄₂₀ in. (1.5mm) hex wrench through the holes in the tensioner.
30. Release the tension from the vise.
31. If a new timing belt is being installed, loosen the timing belt idler bolt.
32. Ensure that the No. 1 cylinder is at TDC on its compression stroke. Check that the camshaft sprocket marks align with the index marks on the upper steel belt cover and that the crankshaft sprocket aligns with the index mark on the oil pump housing.

➡**The timing belt has 3 yellow lines. Each line aligns with the index marks.**

33. Install the timing belt over the crankshaft and camshaft sprockets. The lettering on the belt **KOB** should be readable from the rear of the engine (top of the lettering to the front of the engine). Be sure the yellow lines are aligned with the index marks on the sprockets.

❋❋ WARNING

Do not install the timing belt tensioner with the tensioner rod extended.

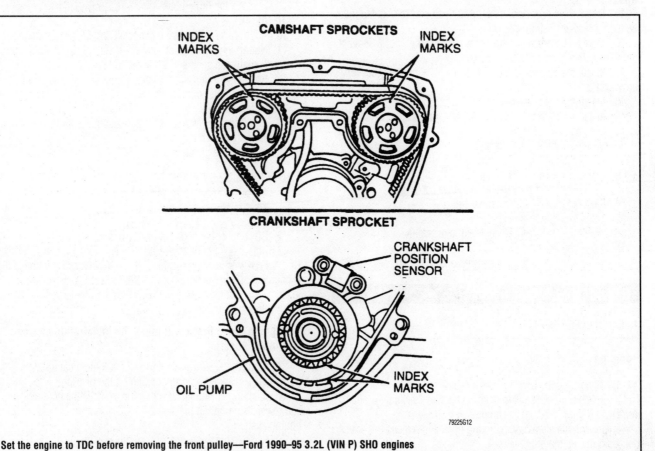

Set the engine to TDC before removing the front pulley—Ford 1990–95 3.2L (VIN P) SHO engines

34. Install the timing belt tensioner on the cylinder block while pushing the timing belt idler toward the timing belt. Install and tighten the tensioner bolts to 12–17 ft. lbs. (16–23 Nm).

35. Install the grommets between the timing belt tensioner and the oil pump.

36. Remove the hex wrench from the timing belt tensioner.

37. If a new timing belt is being installed, perform the following steps:

   a. Remove the hex wrench from the timing belt tensioner, if installed.

   b. Mount Timing Belt Tensioner tool T93P-6254-B or equivalent, using the holes in the power steering pump support.

   c. Hand-tighten the timing belt idler pulley bolt.

   d. Using an in. pound torque wrench with attachment T93P-6254-A or equivalent, rotate the timing belt tensioner clockwise to 4.3 inch lbs. (0.5 Nm).

   e. Tighten the timing belt idler pulley bolt to 27–37 ft. lbs. (36–50 Nm). Remove both timing belt tensioning tools.

38. Raise and safely support the vehicle securely on jackstands.

39. Install the belt guide and lower timing belt cover. Tighten the retaining bolts to 12–17 ft. lbs. (16–23 Nm).

40. Using a suitable tool, install the crankshaft damper. Tighten the damper attaching bolt to 113–126 ft. lbs. (152–172 Nm).

41. Rotate the crankshaft 2 revolutions clockwise until the yellow mark on the damper aligns with the **0** degree mark on the lower timing belt cover.

42. Lower the vehicle.

43. Ensure that the index marks on the camshaft sprockets align with the marks on the rear metal timing belt cover.

44. Route the CKP sensor wiring and connect with the engine wiring harness.

45. Install the center and upper timing belt covers. Tighten the bolts to 12–17 ft. lbs. (16–23 Nm).

46. Install the steering pump pulley. Tighten the nut to 12–17 ft. lbs. (16–23 Nm).

47. Install the accessory drive belt while rotating the accessory drive belt tensioner clockwise.

48. Install the surge tank fitting.

49. Install the intake manifold crossover tube. Tighten the bolts to 11–17 ft. lbs. (15–23 Nm).

50. Install the engine roll damper

51. Install the battery.

52. Connect the wiring to the ignition module.

53. Connect both battery cables, negative cable last.

54. Raise and safely support the vehicle.

55. Install the splash-shield.

56. Install the right front wheel and tire assembly.

57. Lower the vehicle.

58. Run the engine and check for leaks and proper engine operation.

General Motors Corporation

1.6L (VIN 6) ENGINE

1988–93

1. Disconnect the negative battery cable.

2. Rotate the crankshaft to position the No. 1 cylinder on the Top Dead Center (TDC) of its compression stroke.

3. Loosen the alternator mounting bolts and remove the drive belt from the alternator pulley.

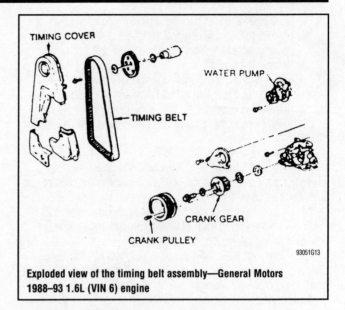

Exploded view of the timing belt assembly—General Motors 1988–93 1.6L (VIN 6) engine

4. If equipped, loosen the air conditioning compressor bolts and remove the drive belt from the compressor pulley.

5. Remove the power steering pump lines and mounting bolts, remove the pump.

6. Unsnap the front cover, the upper half first and remove it from the engine.

7. Mark the timing belt's direction of rotation for reinstallation purposes.

8. Loosen the water pump-to-engine bolts, rotate the water pump to release the tension on the timing belt.

9. Remove the timing belt.

To install:

10. Align the crankshaft and camshaft sprocket timing marks.

11. Noting the direction of rotation, install the timing belt onto the crankshaft, camshaft and water pump sprockets.

12. Rotate the water pump to place tension on the timing belt and tighten the water pump bolts.

13. Install the front cover to the engine.

14. Install the power steering pump, alternator and the air conditioning compressor (if equipped).

15. Connect the negative battery cable.

1.6L (VIN 0, 9, C) ENGINES

1980–87

1. Disconnect the negative battery cable.

2. Rotate the engine to position the No. 1 cylinder at Top Dead Center (TDC) or its compression. The timing mark should be at the **0** degree mark on the timing scale. With the No. 1 cylinder at TDC, insert a 1/8 in. drill bit through the hole in the timing belt upper rear cover into a hole in the camshaft sprocket.

➡**Aligning these holes will make the installation of a new belt much easier.**

3. Remove the distributor cap and mark the location of the rotor in the No. 1 spark plug firing position on the distributor housing.

4. If equipped with air conditioning, remove the compressor and lower its mounting bracket; DO NOT discharge the air conditioning system.

5. Remove the engine accessory drive belts.

6. Remove the engine fan and pulley.

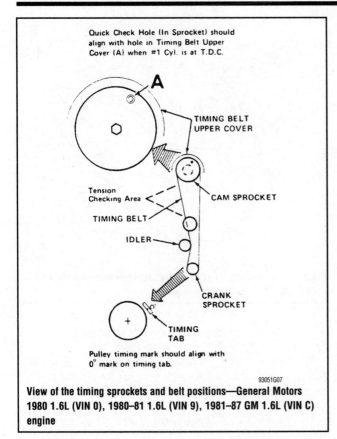

View of the timing sprockets and belt positions—General Motors 1980 1.6L (VIN 0), 1980–81 1.6L (VIN 9), 1981–87 GM 1.6L (VIN C) engine

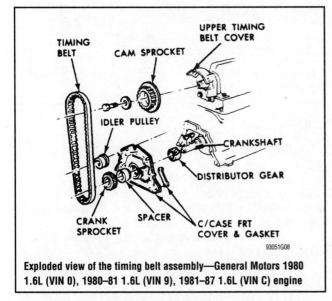

Exploded view of the timing belt assembly—General Motors 1980 1.6L (VIN 0), 1980–81 1.6L (VIN 9), 1981–87 1.6L (VIN C) engine

7. Remove the crankshaft damper bolt and damper pulley.

8. Remove the timing belt cover(s) retaining screws, nuts and the cover(s).

9. Mark the direction of rotation on the timing belt, if to be reused.

10. Loosen the idler pulley adjustment bolt and loosen the tension of the timing belt.

11. Remove the timing belt from the sprockets.

To install:

12. Align the crankshaft pulley timing mark with the **0** mark on the timing scale and the distributor rotor with the scribed mark on the distributor housing.

13. Align the hole in the camshaft sprocket with the hole in the upper rear timing belt cover. Then, insert a ⅛ in. drill bit to hold the sprocket in alignment.

14. Noting the timing belt's direction of rotation, install it onto the crankshaft and camshaft sprockets.

15. Install the crankshaft damper pulley and bolt.

16. Adjust the timing belt tension by performing the following procedure:

   a. Using a Timing Belt Tension Gauge, place it on the belt on the same side as the idler pulley midway between the camshaft sprocket and the idler pulley.

➡**Be sure that the center finger of the gauge extension fits in a notch between the teeth on the belt.**

   b. Rotate the idler pulley to place 70 lbs. (32 kg) of tension on the belt; then, tighten the idler pulley bolt to 15 ft. lbs. (20 Nm).

17. Install the distributor cap.

18. If equipped with air conditioning, install the compressor bracket and compressor.

19. Install the timing belt covers, the engine fan and pulley.

20. Install and adjust the accessory drive belts.

21. Connect the negative battery cable.

1.8L (VIN D) DIESEL ENGINE

1981–86

1. Disconnect the negative battery cable.

2. Drain the cooling system.

3. Remove the fan shroud, cooling fan V-belt and pulley.

4. Disconnect the bypass hose and then remove the upper half of the front timing belt cover.

5. Rotate the crankshaft to position the No. 1 piston at Top Dead Center (TDC) of its compression stroke. Make sure that the notch on the injection pump gear is aligned with the index mark on the front plate. Thread a 8mm x 1.25mm lock bolt through the sprocket and into the front plate.

6. Remove the valve cover and install a Fixing Plate tool J-29761, or equivalent, in the slot at the rear of the camshaft; this will prevent the camshaft from rotating during the procedure.

7. Remove the crankshaft damper pulley and check to make sure that the No. 1 piston is still at TDC.

8. Remove the lower half of the front cover and then remove the timing belt holder from the bottom of the front plate.

9. Mark the direction of rotation on the timing belt for reinstallation purposes.

10. Remove the tension spring from behind the front plate, near to the injection pump.

11. Loosen the tension pulley and slide the timing belt off of the sprockets.

To install:

12. Noting the direction of rotation, slide the timing belt over the sprockets. The belt should be properly tensioned between the sprockets, the cogs on the belt; also, the sprockets should properly engaged and the crankshaft should not be turned.

13. Partially, tighten the bolts in numerical order to prevent movement of the tensioner pulley.

14. Remove the injection pump sprocket lock bolt and the fixing plate on the end of the camshaft.

15. Make sure the mark on the injection pump pulley is aligned with the mark on the plate. The fixing plate should fit smoothly into the slot at the rear of the camshaft, then remove the fixing plate.

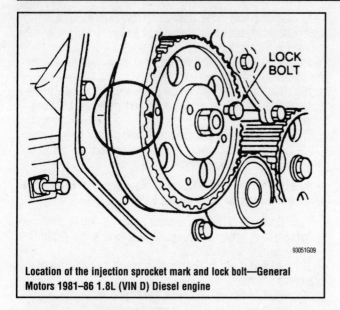

Location of the injection sprocket mark and lock bolt—General Motors 1981–86 1.8L (VIN D) Diesel engine

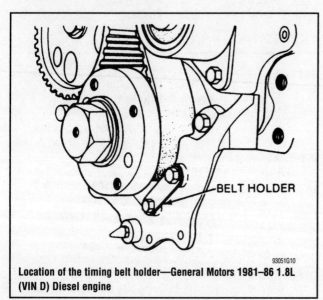

Location of the timing belt holder—General Motors 1981–86 1.8L (VIN D) Diesel engine

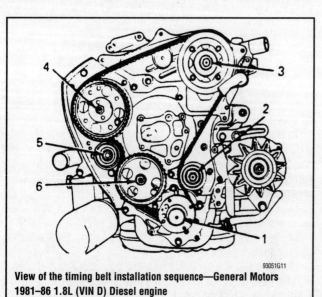

View of the timing belt installation sequence—General Motors 1981–86 1.8L (VIN D) Diesel engine

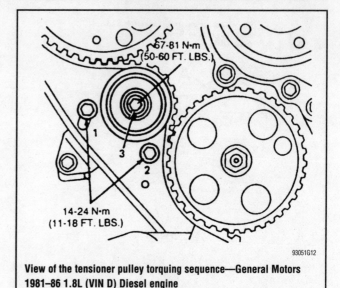

View of the tensioner pulley torquing sequence—General Motors 1981–86 1.8L (VIN D) Diesel engine

16. Install the damper pulley on the hub and check that the No. 1 cylinder is at TDC; DO NOT turn the crankshaft.

17. Loosen the tensioner pulley and plate attaching bolts. Concentrate looseness of the belt on the tensioner, then tighten the bolts in the numerical order to 11–18 ft. lbs. (14–24 Nm) for bolts No. 1 and 2; then 50–60 ft. lbs. (67–81 Nm) for No. 3.

18. Using tool J-29771 or equivalent, check the belt tension between the camshaft sprocket and the injection pump sprocket. Remove the damper pulley and install the belt holder in position away from the timing bolt.

19. Complete the installation by reversing the removal procedures.

20. Connect the negative battery cable.

1.8L (VIN J AND O) ENGINES
2.0L (VIN K AND M) ENGINES

1983–91

1. Disconnect the negative battery cable and drain the engine cooling system to a level below the water pump.

2. Remove the drive belt(s) from the engine.

3. Loosen the drive belt tensioner bolt and the tensioner will swing downward. If necessary, remove the bolt and tensioner from the engine.

4. Remove the cover attaching bolts, then remove the timing belt cover from the engine.

5. Turn the engine to align the timing marks on the gears with the marks on the timing belt rear cover.

6. Loosen the water pump bolts and release timing belt tension with Tension Adjusting tool J-33039 or equivalent. If equipped, remove the air conditioning drive belt.

7. Raise and support the vehicle safely, then remove the right splash shield.

8. Remove the pulley retaining bolt, then remove the pulley from the end of the crankshaft.

9. Lower the vehicle and remove the timing belt. If necessary, remove the retainers and the belt tensioner from the engine.

To install:

10. If removed, install the timing belt tensioner and tighten the retainers to 18 ft. lbs. (25 Nm).

11. If necessary, turn the crankshaft and/or the camshaft gears

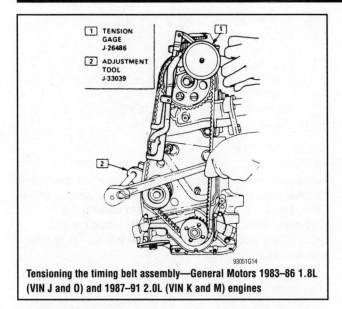

Tensioning the timing belt assembly—General Motors 1983–86 1.8L (VIN J and O) and 1987–91 2.0L (VIN K and M) engines

15. Turn the water pump eccentric counterclockwise until the hole in the tensioner arm is aligned with the hole in the base. In order for the holes to properly align, the engine must be at or near room temperature of 68° F (20° C).

16. Tighten the water pump screws to 18 ft. lbs. (25 Nm) while checking that the tensioner holes remain as adjusted in the prior step.

17. Raise and support the vehicle safely, then install the crankshaft pulley. Coat the pulley retainer bolt threads with Loctite® 242 or equivalent threadlock. Install the retainer and tighten to 20 ft. lbs. (27 Nm).

18. Install the right splash shield and lower the vehicle.

19. Install the timing belt cover and attaching bolts. Tighten to 89 inch lbs. (10 Nm).

20. Pivot or install the timing belt tensioner into position and tighten the retaining bolt to 40 ft. lbs. (54 Nm).

21. Install the drive belt(s).

22. Install fuel vapor pipes.

23. Install the drive belt and, if equipped with air conditioning.

24. Connect the negative battery cable and properly fill the engine cooling system.

2.0L (VIN H) ENGINE

1992–94

1. Disconnect the negative battery cable and drain the engine cooling system to a level below the water pump.

2. Remove the coolant reservoir tank.

3. Remove the serpentine belt from the engine. Remove the fuel vapor pipe assembly.

clockwise to align the timing marks on the gears with the timing marks on the rear cover.

12. Install the timing belt, making sure the portion between the camshaft gear and crankcase gear is in tension.

13. Using tool J-33039 or equivalent, turn the water pump eccentric clockwise until the tensioner contacts the high torque stop. Tighten the water pump screws slightly.

14. Turn the engine clockwise 720 degrees using the crankshaft gear bolt in order to fully seat the belt into the gear teeth.

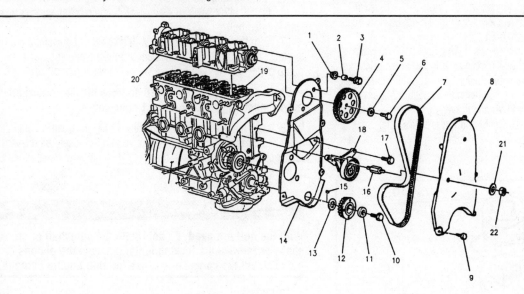

1 GROMMET	12 CRANKSHAFT SPROCKET
2 SLEEVE	13 WASHER
3 BOLT – 10 N·m (89 LBS. IN.)	14 REAR COVER
4 CAMSHAFT SPROCKET	15 KEYWAY
5 WASHER	16 STUD – 48 N·m (35 LBS. FT.)
6 BOLT – 45 N·m (22 LBS. FT.)	17 BOLT – 48 N·m (35 LBS. FT.)
7 TIMING BELT	18 TENSIONER
8 FRONT COVER	19 ENGINE
9 BOLT – 9 N·m (80 MLBS. IN.)	20 CAMSHAFT
10 BOLT – 155 N·m (144 LBS. FT.)	21 WASHER
11 WASHER	22 NUT

Exploded view of the timing covers, gears and belt assembly—General Motors 1992–94 2.0L (VIN H) engine

4. Loosen the drive belt tensioner bolt and the tensioner will swing downward. If necessary, remove the bolt and tensioner from the engine.

5. Remove the cover attaching bolts, then remove the timing belt cover from the engine.

6. Turn the engine to align the timing marks on the gears with the marks on the timing belt rear cover.

7. Loosen the water pump bolts and release timing belt tension with Tension Adjusting tool J-33039 or equivalent. If equipped, remove the air conditioning drive belt.

8. Raise and support the vehicle safely, then remove the right splash shield.

9. Remove the pulley retaining bolt, then remove the pulley from the end of the crankshaft.

10. Lower the vehicle and remove the timing belt. If necessary, remove the retainers and the belt tensioner from the engine.

To install:

11. If removed, install the timing belt tensioner and tighten the retainers to 18 ft. lbs. (25 Nm).

12. If necessary, turn the crankshaft and/or the camshaft gears clockwise to align the timing marks on the gears with the timing marks on the rear cover.

13. Install the timing belt, making sure the portion between the camshaft gear and crankcase gear is in tension.

14. Using tool J-33039 or equivalent, turn the water pump eccentric clockwise until the tensioner contacts the high torque stop. Tighten the water pump screws slightly.

15. Turn the engine clockwise 720 degrees using the crankshaft gear bolt in order to fully seat the belt into the gear teeth.

16. Turn the water pump eccentric counterclockwise until the hole in the tensioner arm is aligned with the hole in the base. In order for the holes to properly align, the engine must be at or near room temperature of 68° F (20° C).

17. Tighten the water pump screws to 18 ft. lbs. (25 Nm) while checking that the tensioner holes remain as adjusted in the prior step.

18. Raise and support the vehicle safely, then install the crankshaft pulley. Coat the pulley retainer bolt threads with Loctite® 242 or equivalent threadlock. Install the retainer and tighten to 15 ft. lbs. (21 Nm).

19. Install the right splash shield and lower the vehicle.

20. Install the timing belt cover and attaching bolts. Tighten to 62 inch lbs. (7 Nm).

21. Pivot or install the timing belt tensioner into position and tighten the retaining bolt to 35 ft. lbs. (48 Nm).

22. Install the serpentine belt.

23. Install fuel vapor pipes.

24. Install the serpentine belt and, if equipped, the air conditioning drive belt. Install the coolant reservoir.

25. Connect the negative battery cable and properly fill the engine cooling system.

3.0L (VIN R) ENGINE

➡**The steps in this procedure are critical in preventing catastrophic engine damage, adhering to this sequence is imperative. There are special tools needed to perform this procedure. It is a good idea to read this procedure several times before attempting to perform this job. This is an interference engine.**

Always turn the crankshaft in the direction of rotation (clockwise), never against engine rotation. Never remove the timing belt without first setting the camshaft gears and crankshaft drive gear to TDC and locking them in place with tool J42069 or equivalent.

The "Timing Belt Installation and Adjustment Table" provides an overview of the steps needed to properly install and adjust the timing belt. Use this table as a reference, not as a substitution, for the steps in this procedure.

1. Disconnect the negative battery cable.

2. Remove the resonance chamber.

3. Remove the front timing belt cover.

4. Remove the harmonic balancer from the crankshaft.

5. Rotate the crankshaft clockwise to 60 degrees BTDC.

6. Install J42069-10, or equivalent, to the crankshaft drive gear with knurled bolt.

7. Turn the engine clockwise, with J42098 or equivalent, until the lever of J42069-10 or equivalent, firmly contacts the water pump pulley flange. Secure the lever to the water pump.

➡**Be sure the engine is not 180 degrees off. The camshaft marks must align with the rear timing cover.**

8. Lock the camshaft gears, using J42069-1, or equivalent, and J42069-2 or equivalent. It may be necessary to loosen the relevant idler pulley to lock the gears, then tighten the idler pulley bolt to 30 ft. lbs. (40 Nm).

9. Loosen the timing belt tensioner and remove the belt.

✴✴ WARNING

With the belt removed, do not rotate the camshaft or crankshaft, or remove the locking tools, because the pistons may contact the valves and cause internal engine damage.

To install:

10. Remove J42069-10 or equivalent.

11. Raise and safely support the vehicle securely on jackstands.

12. Install the timing belt, starting at the crankshaft gear and aligning the double dash (TDC) mark on the belt with the oil pump and belt drive gear.

13. Using J42069-30 or equivalent, secure the belt to prevent the splines from jumping.

14. Lower the vehicle.

15. Route the belt between the idler pulley for camshafts 3 and 4, then between the gears for 3 and 4.

16. If the dash marks on the belt do not align with the camshaft

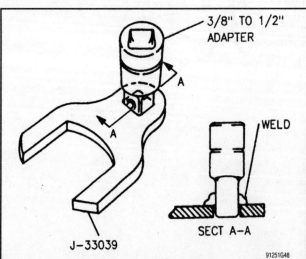

3/8" TO 1/2" ADAPTER

WELD

SECT A-A

J-33039

91251G48

The timing belt tension adjuster may be fabricated or modified to suit the engine—General Motors 1992–94 2.0L (VIN H) engine

and rear timing cover marks, loosen the idler pulley, or move the cam gears slightly, with the camshaft gears still locked in place, until the timing belt can be properly installed. Temporally tighten the idler pulley locking bolt; the locking bolt will be tightened to specification after final adjustments are made.

➡ **The timing belt deflection must be no more than 0.4 in. (10mm) between camshaft gear 4 and the idler pulley. To adjust the deflection, rotate the timing belt idler pulley for camshafts 3 and 4 counterclockwise with tool J42069-40 or equivalent.**

17. Route the belt between the idler pulley for camshafts 1 and 2, then between the gears for 1 and 2.

18. If the dash marks on the belt do not align with the camshaft and rear timing cover marks, loosen the idler pulley, or move the cam gears slightly, with the camshaft gears still locked in place, until the timing belt can be installed. Temporally tighten the idler pulley; the locking bolt will be tightened to specification after final adjustments are made.

19. Apply tension to the belt to keep it from slipping off the gears by turning the timing belt idler pulley for camshafts 1 and 2 counterclockwise with J42069-40 or equivalent. Temporally tighten the idler pulley; the locking bolt will be tightened to specification after final adjustments are made.

20. Complete the routing of the timing belt through the belt tensioner.

21. Apply initial tension by turning the timing belt tensioner counterclockwise, with a 5mm Allen wrench, until the marks are set as shown in the initial timing belt adjustment illustration.

22. Tighten the tensioner locking nut to 15 ft. lbs. (20 Nm).

23. Ensure that the alignment marks are at their specific reference points.

24. Remove J42069-30, J42069-1 and J42069-2.

25. Rotate the engine 2 revolutions clockwise, stopping at 60 degrees BTDC.

26. Install J42069-10 to the crankshaft gear with the knurled bolt.

27. Turn the engine clockwise, with J42098 or equivalent, until

TIMING BELT INSTALLATION AND ADJUSTMENT TABLE

Step	Action	Value	Yes	No
1	Install the timing belt and align marks on the belt with the marks on the camshaft gears and the crankshaft gear. Check the timing belt deflection between the idler pulley for camshafts 3 & 4 and camshaft number 4. Is the timing belt installed, the marks aligned and the timing belt deflection adjusted?	1 cm (0.4 in) maximum	Go to Step 2	—
2	Set the initial timing belt tension at the timing belt tensioner. Is the initial timing belt tension set?	—	Go to Step 3	—
3	Rotate the engine two complete revolutions and secure the crankshaft at Top Dead Center (TDC) with the J 42069-10. Has the engine been rotated and the crankshaft secured to TDC?	—	Go to Step 4	—
4	Starting with camshafts 3 and 4, check the alignment of the marks on the camshaft gears with the marks on the J 42069-20 checking gauge. Do the marks on the camshaft gears align exactly with the marks on J 42069-20?	—	Go to Step 5	Go to Step 6
5	Check the alignment of the marks on camshafts gears 1 and 2 with the marks on the J 42069-20 checking gauge. Do the marks on the camshaft gears align exactly with the marks on J 42069-20?	—	Go to Step 14	Go to Step 10
6	Do the camshaft gear marks line up to the left (BTDC) of the marks on the J 42069-20 checking gauge?	—	Go to Step 8	Go to Step 7
7	Do the camshaft gear marks line up to the right (ATDC) of the marks on the J 42069-20 checking gauge?	—	Go to Step 9	—
8	Turn the idler pulley eccentric, for camshafts 3 and 4, counterclockwise until the marks on the camshaft gear align exactly with the marks on J 42069-20. Rotate the engine two complete revolutions, lock the crankshaft at TDC with J 42069-10 and recheck the alignment of the camshaft gear marks to the marks on J 42069-20. Do the marks on the camshaft gears align exactly with the marks on J 42069-20?	—	Go to Step 5	Go to Step 6
9	Turn the idler pulley eccentric, for camshafts 3 and 4, clockwise until the marks on the camshaft gear align exactly with the marks on J 42069-20. Rotate the engine two complete revolutions, lock the crankshaft at TDC with J 42069-10 and recheck the alignment of the camshaft gear marks to the marks on J 42069-20. Do the marks on the camshaft gears align exactly with the marks on J 42069-20?	—	Go to Step 5	Go to Step 6

Timing belt installation and adjustment table—General Motors 3.0L (VIN R) engine

79225G35

Step	Action	Value	Yes	No
10	Do the camshaft gear marks line up to the left (BTDC) of the marks on the J 42069-20 checking gauge?	—	Go to Step 12	Go to Step 11
11	Do the camshaft gear marks line up to the right (ATDC) of the marks on the J 42069-20 checking gauge?	—	Go to Step 13	—
12	Turn the idler pulley eccentric, for camshafts 1 and 2, counterclockwise until the marks on the camshaft gear align exactly with the marks on J 42069-20. Rotate the engine two complete revolutions, lock the crankshaft at TDC with J 42069-10 and recheck the alignment of the camshaft gear marks to the marks on J 42069-20. Do the marks on the camshaft gears align exactly with the marks on J 42069-20?	—	Go to Step 14	Go to Step 10
13	Turn the idler pulley eccentric, for camshafts 1 and 2, clockwise until the marks on the camshaft gear align exactly with the marks on J 42069-20. Rotate the engine two complete revolutions, lock the crankshaft at TDC with J 42069-10 and recheck the alignment of the camshaft gear marks to the marks on J 42069-20. Do the marks on the camshaft gears align exactly with the marks on J 42069-20?	—	Go to Step 14	Go to Step 10
14	Set the final timing belt tension at the timing belt tensioner. Is the final timing belt tension set?	—	Go to Step 15	—
15	Again, rotate the engine two complete revolutions and lock the crankshaft at TDC. Do a final inspection of the camshaft gear marks' relationship to the J 42069-20 marks. The marks must align exactly. Do the marks on the camshaft gears align exactly with the marks on the J 42069-20?	—	Go to Step 16	Go to Step 2
16	Remove all checking tools and ensure all idler pulleys and the tensioner locking nut are tightened to specifications. Continue with re-assembly of the engine.	—	—	—

79225G36

Timing belt installation and adjustment table (continued)—GM 3.0L (VIN R) engine

the lever of J42069-10 or equivalent firmly contacts the water pump pulley flange. Secure the lever to the water pump.

➡ **The alignment marks on the timing belt will no longer match the marks on the camshaft gears after 1 or more revolutions. The marks on the camshaft gears must align with the notches on the rear timing cover, and the crankshaft drive gear and the oil pump housing should match up to their mark.**

28. If timing belt adjustment is necessary, first adjust camshafts 3 and 4.

29. Check the alignment of camshafts 3 and 4, 1 and 2 with gauge J42069-20.

30. If alignment is OK, then set final belt tension as follows:
   a. Loosen the timing belt tensioner locknut.
   b. Adjust the timing belt tensioner by turning the eccentric cam, with a 5mm hex wrench, until the marks are set.

31. Tighten the tensioner locknut to 15 ft. lbs. (20 Nm).

32. Remove J42069-10 and J42069-20.

33. Rotate the engine clockwise 2 revolutions, stopping at 60 degrees BTDC.

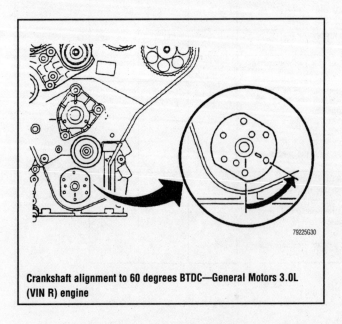

Crankshaft alignment to 60 degrees BTDC—General Motors 3.0L (VIN R) engine

79225G30

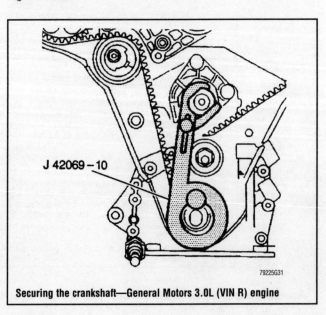

J 42069 – 10

79225G31

Securing the crankshaft—General Motors 3.0L (VIN R) engine

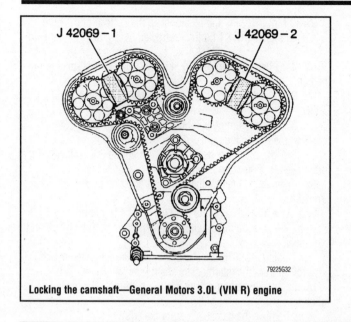

Locking the camshaft—General Motors 3.0L (VIN R) engine

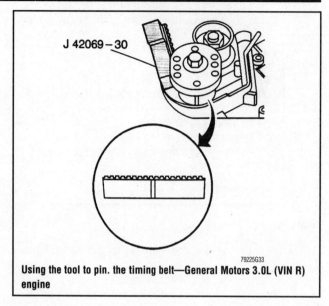

Using the tool to pin. the timing belt—General Motors 3.0L (VIN R) engine

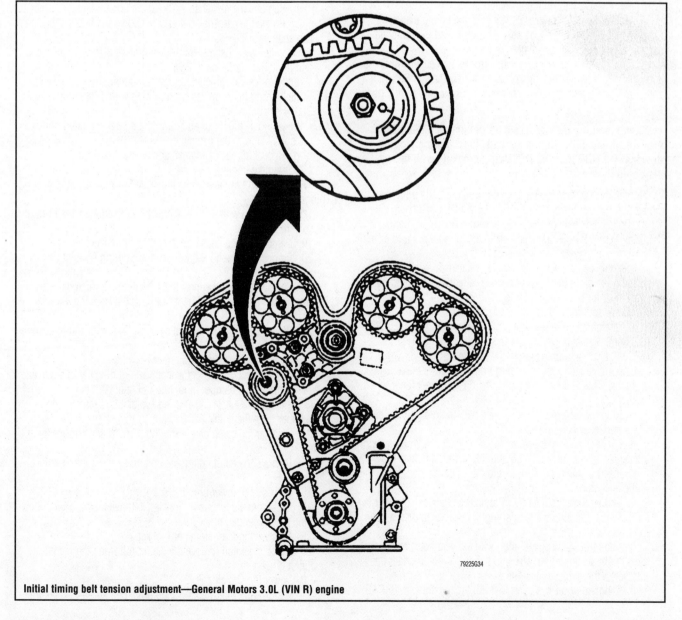

Initial timing belt tension adjustment—General Motors 3.0L (VIN R) engine

34. If further adjustment is needed, then repeat the applicable steps.

35. Tighten the idler pulley bolts for the camshafts to 30 ft. lbs. (40 Nm).

36. Be sure all tools are removed from the engine.

37. Install the harmonic balancer and tighten the bolts to 15 ft. lbs. (20 Nm).

38. Install the timing belt covers.

39. Install the resonance chamber.

40. Connect the negative battery cable, and reprogram applicable accessories.

3.4L (VIN X) ENGINE

1991–97

The 3.4L (VIN X) engine uses a timing chain and camshaft timing belts.

1. Disconnect the negative battery cable.

2. Disconnect and remove the power steering pump from the pump mounting bracket.

3. Remove the left, right and center timing belt covers.

4. Rotate the engine clockwise to align the timing marks, Top Dead Center (TDC) on the No. 1 exhaust stroke, on the camshaft sprockets and intermediate shaft.

5. Loosely clamp the 2 camshaft sprockets on each side of the engine together using clamping pliers or the equivalent. Secure the belt to the right-side cam sprocket with a C-clamp and a wide pad on the belt.

➡**When clamping the sprockets no deflection should be noticed. If any deflection is noticed, loosen the clamping devices. DO NOT mar the camshaft sprockets with the clamping device.**

6. Remove the tensioner side plate retaining bolts from the tensioner and remove the side plate from the actuator and base.

7. Rotate the actuator assembly around the arm pivot and out of the base. Removal of the tensioner from the base allows it to extend to its maximum travel.

8. Set the actuator aside on a table in a vertical position to allow the oil to drain into the boot end. The tensioner should be allowed to sit for 5 minutes prior to refilling with oil.

9. Reset the timing belt actuator as follows:

   a. Straighten out a paper clip or a piece of stiff wire 0.032 in. (0.75mm) diameter to a minimum straight length of 1.85 in. (47mm). Form a double loop in the remaining end.

   b. Remove the rubber end plug from the rear of the tensioner assembly. This will aid in allowing the oil in the tensioner to escape.

   c. Hold the tensioner in your hand with the rubber boot end of the tensioner pointing down.

   d. DO NOT remove the vent plug. Push the paper clip through the center hole in the vent plug and into the pilot hole.

   e. Insert a small screwdriver into the screw slot inside the end of the tensioner.

   f. Retract the tensioner by rotating the tensioner plunger in a clockwise direction while pushing the rod tip against a table top.

   g. Align the screw slot to align with the vent hole, and push the straight section of the wire into the screw slot to retain the plunger in the retracted position.

   h. If tensioner oil has been lost, fill the tensioner with SAE 5W30 Mobil 1®. Fill the tensioner to the bottom of the plug. The tensioner **MUST** be fully retracted before being filled with oil.

10. If the belt is being reused, mark the direction of rotation on the belt.

11. Remove the timing belt tensioner pulley mounting bolt and pulley.

12. Remove the timing belt after first removing the C-clamp retaining the belt to the right-side camshaft sprocket.

13. Remove the Torx® head bolts securing the idler pulleys, if the idlers need to be replaced.

14. Remove the intermediate shaft sprocket using the following procedure:

   a. Use a suitable tool to hold the engine from turning.

   b. Remove the intermediate shaft sprocket mounting bolt and washer.

   c. Using J-38616 or equivalent sprocket puller, remove the sprocket from the intermediate shaft.

15. If the camshaft sprockets need to be removed, proceed as follows:

   a. Remove the camshaft carrier cover(s).

   b. Remove the camshaft sprocket clamping pliers.

   c. Rotate the camshaft being serviced so the flats on the camshaft are face up.

   d. Install a Camshaft Hold-down tool J-38613 or equivalent, and tighten to 22 ft. lbs. (30 Nm).

   e. Remove the camshaft sprocket mounting bolt and washer while holding the camshaft from turning with J-38613 and J-38614 or equivalent

   f. Using J-38616, or an equivalent sprocket puller remove the sprocket from the camshaft.

   g. Repeat for each camshaft sprocket as necessary.

16. Drain the cooling system.

17. Disconnect the lower radiator hose from water pump inlet pipe.

18. If equipped with a manual transaxle, disconnect the front AIR hose at the AIR pipe.

19. Disconnect the heater hose at the front cover.

20. Remove the heater pipe bracket mounting bolts at the frame.

21. Raise and safely support the vehicle on jack stands. Remove the right front tire and wheel assembly and the right inner fender splash-shield.

22. Remove the crankshaft pulley mounting bolts and remove the pulley from the damper.

23. Remove the crankshaft damper as follows:

   a. While holding the crankshaft from turning using a suitable tool, remove the damper mounting bolt and washer.

   b. Install tool J-24420-B or equivalent, and remove the damper from the crankshaft.

24. Place an oil catch pan under the oil filter and remove the oil filter.

25. Remove the air conditioning compressor mounting bracket bolts.

26. Remove the lower front cover bolts.

27. On automatic transaxle vehicles, remove the halfshaft following the recommended procedure.

28. Remove the rear alternator bracket.

29. Disconnect and remove the starter following the recommended procedure.

30. Lower the vehicle.

31. Remove the intermediate shaft drive belt sprocket retaining

A LOCATION OF TIMING MARKS WITH CAM HOLD
  DOWN TOOLS J 38613 INSTALLED
  (#4 TDC COMPRESSION STROKE)
B FRONT COVER TIMING MARK
C LOCATION OF TIMING MARKS WITH DRIVE
  BELT INSTALLED
D LOCATION WHERE CAM HOLD DOWN TOOLS
  ARE INSTALLED
1 RH EXHAUST CAMSHAFT SPROCKET
2 RH INTAKE CAMSHAFT SPROCKET
3 LH INTAKE CAMSHAFT SPROCKET
4 LH EXHAUST CAMSHAFT SPROCKET
5 PERMANENT MARKS PAINTED DOTS REMOVE
  PREVIOUS MARKS IF TIMING IS BEING CHANGED
  AND MARKS AGAIN IN THESE LOCATIONS
6 CRANKSHAFT BALANCER
7 INTERMEDIATE SHAFT SPROCKET

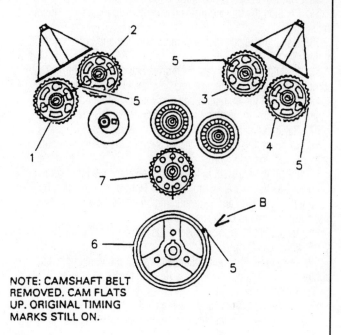

NOTE: CAMSHAFT BELT
REMOVED. CAM FLATS
UP. ORIGINAL TIMING
MARKS STILL ON.

79225G15

Timing marks with hold-down tool in place—General Motors 1991–97 3.4L (VIN X) engines

bolt and remove the intermediate shaft drive belt sprocket using J38616 or equivalent puller.

32. Remove the upper alternator mounting bolts.

33. Remove the forward light relay center screws and position the relay center aside.

34. Disconnect the oil cooler hose from the front cover.

35. Remove the water pump pulley.

36. Remove the upper front cover bolts and remove the front cover.

37. Mark the intermediate shaft sprocket, chain link, crankshaft sprocket and cylinder block for assembly reference. The marks should be made with paint so they won't be lost when the components are removed.

38. Retract the timing chain tensioner shoe as follows:

  a. Insert J-33875 or an equivalent, on both sides of the tensioner.

  b. Pull on the through-pin in the tensioner arm to retract the spring located in the tensioner arm.

  c. While compressing the spring, use a suitable tool, a cotter pin or nail, and insert the pin in the hole in the tensioner assembly to hold the tensioner compressed. The tool used must be strong enough to hold the tensioner compressed.

➡The timing chain, crankshaft sprocket and intermediate shaft sprocket will be removed at the same time. If, when removing the assembly, the intermediate shaft sprocket does not easily come off the intermediate shaft, rotate the crankshaft back and forth to loosen the intermediate shaft sprocket.

39. Install a suitable puller, J-38611 and J-8433 or their equivalents.

40. Tighten the bolt on the puller and slowly pull the crankshaft sprocket off the crankshaft. Be sure the intermediate shaft sprocket is moving along with the crankshaft sprocket.

41. Remove the timing chain and sprockets.

42. Remove the tensioner mounting bolts and remove the tensioner assembly.

To install:

43. Install the tensioner assembly and tensioner assembly mounting bolts finger-tight first. Tighten the bolt in the slotted hole first to 18 ft. lbs. (25 Nm), then tighten the remainder of the bolts to 18 ft. lbs. (25 Nm).

44. Check to ensure that the crankshaft key is fully seated in the crankshaft cutout and the tensioner assembly is fully retracted.

45. Assemble the timing chain, intermediate shaft sprocket and crankshaft sprocket on a work bench. The timing marks made should be in alignment. The large chamfer and counterbore of the crankshaft sprocket are installed facing toward the engine and the intermediate shaft spline sockets are installed facing away from the engine.

46. Install the sprocket and chain assembly onto the engine. As the sprockets are installed, parallel alignment must be maintained.

47. The crankshaft sprocket will have to be pressed on the final 0.31 in. (8mm). This can be done using J-38612, or an equivalent puller.

48. Ensure timing was maintained.

49. Remove the retaining pin from tensioner. Clean all gasket surfaces completely.

50. Apply GM Sealer 1052080 or equivalent, to the lower edges of the sealing surface of the front cover. Install a new gasket on the front cover.

51. Install the front cover on the engine. Apply thread sealant to the large bolts and tighten the bolts enough to pull the front cover against the engine block.

52. Install the water pump pulley.

53. Connect the oil cooler hose to the front cover.

54. Position the forward light relay center and install the mounting screws.

55. Install the upper alternator mounting bolt and tighten it to 22 ft. lbs. (30 Nm).

56. Install the intermediate shaft drive belt sprocket. The sprocket must lock into the intermediate shaft timing chain sprocket. Install the mounting bolt and washer. While holding the engine from turning, tighten the bolt to 95 ft. lbs. (130 Nm).

57. Raise and safely support vehicle on jackstands.

58. Connect and install the starter. Tighten the starter mounting bolts to 32 ft. lbs. (43 Nm).

59. Install the halfshaft following the recommended procedure.

60. Install the rear alternator bracket. Tighten the mounting bolt to 22 ft. lbs. (30 Nm) and the mounting stud to 41 ft. lbs. (55 Nm) and the lower bolt to 61 ft. lbs. (83 Nm).

61. Install the lower front cover bolts and tighten the small bolts to 18 ft. lbs. (25 Nm).

62. Install the air conditioning compressor mounting bolts and tighten the mounting bolts to 37 ft. lbs. (50 Nm).

63. Install a new oil filter.

64. Install the crankshaft damper as follows:

    a. Coat the seal contact area on the damper with clean engine oil.

    b. Align the notch inside the damper with the crankshaft key and slide the damper on until the key is started into the notch.

    c. Using J-29113 or an equivalent puller, press the damper into position on the crankshaft.

    d. Install the crankshaft pulley and pulley mounting bolts. Tighten the pulley mounting bolts to 37 ft. lbs. (50 Nm).

    e. Install the crankshaft damper mounting bolt and washer, and tighten to 78 ft. lbs. (105 Nm).

65. Install the right-side inner fender splash-shield.

66. Install the tire and wheel assembly.

67. Lower the vehicle.

68. Tighten the upper front cover small bolts to 18 ft. lbs. (25 Nm) and the large bolts to 35 ft. lbs. (47 Nm).

69. Connect the heater hose to the front cover.

70. Install the heater pipe bracket mounting bolts.

71. If equipped with a manual transaxle, connect the front AIR hose to the AIR pipe.

72. Connect the lower radiator hose to the water pump.

73. Install the right-side cooling fan. Install the upper radiator support.

74. Position the torque strut mounting bracket and install the mounting bolts and tighten to 52 ft. lbs. (70 Nm).

75. Install the front engine lift hook and tighten the mounting bolt to 52 ft. lbs. (70 Nm).

76. To install the camshaft sprockets, proceed as follows:

    a. Wipe the camshaft noses with clean engine oil.

    b. Install the camshaft sprocket onto the nose of the camshaft.

    c. Install the lockring and shim ring.

    d. Install, but DO NOT tighten, the camshaft sprocket mounting bolts at this time.

77. Install the intermediate shaft sprocket as follows:

    a. Lubricate the seal contact area on the intermediate shaft sprocket with clean engine oil.

    b. Slide the sprocket through the intermediate shaft sprocket seal and engage the locking tangs into the sockets of the chain sprocket.

    c. Lightly lubricate the shaft seal and place it in position on the end of the intermediate shaft.

    d. Install the intermediate shaft sprocket mounting bolt and

washer. Tighten the bolt to 96 ft. lbs. (130 Nm) while holding the crankshaft from turning.

78. Install the timing belt idler pulleys and tighten the Torx® bolts to 37 ft. lbs. (50 Nm).

79. Install the actuator assembly and side plate. Tighten the actuator mounting bracket bolts to 37 ft. lbs. (50 Nm).

80. Install the belt, taking note of direction of rotation if the old belt was used.

81. Install the tensioner pulley to the mounting base. Tighten the bolt to 37 ft. lbs. (50 Nm).

82. Rotate the tensioner pulley counterclockwise into the belt using the cast square lug on the body and engage the ball end of the actuator into the socket on the pulley arm.

83. Remove the tensioner lockpin allowing the tensioner shaft to extend and the pulley to move into the belt.

84. Rotate the tensioner pulley counterclockwise, applying 14 ft. lbs. (18 Nm) of torque.

85. Rotate the engine clockwise 3 times to seat the belt. Align the crankshaft reference marks during the final rotation to TDC. Do not allow the crankshaft to spring back or reverse its direction of rotation.

➡**The timing flats on the camshafts should be 180 degrees apart from the left-side to the right-side. Both camshafts on the same side should be the same.**

86. To perform the camshaft timing procedure, proceed as follows:

    a. Rotate the camshaft flats up on the right-side camshafts and install a Camshaft Hold-down tool J-38613 or equivalent, and tighten to 22 ft. lbs. (30 Nm).

    b. Seat the lockring on the right exhaust and intake camshaft sprockets by threading in the mounting bolt and washer.

    c. Hold the sprocket from turning using tool J-38614 or equivalent.

➡**Running torque of the bolts before seating should be 55 ft. lbs. (75 Nm).**

    d. If less torque is required, replace the shim ring and lockring.

    e. If more torque is required, replace the shim ring and lockring and inspect the bolts for burring.

    f. Seating of the lockring is accomplished when the edge is flush with the sprocket hub.

    g. With the lockring seated tighten the bolt to final torque of 81 ft. lbs. (110 Nm).

    h. Remove J-38613 or equivalent.

    i. Rotate the engine clockwise 1 full revolution or any number of odd revolutions. **DO NOT** rotate the engine backward.

    j. Be sure the timing mark on the damper lines up with the mark on the front cover.

    k. Repeat Substeps **a** through **i** for the left-side.

87. Install the camshaft carrier covers.

88. Install timing the belt left, right and center covers and retaining bolts.

89. Connect the negative battery cable.

90. Start the engine and verify proper operation and engine performance.

GEO/Chevrolet

➡**During these procedures, identify all components removed from the engine so that they may be reinstalled in their original positions. If discarding the old components so new components can be installed, identifying the old items is not necessary.**

1.0L (VIN 2, 5, 6) & 1.3 (VIN 9) ENGINES

➡ **Timing belts must always be completely free of dirt, grease, fluids and lubricants. This includes the sprockets and contact surfaces on which the belt rides. The belt must never be crimped, twisted or bent. Never use tools to pry or wedge the belt.**

1. Disconnect the negative battery cable.
2. Raise and safely support the vehicle.
3. Remove the clips and right-side splash-shield.
4. Remove the lower alternator cover plate.
5. Remove the alternator drive belt, if equipped, the air conditioning drive belt.
6. Remove the water pump pulley.

➡ **It is not necessary to remove the crankshaft timing sprocket bolt (center bolt) to remove the crankshaft pulley.**

7. Remove the crankshaft pulley bolts and crankshaft pulley.
8. Remove the retaining bolts and nut from the timing belt outside cover.
9. Remove the timing belt outside cover.
10. Turn the crankshaft to align the timing marks. The mark on the crankshaft sprocket should align with the arrow mark on the oil pump housing. The mark on the camshaft sprocket should align with the **V** mark on the timing belt inner cover or cylinder head cover.
11. If the timing belt is to be reused, mark the direction of rotation on the belt.
12. Remove the timing belt tensioner, tensioner plate, tensioner spring, spring damper and timing belt.

➡ **Never turn the camshaft or crankshaft independently after the timing belt has been removed. Interference may occur between the pistons and valves, and parts may be damaged.**

13. Inspect the timing belt for wear or cracks, and replace as necessary. Check the tensioner for smooth rotation.
14. If the timing belt sprockets are to be removed, proceed as follows:

    a. Using a 0.39 in. (10mm) rod inserted into the camshaft, hold the camshaft and remove the retaining bolt and camshaft sprocket.

✳✳ WARNING

Be careful not to damage the cylinder head or cylinder head cover mating surfaces. Place a clean shop cloth between the rod and cylinder head. Do not bump the rod hard against the cylinder head when loosening the bolt.

    b. Raise and safely support the vehicle on jack stands.
    c. If equipped with a manual transaxle, lock the crankshaft in position by inserting a suitable flat-bladed tool into the hole in the bottom of the bell housing to engage the flywheel teeth.
    d. If equipped with an automatic transaxle, lock the crankshaft in position by inserting a suitable flat-bladed tool between the flywheel teeth and against the engine block.
    e. Remove the crankshaft sprocket bolt and crankshaft sprocket.

To install:

15. If the timing belt sprockets were removed, proceed as follows:

    a. Install the crankshaft sprocket, aligning the keyway. Lock the crankshaft in place and tighten the crankshaft sprocket bolt to 81 ft. lbs. (110 Nm).
    b. Lower the vehicle.

    c. Install the camshaft sprocket and retaining bolt. Lock the camshaft in place using the rod, and tighten the bolt to 44 ft. lbs. (60 Nm). Remove the locking rod.
16. Install the tensioner plate to the tensioner.
17. Insert the lug of the tensioner plate into the hole of the tensioner.
18. Install the tensioner, tensioner plate and spring. Do not fully tighten the tensioner bolt and stud at this time.
19. Move the tensioner plate in a counterclockwise direction. This should cause the tensioner to move in the same direction. If it does not, remove the tensioner and tensioner plate, and reinsert the tensioner plate lug in the timing plate tensioner hole.
20. Check that the camshaft timing marks are aligned. If not, align the 2 marks by turning the camshaft.
21. Check that the punch mark on the crankshaft timing belt sprocket is aligned with the arrow mark on the oil pump case. If not, align the 2 marks by turning the crankshaft.
22. With the timing marks aligned, install the timing belt on the 2 sprockets. If the old belt is being reused, be sure to install it running in the same direction of original rotation.

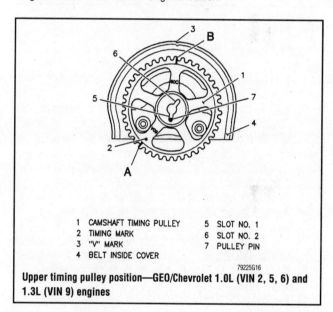

1	CAMSHAFT TIMING PULLEY	5	SLOT NO. 1
2	TIMING MARK	6	SLOT NO. 2
3	"V" MARK	7	PULLEY PIN
4	BELT INSIDE COVER		

79225G16

Upper timing pulley position—GEO/Chevrolet 1.0L (VIN 2, 5, 6) and 1.3L (VIN 9) engines

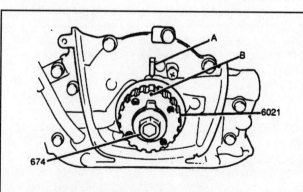

A   ARROW MARK ON OIL PUMP CASE
B   PUNCH MARK ON CRANKSHAFT TIMING GEAR
674   CRANKSHAFT PULLEY TIMING GEAR BOLT
6021   CRANKSHAFT TIMING GEAR

79225G17

Crankshaft timing mark—GEO/Chevrolet 1.0L (VIN 2, 5, 6) and 1.3L (VIN 9) engines

23. Install the tensioner spring and spring damper. Turn the timing belt 2 rotations clockwise after installing the tensioner spring and damper to remove any belt slack. Tighten the tensioner stud to 96 inch lbs. (11 Nm), then the tensioner bolt to 20 ft. lbs. (27 Nm).

➡**Confirm that both sets of timing marks are aligned properly.**

24. Using a new seal, install the timing belt cover and tighten the bolts and nut to 97 inch lbs. (11 Nm).

25. Install the crankshaft pulley. Fit the keyway on the pulley to the crankshaft timing belt sprocket and tighten the bolts to 8–12 ft. lbs. (11–16 Nm).

26. Install the alternator drive belt, if equipped, air conditioning drive belt.

27. Install the lower alternator cover plate. Tighten the bolts to 89 inch lbs. (10 Nm).

28. Install the right-side splash-shield and clips.

29. Lower the vehicle and connect the negative battery cable.

30. Run the engine. Check for leaks.

1.5L (VIN 7, 9) ENGINES

1985–89

1. Disconnect the negative battery cable and drain the cooling system.

2. Remove the engine and mount the engine on an engine stand.

3. Remove the accessory drive belts.

4. Remove the engine mount bracket from the timing cover.

5. Rotate the crankshaft until the notch on the crankshaft pulley aligns with the **0** degree mark on the timing cover and the No. 4 cylinder is on Top Dead Center (TDC) of the compression stroke.

6. Remove the starter and install the Flywheel Holding tool No. J-35271 or equivalent.

7. Remove the crankshaft bolt and pulley.

8. Remove the timing cover bolts and the timing cover.

9. Loosen the tension pulley bolt.

10. Insert an Allen wrench into the tension pulley hexagonal hole and loosen the timing belt by turning the tension pulley clockwise.

11. If reusing the timing belt, mark the direction of rotation on the back of the belt.

12. Remove the timing belt.

To install:

13. Align the groove on the crankshaft timing gear with the mark on the oil pump.

14. Align the camshaft timing gear mark with the upper surface of the cylinder head and the dowel pin in its uppermost position.

15. Place the timing belt arrow in the direction of the engine rotation and install the timing belt. Tighten the tension pulley bolt.

16. Turn the crankshaft 2 complete revolutions and realign the crankshaft timing gear groove with the mark on the oil pump.

17. Loosen the tension pulley bolt and apply tension to the belt with an Allen wrench. Torque the pulley bolt to 37 ft. lbs. (50 Nm) while holding the pulley stationary.

18. To complete the installation, reverse the removal procedures. Torque the crankshaft pulley-to-crankshaft bolt to 109 ft. lbs. (148 Nm).

1.6L (VIN 4) SOHC ENGINE

1985–88 Nova

1. Disconnect the negative battery cable.

2. Loosen the water pump pulley bolts and remove the alterna-

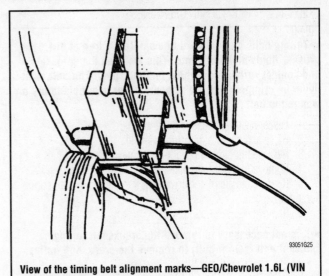

View of the timing belt alignment marks—GEO/Chevrolet 1.6L (VIN 4) SOHC engine—1985–88 Nova

tor/water pump drive belt. If equipped with power steering, remove the power steering pump.

3. Remove the water pump pulley bolts and pulley. Drain the cooling system.

4. Disconnect the upper radiator hose from the water pump outlet. Label and disconnect all vacuum hoses that may be in the way.

5. Remove the upper timing belt cover front cover-to-engine bolts.

6. Remove the upper timing belt front cover and gasket.

7. If equipped with air conditioning, loosen the idler pulley mounting bolt. Loosen the adjusting nut, then remove the air conditioning drive belt and the idler pulley with the adjusting bolt.

8. Remove the alternator bolts and move the alternator aside.

9. Remove the middle timing belt front cover-to-engine bolts and the cover.

10. Raise and safely support the vehicle.

11. Remove the right-side undercover, the flywheel cover, the crankshaft pulley-to-crankshaft bolt and the crankshaft pulley.

12. Remove the lower timing belt front cover-to-engine bolts and the cover.

13. Remove the No. 1 spark plug. Using a socket wrench on the crankshaft pulley bolt, rotate the engine (clockwise) to position the No. 1 cylinder on Top Dead Center (TDC) on its compression stroke.

➡**The TDC of the No. 1 cylinder is located when air is expelled from the cylinder.**

14. If reusing the timing belt, mark an arrow showing the direction of rotation and matchmark the belt to both pulleys.

15. Loosen the idler pulley mounting bolt and push the idler pulley to relieve the belt tension, then retighten the mounting bolt.

16. Remove the timing belt.

To install:

17. Noting the direction of rotation, install the timing belt.

18. Loosen the idler pulley mounting bolt and move the idler pulley to tension the timing belt.

19. Using finger pressure on the longest span between the pulleys, measure the timing belt deflection; it should be 0.24–0.28 in. (6–7mm) at 4.4 lbs. (2 kg). Torque the idler pulley mounting bolt to 27 ft. lbs. (37 Nm).

20. Rotate the crankshaft 2 complete revolutions and realign the timing marks.

21. If adjustment is not correct, loosen the idler pulley bolt and correct the belt tension.

22. To complete the installation, reverse the removal procedures.

1.6L (VIN 5) DOHC ENGINE

1988 Nova and 1991–93 Prizm

1. Disconnect the negative battery cable.
2. Raise the vehicle and safely support it on jackstands.
3. Remove the right front wheel.
4. Remove the splash shield from under the car.
5. Drain the coolant into clean containers. Close the draincocks when the system is empty.
6. Lower the car to the ground. Disconnect the accelerator cable and, if equipped, the cruise control cable.
7. Remove the cruise control actuator, if equipped.
8. Carefully remove the ignition coil.
9. Disconnect the radiator hose at the water outlet.
10. Remove the power steering drive belt and the alternator drive belt.

11. Remove the spark plugs.
12. Rotate the crankshaft clockwise and set the engine to Top Dead Center (TDC) on the compression stroke on No. 1 cylinder. Align the crankshaft marks at zero; look through the oil filler hole and make sure the small hole in the end of the camshaft can be seen.
13. Raise and safely support the vehicle. Disconnect the center engine mount.
14. Lower the vehicle to the ground.
15. Support the engine either from above or below. Disconnect the right engine mount from the engine.
16. Raise the engine and remove the mount.
17. Remove the water pump pulley.
18. Remove the crankshaft pulley.
19. Remove the 10 bolts and remove the timing belt covers with their gaskets.

➡**The bolts are different lengths; they must be returned to their correct location at reassembly. Label or diagram the bolts during removal.**

20. Remove the timing belt guide from the crankshaft pulley.
21. Loosen the timing belt idler pulley, move it to the left (to take tension off the belt) and tighten its bolt.

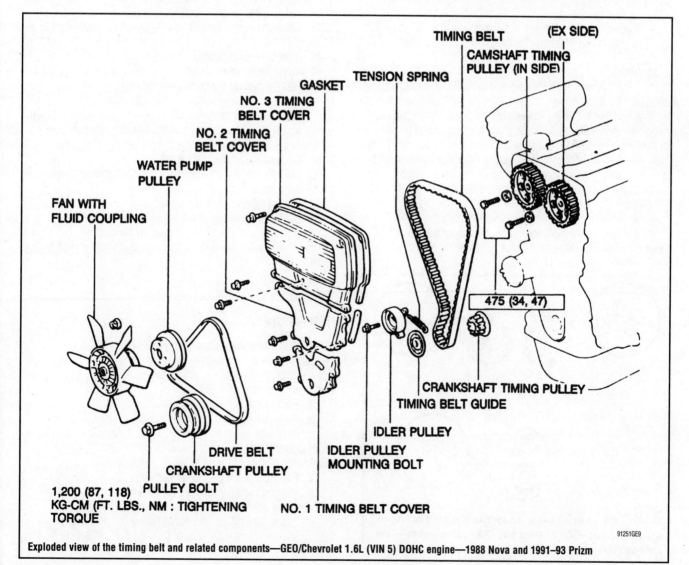

Exploded view of the timing belt and related components—GEO/Chevrolet 1.6L (VIN 5) DOHC engine—1988 Nova and 1991–93 Prizm

22. Make matchmarks on the belt and all pulleys showing the exact placement of the belt. Mark an arrow on the belt showing its direction of rotation.

23. Carefully slip the timing belt off the pulleys.

➡**Do not disturb the position of the camshafts or the crankshaft during removal.**

24. Remove the idler pulley bolt, pulley and return spring.

25. Remove the PCV hose and the valve cover.

26. Use an adjustable wrench to counterhold the camshaft. Be careful not to damage the cylinder head. Loosen the center bolt in each camshaft pulley and remove the pulley. Label the pulleys and keep them clean.

27. Check the timing belt carefully for any signs of cracking or deterioration. Pay particular attention to the area where each tooth or cog attaches to the backing of the belt. If the belt shows signs of damage, check the contact faces of the pulleys for possible burrs or scratches.

28. Check the idler pulley by holding it in your hand and spinning it. It should rotate freely and quietly. Any sign of grinding or abnormal noise indicates the pulley should be replaced.

29. Check the free length of the tension spring. Correct length is 43.5mm measured at the inside faces of the hooks. A spring which has stretched during use will not apply the correct tension to the pulley; replace the spring.

30. If you can test the tension of the spring, look for 22 lbs. (10 kg) of tension at 2 in. (50mm) of length. If in doubt, replace the spring.

To install:

31. Align the camshaft knock pin and the pulley. Reinstall the camshaft timing belt pulleys, making sure the pulley fits properly on the shaft and that the timing marks align correctly. Tighten the center bolt on each pulley to 43 ft. lbs. (58.5 Nm). Be careful not to damage the cylinder head during installation.

32. Before reinstalling the belt, double check that the crank and camshafts are exactly in their correct positions. The alignment marks on the pulleys should align with the cast marks on the head and oil pump.

33. Reinstall the valve covers and the PCV hose.

34. Install the timing belt idler pulley and its tensioning spring. Move the idler to the left and temporarily tighten its bolt.

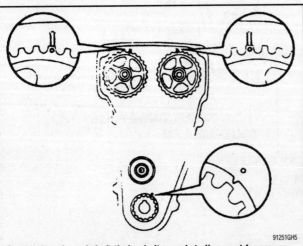

91251GH5

Camshaft and crankshaft timing belt sprocket alignment for proper belt replacement—GEO/Chevrolet 1.6L (VIN 5) DOHC engine—1988 Nova and 1991–93 Prizm

35. Carefully observing the matchmarks made earlier, install the timing belt onto the pulleys.

36. Slowly release tension on the idler pulley bolt and allow the idler to take up tension on the timing belt. DO NOT allow the idler to slam into the belt; the belt may become damaged.

37. Temporarily install the crankshaft pulley bolt. Turn the engine clockwise through 2 complete revolutions, stopping at TDC. Check that each pulley aligns with its marks.

38. Check the tension of the timing belt at a top point halfway between the 2 camshaft sprockets. The correct deflection is 0.16 in. (4mm) at 4.4 lbs. (2 kg) pressure. If the belt tension is incorrect, readjust it by repeating steps 18–20. If the tension is correct, tighten the idler pulley bolt to 27 ft. lbs. (37 Nm).

39. Remove the crankshaft pulley bolt.

40. Install the timing belt guide onto the crankshaft timing pulley.

41. When reinstalling, make certain that the gaskets and their mating surfaces are clean and free from dirt and oil. The gasket itself must be free of cuts and deformations and must fit securely in the grooves of the covers.

42. Reinstall the covers and their gaskets and the 10 bolts in their proper positions.

43. Install the crankshaft pulley, again using the counterholding tool. Tighten the bolt to specifications.

44. Install the water pump pulley.

45. Install the right engine mount. Tighten the through-bolt to 58 ft. lbs. (79 Nm).

46. Reinstall the spark plugs and their wires.

47. Install the alternator drive belt and the power steering drive belt. Adjust the belts to the correct tension.

48. Connect the radiator hose to the water outlet port.

49. Install the ignition coil.

50. Install the cruise control actuator and the cruise control cable, if equipped.

51. Connect the accelerator cable.

52. Refill the cooling system with the correct amount of antifreeze and water.

53. Connect the negative battery cable.

54. Start the engine and check for leaks. Allow the engine to warm up and check the work areas carefully for seepage.

55. Install the splash shield under the car. Install the right front wheel.

1.6L (VIN 6) SOHC ENGINE

1990–92 Storm

1. Disconnect the negative battery cable.

2. Remove the alternator belt, then the power steering belt.

3. Using the Engine Support Fixture tool J-28467-A or equivalent, support the engine.

4. Remove the right engine mount.

5. Remove the timing belt cover and the timing belt cover.

6. Rotate the crankshaft so that the No. 4 cylinder is at Top Dead Center (TDC) on its compression stroke by aligning the camshaft pulley timing mark to the 9 o'clock position.

7. Raise and safely support the vehicle.

8. Remove the crankshaft pulley, the 4 crankshaft pulley side bolts and the crankshaft pulley.

9. Lower the vehicle. Loosen the timing belt tensioner pulley bolt, then move the tensioner away from the belt and secure it with the bolt.

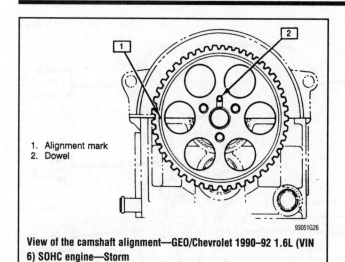

1. Alignment mark
2. Dowel

View of the camshaft alignment—GEO/Chevrolet 1990–92 1.6L (VIN 6) SOHC engine—Storm

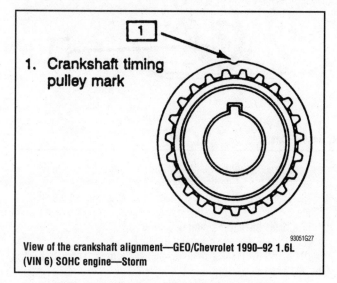

1. **Crankshaft timing pulley mark**

View of the crankshaft alignment—GEO/Chevrolet 1990–92 1.6L (VIN 6) SOHC engine—Storm

10. If reusing the timing belt, mark the rotational direction on the back of the belt.

11. Remove the timing belt.

To install:

12. Noting the timing belt's direction of rotation, install the belt on the crankshaft sprocket, tensioner pulley, water pump sprocket and the camshaft sprocket.

13. Adjust the timing belt tension by performing the following procedure:

    a. Loosen the belt tensioner bolt and move it toward the timing belt.

14. Using an Allen wrench, insert it into the hexagonal hole of the tension pulley. Hold the pulley stationary and temporarily tighten the tension pulley-to-engine bolt.

    a. Rotate the crankshaft 2 complete revolutions and align the crankshaft timing pulley groove with the mark on the oil pump.

    b. Loosen the tension pulley-to-engine bolt.

    c. Using the Allen wrench and a timing belt tension gauge, apply 31 ft. (42 N) of tension on the timing belt.

    d. Hold the pulley stationary and torque the tension pulley-to-engine bolt to 31 ft. lbs. (42 Nm).

15. Check that the camshaft is still aligned to the 9 o'clock position.

16. Install the timing belt cover, secure it with the 2 lower mounting bolts and torque the bolts to 89 inch lbs. (10 Nm).

17. Install the crankshaft pulley to the crankshaft damper and secure the center bolt and the 4 side bolts. Torque the center bolt to 87 ft. lbs. (118 Nm) and the 4 side bolts to 17 ft. lbs. (23 Nm).

18. Install the right side undercover. Lower the vehicle. Install the 4 upper timing belt cover retaining bolts.

19. Install the right engine mount.

20. Install the power steering pump drive belt, alternator drive belt and any other drive belts that may have been removed.

21. Remove the Engine Support Fixture tool J-28467-A or equivalent from the engine.

22. Connect the negative battery cable.

1.6L (VIN 5) DOHC ENGINE

1990–92 Storm

1. Disconnect the negative battery cable.
2. Remove the alternator belt, then the power steering belt.

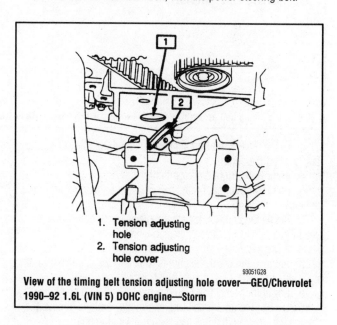

1. Tension adjusting hole
2. Tension adjusting hole cover

View of the timing belt tension adjusting hole cover—GEO/Chevrolet 1990–92 1.6L (VIN 5) DOHC engine—Storm

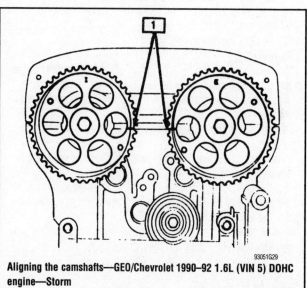

Aligning the camshafts—GEO/Chevrolet 1990–92 1.6L (VIN 5) DOHC engine—Storm

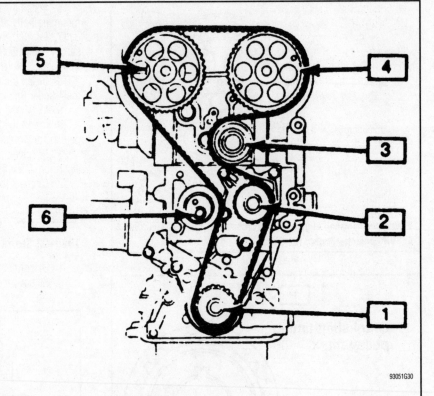

1. **Crankshaft**
2. **Coolant pump**
3. **Idler pulley**
4. **Exhaust camshaft pulley**
5. **Intake camshaft pulley**
6. **Belt tensioner**

View of the timing belt assembly—GEO/Chevrolet 1990–92 1.6L (VIN 5) DOHC engine—Storm

3. Using the Engine Support Fixture tool J-28467-A or equivalent, support the engine.

4. Remove the right engine mount.

5. Remove the upper timing belt cover and the upper timing belt cover.

6. Rotate the crankshaft so that the No. 1 cylinder is at Top Dead Center (TDC) on its compression stroke.

7. Raise and safely support the vehicle.

8. Remove the crankshaft pulley bolt and the crankshaft pulley.

9. Lower the vehicle and remove the lower timing belt cover.

10. Loosen the timing belt tensioner pulley bolt, then move the tensioner away from the belt and secure it with the bolt.

11. If reusing the timing belt, mark the rotational direction on the back of the belt.

12. Remove the timing belt.

To install:

13. Noting the timing belt's direction of rotation, install the belt on the crankshaft sprocket, water pump sprocket, the idler pulley, the exhaust camshaft sprocket, the intake camshaft sprocket and tensioner pulley,.

14. Adjust the timing belt tension by performing the following procedure:

   a. Loosen the belt tensioner bolt and move it toward the timing belt.

15. Using an Allen wrench, insert it into the hexagonal hole of the tension pulley. Hold the pulley stationary and temporarily tighten the tension pulley-to-engine bolt.

   a. Rotate the crankshaft 2 complete revolutions and align the crankshaft timing pulley and the camshaft sprocket marks.

   b. Loosen the tension pulley-to-engine bolt.

   c. Using the Allen wrench and a timing belt tension gauge, apply 31 ft. (42 N) of tension on the timing belt.

   d. Hold the pulley stationary and torque the tension pulley-to-engine bolt to 31 ft. lbs. (42 Nm).

16. Check that the crankshaft and camshaft sprockets are still aligned.

17. Install the lower timing belt cover.

18. Raise and safely support the vehicle.

19. Install the crankshaft pulley bolt and torque the bolt to 87 ft. lbs. (118 Nm).

20. Lower the vehicle. Install the upper timing belt cover.

21. Install the right engine mount.

22. Install the power steering pump drive belt, alternator drive belt and any other drive belts that may have been removed.

23. Remove the Engine Support Fixture tool J-28467-A or equivalent from the engine.

24. Connect the negative battery cable.

1.6L (VIN 6) & 1.8L (VIN 8) ENGINES

1989–90

1. Disconnect the negative battery cable.

2. Elevate the vehicle and safely support it.

3. Remove the right splash shield under the vehicle.

4. Lower the vehicle. Remove the wiring harness from the upper timing belt cover.

5. Depending on equipment, loosen the air conditioner compressor, the power steering pump and the alternator on their adjusting bolts. Remove the drive belts.

6. Remove the crankshaft pulley. The use of a counter holding tool such as J-8614-01 or similar is highly recommended.

7. Remove the valve cover.

8. Remove the windshield washer reservoir.

9. Elevate and safely support the vehicle.

10. Support the engine either from above (tool 28467-A or chain hoist) or below (floor jack and wood block) and remove the through bolt at the right engine mount.

11. Remove the protectors on the mount nuts and studs for the center and rear transaxle mounts.

12. Remove the 2 rear transaxle mount-to-main crossmember nuts. Remove the 2 center transaxle mount-to-center crossmember nuts.

13. Carefully elevate the engine enough to gain access to the water pump pulley.

14. Remove the water pump pulley. Lower the engine to its normal position.

15. Remove the 4 bolts and the lower timing cover. Remove the center timing cover and its bolt, then the upper cover with its 4 bolts.

16. Rotate the crankshaft clockwise to the Top Dead Center (TDC) of the compression stroke for No. 1 cylinder.

17. Loosen the timing belt idler pulley to relieve the tension on the belt, move the pulley away from the belt and temporarily tighten the bolt to hold it in the loose position.

18. Make matchmarks on the belt and both pulleys showing the exact placement of the belt. Mark an arrow on the belt showing its direction of rotation.

19. Carefully slip the timing belt off the pulleys.

➡**Do not disturb the position of the camshafts or the crankshaft during removal.**

20. Remove the idler pulley bolt, pulley and return spring.

21. Use an adjustable wrench mounted on the flats of the camshaft to hold the camshaft from moving. Loosen the center bolt in the camshaft timing pulley and remove the pulley.

22. Check the timing belt carefully for any signs of cracking or deterioration. Pay particular attention to the area where each tooth or cog attaches to the backing of the belt. If the belt shows signs of damage, check the contact faces of the pulleys for possible burrs or scratches.

23. Check the idler pulley by holding it and spinning it. It should rotate freely and quietly. Any sign of grinding or abnormal noise indicates replacement of the pulley.

24. Check the free length of the tension spring. Correct length is 1.5 in. (38.5mm) measured at the inside faces of the hooks. A spring that has stretched during use will not apply the correct tension to the pulley; replace the spring.

To install:

25. Test the tension of the spring, look for 8.4 lbs. (3.8 kg) of tension at 2.0 in. (50mm) of length. If in doubt, replace the spring.

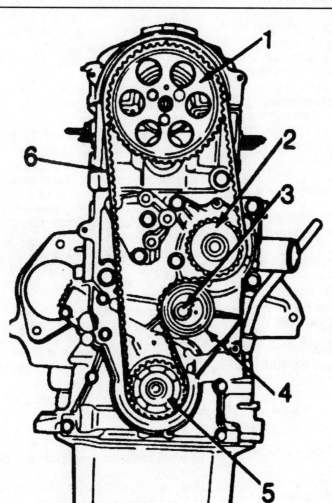

1. Camshaft timing pulley
2. Water pump timing pulley
3. Bolt
4. Tension pulley
5. Crankshaft timing pulley
6. Timing belt

91251G45

View of the timing belt assembly—GEO/Chevrolet 1985–89 1.5L (VIN 7, 9), 1990 1.6L (VIN 6) and 1.8L (VIN 8) engines

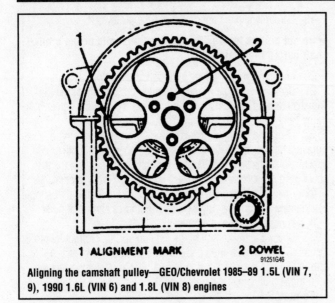

Aligning the camshaft pulley—GEO/Chevrolet 1985–89 1.5L (VIN 7, 9), 1990 1.6L (VIN 6) and 1.8L (VIN 8) engines

1 ALIGNMENT MARK   2 DOWEL

91251G46

26. Reinstall the camshaft timing belt pulley, making sure the pulley fits properly on the shaft and that the timing marks align correctly. Tighten the center bolt to 43 ft. lbs. (58 Nm).

27. Before reinstalling the belt, double check that the crank and camshafts are exactly in their correct positions. The alignment mark on the end of the camshaft bearing cap should show through the small hole in the camshaft pulley and the small mark on the crankshaft timing belt pulley should align with the mark on the oil pump.

28. Reinstall the timing belt idler pulley and the tension spring. Pry the pulley to the left as far as it will go and temporarily tighten the retaining bolt. This will hold the pulley in its loosest position.

29. Install the timing belt, observing the matchmarks made earlier. Make sure the belt is fully and squarely seated on the upper and lower pulleys.

30. Loosen the retaining bolt for the timing belt idler pulley and allow it to tension the belt.

31. Temporarily install the crankshaft pulley bolt and turn the crank clockwise 2 full revolutions from TDC-to-TDC. Insure that each timing mark realigns exactly.

32. Tighten the timing belt idler pulley retaining bolt to 27 ft. lbs. (37 Nm).

33. Measure with the Timing Belt Deflection tool 23600-B or similar, looking for 0.20–0.24 in. (5–6mm) of deflection at 4.4 pounds of pressure. If the deflection is not correct, readjust the idler pulley by repeating Steps 15 through 18.

34. Remove the bolt from the end of the crankshaft.

35. Install the timing belt guide onto the crankshaft and install the lower timing belt cover.

36. When reinstalling, make certain that the gaskets and their mating surfaces are clean and free from dirt and oil. The gasket itself must be free of cuts and deformations and must fit securely in the grooves of the covers.

37. Install the covers and the bolts; tighten the bolts to 43 inch lbs. (5 Nm).

38. Elevate the engine and install the water pump pulley.

39. Lower the engine to its normal position. Install the through bolt in the right engine mount and tighten it to 64 ft. lbs. (87 Nm) with the bolt secure, the engine lifting apparatus may be removed.

40. Install the valve cover.

41. Install the crankshaft pulley and tighten its bolt to 87 ft. lbs. (118 Nm).

42. Reinstall the air conditioning compressor, the power steering pump and the alternator. Install their belts and adjust them to the correct tension.

43. Reconnect the wiring harness to the upper timing belt cover.

44. Raise the vehicle and safely support it.

45. Install the 2 nuts on the center transaxle mount and the rear transaxle mount. Tighten all the nuts to 45 ft. lbs. (61 Nm).

46. Install the protectors on the nuts and studs.

47. Install the splash shield under the vehicle.

48. Lower the vehicle to the ground.

49. Install the windshield washer reservoir and connect the negative battery cable.

1991–00

1. Disconnect the negative battery cable.

2. Remove the windshield washer reservoir from the engine compartment.

3. If equipped with cruise control, proceed as follows:
  a. Remove the cruise control actuator cover.
  b. Disconnect the cruise control harnesses.
  c. Disconnect the control cable.
  d. Remove the bolts and actuator from the vehicle.

4. Raise and safely support the vehicle on jackstands.

5. Remove the right front wheel.

6. Remove the bolts and plastic clips and the right front wheel housing.

7. Remove the alternator/water pump drive belt.

8. Lower the vehicle.

9. If equipped with air conditioning, proceed as follows:
  a. Remove the air conditioning compressor drive belt.
  b. Disconnect the compressor harness.
  c. Remove the bolts and compressor, without disconnecting the refrigerant lines. Suspend the compressor aside.
  d. Remove the compressor mount bracket.

10. Remove the power steering pump drive belt.

11. Disconnect the wiring from the alternator and oil pressure switch.

12. Remove the engine wiring harness cover.

13. Remove the wiring harness from the cylinder head cover.

14. Disconnect the ignition wires from the spark plugs, then remove the spark plugs.

15. Remove the PCV hoses from the valve cover.

16. Remove the cap nuts, the seal washers and the cylinder head cover with the gasket.

17. Turn the crankshaft to align the timing mark on the crankshaft pulley at **0**, setting the piston in the No. 1 cylinder at Top Dead Center (TDC) on the compression stroke. Check that the valve lash adjusters on the No. 1 cylinder are loose. If not, turn the crankshaft pulley 1 complete revolution (360 degrees).

18. Remove the engine ground wire from the right fender apron.

19. Install a suitable support under the engine and remove the engine mount.

20. Remove the water pump pulley.

21. Remove the crankshaft pulley using a suitable puller.

22. Remove the 9 retaining bolts and the timing belt covers.

23. Slide the timing belt guide from the crankshaft.

24. Be sure the timing belt sprockets are properly aligned.

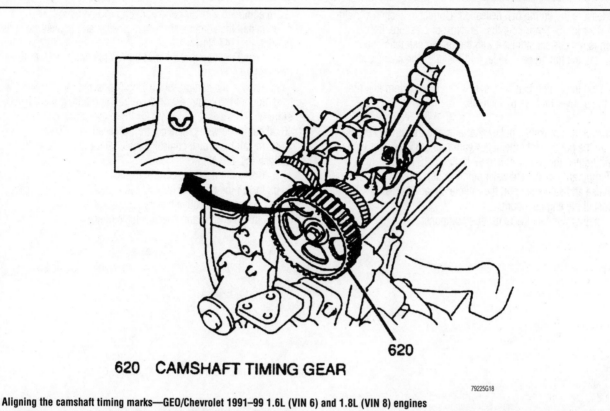

620 CAMSHAFT TIMING GEAR

79225G18

Aligning the camshaft timing marks—GEO/Chevrolet 1991–99 1.6L (VIN 6) and 1.8L (VIN 8) engines

✳✳ WARNING

Do not turn the crankshaft or camshaft independently after removal of the timing belt; binding or damage to engine components could result. If the timing belt is to be reused, mark the belt with an arrow showing the direction of engine revolution.

25. Remove the timing belt tensioner bolt, tensioner and tension spring.

26. Remove the timing belt from the sprockets. Inspect the timing belt for cracked or damaged teeth. Replace as necessary.

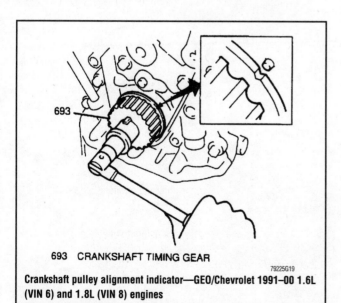

693 CRANKSHAFT TIMING GEAR

79225G19

Crankshaft pulley alignment indicator—GEO/Chevrolet 1991–00 1.6L (VIN 6) and 1.8L (VIN 8) engines

✳✳ WARNING

Do not bend, twist or turn the timing belt.

27. If the camshaft sprocket is to be removed, hold the camshaft stationary using a wrench positioned on the hexagon cast into the camshaft, and remove the sprocket retaining bolt and sprocket.

✳✳ WARNING

Be careful not to damage the cylinder head when holding the camshaft in place.

28. If the crankshaft sprocket is to be removed, pry it from the crankshaft using 2 flat-bladed prybars.
   To install:
29. Align the camshaft key with the groove on the sprocket and slide the sprocket on. Hold the camshaft with the wrench at the hexagonal portion of the camshaft, and tighten the camshaft timing sprocket bolt to 43 ft. lbs. (59 Nm).

30. Be sure the sprocket is still properly aligned.

31. Install the crankshaft timing sprocket. Align the crankshaft key with the groove on the sprocket and slide it on.

32. Reinstall the timing belt tensioner and the tension spring. Pry the tensioner to the left as far as it will go and temporarily tighten the retaining bolt.

33. Install the timing belt. If installing the old belt, observe the matchmarks made during removal.

34. Loosen the retaining bolt for the timing belt tensioner and allow it to tension the belt.

35. Temporarily install the crankshaft pulley bolt and turn the crankshaft clockwise 2 full revolutions. Be sure each timing mark realigns exactly.

36. Tighten the timing belt tensioner bolt to 27 ft. lbs. (37 Nm).

37. Measure the timing belt deflection. Correct deflection should be 0.20–0.24 in. (5–6mm) at 4 lbs. (20 Nm) of pressure. If the deflection is not correct, adjust it with the timing belt tensioner.

38. Install the timing belt guide, with the cup side facing outward.

39. Install the timing belt covers, installing the bottom one first. Tighten the 9 cover bolts to 62 inch lbs.
(7 Nm).

40. Install the crankshaft pulley after aligning the pulley key with the slot on the pulley. Hold the pulley with tool J-8614-01 or equivalent, and tighten the pulley bolt to 87 ft. lbs. (118 Nm).

41. Temporarily install the water pump pulley.

42. Raise and safely support the vehicle on jackstands.

43. Install the engine mount.

44. Install or connect the remaining components.

45. If equipped with air conditioning, proceed as follows:

a. Install the compressor mount bracket and tighten the bolts to 35 ft. lbs. (47 Nm).

b. Install the compressor and tighten the bolts to 18 ft. lbs. (25 Nm).

c. Engage the compressor wiring connector.

d. Install the compressor drive belt and adjust the tension.

46. If equipped with cruise control, proceed as follows:

a. Install the cruise control actuator and tighten the bolts to 89 inch lbs. (10 Nm).

b. Connect the cruise control cable.

c. Install the cruise control actuator cover.

47. Connect the negative battery cable.

48. Start the engine and check vehicle operation.

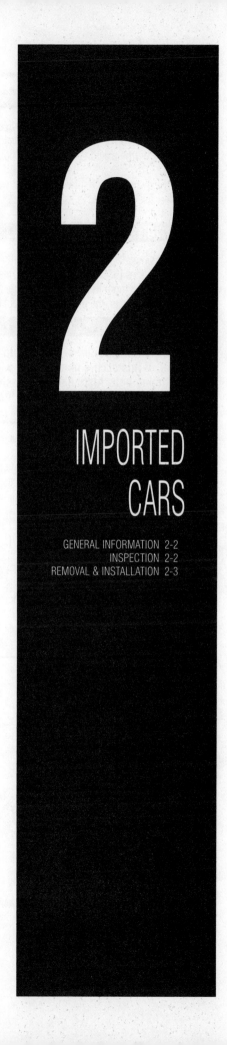

2

IMPORTED CARS

GENERAL INFORMATION

Whenever a vehicle with an unknown service history comes into your repair facility or is recently purchased, here are some points that should be asked to help prevent costly engine damage:
- Does the owner know if, or when the belt was replaced?
- If the vehicle purchased is used, or the condition and mileage of the last timing belt replacement are unknown, it is recommended to inspect, replace or at least inform the owner that the vehicle is equipped with a timing belt.
- Note the mileage of the vehicle. The average replacement interval for a timing belt is approximately 60,000 miles (96,000 km).

INSPECTION

The average replacement interval for a timing belt is approximately 60,000 miles (96,000km). If, however, the timing belt is inspected earlier or more frequently than suggested, and shows signs of wear or defects, the belt should be replaced at that time.

✳✳ WARNING

Never allow antifreeze, oil or solvents to come into with a timing belt. If this occurs immediately wash the solution from the timing belt. Also, never excessively bend or twist the timing belt; this can damage the belt so that its lifetime is severely shortened.

Inspect both sides of the timing belt. Replace the belt with a new one if any of the following conditions exist:
- Hardening of the rubber—back side is glossy without resilience and leaves no indentation when pressed with a fingernail
- Cracks on the rubber backing
- Cracks or peeling of the canvas backing
- Cracks on rib root
- Cracks on belt sides
- Missing teeth or chunks of teeth
- Abnormal wear of belt sides—the sides are normal if they are sharp, as if cut by a knife.

If none of these conditions exist, the belt does not need replacement unless it is at the recommended interval. The belt MUST be replaced at the recommended interval.

✳✳ WARNING

On interference engines, it is very important to replace the timing belt at the recommended intervals, otherwise expensive engine damage will likely result if the belt fails.

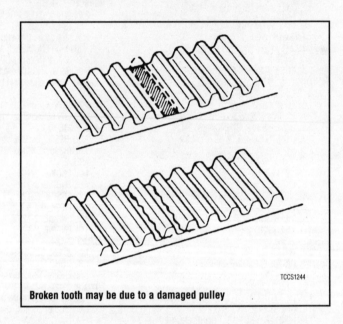

TCCS1244

Broken tooth may be due to a damaged pulley

TCCS1242

Never bend or twist a timing belt excessively, and do not allow solvents, antifreeze, gasoline, acid or oil to come into contact with the belt

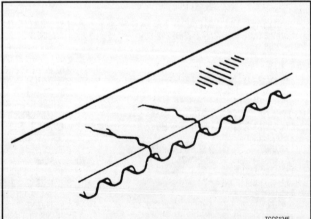

TCCS1245

Back surface worn or cracked from a possible overheated engine or interference with the belt cover

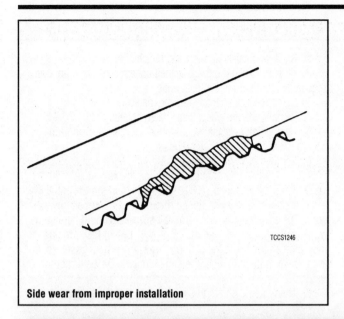

Side wear from improper installation

TCCS1246

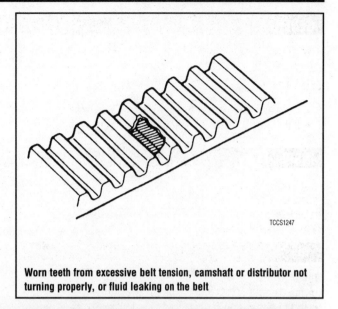

Worn teeth from excessive belt tension, camshaft or distributor not turning properly, or fluid leaking on the belt

TCCS1247

REMOVAL & INSTALLATION

Acura/Sterling

➡The radio may have a coded theft protection circuit. Obtain the code before disconnecting the battery, removing the radio fuse or removing the radio.

1.6L (D16A1), 1.7L (B17A1),
1.8L (B18A1, B18B1 & B18C1) ENGINES

1. Disconnect the negative battery cable.
2. Turn the crankshaft pulley until cylinder No. 1 is set to Top Dead Center (TDC) on the compression stroke. The white crankshaft pulley mark should be aligned with the pointer on the lower timing belt cover.
3. Remove all necessary components to gain access to the cylinder head and timing belt covers.
4. Remove the cylinder head and timing belt covers.
5. Remove the crankshaft pulley bolt and the crankshaft pulley. To remove the crankshaft pulley bolt a pulley holder (holder attachment tool part No. 07MAB-PY3010A and holder handle tool part No. 07JAB-001020A or equivalent) will be needed to keep the crankshaft from turning.

✳ WARNING

Do not use the timing belt covers to store small parts. Grease or oil can transfer from the parts to the cover, then to the belt. Clean the covers thoroughly before installation.

6. Recheck that the No. 1 piston is at TDC on its compression stroke. Align the groove on the toothed side of the crankshaft timing belt drive sprocket to the arrow pointer on the oil pump.
7. To set the camshafts to top dead center for the No. 1 cylinder, align the hole in each camshaft with the holes in the No. 1 camshaft holders, then push 5.0mm pin punches into the holes. Be sure that the **UP** arrows are pointing up and that the TDC marks on the intake and exhaust sprockets are aligned.
8. Loosen the tensioner adjusting bolt 180 degrees (½ turn). Push on the tensioner to remove the tension from the timing belt,

then retighten the bolt. If the timing belt is to be reinstalled, mark the direction of rotation on the belt with a crayon or white paint.
9. Remove the timing belt from the sprockets.

➡Be sure the water pump pulley turns counterclockwise freely. Check for signs of seal leakage; a small amount of weeping from the bleed hole is normal.

10. If necessary, remove the timing belt tensioner by performing the following:
    a. Remove the timing belt tensioner spring.
    b. Remove the bolt from the timing belt tensioner and remove the tensioner.
To install:
11. If the timing belt tensioner was removed, perform the following:
    a. Position the timing belt tensioner on the engine and install the attaching bolt loosely.

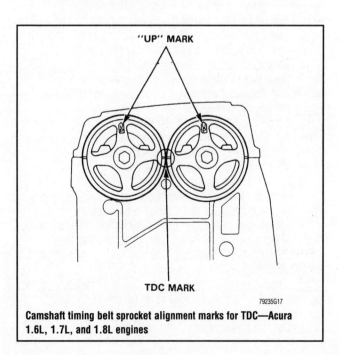

"UP" MARK

TDC MARK

79235G17

Camshaft timing belt sprocket alignment marks for TDC—Acura 1.6L, 1.7L, and 1.8L engines

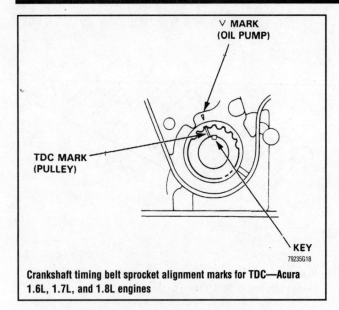

Crankshaft timing belt sprocket alignment marks for TDC—Acura 1.6L, 1.7L, and 1.8L engines

2.2L (F22B1) ENGINES

1. Raise and safely support the vehicle.
2. Be sure to acquire the anti-theft code for the radio and the frequencies for the radio's preset buttons.
3. Disconnect the negative battery cable.
4. Remove the left wheel well splash shield.
5. Remove all necessary components to gain access to the cylinder head (valve) and timing belt covers.
6. Remove the cylinder head and timing belt covers.
7. Turn the engine to align the timing marks and set cylinder No. 1 to Top Dead Center (TDC). The white mark on the crankshaft sprocket should align with the pointer on the timing belt cover. The words **UP** embossed on the camshaft sprocket should be aligned in the upward position. The marks on the edge of the sprocket should be aligned with the cylinder head or the back cover upper edge. Once in this position, the engine must **NOT** be turned or disturbed.
8. There are 2 belts in this system; the belt running to the

b. Install the timing belt tensioner spring.
c. Push the tensioner down, then snug the tensioner bolt to hold this position.

➡**Before reinstallation, check every component for cleanliness. All covers, pulleys, shields, etc. must be completely free of grease and oil.**

➡**Install the timing belt in the correct sequence. Also, if installing the old belt, be sure it is turning the same direction.**

12. Install the timing belt first to the crankshaft pulley, then to the adjuster, then to the water pump pulley, the exhaust camshaft and finally to the intake camshaft pulley.
13. Install the lower belt cover. Install the crankshaft pulley, tightening the bolt to 130 ft. lbs. (177 Nm). Lubricate the threads and the flange of the bolt with engine oil before installation.
14. Loosen the adjusting bolt, allowing the adjuster to tension the belt. Retighten the bolt to 40 ft. lbs. (54 Nm).
15. Remove the pin punches from the camshafts.
16. Rotate the crankshaft 4–6 turns counterclockwise. This allows the belt to equalize tension across all of the pulleys.
17. Once again, set the engine to TDC compression for cylinder No. 1. Check that all timing marks for the camshaft and crankshaft are properly aligned. If any mark is out of alignment, remove the timing belt and reinstall it.
18. Loosen the adjusting bolt 180 degrees (½ turn). Rotate the crankshaft counterclockwise until the camshaft pulleys have moved 3 teeth. Retighten the adjusting bolt to 40 ft. lbs. (54 Nm).
19. Check the torque of the crankshaft pulley bolt.
20. Install the other timing belt covers.
21. Install the rubber seal in the groove of the cylinder head cover. Be sure that the seal and groove are thoroughly clean first.
22. Apply liquid gasket to the rubber seal at the 8 corners of the recesses. Do not install the parts if 20 minutes or more have elapsed since applying the liquid gasket. Instead, reapply liquid gasket after removing the old residue.
23. Install the cylinder head cover and all other applicable components. Tighten the cylinder head cover nuts in 2 steps to 88 inch lbs. (10 Nm).
24. Connect the negative battery cable.

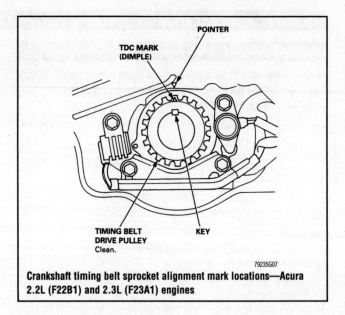

Crankshaft timing belt sprocket alignment mark locations—Acura 2.2L (F22B1) and 2.3L (F23A1) engines

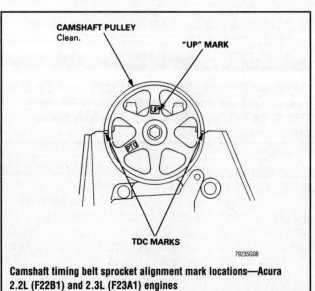

Camshaft timing belt sprocket alignment mark locations—Acura 2.2L (F22B1) and 2.3L (F23A1) engines

camshaft sprocket is the timing belt, the other, shorter belt drives the balance shaft and is referred to as the balancer shaft belt or timing balancer belt. Lock the timing belt adjuster in position, by installing one of the lower timing belt cover bolts in the adjuster arm.

9. Loosen the timing belt and balancer shaft tensioner adjuster nut, do not loosen the nut more than 1 revolution. Push the tensioner for the balancer belt away from the belt to relieve the tension. Hold the tensioner and tighten the adjusting nut to hold the tensioner in place.

10. Carefully remove the balancer belt. Do not crimp or bend the belt; protect it from contact with oil or coolant.

11. Remove the balancer belt sprocket from the crankshaft.

12. Loosen the lockbolt, installed in the timing belt adjuster and the adjusting nut. Push on the timing belt adjuster to remove the tension on the timing belt, then tighten the adjuster nut.

13. Remove the timing belt by sliding it off the sprockets. Do not crimp or bend the belt; protect it from contact with oil or coolant.

14. If necessary, remove the belt tensioners by performing the following:

   a. Remove the springs from the balancer belt and the timing belt tensioners.

   b. Remove the adjusting nut from the belt tensioners.

   c. Remove the bolt from the balancer belt adjuster lever, then remove the lever and the tensioner pulley.

   d. Remove the lockbolt from the timing belt tensioner lever, then remove the tensioner pulley and lever from the engine.

15. This is an excellent time to check or replace the water pump. Even if the timing belt is only being replaced as part of a good maintenance schedule, consider replacing the pump at the same time.

To install:

16. If the water pump is to be replaced, install a new O-ring on the pump and make certain it is properly seated. Install the water pump and tighten the mounting bolts to 106 inch lbs. (12 Nm).

17. If the tensioners were removed, perform the following to install them:

   a. Install the timing belt tensioner lever and the tensioner pulley.

   b. Install the balancer belt pulley and adjuster lever.

   c. Install the adjusting nut and bolt to the balancer belt adjuster lever.

   d. Install the springs to the tensioners.

   e. Install the lockbolt to the timing belt tensioner, then move the tensioner it's full deflection and tighten the lockbolt.

   f. Move the balancer belt tensioner it's full deflection and tighten the adjusting nut to hold its position.

18. The pointer on the crankshaft sprocket should be aligned with the pointer on the oil pump. The camshaft sprocket must be aligned so that the word **UP** is at the top of the sprocket and the marks on the edge of the sprocket are aligned with the surfaces of the head or the back cover upper edge.

19. Install the timing belt in the following sequence:

   a. Start the belt on the crankshaft sprocket.

   b. Then, around the tensioner sprocket.

   c. On the water pump sprocket.

   d. Finally, around the camshaft sprocket.

20. Check the timing marks to be sure that they did not move.

21. Loosen, then retighten the timing belt adjusting nut.

22. Install the timing/balancer belt drive sprocket and the lower timing belt cover.

23. Install the crankshaft pulley and bolt, tighten the bolt to 181 ft. lbs. (245 Nm). Rotate the crankshaft sprocket 5–6 revolutions to properly position the timing belt on the sprockets.

24. Set the No. 1 cylinder to TDC and loosen the timing belt adjusting nut 1 turn. Turn the crankshaft counterclockwise until the cam sprocket has moved 3 teeth; this creates the proper tension on the timing belt.

25. Tighten the timing belt adjusting nut.

26. Set the crankshaft sprocket and the camshaft sprocket to TDC. If the sprockets do not align, remove the belt to realign the marks, then install the belt.

27. Remove the crankshaft pulley and the lower cover.

28. With the timing marks aligned, lock the timing belt adjuster in place with one of the lower cover mounting bolts.

29. Loosen the adjusting nut and ensure the timing balancer belt adjuster moves freely.

30. Align the rear timing balancer sprocket using a 6 x 100mm bolt or rod. Mark the bolt or rod at a point 2.9 in. (74mm) from the end. Remove the bolt from the maintenance hole on the side of the block; insert the bolt/rod into the hole and align the 2.9 in. (74mm) mark with the face of the hole. This will hold the shaft in place during installation.

31. Align the groove on the front balancer shaft sprocket with the pointer on the oil pump.

32. Install the balancer belt. Once the belts are in place, be sure that all the engine alignment marks are still correct. If not, remove the belts, realign the engine and reinstall the belts. Once the belts are properly installed, slowly loosen the adjusting nut, allowing the tensioner to move against the belt. Remove the bolt from the maintenance hole and reinstall the bolt and washer.

33. Install the crankshaft pulley, then turn the crankshaft sprocket 1 turn counterclockwise and tighten the timing belt adjusting nut to 33 ft. lbs. (45 Nm).

34. Remove the crankshaft pulley and the bolt locking the timing belt adjuster in place.

35. Install the lower timing belt cover and tighten the bolts to 106 inch lbs. (12 Nm).

36. Install a new seal around the adjusting nut. Do not loosen the adjusting nut.

37. Install the crankshaft pulley. Coat the threads and seating face of the pulley bolt with engine oil, then install and tighten the bolt to 181 ft. lbs. (250 Nm).

38. Install the dipstick tube, then install the side engine mount. Tighten the bolt and nut attaching the mount to the engine to 40 ft. lbs. (55 Nm). Tighten the through-bolt and nut to 47 ft. lbs. (65 Nm), then remove the jack from under the engine.

39. Install the upper timing belt cover. Tighten the bolt toward the exhaust manifold to 89 inch lbs. (10 Nm) and the bolt toward the intake manifold to 106 inch lbs. (12 Nm).

40. Install the cylinder head cover gasket in the groove of the cylinder head cover. Before installing the gasket, thoroughly clean the seal and the groove. Seat the recesses for the camshaft first, then work it into the groove around the outside edges. Be sure the gasket is seated securely in the corners of the recesses.

41. Apply liquid gasket to the 4 corners of the recesses in the cylinder head cover gasket. Do not install the parts if 5 minutes or more have elapsed since applying liquid gasket. After assembly, wait at least 20 minutes before filling the engine with oil.

42. Install the cylinder head (valve) cover and all other applicable components. Tighten the bolts attaching the cylinder head cover in 2 steps to the proper sequence of 89 inch lbs. (10 Nm).

43. Connect the negative battery cable.

2.3L (F23A1) ENGINE

➡ **Under normal driving conditions, the timing belt and timing balancer belt are to be replaced at 105,000 miles (168,000 km).**

1. Raise and safely support the vehicle.
2. Be sure to acquire the anti-theft code for the radio and the frequencies for the radio's preset buttons.
3. Disconnect the negative battery cable.
4. Remove the left wheel-well splash shield.
5. Rotate the crankshaft pulley so that the No. 1 piston is at Top Dead Center (TDC) of its compression stroke.
6. Loosen the power steering pump's adjusting bolt, locknut and mounting nut; then, remove the pump's belt.
7. Loosen the alternator's adjusting bolt, locknut and mounting nut; then, remove the alternator's belt.
8. Remove the alternator's terminal and connector.
9. To remove the engine's side mount, located at the front of the engine, perform the following procedure:

   a. Using a floor jack, position a cushion between the jack and the oil pan.

   b. Raise the engine slightly to take the weight off of the side mount.

   c. Remove the side mount-to-engine bolts, the side mount-to-chassis bolts and the side mount.

10. Remove the dipstick and the dipstick tube-to-engine bolt; then, pull the tube from its O-ring mount.

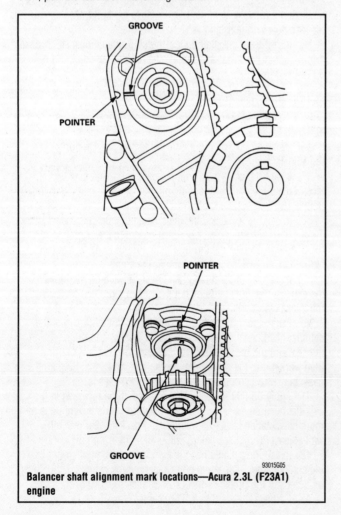

Balancer shaft alignment mark locations—Acura 2.3L (F23A1) engine

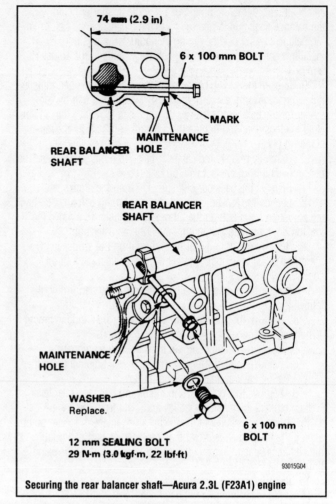

Securing the rear balancer shaft—Acura 2.3L (F23A1) engine

11. Remove the cylinder head cover.
12. Remove the crankshaft pulley by performing the following procedure:

   a. Using the Holder Handle and the 50mm Offset Holder Attachment tool 07MAB-PY3010A or equivalent, and a 19mm socket, secure the crankshaft pulley and remove the crankshaft pulley bolt.

   b. Remove the washer and pull the crankshaft pulley from the engine.

13. From the lower timing belt cover, remove the rubber seal concealing the adjusting nut. Remove the upper timing belt cover-to-engine bolts and the cover.
14. Loosen the adjusting nut ⅔–1 turn. Push the tensioner to relieve the tension from the timing belt and the balancer belt; then, retighten the adjusting nut.
15. Remove the balancer and the timing belts.

To install:

16. Clean the upper and lower timing belt covers.
17. Remove the balancer belt drive pulley from the crankshaft.
18. Position the timing belt pulley so the No. 1 piston is at TDC of its compression stroke. Align the mark on the pulley (near keyway) with the pointer mark on the oil pump.
19. Adjust the camshaft pulley so that the No. 1 piston is the TDC of the compression stroke. Align the TDC marks on the pulley with the upper surface of the back cover and the **UP** mark should be facing upward.
20. Install the timing belt in the following sequence:

a. Crankshaft pulley sprocket.
b. Adjusting pulley.
c. Water pump pulley
d. Camshaft pulley

❋❋ **WARNING**

Make sure that the camshaft and crankshaft pulleys are at TDC.

21. Loosen and retighten the adjusting nut to tension the timing belt.

22. Install the balancer belt drive pulley and the lower timing belt cover.

23. Install the crankshaft pulley and finger-tighten the bolt and washer. Using the Holder Handle and the 50mm Offset Holder Attachment tool 07MAB-PY3010A or equivalent, and a 19mm socket with a torque wrench, tighten the crankshaft pulley bolt to 181 ft. lbs. (245 Nm).

24. Rotate the crankshaft pulley about 5–6 turns counterclockwise to position the timing belt on the pulleys.

25. To adjust the timing belt tension, perform the following procedure:

 a. Make sure that the No. 1 piston is at TDC of its compression stroke.

 b. Loosen the adjusting nut ⅔–1 turn. Rotate the crankshaft counterclockwise 3 teeth on the camshaft pulley.

 c. Tighten the adjusting nut to 33 ft. lbs. (44 Nm).

26. Retighten the crankshaft pulley bolt to 181 ft. lbs. (245 Nm).

27. Make sure the crankshaft and camshaft pulleys are at TDC.

➡**If the camshaft or crankshaft pulley is not at TDC, remove the timing belt and re-perform the adjustment procedure.**

28. Remove the crankshaft pulley and the timing belt lower cover.

29. Install a 6 x 1.0mm bolt to lock the timing belt adjuster arm in place.

30. Loosen the adjusting nut ⅔–1 turn and verify that the balancer belt adjuster moves freely.

31. Position the tensioner to remove tension from the balancer belt, then tighten the adjusting nut.

32. Align the rear balancer shaft pulley by performing the following procedure:

 a. Using a 6 x 100mm bolt, scribe a line 2.9 in. (74mm) from the end of the bolt.

 b. Insert the bolt into the maintenance hole to the scribed line.

 c. Align the groove on the front balancer shaft pulley with the pointer on the oil pump housing.

33. Install the balancer belt. Loosen the adjuster nut ⅔–1 turn to place tension on the balancer belt.

34. Remove the 6 x 100mm bolt from the rear balancer shaft and install the 12mm sealing bolt.

35. Install the crankshaft pulley and tighten the bolt to 181 ft. lbs. (245 Nm).

36. Rotate the crankshaft pulley 1 turn counterclockwise and tighten the adjusting nut to 33 ft. lbs. (44 Nm).

37. Remove the 6 x 1.00mm bolt from the timing belt adjuster arm.

38. Remove the crankshaft pulley and install the lower timing belt cover.

39. Install the rubber seal over the adjusting nut.

➡**DO NOT loosen the adjusting nut.**

40. Install the crankshaft pulley and tighten the bolt to 181 ft. lbs. (245 Nm).

41. To complete the installation, reverse the removal procedures. Adjust the tension of the drive belts.

42. Connect the negative battery cable.

2.5L (G25A1 & G25A4) ENGINES

1. Disconnect the negative battery cable.

2. Set the No. 1 piston at Top Dead Center (TDC) on the compression stroke. The white TDC mark on the crankshaft pulley must align with the pointers on the lower belt cover.

3. Remove all necessary components to gain access to the timing belt covers.

4. Remove the timing belt upper cover.

5. Be sure the **UP** mark and the TDC marks on the camshaft sprocket are correctly positioned.

6. Rotate the crankshaft to align the white timing mark on the crankshaft pulley with the pointer on the lower cover. There are similar alignment marks on the crankshaft sprocket and oil pump housing.

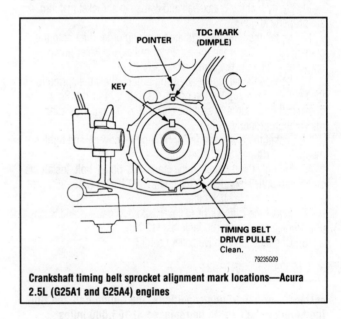

Crankshaft timing belt sprocket alignment mark locations—Acura 2.5L (G25A1 and G25A4) engines

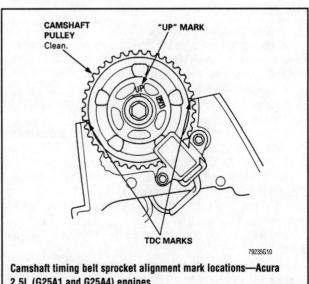

Camshaft timing belt sprocket alignment mark locations—Acura 2.5L (G25A1 and G25A4) engines

7. Remove the timing belt lower cover.

8. Loosen the adjusting bolt 180 degrees (½ turn), then push the tensioner down to relieve the belt tension. Retighten the adjusting bolt to 33 ft. lbs. (44 Nm).

➡ **Do not remove the adjusting bolt and tensioner pulley unless they are to be replaced. The bolt is only loosened and tightened in this procedure to tension the timing belt.**

9. Remove the timing belt.

10. Inspect the timing belt tensioner pulley and tension spring for signs of wear. Remove and replace the tensioner assembly as necessary.

To install:

✷✷ WARNING

Replace the timing belt if it shows any signs of wear, damage or contamination from oil or coolant. The source of any oil or coolant contamination must be determined and corrected before the new timing belt may be installed.

11. Verify that the crankshaft and camshaft sprocket matchmarks are properly aligned.

12. Install the timing belt in the following order: first onto the crankshaft sprocket, then the tensioner pulley, the water pump sprocket and finally the camshaft sprocket.

13. Loosen the tensioner adjusting bolt to allow the spring to set the tension. Then, tighten the bolt to 33 ft. lbs. (44 Nm). Rotate the crankshaft 6 full turns to seat the belt and verify that the timing marks align properly.

14. Install the lower and upper timing belt covers, and tighten the bolts to 106 inch lbs. (12 Nm).

15. Oil only the threads on the crankshaft pulley bolt. Install the crankshaft pulley and tighten the bolt to 181 ft. lbs. (245–250 Nm).

16. Install the dipstick tube.

17. Install the cylinder head cover with new gaskets and washers. Tighten the cap to 106 inch lbs. (12 Nm).

18. Connect the negative battery cable.

3.2L (C32A6) ENGINE—3.2TL

➡ **Under normal driving conditions, the timing belt and timing balancer belt are to be replaced at 105,000 miles (168,000 km).**

1. Raise and safely support the vehicle.

2. Be sure to acquire the anti-theft code for the radio and the frequencies for the radio's preset buttons.

3. Disconnect the negative battery cable.

4. Remove the left wheel well splash shield.

5. Loosen the power steering pump's adjusting bolt, locknut and mounting nut; then, remove the pump's belt.

6. Loosen the alternator's adjusting bolt, locknut and mounting nut; then, remove the alternator's belt.

7. Remove the alternator's terminal and connector.

8. To remove the engine-to-chassis center bracket, located at the front of the engine, perform the following procedure:

   a. Using a floor jack, position a cushion between the jack and the oil pan.

   b. Raise the engine slightly to take the weight off of the center bracket.

   c. Remove the center bracket-to-engine bolts, the center bracket-to-chassis bolts and the center bracket.

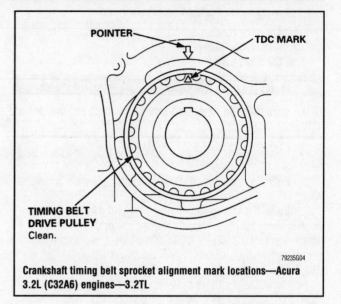
Crankshaft timing belt sprocket alignment mark locations—Acura 3.2L (C32A6) engines—3.2TL

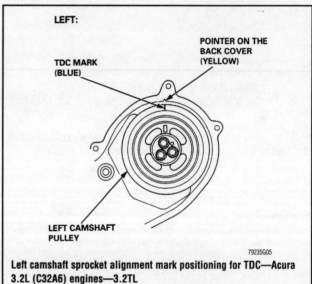
Left camshaft sprocket alignment mark positioning for TDC—Acura 3.2L (C32A6) engines—3.2TL

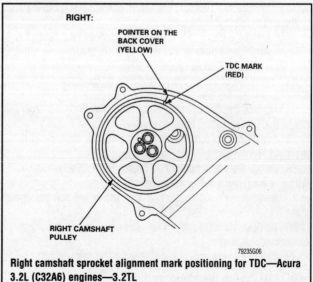
Right camshaft sprocket alignment mark positioning for TDC—Acura 3.2L (C32A6) engines—3.2TL

9. Remove the TCS upper and lower brackets.
10. Disconnect the TCS throttle sensor and actuator connectors and remove the TCS control valve assembly.
11. From the front of the engine, remove the oil pressure switch connector, the engine ground cable and the engine wire harness cover.
12. Remove the dipstick and the dipstick tube-to-engine bolt; then, pull the tube from its O-ring mount. Discard the O-ring.
13. Turn the engine to align the timing marks and set cylinder No. 1 to Top Dead Center (TDC) on the compression stroke. The white mark on the crankshaft pulley should align with the pointer on the timing belt cover. Remove the inspection caps on the upper timing belt covers to check the alignment of the timing marks. The pointers for the camshafts should align with the marks on the camshaft sprockets.
14. Remove the crankshaft pulley by performing the following procedure:
    a. Using the Holder Handle and the 50mm Offset Holder Attachment tool 07MAB-PY3010A or equivalent, and a 19mm socket, secure the crankshaft pulley and remove the crankshaft pulley bolt.
    b. Remove the washer and pull the crankshaft pulley from the engine.
15. Remove all necessary components for access to the timing belt covers.
16. Remove the upper and lower timing belt covers. Clean any dirt, oil or grease from the covers. Do not use the covers for storing removed items.
17. Loosen the timing belt tensioner adjusting bolt 180 degrees (½ turn). Push on the tensioner to remove tension from the timing belt, then tighten the adjusting bolt.
18. Remove the timing belt.
19. If necessary, remove the timing belt tensioner by performing the following:
    a. Remove the spring from the tensioner.
    b. Remove the bolt mounting the tensioner, then remove the tensioner.
To install:

❊❊ WARNING

Do not rotate the crankshaft pulley or camshaft pulleys with the timing belt removed. The pistons may hit the valves and cause damage.

20. If necessary, install the timing belt tensioner by performing the following:
    a. Install the tensioner and the attaching bolt.
    b. Move the tensioner its full deflection to the left and tighten the bolt.
    c. Install the spring to the tensioner.
21. Remove the spark plugs.
22. Set the timing belt drive (crankshaft) sprocket so that the No. 1 piston is at TDC. Align the TDC mark on the tooth of the timing belt drive sprocket with the pointer on the oil pump.
23. Set the camshaft pulleys so that the No. 1 piston is at TDC. Align the TDC mark on the camshaft pulleys to the pointers on the back covers.
24. Install the timing belt on the sprockets in the following sequence: drive sprocket (crankshaft), tensioner pulley, left camshaft sprocket, water pump pulley, right camshaft sprocket.
25. Loosen, then retighten the timing belt adjuster bolt to tension the timing belt.

26. Install the lower timing belt cover.
27. Install the crankshaft pulley and finger-tighten the bolt and washer. Using the Holder Handle and the 50mm Offset Holder Attachment tool 07MAB-PY3010A or equivalent, and a 19mm socket with a torque wrench, tighten the crankshaft pulley bolt to 181 ft. lbs. (245 Nm).
28. Rotate the crankshaft 5–6 turns clockwise so that the timing belt positions itself properly on the sprockets.
29. Set cylinder No. 1 to TDC by aligning the timing marks. If the timing marks do not align, remove the timing belt, then adjust the components and reinstall the timing belt.
30. Rotate the crankshaft clockwise enough to move the camshaft pulley 9 teeth (the **blue** mark on the crankshaft pulley should align with the pointer on the lower cover).
31. Loosen the timing belt adjusting bolt 180 degrees (½ turn), then tighten the bolt to 31 ft. lbs. (42 Nm).
32. Install the upper timing belt covers, then install all applicable components. When installing the center bracket, tighten the bolts attaching the brackets to 40 ft. lbs. (54 Nm), then the mount through-bolt to 40 ft. lbs. (54 Nm).
33. To complete the installation, reverse the removal procedures. Adjust the tension of the drive belts.
34. Connect the negative battery cable.

3.0L (J30A1) & 3.2L (J32A1) ENGINES

1. Raise and safely support the vehicle.
2. Disconnect the negative battery cable.
3. Remove the ignition coil cover.
4. Remove the front tire/wheel assemblies. Remove the splash shield from under the vehicle.
5. Move the alternator's drive belt tensioner to relieve the tension; then, remove the alternator belt.
6. Loosen the power steering pump's adjusting bolt, locknut and mounting nut; then, remove the pump's belt.
7. To remove the engine-to-chassis side mount, located at the front of the engine, perform the following procedure:
    a. Using a floor jack, position a cushion between the jack and the oil pan.
    b. Raise the engine slightly to take the weight off of the side mount.
    c. Remove the side mount-to-engine bracket bolt, the side mount-to-chassis bolts and the side mount.
8. Remove the dipstick and the dipstick tube-to-engine bolt; then, pull the tube from its O-ring mount. Discard the O-ring.
9. Turn the engine to align the timing marks and set cylinder No. 1 to Top Dead Center (TDC) on the compression stroke. The white mark on the crankshaft pulley should align with the pointer on the timing belt cover. Remove the inspection caps on the upper timing belt covers to check the alignment of the timing marks. The pointers for the camshaft pulleys should align with the marks on rear upper cover mark.
10. Remove the crankshaft pulley by performing the following procedure:
    a. Using the Holder Handle and the 50mm Offset Holder Attachment tool 07MAB-PY3010A or equivalent, and a 19mm socket, secure the crankshaft pulley and remove the crankshaft pulley bolt.
    b. Remove the washer and pull the crankshaft pulley from the engine.
11. Remove all necessary components for access to the timing belt covers.

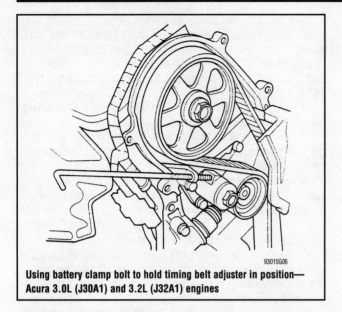

Using battery clamp bolt to hold timing belt adjuster in position—Acura 3.0L (J30A1) and 3.2L (J32A1) engines

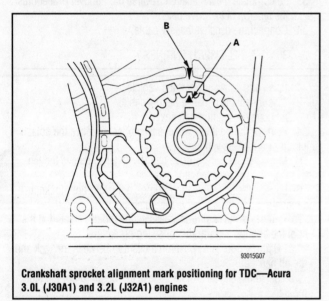

Crankshaft sprocket alignment mark positioning for TDC—Acura 3.0L (J30A1) and 3.2L (J32A1) engines

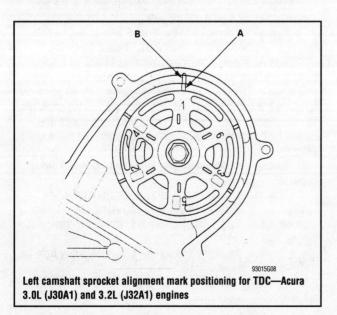

Left camshaft sprocket alignment mark positioning for TDC—Acura 3.0L (J30A1) and 3.2L (J32A1) engines

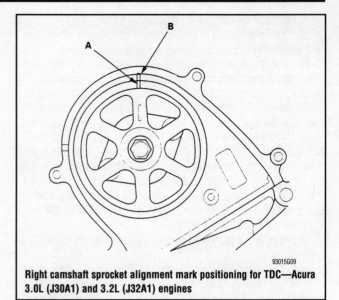

Right camshaft sprocket alignment mark positioning for TDC—Acura 3.0L (J30A1) and 3.2L (J32A1) engines

12. Remove the upper and lower timing belt covers. Clean any dirt, oil or grease from the covers. Do not use the covers for storing removed items.

13. Remove one of the battery clamp bolts and grind a 45 degree bevel on the threaded end. Screw the battery clamp bolt into hole provided at the base of the right cylinder head to hold the timing belt adjuster in it's current position; tighten the bolt by hand, DO NOT use a wrench.

14. Remove the engine mount bracket.

15. At the base of the left cylinder head, loosen the idler pulley bolt about 5–6 turns; then, remove the timing belt.

To install:

✳✳ WARNING

Do not rotate the crankshaft pulley or camshaft pulleys with the timing belt removed. The pistons may hit the valves and cause damage.

16. Remove the spark plugs.

17. Set the timing belt drive (crankshaft) sprocket so that the No. 1 piston is at TDC. Align the TDC mark on the tooth of the timing belt drive sprocket with the pointer on the oil pump.

18. Set the camshaft pulleys so that the No. 1 piston is at TDC. Align the TDC mark on the camshaft pulleys to the pointers on the back covers.

19. Remove the battery clamp bolt from the back cover. Remove the auto-tensioner.

20. Service the auto-tensioner by performing the following procedure:

    a. Position the auto-tensioner in a soft jawed vise with the maintenance bolt facing upward. DO NOT grip the body of the auto-tensioner.

    b. Remove the maintenance bolt.

    c. Be careful not to spill the oil from inside the tensioner. If oil is spilled, replenish it; the total capacity is 0.22 fl. oz. (6.5 ml).

    d. Using Stopper tool 14540-P8A-A01 or equivalent, position it on the auto-tensioner while turning the internal screw.

    e. Insert a flat-blade screwdriver into the maintenance hole and turn it clockwise to compress the bottom.

Be careful not to damage the threads or the gasket contact surface with the screwdriver.

f. Using a new gasket, reinstall the maintenance bolt and torque it to 72 inch lbs. (8 Nm).

g. Make sure that no oil is leaking around the maintenance bolt and install the auto-tensioner; torque the bolts 33 ft. lbs. (44 Nm).

➡**Make sure that the Stopper tool 14540-P8A-A01 or equivalent stays in place.**

21. Install the timing belt on the sprockets in the following sequence: drive sprocket (crankshaft), idler pulley, left camshaft sprocket, water pump pulley, right camshaft sprocket and adjusting pulley.

22. Torque the idler pulley bolt to 33 ft. lbs. (44 Nm).

23. Remove the Stopper tool from the auto-tensioner.

24. Install the engine mount-to-engine and torque the No. 10 bolts to 33 ft. lbs. (44 Nm) and the No. 6 bolt to 104 inch lbs. (12 Nm).

25. Install the lower and upper timing belt covers.

26. Install the crankshaft pulley and finger-tighten the bolt and washer. Using the Holder Handle and the 50mm Offset Holder Attachment tool 07MAB-PY3010A or equivalent, and a 19mm socket with a torque wrench, tighten the crankshaft pulley bolt to 181 ft. lbs. (245 Nm).

27. Rotate the crankshaft 5–6 turns clockwise so that the timing belt positions itself properly on the sprockets.

28. Set cylinder No. 1 to TDC by aligning the timing marks. If the timing marks do not align, remove the timing belt, then adjust the components and reinstall the timing belt.

29. Install all applicable components.

30. To complete the installation, reverse the removal procedures. Adjust the tension of the drive belts.

31. Disconnect the negative battery cable.

**2.5L (C25A1), 2.7L (C27A1),
3.2L (C32A1) & 3.5L (C35A1) ENGINES**

➡**Under normal driving conditions, the timing belt and timing balancer belt are to be replaced at 105,000 miles (168,000 km).**

1. Raise and safely support the vehicle.

2. Be sure to acquire the anti-theft code for the radio and the frequencies for the radio's preset buttons.

3. Disconnect the negative battery cable.

4. If necessary, remove the strut brace located at the top rear of the engine.

5. Rotate the crankshaft pulley so that the No. 1 piston is at Top Dead Center (TDC) of its compression stroke.

6. Remove the top engine cover-to-engine bolts and the cover.

7. Remove the air intake duct and air cleaner housing.

8. Loosen the alternator's adjusting rod, locknut and mounting bolt; then, remove the alternator's belt.

9. Loosen the idler pulley center nut and adjusting bolt; then, remove the air conditioning compressor belt.

10. Remove the TCS control valve upper and lower brackets.

11. Loosen the power steering pump's adjusting bolt, locknut and mounting bolt; then, remove the pump's belt.

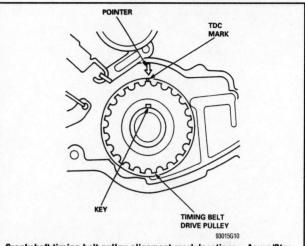

Crankshaft timing belt pulley alignment mark locations—Acura/Sterling 2.5L (C25A1), 2.7L (C27A1), 3.2L (C32A1) and 3.5L (C35A1) engines

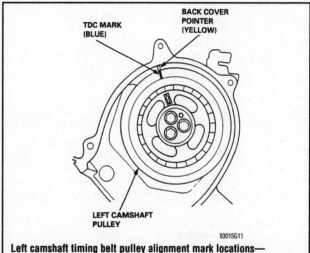

Left camshaft timing belt pulley alignment mark locations—Acura/Sterling 2.5L (C25A1), 2.7L (C27A1), 3.2L (C32A1) and 3.5L (C35A1) engines

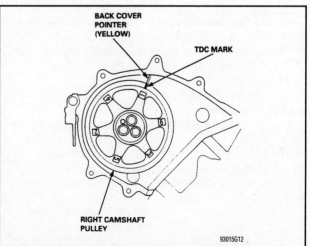

Right camshaft timing belt pulley alignment mark locations—Acura/Sterling 2.5L (C25A1), 2.7L (C27A1), 3.2L (C32A1) and 3.5L (C35A1) engines

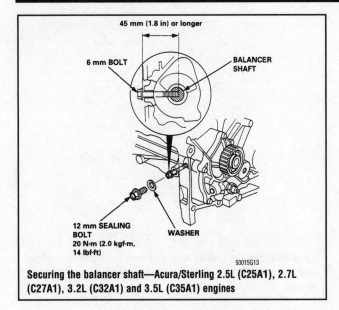

Securing the balancer shaft—Acura/Sterling 2.5L (C25A1), 2.7L (C27A1), 3.2L (C32A1) and 3.5L (C35A1) engines

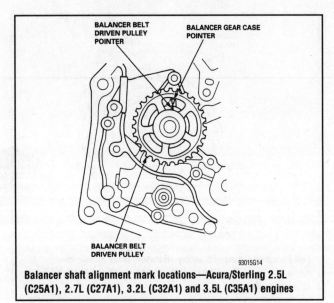

Balancer shaft alignment mark locations—Acura/Sterling 2.5L (C25A1), 2.7L (C27A1), 3.2L (C32A1) and 3.5L (C35A1) engines

12. Disconnect the TCS throttle sensor connector, the TCS throttle actuator connector and the Throttle Position (TP) sensor; then, remove the TCS control valve assembly.

13. Disconnect the Vehicle Speed Sensor (VSS) harness connector and remove the wire harness holder.

14. Remove the breather and vacuum hoses.

15. From the right timing belt cover, remove the Ignition Control Module (ICM) bracket.

16. From the left side of the crankshaft pulley, remove the idler pulley bracket.

17. Remove the dipstick and the dipstick tube-to-engine bolt; then, pull the tube from its O-ring mount. Discard the O-ring.

18. Remove the crankshaft pulley by performing the following procedure:

    a. Using the Holder Handle and the 50mm Offset Holder Attachment tool 07MAB-PY3010A or equivalent, and a 19mm socket, secure the crankshaft pulley and remove the crankshaft pulley bolt.

    b. Remove the washer and pull the crankshaft pulley from the engine.

19. From the upper and lower timing belt cover.

20. Loosen the balancer belt adjusting nut 180 degrees (½) turn. Push the tensioner to relieve the tension from the balancer belt; then, retighten the adjusting bolt. Remove the balancer belt.

21. Loosen the timing belt adjusting nut 180 degrees (½) turn. Push the tensioner to relieve the tension from the timing belt; then, retighten the adjusting bolt. Remove the timing belt.

To install:

22. Remove the spark plugs.

23. Remove the balancer belt drive pulley and the timing belt guide plate from the crankshaft.

24. Clean the upper and lower timing belt covers.

25. Position the timing belt pulley so the No. 1 piston is at TDC of its compression stroke. Align the mark on the pulley (near keyway) with the pointer mark on the oil pump.

26. Adjust the camshaft pulley so that the No. 1 piston is the TDC of the compression stroke. Align the TDC marks on the pulley with the upper surface of the back cover; the arrow mark on the left camshaft pulley and the **1** on the right camshaft pulley should be facing the back cover pointer.

27. Install the timing belt in the following sequence:

    a. Crankshaft timing belt pulley sprocket.

    b. Adjusting pulley.

    c. Left camshaft pulley

    d. Water pump pulley

    e. Right camshaft pulley

❊❊ WARNING

Make sure that the camshaft and crankshaft pulleys are at TDC.

➡**For easier installation, turn the right camshaft pulley about ½ tooth from TDC.**

28. Loosen and retighten the timing belt adjusting bolt to tension the timing belt.

29. Install the lower cover and the crankshaft pulley.

30. Rotate the crankshaft pulley about 5–6 turns clockwise to position the timing belt on the pulleys.

31. To adjust the timing belt tension, perform the following procedure:

    a. Make sure that the No. 1 piston is at TDC of its compression stroke.

    b. Rotate the crankshaft clockwise 10 teeth on the camshaft pulley; the **blue** mark on the crankshaft pulley should align with the lower cover pointer.

    c. Loosen the adjusting nut 180 degrees (½ turn).

    d. Tighten the adjusting nut to 31 ft. lbs. (42 Nm).

32. Remove the crankshaft pulley and the lower cover; then, install the timing belt guide plate and the balancer belt drive pulley.

33. Align the balancer shaft pulley by performing the following procedure:

    a. Using a 6 x 45mm bolt, insert it into the maintenance hole and the balancer shaft.

    b. Align the pointer on the balancer belt pulley with the pointer on the balancer gear case.

34. Adjust the timing belt drive pulley so that the No. 1 piston is at TDC of the compression stroke.

35. Install the balancer belt drive pulley and the balancer belt.

36. Loosen and retighten the balancer adjuster bolt to place tension on the balancer belt.

37. Remove the 6mm bolt and install the sealing bolt in the maintenance hole using a new washer.

38. Install the crankshaft pulley. Rotate the crankshaft pulley about 5–6 turns clockwise to position the timing belt on the pulleys.

39. Loosen the balancer belt adjuster bolt 180 degrees (½ turn) and retighten the bolt to 33 ft. lbs. (44 Nm).

40. Remove the crankshaft pulley.

41. Install the upper and lower timing belt covers and the crankshaft pulley.

42. Install the crankshaft pulley and finger-tighten the bolt and washer. Using the Holder Handle and the 50mm Offset Holder Attachment tool 07MAB-PY3010A or equivalent, and a 19mm socket with a torque wrench, tighten the crankshaft pulley bolt to 181 ft. lbs. (245 Nm).

43. Make sure the crankshaft and camshaft pulleys are at TDC.

➡**If the camshaft or crankshaft pulley is not at TDC, remove the timing belt and re-perform the adjustment procedure.**

44. To complete the installation, reverse the removal procedures. Adjust the tension of the drive belts.

45. Connect the negative battery cable.

3.0L (C30A1) & 3.2L (C32B1) ENGINES

1. Disconnect the negative battery cable.

2. Raise and safely support the vehicle. Remove the right rear wheel.

3. Remove the engine oil cooler base assembly without disconnecting the hoses.

4. Lower the vehicle to a comfortable working height and remove the rear strut brace.

5. Remove the intake manifold plate.

6. Remove the top cover.

7. Remove the injector cover and the wire harness covers.

8. Remove the ignition coil covers and the ignition coils.

9. Remove the coolant expansion tank and the air cleaner case.

10. Remove the alternator.

11. Disconnect the side engine mount and position it back into the body cavity.

12. Remove the transmission mount.

13. Remove the cylinder head covers.

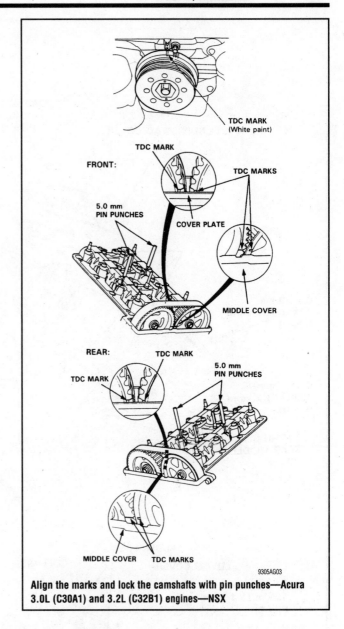

9305AG03

Align the marks and lock the camshafts with pin punches—Acura 3.0L (C30A1) and 3.2L (C32B1) engines—NSX

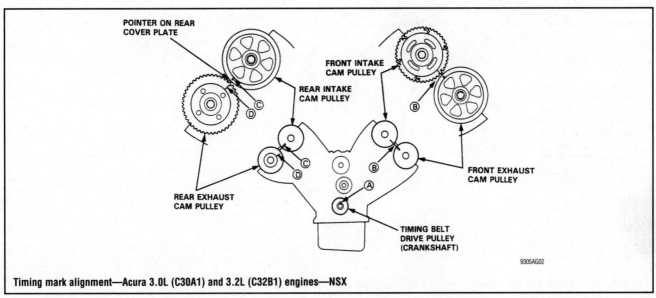

9305AG02

Timing mark alignment—Acura 3.0L (C30A1) and 3.2L (C32B1) engines—NSX

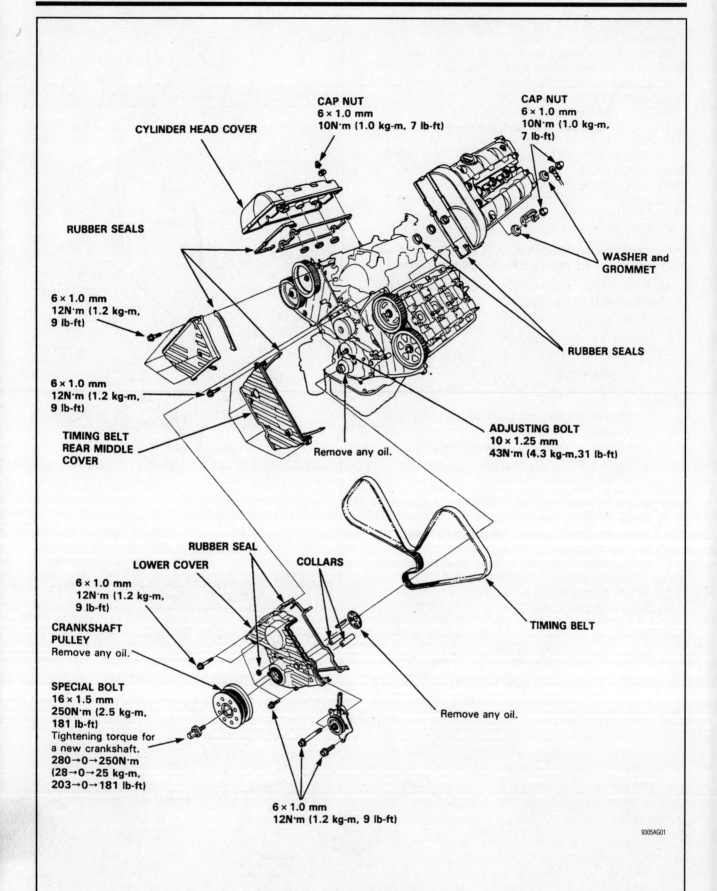

CYLINDER HEAD COVER

CAP NUT
6 × 1.0 mm
10N·m (1.0 kg-m, 7 lb-ft)

CAP NUT
6 × 1.0 mm
10N·m (1.0 kg-m, 7 lb-ft)

RUBBER SEALS

WASHER and GROMMET

6 × 1.0 mm
12N·m (1.2 kg-m, 9 lb-ft)

RUBBER SEALS

6 × 1.0 mm
12N·m (1.2 kg-m, 9 lb-ft)

TIMING BELT REAR MIDDLE COVER

Remove any oil.

ADJUSTING BOLT
10 × 1.25 mm
43N·m (4.3 kg-m, 31 lb-ft)

RUBBER SEAL

LOWER COVER

COLLARS

6 × 1.0 mm
12N·m (1.2 kg-m, 9 lb-ft)

TIMING BELT

CRANKSHAFT PULLEY
Remove any oil.

SPECIAL BOLT
16 × 1.5 mm
250N·m (2.5 kg-m, 181 lb-ft)
Tightening torque for a new crankshaft.
280→0→250N·m
(28→0→25 kg-m, 203→0→181 lb-ft)

Remove any oil.

6 × 1.0 mm
12N·m (1.2 kg-m, 9 lb-ft)

9305AG01

Exploded view of the timing belt assembly—Acura 3.0L (C30A1) and 3.2L (C32B1) engines—NSX

14. Rotate the crankshaft so that the No. 1 cylinder is at Top Dead Center (TDC), and the timing marks are aligned and the camshaft pulley **UP** marks are at the top.

15. Align the holes in the camshaft holders to the holes in the camshafts, and insert 5mm pin punches to lock the camshafts in position while the belt is removed.

16. Using a jack, raise the engine slightly.

17. Remove the alternator bracket stiffener.

18. Remove the air conditioning adjusting pulley and belt.

19. Remove the dipstick tube mounting bolt.

20. Remove the front and rear timing belt middle covers.

21. Remove the crankshaft pulley and remove the lower timing belt cover.

22. Loosen the timing belt adjuster pulley bolt and push on the tensioner to relieve belt tension. Re-tighten the adjuster bolt.

➡**If reusing the timing belt, mark its direction of rotation on the back of the belt.**

23. Remove the timing belt.

To install:

➡**Do not rotate the crankshaft or camshafts before installing the timing belt.**

24. Noting the direction of rotation, install the timing belt. Route the belt in the following order:
- Crankshaft pulley
- Adjusting pulley
- Front exhaust cam pulley
- Front intake cam pulley
- Water pump pulley
- Rear intake cam pulley
- Rear exhaust cam pulley.

25. Install the lower timing belt cover and the crankshaft pulley.

26. Remove the pin punches.

27. Adjust the timing belt tension by performing the following procedure:

a. Loosen the tensioner bolt.

b. While holding the crankshaft from turning, remove slack in the timing belt by rotating the camshafts slightly and in the order of belt installation.

c. Tighten the tensioner bolt.

d. Rotate the crankshaft clockwise 9 teeth on a camshaft pulley. The **blue** mark on the crankshaft pulley should align with the pointer on the lower timing cover.

e. Loosen the tensioner bolt.

f. Tighten the tensioner bolt to 31 ft. lbs. (43 Nm).

28. Rotate the crankshaft by hand 2 complete revolutions and insure the timing marks align.

29. Install the front and rear timing belt covers.

30. Install the dipstick tube mounting bolt.

31. Install the air conditioning adjustment pulley and belt.

32. Install the alternator bracket stiffener.

33. Install the cylinder head covers.

34. Lower the engine and install the transmission mount.

35. Install the side engine mount.

36. Install the alternator.

37. Install the air cleaner case and coolant expansion tank.

38. Install the ignition coils.

39. Install the injector cover and the wire harness covers.

40. Install the ignition coil covers.

41. Install the top cover and the intake manifold plate.

42. Install the rear strut brace.

43. Install the engine oil cooler base assembly.

44. Install the right rear wheel.

45. Lower the vehicle.

46. Connect the negative battery cable.

Audi

1.6L 4-CYLINDER ENGINES

1980–87

1. Disconnect the negative battery cable.

2. Raise and safely support the vehicle.

3. Using the large bolt on the crankshaft sprocket, rotate the engine until the No. 1 cylinder is at Top Dead Center (TDC) of the compression stroke. At this point, both valves will be closed and the **0** mark on the flywheel will be aligned with the pointer on the bell housing.

4. If the belt hasn't jumped, the timing mark on the rear face of

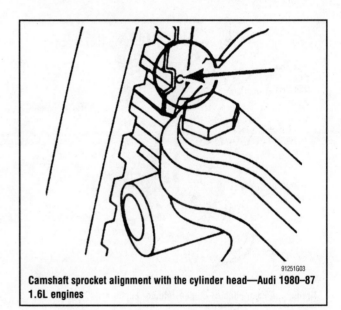

Camshaft sprocket alignment with the cylinder head—Audi 1980–87 1.6L engines

91251G03

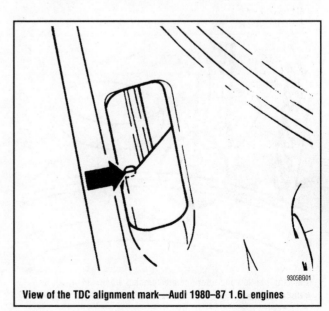

View of the TDC alignment mark—Audi 1980–87 1.6L engines

9305BG01

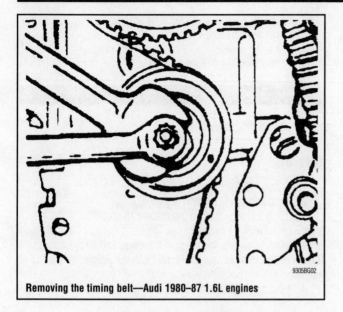

Removing the timing belt—Audi 1980–87 1.6L engines

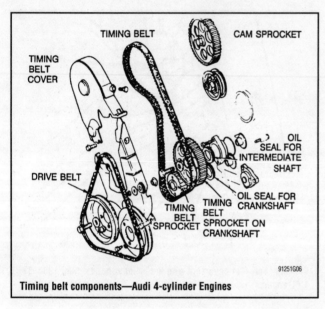

Timing belt components—Audi 4-cylinder Engines

the camshaft sprocket should be aligned with the upper left edge of the valve cover.

5. Loosen the alternator adjusting bolts, pivot the alternator over and slip the drive belt off.

6. Loosen the air conditioning compressor mounting bolts and remove the drive belt.

7. Remove the valve cover nuts and remove the valve cover and retaining straps.

8. Remove the upper timing belt cover nuts. Note the position of the washers and spacers while removing the cover.

9. Remove the crankshaft pulley retaining bolts. If the sprocket or rear cover is to be serviced, loosen the crankshaft sprocket bolt.

➡ **To remove the crankshaft sprocket bolt on manual transaxle vehicles, place the vehicle in 5th gear and have an assistant apply the brake. The will stop the engine from rotating while loosening the bolt. On automatic transaxle vehicles, remove the starter and hold the flywheel from turning using a flywheel holding tool VW 10-201 or equivalent.**

10. Remove the crankshaft pulley.

11. Remove the water pump pulley retaining bolts and remove the pulley.

12. Remove the lower cover retaining nuts and remove the cover.

13. While holding the large hex nut on the tensioner pulley, loosen the pulley locknut.

14. Turn the tensioner counterclockwise to relieve the tension on the timing belt.

15. Carefully slide the timing belt off the sprockets and remove the belt.

To install:

16. If the engine has moved or jumped timing, use the large bolt on the crankshaft sprocket to rotate the engine until the No. 1 cylinder is at TDC of the compression stroke. At this point, both valves will be closed and the **0** mark on the flywheel will be aligned with the pointer on the bell housing. Rotate the camshaft until the timing mark on the rear face of the camshaft sprocket is aligned with the upper left edge of the valve cover.

17. Install the crankshaft pulley and check that the notch on the pulley is aligned with the mark on the intermediate shaft sprocket. If not, rotate the intermediate shaft until they align.

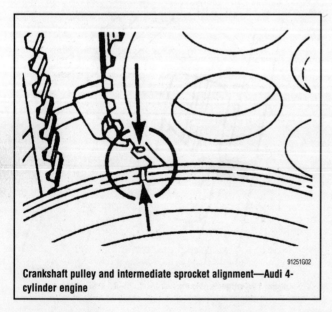

Crankshaft pulley and intermediate sprocket alignment—Audi 4-cylinder engine

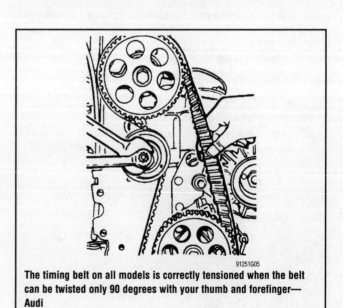

The timing belt on all models is correctly tensioned when the belt can be twisted only 90 degrees with your thumb and forefinger—Audi

If the timing marks are not correctly aligned with the No. 1 piston at TDC of the compression stroke when the belt is installed, valve timing will be incorrect. Poor performance and possible engine damage can result from the improper valve timing.

18. Remove the crankshaft pulley. Note the pulley location on the crankshaft sprocket so it can be replaced in the same position.

19. Hold the large nut on the tensioner pulley and loosen the smaller locknut. Turn the tensioner counterclockwise to loosen and install the timing belt.

20. Slide the timing belt onto the sprockets and adjust the belt tension. The timing belt tension is correct when the belt can be twisted 90 degrees midway between the camshaft and the intermediate shaft drive sprockets.

21. Torque the tensioner locknut to 33 ft. lbs. (45 Nm).

22. Install the upper and lower timing belt covers.

23. Install the crankshaft pulley. Torque the retaining bolts to 15 ft. lbs. (20 Nm).

24. If removed, install and torque the crankshaft sprocket bolt to 66 ft. lbs. (89 Nm) plus ½ additional turn.

25. Install the water pump pulley and torque the bolts to 18 ft. lbs. (25 Nm).

26. Install the belts and adjust to the proper tension.

27. Install any previously removed shields.

28. Lower the vehicle.

29. Connect the negative battery cable.

30. Road test the vehicle for proper operation.

1.8L (AEB) 4-CYLINDER ENGINES

1. Disconnect the negative battery cable.

2. Raise and safely support the vehicle.

3. From under the vehicle, remove the splash shield.

4. Remove the front bumper.

5. Remove the intake air duct between the grille/front end assembly and the air cleaner housing.

6. Remove the grille/front end assembly-to-chassis bolts and the grille/front end assembly-to-vehicle fasteners.

7. If installed, remove the wiring harness retaining clamps from the left side of the radiator frame.

8. Install Support tools 3369 or equivalent bolts into the grille/front end assembly-to-chassis holes; then, pull the grille/front end assembly forward until it hits the stops.

➡**If necessary to secure the grille/front end assembly, install M6 bolts into the rear bored holes of the grille/front end assembly and the fender.**

9. Loosen the air conditioning compressor's serpentine belt tensioner bolts; release the belt tension and remove the belt.

10. Place an open-end wrench on the alternator's belt tensioner and rotate it clockwise toward the alternator to release the belt's tension. Remove the alternator's serpentine drive belt and release the tensioner.

➡**If necessary to lock the alternator tensioner in position, align the housing holes and insert an Allen wrench into the holes to secure its movement.**

11. Using a 5 x 60mm bolt, secure the viscous fan pulley. Using

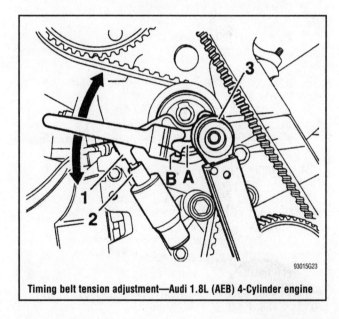
Timing belt tension adjustment—Audi 1.8L (AEB) 4-Cylinder engine

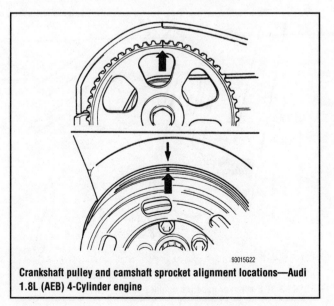
Crankshaft pulley and camshaft sprocket alignment locations—Audi 1.8L (AEB) 4-Cylinder engine

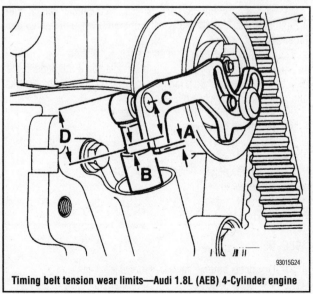
Timing belt tension wear limits—Audi 1.8L (AEB) 4-Cylinder engine

a hex wrench, remove the viscous fan-to-pulley bolts. Remove the viscous fan assembly.

12. Remove the upper timing belt cover.

➡**If reusing the timing belt, mark its rotational direction so it may be installed in its original position.**

13. Using the center bolt, rotate the crankshaft in the direction of engine rotation to position the No. 1 cylinder at Top Dead Center (TDC) of its compression stroke.

14. Remove the damper pulley-to-crankshaft bolts and the damper.

15. Remove the lower timing belt cover.

16. Using a Torx Wrench T45 or equivalent, loosen the timing belt tensioner, push the tensioner downward and remove the timing belt.

To install:

17. Align the camshaft sprocket timing mark with the cylinder head cover mark.

18. Install the timing belt on the crankshaft sprocket with the arrow facing the rotational direction.

19. Install the lower timing belt cover.

20. Using a bolt, secure the damper/belt pulley on the crankshaft.

21. Align the crankshaft damper/belt pulley with the housing timing mark so that the No. 1 cylinder is at TDC of its compression stroke.

22. Install the timing belt on the camshaft sprocket and belt tensioner.

23. Using a 2-pin Spanner Matra V159 Wrench or equivalent, lift (turn clockwise) the timing belt tensioner cylinder No. 1 until it is fully extended and tensioner cylinder No. 2 is raised approx. 1mm; then, hand-tighten the mounting bolt.

24. Rotate the crankshaft 2 complete rotation in the running direction.

25. Inspect area "A" for proper alignment with the upper edge of piston No. 2 and adjust if necessary.

- Area "A": adjustment OK
- Area "B": wear limit
- Area "C": re-adjust and check belt drive including tensioner for wear.

➡**If the piston edge is located in area "A", measurement "D" is 0.984–1.142 in. (25–29mm).**

26. After adjustment has been verified, secure the tensioner with a 2-pin Spanner Matra V159 Wrench or equivalent, and tighten the mounting bolt.

27. Complete the damper to crankshaft installation.

28. Using the center bolt, rotate the crankshaft 2 rotations in the direction of engine rotation until the camshaft and crankshaft marks align with their respective reference points.

29. Install the upper timing belt cover.

30. Install the drive belts.

31. Replace the remaining components by reversing the removal procedures.

32. Install the negative battery cable last.

33. Test drive the vehicle.

2.0L DIESEL, 2.2L & 2.3L 5-CYL. GASOLINE ENGINES

1980–95

1. Disconnect the negative battery cable.
2. Using the large bolt on the crankshaft sprocket, rotate the engine until the No. 1 cylinder is at Top Dead Center (TDC) of the compression stroke. Align the TDC mark **0** with the cast mark on the bell housing.

3. The timing mark on the rear face of the camshaft sprocket should be aligned with the upper left edge of the valve cover. On 2.2L 20 Valve Turbo engines, the camshaft timing mark should be aligned with the notch in the rear belt cover at the 12 o'clock position.

4. Remove the alternator, power steering, and air conditioner compressor drive belts.

5. Remove the retaining nuts and remove the upper timing belt cover. Take care not to lose any of the washers or spacers.

6. Remove the crankshaft balancer center bolt.

➡**To remove the crankshaft balancer bolt on manual transaxle vehicles, place the vehicle in 5th gear and have an assistant apply the brake. The will stop the engine from rotating while loosening the bolt. On automatic transaxle vehicles, remove the starter and hold the flywheel from turning, using a flywheel holding tool VW 10-201 or equivalent. This bolt is extremely tight.**

7. Remove the lower timing belt cover bolts and remove the cover.

8. Loosen the water pump bolts only enough to turn the pump clockwise.

➡**By loosening the water pump bolts, the coolant may drain from the engine at the water pump. If necessary, drain the cooling system, remove the water pump and reinstall it with a new O-ring.**

9. Slide the timing belt off the sprockets.

To install:

10. Align crankshaft and camshaft timing marks.

11. Install the timing belt and turn the water pump counterclockwise to tighten the belt.

12. Torque the water pump bolts to 15 ft. lbs. (20 Nm).

➡**The timing belt is correctly tensioned when it can be twisted 90 degrees along the straight run between the camshaft sprocket and water pump. The belt must not be jammed between the oil pump and sprocket when installing the vibration damper.**

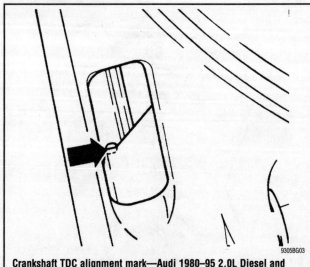

Crankshaft TDC alignment mark—Audi 1980–95 2.0L Diesel and 2.2L, 2.3L 5-cylinder gasoline engines

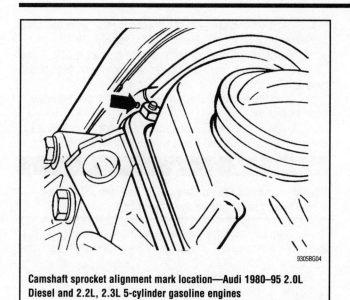

Camshaft sprocket alignment mark location—Audi 1980–95 2.0L Diesel and 2.2L, 2.3L 5-cylinder gasoline engines

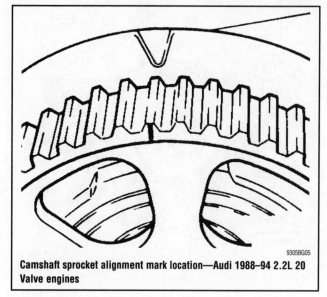

Camshaft sprocket alignment mark location—Audi 1988–94 2.2L 20 Valve engines

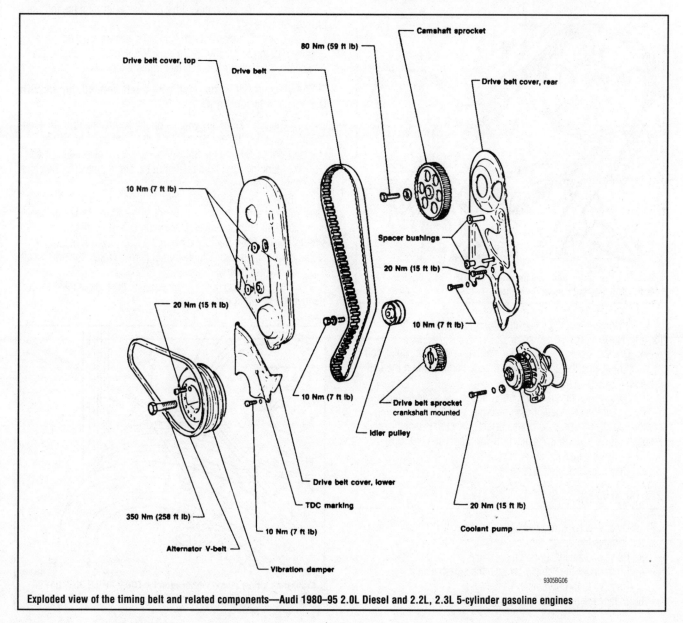

Exploded view of the timing belt and related components—Audi 1980–95 2.0L Diesel and 2.2L, 2.3L 5-cylinder gasoline engines

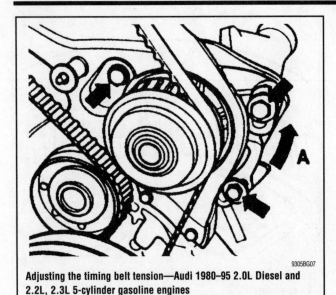

Adjusting the timing belt tension—Audi 1980–95 2.0L Diesel and 2.2L, 2.3L 5-cylinder gasoline engines

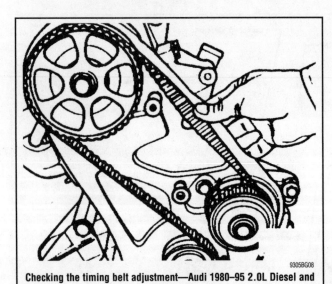

Checking the timing belt adjustment—Audi 1980–95 2.0L Diesel and 2.2L, 2.3L 5-cylinder gasoline engines

13. Install the timing belt covers and torque the bolts to 89 inch lbs. (10 Nm).

14. Use the same procedure to install the crankshaft center bolt as when removing the bolt. Apply a locking compound on the bolt threads.

15. Torque the bolt to 258 ft. lbs. (350 Nm) in several steps.

16. Install the alternator, power steering, and air conditioning compressor belts. These belts are correctly tensioned when they can be depressed ⅜ in. (9.5mm) along their longest straight run.

17. Connect negative battery cable.

18. Check and set ignition timing.

2.8L (AHA) V6 ENGINES

1. Disconnect the negative battery cable.
2. Remove the upper engine cover.
3. Raise and safely support the vehicle.
4. From under the vehicle, remove the splash shield.
5. Remove the front bumper.
6. Disengage the hood lock cable

7. Remove the intake air duct between the lock carrier and the air cleaner housing at the grille/front end assembly.

8. Remove the grille/front end assembly-to-chassis bolts.

9. Disconnect the electrical connectors from the grille/front end assembly.

10. Drain the engine coolant and disconnect the coolant hoses from the radiator.

11. Detach the air conditioning condenser from the grille/front end assembly and suspend it on a wire at the front wheel.

⁂ WARNING

DO NOT suspend the condenser by its lines. The condenser lines must not be bent or kinked.

12. Drain the automatic transmission fluid from the transmission and the transmission cooler. Disconnect the hydraulic lines from the transmission cooler.

13. If equipped, remove the charge air cooler.

14. Remove the grille/front end assembly-to-vehicle fasteners and the grille/front end assembly from the vehicle.

15. Remove the serpentine drive belt by performing the following procedure:

  a. Using Spanner Wrench No. 3212, secure the viscous fan pulley. Using an Open-end Spanner Wrench 3212, remove the viscous fan bearing housing by turning it clockwise.

➡ **The viscous fan is mounted with a left-handed thread; turn it clockwise to loosen it.**

  b. Place a 17mm box wrench on the serpentine drive belt tensioner and turn the tensioner clockwise until the 2 holes align; insert drift 3204 into the holes to secure the tensioner in place.

  c. Mark the running direction of the serpentine drive belt and remove it from the pulleys.

16. Rotate the crankshaft by hand to align the crankshaft pulley mark with the arrow on the engine housing and the large hole in each camshaft sprocket must face inward and must align; this should be Top Dead Center (TDC) of the No. 1 cylinder's compression stroke. If these conditions are not correct, rotate the crankshaft 1 complete revolution and realign.

17. On the left side of the cylinder block near the crankshaft, remove the sealing plug.

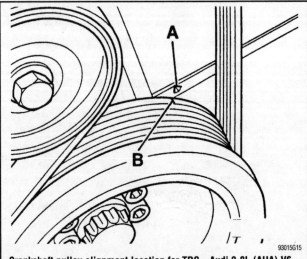

Crankshaft pulley alignment location for TDC—Audi 2.8L (AHA) V6 engine

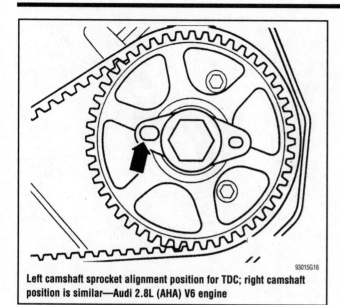

Left camshaft sprocket alignment position for TDC; right camshaft position is similar—Audi 2.8L (AHA) V6 engine

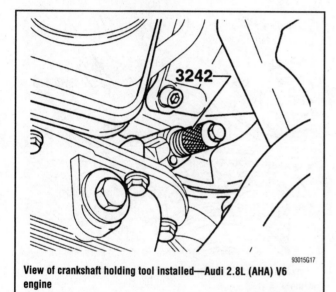

View of crankshaft holding tool installed—Audi 2.8L (AHA) V6 engine

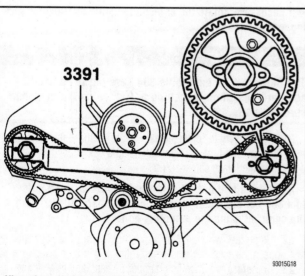

View of camshaft locator bar installed—Audi 2.8L (AHA) V6 engine

18. Insert Crankshaft Holder tool No. 3242 into the sealing plug hole to secure the crankshaft.

19. Using a 8mm Allen® wrench, rotate the timing belt tensioner roller clockwise until the tensioner is compressed; then, insert a 2mm spring pin through the tensioner housing and tensioner plunger to secure it in place. When the plunger is secure, release the wrench tension.

20. Remove the damper-to-crankshaft bolts and the damper.

➡️**It is not necessary to remove the center bolt when removing the crankshaft damper.**

21. Remove the serpentine belt idler and the crankshaft damper guard.

22. Mark the running direction of the timing belt and remove it from the pulleys.

To install:

23. Make sure that the camshaft pulleys and the crankshaft pulley are in alignment with TDC of the No. 1 cylinder's compression stroke.

24. Install the timing belt; make sure the timing belt is installed in the correct running direction from which it was removed.

25. Using a 8mm Allen® wrench, rotate the timing belt tensioner roller clockwise until the tensioner is compressed; then, remove the 2mm spring pin from the tensioner housing. Slowly, release the tensioner's spring pressure to put pressure on the timing belt.

26. Install the crankshaft damper guard and the serpentine belt idler pulley; torque the idler pulley bolts to 33 ft. lbs. (45 Nm).

27. Install the crankshaft damper and torque the damper-to-crankshaft bolts to 15 ft. lbs. If the damper-to-crankshaft center bolt was removed, torque it to 147 ft. lbs. (200 Nm) plus 180° ½ turn).

28. Remove the Crankshaft Holder tool No. 3242 and install the sealing plug.

29. Replace the remaining components by reversing the removal procedures.

30. Refill the cooling system and the automatic transaxle. Connect the electrical connectors. Install the negative battery cable last.

31. Test drive the vehicle.

3.7L & 4.2L V8 ENGINES

1. Obtain the code for the anti-theft radio.

2. Disconnect the negative battery cable; the battery is under the cover located at the right side of the trunk.

3. Remove the noise insulation panel and the air intake grille from the bumper.

4. Drain the engine coolant from the radiator. Mark running direction on the serpentine drive belt to aid in its reinstallation.

5. Place box-end wrench on the serpentine drive belt tensioner and slowly turn the tensioner counterclockwise to release the belt tension and remove the serpentine drive belt.

6. Remove the upper engine cover, the coolant hose clamp from the right belt guard and air cleaner-to-throttle body air intake duct.

7. On the upper left of the radiator, remove air shroud for the viscous fan and electric coolant fan; then, move the air shroud aside with the wiring connected.

➡️**It may be necessary to remove the viscous fan with the air shroud.**

8. Using the Two-Hole Nut Driver tool 3212, or equivalent, counterhold the viscous fan pulley and remove the viscous fan from the pulley.

➡ **The viscous fan has a left-hand thread which must be loosened clockwise.**

9. Remove the right front engine support mount.

10. Disconnect and remove the coolant hoses from the front of the engine; then, loosen the coolant hose at the upper right side of the radiator and move it to the right.

11. Using Special tool 3197 to hold the crankshaft pulley, loosen its center bolt approximately 1 turn.

❊ WARNING

When the crankshaft pulley center bolt has been loosened, it must be replaced.

12. Rotate the crankshaft and align the Top Dead Center (TDC) marks on the crankshaft pulley.

➡ **If the crankshaft pulley center bolt has been loosened, use Special Wrench tool 3197 or equivalent, to turn the crankshaft.**

13. At the rear of the left cylinder head, remove the Camshaft Position (CMP) sensor housing.

➡ **If the CMP is not positioned behind the sensor plate window, rotate the crankshaft 360° (1 full turn).**

14. Remove the camshaft flange from the right rear cylinder head.

15. Disconnect the electrical connector from the intake manifold change-over valve switch.

16. Remove the left timing belt cover, the serpentine drive belt guide pulley cap from the right timing belt cover and the upper part of the air cleaner housing.

17. Remove the right timing belt cover bolts and the cover.

18. Using the Camshaft Sprocket Holder tool 3036 or equivalent, secure the camshaft sprocket; then, loosen the camshaft sprocket bolts.

19. At the left cylinder head, remove the CMP sensor plate-to-camshaft bolt, camshaft sensor housing bolts and the sensor.

20. At the rear of each cylinder head, install and tighten the Camshaft Locking tool 3341 or equivalent.

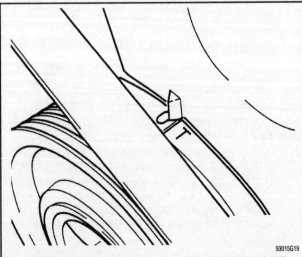

Crankshaft pulley alignment location for TDC—Audi 3.7L and 4.2L V8 engines

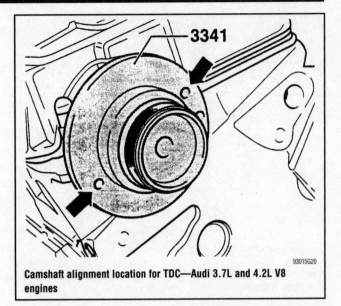

Camshaft alignment location for TDC—Audi 3.7L and 4.2L V8 engines

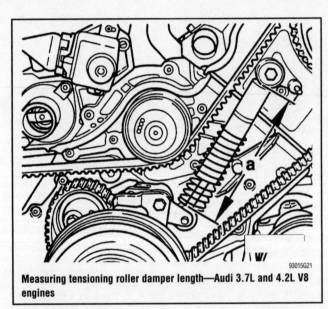

Measuring tensioning roller damper length—Audi 3.7L and 4.2L V8 engines

➡ **If necessary, rotate the camshaft slightly to allow the locking tool to engage.**

❊ WARNING

The Camshaft Locking tools 3341 are not to be used as counterholds when loosening or tightening the camshaft sprocket bolts.

21. If necessary to rotate the camshaft(s), remove the cylinder head cover(s) and turn the camshaft using a hex-shaped end wrench on the intake camshaft nut.

➡ **If the drive belt or belt sprocket is not loose, use Holder tool 3036 or equivalent to turn the camshaft; DO NOT turn the sprocket using the mounting bolt.**

22. Using the Two-Hole Nut Turner tool Matra V/159 or equivalent, loosen the eccentric tensioning roller and turn its lowest point.

23. Compress the drive belt damper by hand and remove the tensioning roller.

Be sure to mark the running direction of the timing belt, if the belt is to be reused.

24. Remove the timing belt from the camshaft sprocket.
25. Using a plastic hammer, lightly strike the edge of the camshaft sprockets to loosen them from the tapered ends of the camshafts.
26. Remove the 4 vibration damper-to-drive belt sprocket screws, the center bolt and the vibration damper.

➡**DO NOT remove the crankshaft sprocket from the crankshaft.**

27. Remove the timing belt.
To install:
28. Position the timing belt over the crankshaft sprocket and install the vibration damper on the crankshaft.
29. Install the vibration damper and the 4 vibration damper-to-sprocket bolts; hand-tighten only.
30. Using a new center bolt, secure the vibration damper and hand-tighten.
31. Install the camshaft sprockets on the camshafts finger-tight.
32. Position the timing belt over the sprockets, guide pulley and water pump pulley.
33. Push the belt over the tensioning roller with the eccentric insert and tighten the nut; make sure eccentric insert is still able to turn.
34. Torque the vibration damper-to-crankshaft sprocket bolts to 18 ft. lbs. (25 Nm).
35. Using the Two-Hole Nut Turner tool Matra V/159 or equivalent, turn the tensioning roller clockwise to adjust the damper length of the tensioning roller to 5.35–5.47 in. (136–139mm) cold or 4.96–5.08 in. (126–129mm) warm. Then, tighten the tensioning roller to 18 ft. lbs. (25mm).
36. Use Holder tool 3036, or equivalent, to secure the camshaft sprockets, tighten the sprocket bolts to 41 ft. lbs. (55 Nm).
37. From the rear of the cylinder heads, remove the camshaft locking devices.
38. Install the CMP sensor housing behind the CMP sensor and torque the bolts to 15 ft. lbs. (20 Nm).
39. Install the CMP sensor plate and engage in the camshaft catch.
40. Rotate the crankshaft at least 2 revolutions.
41. Torque the crankshaft center bolt to 258 ft. lbs. (350 Nm) with Special tool 3197 or to 332 ft. lbs. (450 Nm) without special tool.
42. Replace the remaining components by reversing the removal procedures.
43. Refill the cooling system and the automatic transaxle. Connect the electrical connectors. Install the negative battery cable last.
44. Test drive the vehicle.

BMW

2.4L (524 TD) DIESEL ENGINE

1985–86

1. Disconnect the negative battery cable.
2. Drain the cooling system.
3. Remove the accessory drive belts.
4. Remove the fan assembly and the water pump assembly.
5. Remove the vibration damper and pulley.

6. Disconnect the heater hose from the thermostat housing.
7. Remove the 4 bolts attaching the camshaft drive belt cover to the crankcase and remove the cover.
8. Remove the rocker arm cover.
9. Rotate the crankshaft until the No. 1 cylinder is at Top Dead Center (TDC) on its compression stroke (valves of No. 6 overlapping).
10. Install TDC Aligning Pin 13-5-340 or equivalent.

➡**Flat side of nut or cam position tool should be facing down.**

11. Using a piece of chalk or similar marker, mark the direction of engine rotation on the drive belt, unless a new belt is to be installed.
12. Loosen both belt tensioner bolts.
13. Remove the camshaft drive belt.
To install:
14. Insert a 0.098 in. (2.5mm) thick feeler gauge blade between the Cam Positioning tool No. 11-3-090 or equivalent, at the right front corner of the gasket mating surface of the cylinder head if

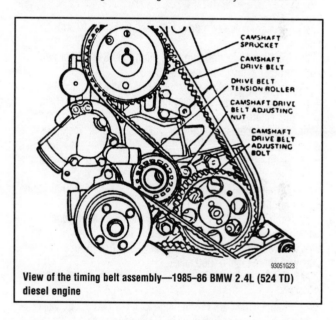

View of the timing belt assembly—1985–86 BMW 2.4L (524 TD) diesel engine

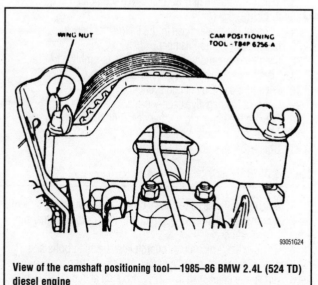

View of the camshaft positioning tool—1985–86 BMW 2.4L (524 TD) diesel engine

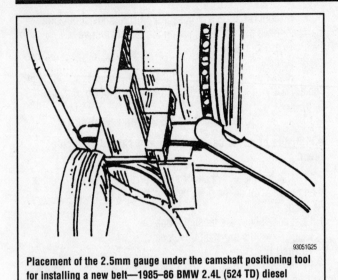

Placement of the 2.5mm gauge under the camshaft positioning tool for installing a new belt—1985–86 BMW 2.4L (524 TD) diesel engine

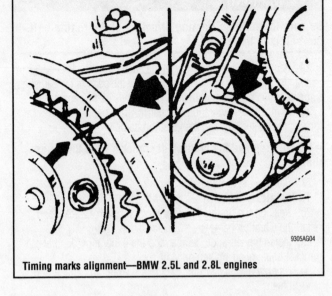

Timing marks alignment—BMW 2.5L and 2.8L engines

using a new drive belt or a drive belt used less than 10,000 miles.

15. Install the Injection Pump Aligning Pin 13-5-340 or equivalent, through the injection pump sprocket.

16. Rotate the injection pump sprocket clockwise against the pin.

17. Install the camshaft drive belt. Starting at the crankshaft, route the belt around the intermediate shaft sprocket, injection pump sprocket, camshaft sprocket and tension roller, keeping slack to a minimum.

18. Hand-tighten the belt with the belt tensioner until all slack is gone.

19. Remove the Injection Pump Aligning Pin 13-5-340 or equivalent, from the injection pump sprocket.

20. Adjust the belt tension by tightening the belt tensioner. Tighten the belt tensioner to 30.5–32.5 ft. lbs. (41.5–44 Nm)) on belts with less than 10,000 miles or 22–25 ft. lbs. (30–34 Nm) for belts with more than 10,000 miles.

21. Tighten both belt tensioner holding bolts to 15–18 ft. lbs. (20–24 Nm).

22. Remove the Cam Positioning tool 11-3-090 or equivalent.

23. Install the camshaft drive belt cover and tighten bolts to 72–84 ft. lbs. (8–9.5 Nm).

24. Connect the heater hose to the thermostat housing.

25. Install the vibration damper.

26. Install the fan and water pump pulley assembly.

27. Install and adjust the accessory drive belts.

28. Fill and bleed the cooling system.

29. Connect the negative battery cable.

30. Operate the engine to normal operating temperatures and check for oil and/or coolant leaks.

31. Check the injection pump timing.

2.5L & 2.8L ENGINES

1. Disconnect the negative battery cable. Drain the cooling system.

2. Remove the distributor cap and rotor. Remove the inner distributor cover and seal.

3. Remove the 2 distributor guard plate attaching bolts and 1 nut. Remove the rubber guard and take out the guard plate.

4. Rotate the crankshaft to set No. 1 piston at Top Dead Center (TDC) of the compression stroke.

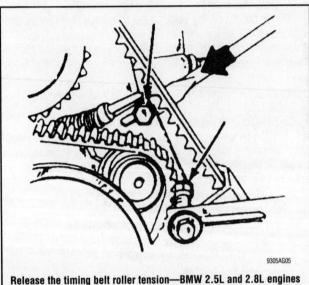

Release the timing belt roller tension—BMW 2.5L and 2.8L engines

➤ At TDC of No. 1 piston compression stroke, the camshaft sprocket arrow should align directly with the mark on the cylinder head.

5. Remove the radiator.

6. Remove the lower splash guard and take off the alternator, power steering and air conditioning belts.

7. Remove the crankshaft pulley and vibration damper.

8. Hold the crankshaft hub from rotating with tool 11-2-150 or equivalent. Remove the crankshaft hub bolt.

9. Install the hub bolt into the crankshaft about 3 turns and use the proper gear puller, to remove the crankshaft hub.

10. Remove the bolt from the engine end of the alternator bracket. Loosen the alternator adjusting bolt and swing the bracket aside.

11. Lift out the Top Dead Center (TDC) transmitter and set it aside.

12. Remove the remaining bolt and lift off the lower timing belt cover.

13. Loosen the tensioner pulley bolts.

14. Release the tension on the belt by pushing on the tensioner pulley bracket and tightening the upper tensioner bolt.

15. Mark the running direction of the timing belt if it is to be reused and remove the belt.

To install:

16. Check the camshaft and crankshaft alignment marks and install the timing belt.

17. Loosen the upper tensioner pulley bolt.

18. Rotate the crankshaft with a wrench to tension the timing belt.

19. Tighten the tensioner pulley bolts, upper bolt first.

20. Rotate the crankshaft 2 complete revolutions and insure that the timing marks align.

21. Install the lower timing belt cover and the TDC transmitter.

22. Replace the alternator bracket.

23. Install the crankshaft hub, pulley and vibration damper.

24. Install the accessory drive belts and the lower splash guard.

25. Install the radiator.

26. Install the distributor guard plate.

27. Install the distributor inner cover and seal.

28. Install the cap and rotor.

29. Connect the negative battery cable and refill the cooling system.

Chrysler Imports

1.4L & 1.5L ENGINES

1980–87

1. Disconnect the negative battery cable.

2. Rotate the crankshaft to position the No. 1 cylinder on Top Dead Center (TDC) of the compression stroke.

3. Remove the fan drive belt, the fan blades, spacer and water pump pulley.

4. Remove the timing belt cover.

5. Loosen the timing belt tensioner mounting bolt and move the tensioner toward the water pump. Temporarily secure the tensioner.

6. Remove the crankshaft pulley and slide the belt off of the camshaft and crankshaft sprockets.

7. Inspect the drive sprockets for abnormal wear, cracks or damage and replace as necessary. Remove and inspect the tensioner. Check for smooth pulley rotation, excessive play or noise. Replace the tensioner, if necessary.

8. If reusing the timing belt, mark the rotational direction on the back of the belt.

9. Remove the timing belt.

To install:

10. Make sure that the timing mark on the camshaft sprocket is aligned with the pointer on the cylinder head and that the crankshaft sprocket mark is aligned with the mark on the engine case.

11. Noting the direction of rotation, install the timing belt; make sure that there is no play between the camshaft and crankshaft sprockets.

12. Loosen the tensioner from its temporary position so that the spring pressure will allow it to contact the timing belt.

13. Rotate the crankshaft 2 complete turns in the normal rotation direction to remove any belt slack. Turn the crankshaft until the timing marks are aligned. If the timing has slipped, remove the belt and repeat the procedure.

14. Once again, rotate the engine 2 complete revolutions until the timing marks are aligned. Recheck the belt tension.

➡**When the tension side of the timing belt and the tensioner are pushed in horizontally with a moderate force (about 11 lbs. [5 kg]), and the cogged side of the belt covers about a ¼ in. of the tensioner right side mounting bolt head (across flats), the tension is correct.**

15. Install the timing belt cover, the water pump pulley, spacer, fan blades and drive belt.

16. Connect the negative battery cable.

1.5L & 1.8L ENGINES

1988–94

1. Disconnect the negative battery cable. Remove the engine under cover.

2. Rotate crankshaft clockwise and position the No. 1 cylinder at Top Dead Center (TDC) of the compression stroke.

3. Raise and safely support the weight of the engine using the appropriate equipment. Remove the front engine mount bracket and accessory drive belts.

4. On Summit Wagon, remove the coolant reservoir tank.

5. Remove the drive belts, tension pulley brackets, water pump pulley and crankshaft pulley.

6. Remove all attaching screws and remove the upper and lower timing belt covers.

7. Make a mark on the back of the timing belt indicating the direction of rotation so it may be reassembled in the same direction if it is to be reused. Loosen the timing belt tensioner and remove the timing belt.

➡**If coolant or engine oil comes in contact with the timing belt, they will drastically shorten its life. Also, do not allow engine oil or coolant to contact the timing belt sprockets or tensioner assembly.**

8. Remove the tensioner spacer, tensioner spring and tensioner assembly.

9. Inspect the timing belt for cracks on back surface, sides, bottom and check for separated canvas. Check the tensioner pulley for smooth rotation.

To install:

10. Position the tensioner, tensioner spring and tensioner spacer on engine block.

11. Align the timing marks on the camshaft sprocket and crankshaft sprocket. This will position No. 1 piston on TDC on the compression stroke.

12. Position the timing belt on the crankshaft sprocket and keeping the tension side of the belt tight, set it on the camshaft sprocket.

13. Apply counterclockwise force to the camshaft sprocket to give tension to the belt and make sure all timing marks are aligned.

14. Loosen the pivot side tensioner bolt and the slot side bolt. Allow the spring to remove the slack.

15. Tighten the slot side tensioner bolt and then the pivot side bolt. If the pivot side bolt is tightened first, the tensioner could turn with bolt, causing over tension.

16. Turn the crankshaft clockwise. Loosen the pivot side tensioner bolt and then the slot side bolt to allow the spring to take up any remaining slack. On 1.8L engine, tighten the adjuster bolt to 18 ft. lbs. (24 Nm). On 1.5L engine, tighten the slot bolt and then the pivot side bolt to 14–20 ft. lbs. (20–27 Nm).

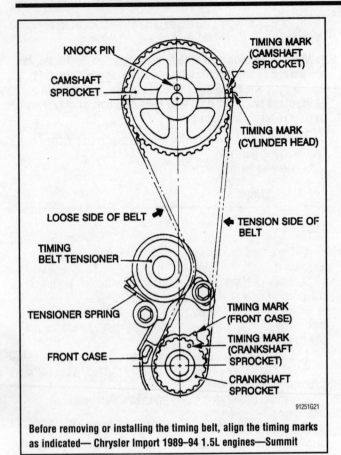

Before removing or installing the timing belt, align the timing marks as indicated— Chrysler Import 1989–94 1.5L engines—Summit

17. Install the timing belt covers and all related items.
18. Connect the negative battery cable.

1995–00

1. Rotate the crankshaft clockwise and position the engine at Top Dead Center (TDC) on the compression stroke for the No. 1 cylinder.

2. Remove the drive belts, tension pulley brackets, water pump pulley and crankshaft pulley.

3. Remove all attaching screws and remove the upper and lower timing belt covers.

4. Make a mark on the back of the timing belt indicating the direction of rotation so it may be reassembled in the same direction if it is to be reused. Loosen the timing belt tensioner and remove the timing belt.

5. For the 1.5L engine, loosen the timing belt tensioner and move the tensioner to provide slack to the timing belt. Tighten the tensioner in this position.

6. For the 1.8L engine, loosen the timing belt tensioner, insert a thin prytool into the tensioner and release tension by prying against the spring tension. Temporarily tighten the tensioner bolt to provide slack.

7. Remove the timing belt.

✳✳ WARNING

Coolant and engine oil will damage the rubber in the timing belt, drastically reducing its life. Do not allow engine oil or coolant to contact the timing belt, the sprockets or tensioner assembly.

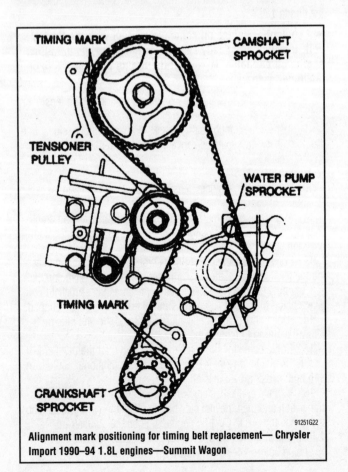

Alignment mark positioning for timing belt replacement— Chrysler Import 1990–94 1.8L engines—Summit Wagon

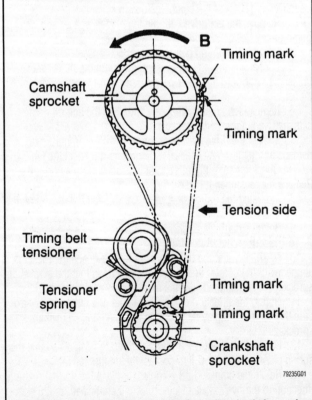

Before removing or installing the timing belt, align the timing marks as indicated—Chrysler Import 1.4L and 1.5L engines

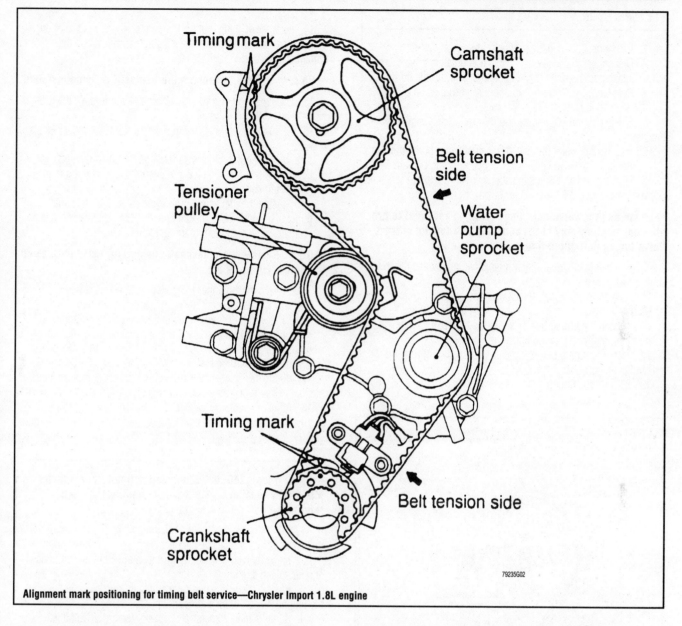

Timing mark

Camshaft sprocket

Belt tension side

Tensioner pulley

Water pump sprocket

Timing mark

Belt tension side

Crankshaft sprocket

79235G02

Alignment mark positioning for timing belt service—Chrysler Import 1.8L engine

8. If defective, remove the tensioner spacer, tensioner spring and tensioner assembly.

➡**It is recommended that the timing belt be replaced at least every 60,000 miles (96,000 km).**

9. Inspect the timing belt for cracks or wear. Check the tensioner pulley for smooth rotation.

To install:

10. If removed, position the tensioner, tensioner spring and tensioner spacer on the engine block.

11. Align the timing marks on the camshaft sprocket and crankshaft sprocket. This will position the No. 1 piston at TDC on its compression stroke.

12. For the 1.5L engine, position the timing belt on the crankshaft sprocket, keeping the tension side of the belt tight, set it on the camshaft sprocket, then the tensioner sprocket.

13. For the 1.8L engine, position the timing belt on the crankshaft sprocket, water pump sprocket, camshaft sprocket and the tensioner sprocket, keeping the tension side of the belt tight.

14. Apply a slight counterclockwise force to the camshaft sprocket to give tension to the belt, and be sure all timing marks are aligned.

15. Loosen the pivot side tensioner bolt and the slot side bolt. Allow the spring to remove any slack in the timing belt.

16. For the 1.5L engine, tighten the slot side tensioner bolt, then the pivot side bolt. If the pivot side bolt is tightened first, the tensioner could turn with the bolt, causing over tension.

17. For the 1.8L engine, turn the crankshaft clockwise 2 rotations, then tighten the adjuster bolt to 18 ft. lbs. (24 Nm) and the pivot (spring) bolt to 35 ft. lbs. (45 Nm).

18. Turn the crankshaft clockwise. Loosen the pivot side tensioner bolt, then the slot side bolt to allow the spring to take up any remaining slack. Tighten the slot bolt, then the pivot side bolt to 17 ft. lbs. (24 Nm).

19. Install the timing belt covers, and tighten the cover bolts to 96 inch lbs. (11 Nm). Install all applicable components.

1.6L SOHC ENGINE

1. Disconnect the negative battery cable.
2. Disconnect the accelerator cable and breather hose. Remove the air intake duct between the turbocharger and throttle body.
3. Remove the distributor cap/spark plug wire as an assembly.
4. Remove the accessory drive belts.
5. Safely support the engine and remove the left engine mount bracket.
6. Remove the water pump and power steering pump pulleys.
7. Remove the valve cover and the upper timing belt cover.
8. Loosen the valve adjusting screws until the tip of each screw protrudes less than 0.08 inch (2mm) from the rocker arm.

➡**Loosening the valve adjusting screws is essential to provide enough free-play at the camshaft to allow for correct valve timing during the timing belt installation.**

9. Remove the damper pulley, crankshaft pulley and the lower timing belt cover.
10. Rotate the crankshaft (clockwise) until the timing marks are aligned.
11. Loosen the timing belt tensioner bolts, move the tensioner toward the water pump and retighten the bolts.
12. Mark the timing belt to indicate its direction of rotation for reinstallation purposes.
13. Remove the timing belt.

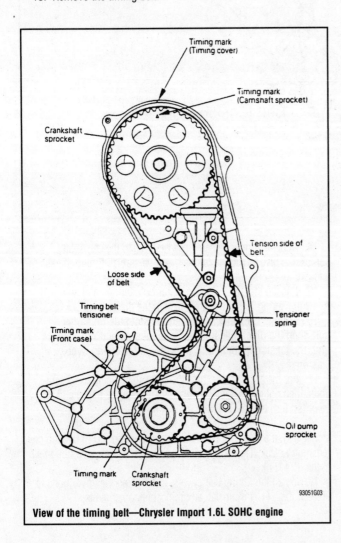

View of the timing belt—Chrysler Import 1.6L SOHC engine

93051G03

To install:
14. Install the timing belt (in the correct direction of rotation) over the crankshaft sprocket, the oil pump sprocket and the camshaft sprocket

➡**Be sure to keep tension on the belt as it is being installed.**

15. Loosen the timing belt tensioner bolts and move the tensioner toward the timing belt.
16. To adjust the timing belt tension, perform the following procedure:
   a. Apply counterclockwise force on the camshaft sprocket to tighten the belt tension side and ensure that the timing marks remain aligned.
   b. Install the crankshaft pulley and rotate the crankshaft (clockwise) until the camshaft sprocket timing mark is 2 teeth away from the alignment mark on the rear cover.

➡**DO NOT rotate the crankshaft counterclockwise or push on the belt to check its tension.**

   c. Move the tensioner to apply pressure on the belt and tighten the tensioner nut; then, tighten the pivot bolt.
   d. Place your thumb on the belt between the oil pump and camshaft sprockets and press the belt outward. The flex between timing belt and the cover should be 0.23 in., if not, loosen the pivot and tension the belt.
17. Rotate the crankshaft 2 revolutions and realign the timing marks to ensure that the engine is in time.
18. Adjust the valve clearances.
19. Complete the installation by reversing the removal procedure.

1.6L DOHC ENGINE

➡**Special tools MD998752 tension pulley torque adapter and MD998738 tension pulley locker or equivalents, are required.**

1. Bring the engine to No. 1 piston at Top Dead Center (TDC) of the compression stroke. Disconnect the negative battery cable.
2. Raise the vehicle and support it safely. Remove the under engine splash shield.
3. Place a piece of wood on a suitable floor jack and support the engine. Remove the engine mount bracket.
4. Remove the alternator and power steering drive belts. Remove the air conditioner drive belt and tensioner assembly.
5. Remove the water pump pulley and the crankshaft pulley.
6. Remove the upper and lower timing belt covers.
7. Remove the engine center cover. Remove the breather hose from the rear of the rocker cover. Remove the PCV hose. Disconnect the spark plug cables from the plugs.
8. Remove the rocker cover and rear half-moon seal.
9. Confirm the engine is still at No. 1 TDC. The timing marks on the camshaft sprocket and the upper surface of the cylinder head should coincide. The dowel pin on the front of the camshafts should be in the 12 o'clock position. Remove the automatic belt tensioner. Loosen the tensioner pulley center bolt.
10. If the timing belt is to be reused, mark an arrow, on the belt, in the direction of rotation, for installation reference. Remove the timing belt.
To install:
11. Install the automatic tensioner, after reset.

➡**To reset the tensioner: Keep the adjuster level and clamp it in a soft jawed vise. Clamp with the extended adjuster on one**

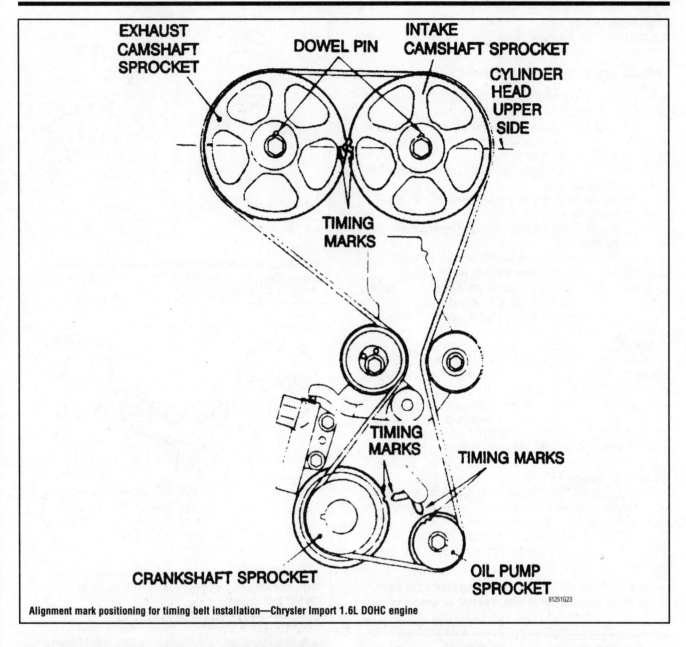

Alignment mark positioning for timing belt installation—Chrysler Import 1.6L DOHC engine

side and the end mounting a plug on the other side. If the plug extends out of the adjuster body, place a suitable hole sized washer over the plug so the vise jaw pushes on the washer, not the plug. Close the vise slowly, forcing the adjuster back into the body. When the hole in the adjuster boss aligns with the adjuster rod, insert a snug fitting pin or wire into the holes to keep the rod in the compressed position. With the locking pin or wire in place, install the tensioner.

12. Align the timing marks on the camshaft sprockets. Align the crankshaft timing marks. Align the oil pump timing marks. Place the timing belt around the intake camshaft and secure it to the sprocket with a stationary binder spring clip. Install the timing belt around the exhaust camshaft sprocket, check sprocket marks for alignment and secure the belt with a second binder clip on the exhaust sprocket.

13. Install the timing belt around the idler pulley, oil pump sprocket, crankshaft sprocket and the tensioner pulley.

14. Lift up the tensioner pulley against the belt and tighten the center bolt to hold it in position.

15. Check to see that all of the timing marks are aligned. Remove the binder clips. Rotate the crankshaft a quarter turn counter clockwise. Then turn the crankshaft clockwise until the timing marks are aligned.

16. Place special tool MD998752 or equivalent, on a torque wrench. Insert the tool into the place provided on the tension pulley. Loosen the center pulley bolt and apply 26 inch lbs. (3.0 Nm) of pressure against the timing belt with the tension pulley. While holding the required torque, tighten the center bolt. Screw in special tool MD998738 or equivalent, through the left engine support bracket until it contacts the tensioner arm bracket. Turn the tool a little more to secure the tensioner and remove the locking wire place into the automatic adjuster when it was reset.

17. Remove the special tool. Rotate the crankshaft 2 complete turns clockwise and allow it to sit for about 15 minutes. Then measure the protrusion of the automatic adjuster. It should be 0.015–0.018 in. (0.381–0.457mm). If the proper amount of protrusion is not present, repeat the tensioning process.

1.8L (LASER/TALON),
2.0L (VIN V) & 2.4L ENGINES

1990–95, Except 1993–95 2.4L Engine

1. Position the engine so the No. 1 piston is at Top Dead Center (TDC) of the compression stroke.

2. Disconnect the negative battery cable. On Summit Wagon with 2.4L engine, remove the coolant reservoir and the power steering and air conditioner hose clamp bolt.

3. Remove the drive belts, tension pulley brackets, water pump pulley and crankshaft pulley.

4. Remove all attaching screws and remove the upper and lower timing belt covers.

5. Remove the timing belt tensioner pulley, tensioner arm, idler pulley and the timing belt.

6. Locate the access plug on the side of block. Remove the plug and install a Phillips screwdriver. Remove the oil pump sprocket nut, oil pump sprocket, special washer, flange and spacer.

7. Remove the outer crankshaft sprocket and flange.

8. Remove the silent shaft (inner) belt tensioner and remove the inner belt.

To install:

9. Align the timing marks of the silent shaft sprockets and the crankshaft sprocket with the timing marks on the front case. Wrap the timing belt around the sprockets so there is no slack in the upper span of the belt and the timing marks are still aligned.

10. Install the tensioner pulley and move the pulley by hand so the long side of the belt deflects 0.20–0.28 in. (5–7mm).

11. Hold the pulley tightly so the pulley cannot rotate when the bolt is tightened. Tighten the bolt to 15 ft. lbs. (20 Nm) and recheck the deflection amount.

12. Install the timing belt tensioner fully toward the water pump and tighten the bolts. Place the upper end of the spring against the water pump body.

13. Align the timing marks of the camshaft, crankshaft and oil pump sprockets with their corresponding marks on the front case or rear cover.

➡ There is a possibility to align all timing marks and have the oil pump sprocket out of time, causing an engine vibra-

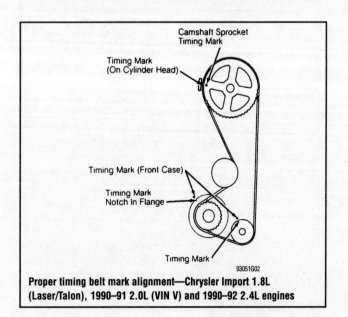

Proper timing belt mark alignment—Chrysler Import 1.8L (Laser/Talon), 1990–91 2.0L (VIN V) and 1990–92 2.4L engines

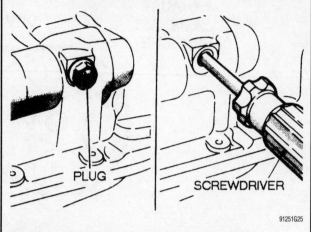

Checking the oil pump sprocket shaft for proper positioning—Chrysler Import 1.8L (Laser/Talon), 1990–91 2.0L (VIN V) and 1990–92 2.4L engines

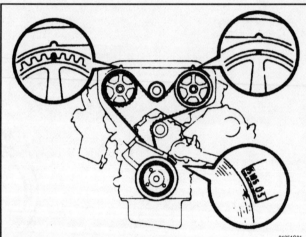

Silent shaft sprocket alignment marks for inner belt replacement—Chrysler Import 1.8L (Laser/Talon), 1990–91 2.0L (VIN V) and 1990–92 2.4L engines

tion during operation. If the following step is not followed exactly, there is a 50 percent chance that the oil pump shaft alignment will be 180 degrees off.

14. Before installing the timing belt, ensure that the oil pump sprocket is in the correct position as follows:

a. Remove the plug from the rear side of the block and insert a Phillips screwdriver with shaft diameter of 0.31 in. (8mm) into the hole.

b. With the timing marks still aligned, the shaft of the tool must be able to go in at least 2.36 in. (60mm). If the tool can only go in 0.79–0.98 in. (20–25mm), the shaft is not in the correct orientation and will cause a vibration during engine operation. Remove the tool from the hole and turn the oil pump sprocket 1 complete revolution. Realign the timing marks and insert the tool. The shaft of the tool must go in at least 2.36 in. (60mm).

c. Recheck and realign the timing marks.

d. Leave the tool in place to hold the oil pump shaft while continuing.

15. Install the belt to the crankshaft sprocket, oil pump sprocket, then camshaft sprocket. While doing so, make sure there is no slack between the sprocket except where the tensioner is installed.

16. Tighten oil pump sprocket bolt to 26–29 ft. lbs. (34–40 Nm) and tighten crankshaft bolt to 80–94 ft. lbs. (110–130 Nm).

17. Recheck the timing mark alignment. If all are aligned, loosen the tensioner mounting bolt and allow the tensioner to apply tension to the belt.

18. Remove the tool that is holding the silent shaft and rotate the crankshaft a distance equal to 2 teeth on the camshaft sprocket. This will allow the tensioner to automatically apply the proper tension on the belt. Do not manually overtighten the belt or it will howl.

19. Tighten the lower mounting bolt first, then the upper spacer bolt.

20. To verify correct belt tension, check that the deflection at the longest span of the belt has 0.40 in. (12mm) clearance from the belt cover.

21. Install the timing belt covers and all related items.

22. Connect the negative battery cable.

1993–95

1. Disconnect the negative battery cable.

2. Remove the timing belt upper and lower covers.

3. Rotate the crankshaft clockwise and align the timing marks so No. 1 piston will be at Top Dead Center (TDC) of the compression stroke. At this time the timing marks on the camshaft sprocket and the upper surface of the cylinder head should coincide, and the dowel pin of the camshaft sprocket should be at the upper side.

➡**Always rotate the crankshaft in a clockwise direction. Make a mark on the back of the timing belt indicating the direction of rotation so it may be reassembled in the same direction if it is to be reused.**

4. Remove the auto tensioner and remove the outermost timing belt.

5. Remove the timing belt tensioner pulley, tensioner arm, idler pulley.

6. Locate the access plug on the side of block. Remove the

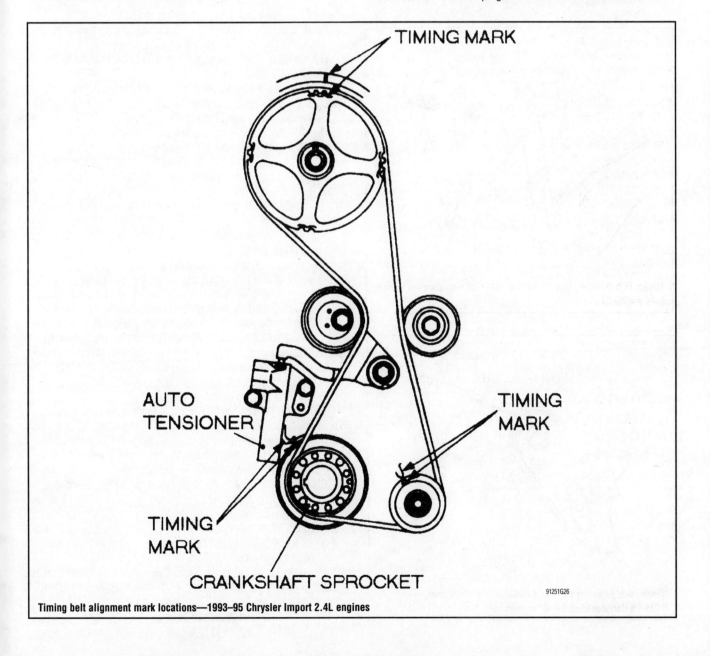

TIMING MARK

AUTO
TENSIONER

TIMING
MARK

TIMING
MARK

CRANKSHAFT SPROCKET

91251G26

Timing belt alignment mark locations—1993–95 Chrysler Import 2.4L engines

plug and install a Phillips screwdriver. Remove the oil pump sprocket nut, oil pump sprocket, special washer, flange and spacer.

7. Remove the silent shaft (inner) belt tensioner and remove the belt.

To install:

8. Align the timing marks on the crankshaft sprocket and the silent shaft sprocket. Fit the inner timing belt over the crankshaft and silent shaft sprocket. Ensure that there is no slack in the belt.

9. While holding the inner timing belt tensioner, adjust the timing belt tension by applying a force towards the center of the belt, until the tension side of the belt is taut. Tighten the tensioner bolt.

➡**When tightening the bolt of the tensioner, ensure that the tensioner pulley shaft does not rotate with the bolt. Allowing it to rotate with the bolt can cause excessive tension on the belt.**

10. Check belt for proper tension by depressing the belt on long side and noting the belt deflection. The desired reading is 0.20–0.28 in. (5–7mm). If tension is not correct, readjust and check belt deflection.

11. Install the flange, crankshaft and washer to the crankshaft. The

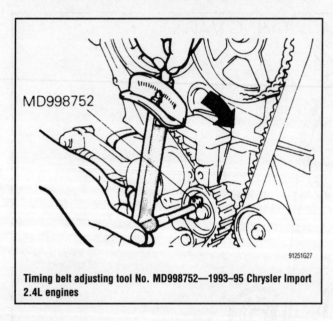

Timing belt adjusting tool No. MD998752—1993–95 Chrysler Import 2.4L engines

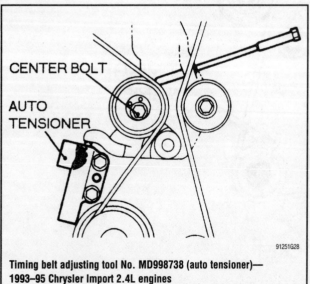

Timing belt adjusting tool No. MD998738 (auto tensioner)— 1993–95 Chrysler Import 2.4L engines

flange on the crankshaft sprocket must be installed towards the inner timing belt sprocket. Tighten bolt to 80–94 ft. lbs. (110–130 Nm).

➡**There is a possibility to align all timing marks and have the oil pump sprocket out of time, causing an engine vibration during operation. If the following step is not followed exactly, there is a 50 percent chance that the oil pump shaft alignment will be 180 degrees off.**

12. Before installing the timing belt, ensure that the oil pump sprocket is in the correct position as follows:

a. Remove the plug from the rear side of the block and insert a Phillips screwdriver with shaft diameter of 0.31 in. (8mm) into the hole.

b. With the timing marks still aligned, the shaft of the tool must be able to go in at least 2.36 in. (60mm). If the tool can only go in 0.79–0.98 in. (20–25mm), the shaft is not in the correct orientation and will cause a vibration during engine operation. Remove the tool from the hole and turn the oil pump sprocket 1 complete revolution. Realign the timing marks and insert the tool. The shaft of the tool must go in at least 2.36 in. (60mm).

c. Recheck and realign the timing marks.

d. Leave the tool in place to hold the silent shaft while continuing.

13. To install the oil pump sprocket and tighten the nut to 36–43 ft. lbs. (50–60 Nm).

14. Position the auto-tensioner into a vise with soft jaws. The plug at the rear of tensioner protrudes, be sure to use a washer as a spacer to protect the plug from contacting vise jaws.

15. Slowly push the rod into the tensioner until the set hole in rod is aligned with set hole in the auto-tensioner.

16. Insert a 0.055 in. (1.4mm) wire into the aligned set holes. Unclamp the tensioner from vise and install to vehicle. Tighten tensioner to 17 ft. lbs. (24 Nm).

17. When installing timing belt, the camshaft sprocket dowel pin should be located on top. Align all timing marks.

18. Align the crankshaft sprocket, camshaft sprocket and oil pump sprocket timing marks.

19. Install the timing belt as follows:

a. Install the timing belt around the idler pulley, oil pump sprocket, crankshaft sprocket, camshaft and the tensioner pulley.

b. Lift upward on the tensioner pulley in a clockwise direction and tighten the center bolt. Make sure all timing marks are aligned.

c. Rotate the crankshaft ¼ turn counterclockwise. Then, turn in clockwise until the timing marks are aligned again.

20. Loosen the center bolt. Using tool No. MD998752 or equivalent and a torque wrench, apply a torque of 22–24 inch lbs. (2.6–2.8 Nm). Tighten the center bolt.

21. Screw the tool No. MD998738 into the engine left support bracket until its end makes contact with the tensioner arm and tighten tensioner pulley to 35 ft. lbs. (48 Nm). At this point, screw the special tool in some more and remove the set wire attached to the auto tensioner. Then remove the special tool.

22. Rotate the crankshaft 2 complete turns clockwise and let it sit for approximately 15 minutes. Then, measure the auto tensioner protrusion (the distance between the tensioner arm and auto tensioner body) to ensure that it is within 0.15–0.18 in. (3.8–4.5mm). If out of specification, repeat belt adjustment procedure until the specified value is obtained.

23. If the timing belt tension adjustment is being performed with the engine mounted in the vehicle, and clearance between the tensioner arm and the auto tensioner body cannot be measured, the following alternative method can be used:

a. Screw in tool No. MD998738 or equivalent, until its end makes contact with the tensioner arm.

b. After the tool makes contact with the arm, screw it in some more to retract the auto tensioner pushrod while counting the number of turns the tool makes until the tensioner arm is brought into contact with the auto tensioner body. Make sure the number of turns the tool makes conforms with the standard value of 2.5–3 turns.

c. Install the rubber plug to the timing belt rear cover.

24. Install the timing belt covers and all related items.

25. Connect the negative battery cable.

1996–00

1. Rotate the crankshaft so that the No. 1 piston is at Top Dead Center (TDC) on its compression stroke.

2. Remove the timing belt covers.

3. To loosen the timing (outer) belt tensioner, install Special Tool MD998738 or equivalent, to the slot, then screw inward to move the tensioner toward the water pump. Once the tension has been relieved, remove the outer timing belt.

→If the timing belts are going to be reused, mark the direction of their rotation on the belts. This will ensure the belt is reinstalled in same direction, extending belt life.

4. Remove the outer crankshaft sprocket and flange.

5. Loosen the silent shaft (inner) belt tensioner and remove the belt.

To install:

6. Turn both tensioner pulleys and check for any signs of bearing wear.

7. Align the timing marks of the silent shaft sprockets and the crankshaft sprocket with the timing marks on the front case. Route the timing belt around the sprockets so there is no slack in the upper span of the belt and the timing marks are still aligned.

8. Install the tensioner pulley and move the pulley by hand so the long side of the belt deflects approximately ¼ in. (6mm).

9. Hold the pulley tightly so the pulley cannot rotate when the bolt is tightened. Tighten the bolt to 14 ft. lbs. (19 Nm) and recheck the deflection amount.

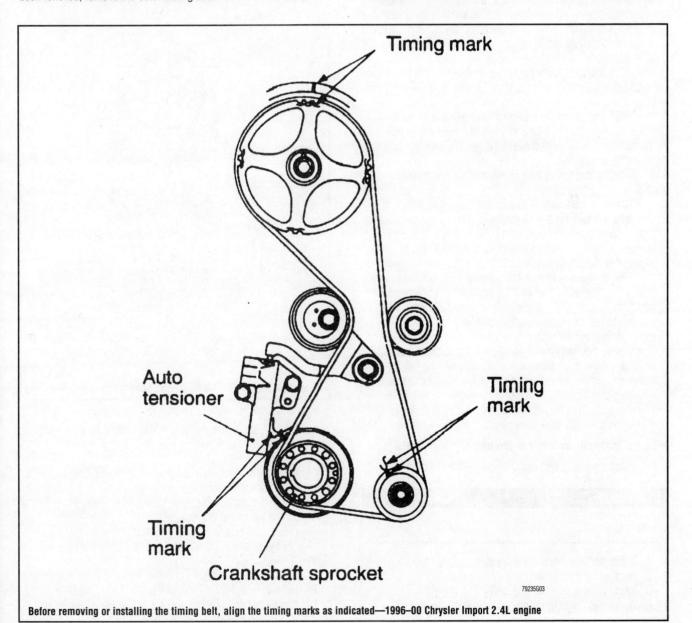

Timing mark

Auto tensioner

Timing mark

Timing mark

Crankshaft sprocket

79235G03

Before removing or installing the timing belt, align the timing marks as indicated—1996–00 Chrysler Import 2.4L engine

10. Align the timing marks of the camshaft, crankshaft and oil pump sprockets with their corresponding marks on the front case or rear cover.

➡There is a possibility to align all timing marks and have the oil pump sprocket and silent shaft out of time, causing an engine vibration during operation. If the following step is not followed exactly, there is a 50 percent chance that the silent shaft alignment will be 180 degrees (½ turn) off.

11. Before installing the timing belt, ensure that the left-side (rear) silent shaft (oil pump sprocket) is in the correct position as follows:

    a. Remove the plug from the rear side of the block and insert a tool with an outer shaft diameter of 0.31 in. (8mm) into the hole.

    b. With the timing marks still aligned, the shaft of the tool must be able to go in at least 2½ in. (63.5mm). If the tool can only go in approximately 1 in. (25mm), the silent shaft is not in the correct orientation and will cause a vibration during engine operation. Remove the tool from the hole and turn the oil pump sprocket 1 complete revolution. Realign the timing marks and insert the tool. The shaft of the tool should now go in at least 2½ in. (63.5mm)

    c. Recheck and realign the timing marks.

    d. Leave the tool in place to hold the silent shaft while continuing.

12. Install the belt on the crankshaft sprocket, the oil pump sprocket, then the camshaft sprocket—in that order. While doing so, be sure there is no slack between the sprockets, except where the tensioner is installed.

13. To adjust the timing (outer) belt perform the following steps:

    a. Turn the crankshaft ¼ turn counterclockwise, then turn it clockwise to move the No. 1 cylinder to TDC.

    b. Loosen the center bolt. Using tool MD998752 or equivalent, and a torque wrench, apply 31 inch lbs. (3.6 Nm) to the tensioner, then tighten the center bolt.

    c. Thread the special tool into the engine left support bracket until its end makes contact with the tensioner arm. At this point, thread the special tool in some more and remove the set wire attached to the auto-tensioner, if the wire was not previously removed. Remove the tool.

    d. Rotate the crankshaft 2 complete turns clockwise and let it sit for approximately 15 minutes. Then, measure the auto-tensioner protrusion (the distance between the tensioner arm and auto-tensioner body) to ensure that it is within 0.15–0.18 in. (3.8–4.5mm). If out of specification, repeat Substeps **a** through **d** until the specified value is obtained.

➡Do not manually overtighten the belt or it will howl.

14. Install the timing belt covers and all related items.

Daihatsu

1.0L ENGINE

1. Disconnect the negative battery cable.
2. Remove the air cleaner.
3. Remove the accessory drive belts.
4. Remove the right front engine undercover.

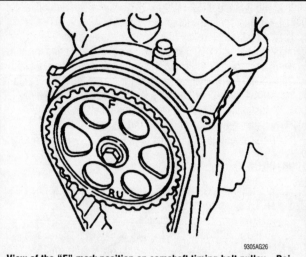

View of the "F" mark position on camshaft timing belt pulley—Daihatsu 1.0L engine—Charade

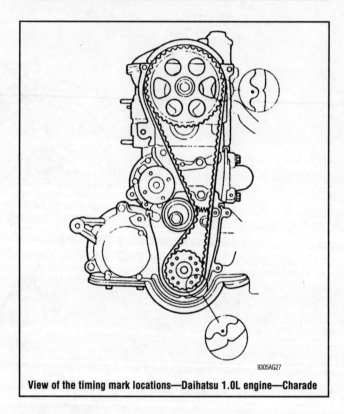

View of the timing mark locations—Daihatsu 1.0L engine—Charade

5. Support the engine with a jack and remove the right motor mount and bracket.
6. Remove the water pump pulley.
7. Remove the crankshaft pulley and remove the timing covers. Remove the crankshaft timing pulley flange.
8. Rotate the crankshaft so that the **F** on the camshaft pulley is facing up and the timing marks align.
9. Loosen the tensioner pulley lockbolt.
10. Pry the tensioner pulley away from the timing belt and tighten the lockbolt.
11. Remove the timing belt.

To install:

12. Ensure that the camshaft and crankshaft timing marks are

aligned with the timing marks on the engine and install the timing belt.

13. Loosen the tensioner lockbolt and allow it to apply tension to the timing belt. Tighten the tensioner lockbolt.

14. Rotate the crankshaft 2 complete turns clockwise and ensure that the timing marks align.

15. Loosen the tensioner lockbolt and allow the tensioner to take up any remaining slack in the timing belt. Tighten the tensioner lockbolt to 25–33 ft. lbs. (33–44 Nm).

16. Install the crankshaft timing pulley flange and the timing covers.

17. Install the crankshaft pulley and tighten the bolt to 65–72 ft. lbs. (88–98 Nm).

18. Install the right motor mount and bracket. Remove the jack.

19. Install the accessory drive belts and the engine undercover.

20. Install the air cleaner.

21. Connect the negative battery cable.

1.3L ENGINE

1. Disconnect the negative battery cable.
2. Remove the accessory drive belts.
3. If equipped with air conditioning, perform the following:
   a. Drain the cooling system.
   b. Remove the radiator and cooling fan.
   c. Remove the idler pulley assembly.
   d. Remove the air conditioning compressor and position aside.
4. If equipped, remove the power steering pump and position aside.
5. Remove the air cleaner.

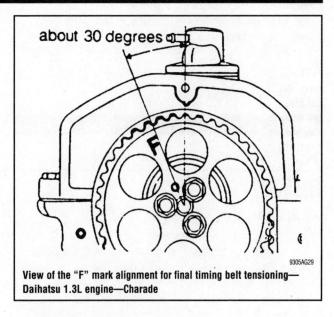

about 30 degrees

View of the "F" mark alignment for final timing belt tensioning— Daihatsu 1.3L engine—Charade

9305AG29

6. Disconnect the oil pressure switch connector. Unbolt and remove the oil pressure switch wiring harness.

7. Remove the water pump pulley.

8. Support the engine with a jack. Remove the right engine mount and bracket.

9. Remove the right inner fender service cover.

10. Remove the crankshaft pulley.

11. Remove the upper and lower timing belt covers.

12. Rotate the crankshaft so that the **F** on the camshaft pulley is facing up and the timing marks align.

13. Loosen the tensioner pulley lockbolt.

14. Pry the tensioner pulley away from the timing belt and tighten the lockbolt.

15. Remove the timing belt.

To install:

16. Ensure that the camshaft and crankshaft timing marks are aligned with the timing marks on the engine and install the timing belt.

17. Loosen the tensioner lockbolt and allow it to apply tension to the timing belt. Tighten the tensioner lockbolt.

18. Rotate the crankshaft clockwise until the **F** mark on the camshaft pulley is 3 teeth away from aligning with the indicator mark on the engine.

19. Loosen the tensioner lockbolt.

20. Continue to rotate the crankshaft clockwise until the **F** mark aligns with the indicator mark. Tighten the tensioner lockbolt to 22–33 ft. lbs. (29–44 Nm).

21. Rotate the crankshaft 2 complete turns clockwise and ensure that the timing marks align.

22. Install the upper and lower timing covers.

23. Install the crankshaft pulley. Tighten the bolts to 15–22 ft. lbs. (20–29 Nm).

24. Install the right inner fender service cover.

25. Install the right engine mount and bracket. Remove the engine support jack.

26. Install the water pump pulley.

27. Route the oil pressure switch wiring harness through the hole in the right engine mount and install the clamp bolt. Connect the oil pressure switch connector.

28. Install the power steering pump, if removed.

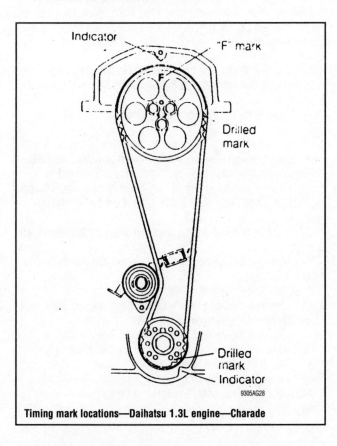

Indicator   "F" mark

Drilled mark

Drilled mark
Indicator

9305AG28

Timing mark locations—Daihatsu 1.3L engine—Charade

29. Install the air conditioning compressor and idler pulley assembly, if removed.

30. Install the radiator and cooling fan, if removed. Fill the cooling system.

31. Install the air cleaner.

32. Connect the negative battery cable.

Honda

4-CYLINDER ENGINES

1980–1995

WITHOUT BALANCE SHAFTS

1. Disconnect the negative battery cable.

2. Rotate the crankshaft to set the engine at Top Dead Center (TDC) on the compression stroke for the No. 1 piston.

3. Raise and support the vehicle.

4. Remove the left front wheel and splash shield.

5. Remove the valve cover.

6. Remove the upper timing cover.

7. Remove the accessory drive belts.

8. If equipped, remove the power steering pump and bracket.

9. If equipped with air conditioning, remove the tensioner pulley and bracket. For Preludes, remove the air conditioning compressor and bracket.

10. If equipped with an accessory belt-driven water pump, remove the water pump pulley.

11. Remove the crankshaft pulley.

12. Remove the lower timing cover.

13. If equipped with an engine side-mount, support the engine and remove the mount.

14. Loosen the tensioner adjusting and pivot bolts and remove the timing belt.

To install:

15. Insure that the camshaft and crankshaft timing marks are aligned and install the timing belt.

➡**Some engines do not have an UP indication on the camshaft pulley. In these cases, align the camshaft marks with the woodruff key facing up.**

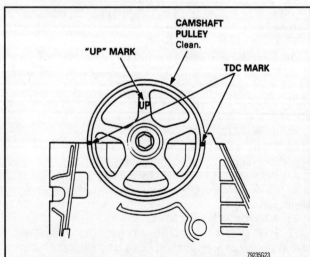

Single camshaft timing belt sprocket TDC mark positioning for timing belt installation—Honda 1980–97 4-cylinder SOHC engines

16. Rotate the crankshaft counterclockwise ¼ turn and tighten the tensioner adjustment bolt.

17. Tighten the tensioner pivot bolt.

18. Install the lower timing cover and the crankshaft pulley.

19. Install the water pump pulley, if removed.

20. If removed, install the air conditioning compressor and bracket.

21. If removed, install the engine side-mount, the power steering pump and bracket, and the air conditioning tensioner pulley and bracket.

22. Install the accessory drive belts and the upper timing cover.

23. Install the valve cover.

24. Install the splash shield and the left front wheel.

25. Lower the vehicle.

26. Connect the negative battery cable.

WITH BALANCE SHAFTS

1. Disconnect the negative battery cable.

2. Turn the crankshaft to align the timing belt matchmarks and set cylinder No. 1 to Top Dead Center (TDC) on the compression stroke. Once in this position, the engine must NOT be turned or disturbed.

3. Remove all necessary components to gain access to the cylinder head and timing belt covers.

4. Remove the cylinder head and timing belt covers.

5. There are 2 belts in this system; the one running to the camshaft pulley is the timing belt. The other, shorter one drives the balance shafts and is referred to as the balancer belt or timing balancer belt. Lock the timing belt adjuster in position by installing one of the lower timing belt cover bolts to the adjuster arm.

6. Loosen the timing belt and balancer shafts tensioner adjuster nut, but do not loosen the nut more than 1 turn. Push the tensioner for the balancer belt away from the belt to relieve the tension. Hold the tensioner and tighten the adjusting nut to hold the tensioner in place.

7. Carefully remove the balancer belt. Do not crimp or bend the belt; protect it from contact with oil or coolant. Slide the belt off the pulleys.

8. Remove the balancer belt drive sprocket from the crankshaft.

9. Loosen the lockbolt installed to the timing belt adjuster and loosen the adjusting nut. Push the timing belt adjuster to remove the tension on the timing belt, then tighten the adjuster nut.

10. Remove the timing belt. Do not crimp or bend the belt; protect it from contact with oil or coolant. Slide the belt off the pulleys.

11. If defective, remove the belt tensioners by performing the following:

   a. Remove the springs from the balancer belt and the timing belt tensioners.

   b. Remove the adjusting nut.

   c. Remove the bolt from the balancer belt adjuster lever, then remove the lever and the tensioner pulley.

   d. Remove the lockbolt from the timing belt tensioner lever, then remove the tensioner pulley and lever from the engine.

12. This is an excellent time to check or replace the water pump. Even if the timing belt is only being replaced as part of a good maintenance schedule, consider replacing the pump at the same time.

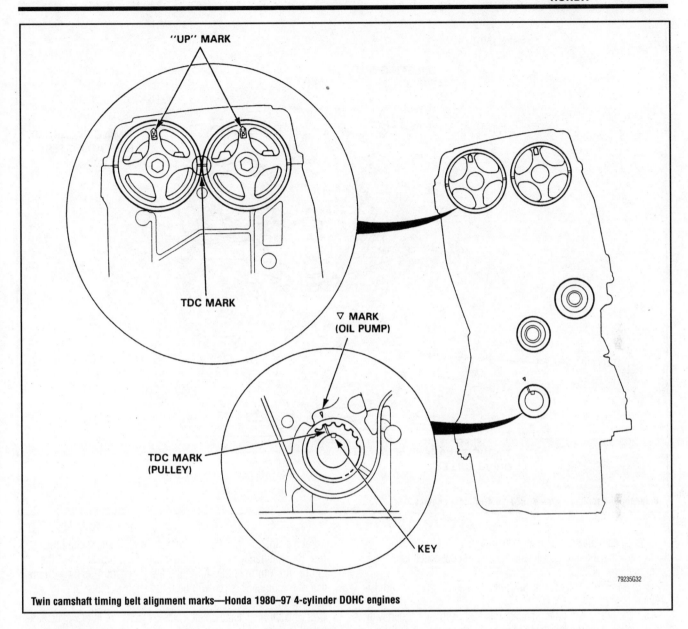

Twin camshaft timing belt alignment marks—Honda 1980–97 4-cylinder DOHC engines

To install:

13. If the water pump is to be replaced, install a new O-ring and make certain it is properly seated. Install the water pump and retaining bolts. Tighten the mounting bolts to 106 inch lbs. (12 Nm).

14. If the tensioners were removed, perform the following to install them:

a. Install the timing belt tensioner lever and tensioner pulley.

➡**The tensioner lever must be properly positioned on its pivot pin located on the oil pump. Be sure that the timing belt lever and tensioner moves freely and does not bind.**

b. Install the lockbolt to the timing belt tensioner, do not tighten the lockbolt at this time.

c. Install the balancer belt pulley and adjuster lever.

d. Install the adjusting nut and the bolt to the balancer belt adjuster lever. Do not tighten the adjuster nut or bolt at this time.

➡**Be sure that the balancer lever and tensioner moves freely and does not bind.**

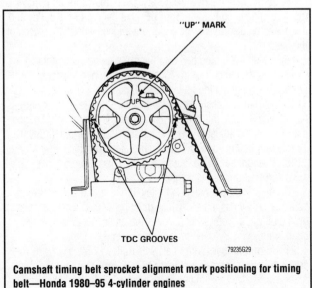

Camshaft timing belt sprocket alignment mark positioning for timing belt—Honda 1980–95 4-cylinder engines

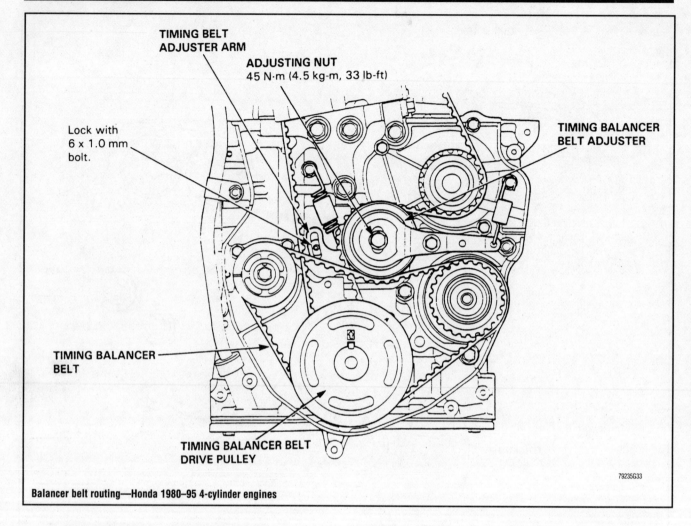

TIMING BELT
ADJUSTER ARM

ADJUSTING NUT
45 N·m (4.5 kg-m, 33 lb-ft)

Lock with
6 x 1.0 mm
bolt.

TIMING BALANCER
BELT ADJUSTER

TIMING BALANCER
BELT

TIMING BALANCER BELT
DRIVE PULLEY

Balancer belt routing—Honda 1980–95 4-cylinder engines

79235G33

e. Install the springs to the tensioners.

f. Move the timing belt tensioner its full deflection and tighten the lockbolt.

g. Move the balancer its full deflection and tighten the adjusting nut.

15. The crankshaft timing pointer must be perfectly aligned with the white mark on the flywheel or flex-plate; the camshaft pulley must be aligned so that the word **UP** is at the top of the pulley and the marks on the edge of the pulley are aligned with the surfaces of the head.

16. Install the timing belt over the pulleys and tensioners.

17. Loosen the bolt used to lock the timing belt tensioner. Loosen, then tighten the timing belt adjusting nut.

18. Turn the crankshaft counterclockwise until the cam pulley has moved 3 teeth; this creates tension on the timing belt. Loosen, then tighten the adjusting nut and tighten it to 33 ft. lbs. (45 Nm). Tighten the bolt used to lock the timing belt tensioner.

19. Realign the timing belt marks, then install the balancer belt drive sprocket on the crankshaft.

20. Align the front balancer pulley; the face of the front timing balancer pulley has a mark, which must be aligned with the notch on the oil pump body. This pulley is the one at 10 o'clock to the crank pulley when viewed from the pulley end.

21. Align the rear timing balancer pulley (2 o'clock from the crank pulley) using a 6 x 100mm bolt or rod. Mark the bolt or rod at a point 2.9 in. (74mm) from the end. Remove the bolt from the maintenance hole on the side of the block; insert the bolt or rod into the hole. Align the 2.9 in. (74mm) mark with the face of the hole. This pin will hold the shaft in place during installation.

22. Install the balancer belt. Once the belts are in place, be sure that all the engine alignment marks are still correct. If not, remove the belts, realign the engine and reinstall the belts. Once the belts are properly installed, slowly loosen the adjusting nut, allowing the tensioner to move against the belt. Remove the pin from the maintenance hole and reinstall the bolt and washer.

23. Turn the crankshaft 1 full turn, then tighten the adjuster nut to 33 ft. lbs. (45 Nm). Remove the bolt used to lock the timing belt tensioner.

24. Install the lower cover, ensuring the rubber seals are in place. Install a new seal around the adjusting nut, DO NOT loosen the adjusting nut.

25. Install the key on the crankshaft and install the crankshaft pulley. Apply oil to the bolt threads and tighten it to 181 ft. lbs. (250 Nm).

26. Install the upper timing belt cover and all applicable components. When installing the side engine mount, tighten the bolt and nut attaching the mount to the engine to 40 ft. lbs. (55 Nm) and the through-bolt and nut to 47 ft. lbs. (65 Nm).

27. Connect the negative battery cable.

1.6L ENGINES

1996–97

1. Disconnect the negative battery cable.
2. Rotate the crankshaft to set the engine at Top Dead Center (TDC) on the compression stroke for the No. 1 piston. The white mark on the crankshaft pulley should align with the pointers on the timing cover. Once the engine is in this position, it must not be disturbed.
3. Remove all necessary components for access to the cylinder head and timing belt covers. Cover the rocker arm and shaft assemblies with a towel or sheet of plastic to keep out dust and foreign objects.
4. Remove the timing belt covers.
5. Loosen the timing belt adjusting bolt 180 degrees (½ turn). Push the tensioner pulley down to release the belt tension. After releasing the tension, retighten the tensioner pulley bolt until snug.

➡**Do not remove the tensioner pulley unless it is to be replaced.**

6. Remove the timing belt. Mark the direction of the belt's rotation if it is to be reinstalled.

To install:

➡**Inspect the water pump when replacing the timing belt; the manufacturer recommends replacing the water pump at the timing belt's service interval. Replace the timing belt if it shows any signs of wear, or if it is contaminated with oil or coolant.**

7. Verify that the timing is set at TDC on the compression stroke for the No. 1 cylinder as follows:
   a. The groove in the crankshaft sprocket must align with the pointer on the oil pump.
   b. The TDC marks on the camshaft sprockets must align with the pointer located between the sprockets. The TDC marks will also be in line with the upper surface of the head.
   c. On other engines, the TDC mark on the camshaft sprocket must align with the pointer on the back cover.
   d. The **UP** mark on the camshaft sprocket must point up.
8. Install the timing belt onto the crankshaft sprocket, then around the adjusting pulley and water pump sprocket, and finally over the camshaft sprocket.
9. Loosen the adjusting pulley bolt 180 degrees (½ turn). Then, tighten the adjusting bolt to 40 ft. lbs. (55 Nm).
10. Install the lower timing belt cover and the crankshaft pulley. Apply a light coat of fresh oil to the pulley bolt threads, then tighten it to 134 ft. lbs. (181 Nm).
11. Rotate the crankshaft 5–6 turns counterclockwise to position the belt on the sprockets.
12. Adjust the timing belt tension, as follows:
   a. Set the No. 1 piston at TDC on the compression stroke for the No. 1 cylinder.
   b. Loosen the adjusting pulley bolt 180 degrees (½ turn).
   c. Rotate the crankshaft counterclockwise so that the camshaft sprocket moves 3 teeth from the TDC/compression position.
   d. Tighten the adjusting bolt to 33 ft. lbs. (45 Nm).
   e. Tighten the crankshaft pulley to 134 ft. lbs. (181 Nm).
13. Verify that the crankshaft and camshaft sprockets will align properly at the TDC/compression position. If the camshaft pulley is

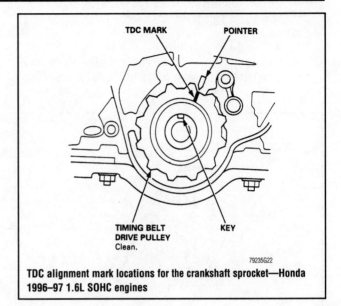

TDC alignment mark locations for the crankshaft sprocket—Honda 1996–97 1.6L SOHC engines

not at TDC/compression, remove the timing belt, adjust the sprocket positions and reinstall the belt.
14. Install the upper timing and cylinder head covers, and all other applicable components. When reattaching the side engine mount, tighten the support nuts to 54 ft. lbs. (75 Nm).
15. Connect the negative battery cable.

2.2L (F22A1) ENGINES

1. Disconnect the negative battery cable.
2. Turn the crankshaft to align the timing belt matchmarks and set cylinder No. 1 to Top Dead Center (TDC) on the compression stroke. Once in this position, the engine must NOT be turned or disturbed.
3. Remove all necessary components to gain access to the cylinder head and timing belt covers.
4. Remove the cylinder head and timing belt covers.
5. There are 2 belts in this system; the one running to the camshaft pulley is the timing belt. The other, shorter one drives the balance shafts and is referred to as the balancer belt or timing balancer belt. Lock the timing belt adjuster in position by installing one of the lower timing belt cover bolts to the adjuster arm.
6. Loosen the timing belt and balancer shafts tensioner adjuster nut, but do not loosen the nut more than 1 turn. Push the tensioner for the balancer belt away from the belt to relieve the tension. Hold the tensioner and tighten the adjusting nut to hold the tensioner in place.
7. Carefully remove the balancer belt. Do not crimp or bend the belt; protect it from contact with oil or coolant. Slide the belt off the pulleys.
8. Remove the balancer belt drive sprocket from the crankshaft.
9. Loosen the lockbolt installed to the timing belt adjuster and loosen the adjusting nut. Push the timing belt adjuster to remove the tension on the timing belt, then tighten the adjuster nut.
10. Remove the timing belt. Do not crimp or bend the belt; protect it from contact with oil or coolant. Slide the belt off the pulleys.
11. If defective, remove the belt tensioners by performing the following:

a. Remove the springs from the balancer belt and the timing belt tensioners.

b. Remove the adjusting nut.

c. Remove the bolt from the balancer belt adjuster lever, then remove the lever and the tensioner pulley.

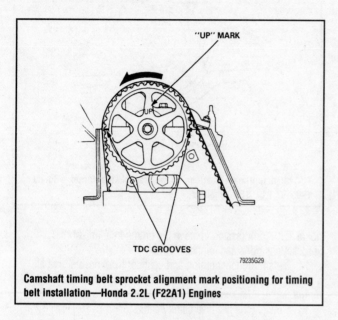

Camshaft timing belt sprocket alignment mark positioning for timing belt installation—Honda 2.2L (F22A1) Engines

d. Remove the lockbolt from the timing belt tensioner lever, then remove the tensioner pulley and lever from the engine.

12. This is an excellent time to check or replace the water pump. Even if the timing belt is only being replaced as part of a good maintenance schedule, consider replacing the pump at the same time.

To install:

13. If the water pump is to be replaced, install a new O-ring and make certain it is properly seated. Install the water pump and retaining bolts. Tighten the mounting bolts to 106 inch lbs. (12 Nm).

14. If the tensioners were removed, perform the following to install them:

a. Install the timing belt tensioner lever and tensioner pulley.

➡The tensioner lever must be properly positioned on its pivot pin located on the oil pump. Be sure that the timing belt lever and tensioner moves freely and does not bind.

b. Install the lockbolt to the timing belt tensioner, do not tighten the lockbolt at this time.

c. Install the balancer belt pulley and adjuster lever.

d. Install the adjusting nut and the bolt to the balancer belt adjuster lever. Do not tighten the adjuster nut or bolt at this time.

➡Be sure that the balancer lever and tensioner moves freely and does not bind.

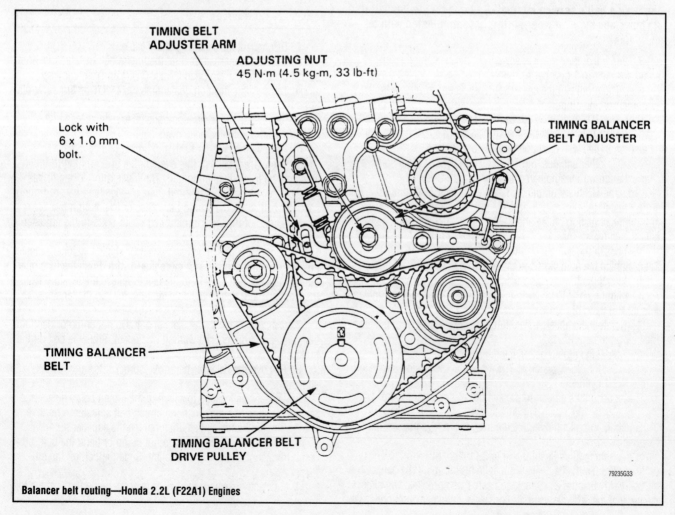

Balancer belt routing—Honda 2.2L (F22A1) Engines

e. Install the springs to the tensioners.

f. Move the timing belt tensioner its full deflection and tighten the lockbolt.

g. Move the balancer its full deflection and tighten the adjusting nut.

15. The crankshaft timing pointer must be perfectly aligned with the white mark on the flywheel or flex-plate; the camshaft pulley must be aligned so that the word **UP** is at the top of the pulley and the marks on the edge of the pulley are aligned with the surfaces of the head.

16. Install the timing belt over the pulleys and tensioners.

17. Loosen the bolt used to lock the timing belt tensioner. Loosen, then tighten the timing belt adjusting nut.

18. Turn the crankshaft counterclockwise until the cam pulley has moved 3 teeth; this creates tension on the timing belt. Loosen, then tighten the adjusting nut and tighten it to 33 ft. lbs. (45 Nm). Tighten the bolt used to lock the timing belt tensioner.

19. Realign the timing belt marks, then install the balancer belt drive sprocket on the crankshaft.

20. Align the front balancer pulley; the face of the front timing balancer pulley has a mark, which must be aligned with the notch on the oil pump body. This pulley is the one at 10 o'clock to the crank pulley when viewed from the pulley end.

21. Align the rear timing balancer pulley (2 o'clock from the crank pulley) using a 6 x 100mm bolt or rod. Mark the bolt or rod at a point 2.9 in. (74mm) from the end. Remove the bolt from the maintenance hole on the side of the block; insert the bolt or rod into the hole. Align the 2.9 in. (74mm) mark with the face of the hole. This pin will hold the shaft in place during installation.

22. Install the balancer belt. Once the belts are in place, be sure that all the engine alignment marks are still correct. If not, remove the belts, realign the engine and reinstall the belts. Once the belts are properly installed, slowly loosen the adjusting nut, allowing the tensioner to move against the belt. Remove the pin from the maintenance hole and reinstall the bolt and washer.

23. Turn the crankshaft 1 full turn, then tighten the adjuster nut to 33 ft. lbs. (45 Nm). Remove the bolt used to lock the timing belt tensioner.

24. Install the lower cover, ensuring the rubber seals are in place. Install a new seal around the adjusting nut, DO NOT loosen the adjusting nut.

25. Install the key on the crankshaft and install the crankshaft pulley. Apply oil to the bolt threads and tighten it to 181 ft. lbs. (250 Nm).

26. Install the upper timing belt cover and all applicable components. When installing the side engine mount, tighten the bolt and nut attaching the mount to the engine to 40 ft. lbs. (55 Nm) and the through-bolt and nut to 47 ft. lbs. (65 Nm).

27. Connect the negative battery cable.

2.2L (F22B1 & F22B2) ENGINES

1. Disconnect the negative battery cable.

2. Remove the cylinder head (valve) and upper timing belt covers.

3. Turn the engine to align the timing marks and set cylinder No. 1 to Top Dead Center (TDC) on the compression stroke. The white mark on the crankshaft sprocket should align with the pointer on the timing belt cover. The words **UP** embossed on the camshaft sprocket should be aligned in the upward position. The marks on the edge of the sprocket should be aligned with the cylinder head or

the back cover upper edge. Once in this position, the engine must NOT be turned or disturbed.

4. Remove all necessary components for access to the lower timing belt cover, then remove the cover.

5. There are 2 belts in this system; the one running to the camshaft sprocket is the timing belt. The other, shorter one drives the balance shaft and is referred to as the balancer shaft belt or timing balancer belt. Lock the timing belt adjuster in position, by installing one of the lower timing belt cover bolts to the adjuster arm.

6. Loosen the timing belt and balancer shafts tensioner adjuster nut, do not loosen the nut more than 1 turn. Push the tensioner for the balancer belt away from the belt to relieve the tension. Hold the tensioner and tighten the adjusting nut to hold the tensioner in place.

7. Carefully remove the balancer belt. Do not crimp or bend the belt; protect it from contact with oil or coolant.

8. Remove the balancer belt sprocket from the crankshaft.

9. Loosen the lockbolt installed to the timing belt adjuster and loosen the adjusting nut. Push the timing belt adjuster to remove the tension on the timing belt, then tighten the adjuster nut.

10. Remove the timing belt by sliding it off the sprockets. Do not crimp or bend the belt; protect it from contact with oil or coolant.

11. If defective, remove the belt tensioners by performing the following:

a. Remove the springs from the balancer belt and the timing belt tensioners.

b. Remove the adjusting nut from the belt tensioners.

c. Remove the bolt from the balancer belt adjuster lever, then remove the lever and the tensioner pulley.

d. Remove the lockbolt from the timing belt tensioner lever, then remove the tensioner pulley and lever from the engine.

12. This is an excellent time to check or replace the water pump. Even if the timing belt is only being replaced as part of a good maintenance schedule, consider replacing the pump at the same time.

To install:

13. If the water pump is to be replaced, install a new O-ring and make certain it is properly seated. Install the water pump and tighten the mounting bolts to 106 inch lbs. (12 Nm).

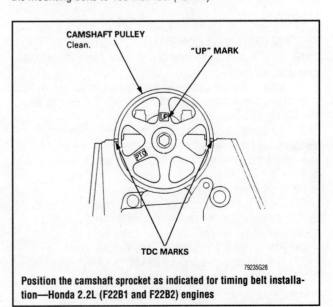

Position the camshaft sprocket as indicated for timing belt installation—Honda 2.2L (F22B1 and F22B2) engines

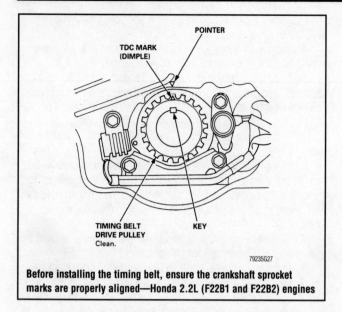

POINTER

TDC MARK
(DIMPLE)

TIMING BELT
DRIVE PULLEY
Clean.

KEY

79235G27

Before installing the timing belt, ensure the crankshaft sprocket marks are properly aligned—Honda 2.2L (F22B1 and F22B2) engines

14. If the tensioners were removed, perform the following procedures:

    a. Install the timing belt tensioner lever and the tensioner pulley.

    b. Install the balancer belt pulley and adjuster lever.

    c. Install the adjusting nut and the bolt to the balancer belt adjuster lever.

    d. Install the springs to the tensioners.

    e. Install the lockbolt to the timing belt tensioner, then move it its full deflection and tighten the lockbolt.

    f. Move the balancer it's full deflection and tighten the adjusting nut to hold its position.

15. The pointer on the crankshaft sprocket should be aligned with the pointer on the oil pump. The camshaft sprocket must be aligned so that the word **UP** is at the top of the sprocket and the marks on the edge of the sprocket are aligned with the surfaces of the head or the back cover upper edge.

16. Install the timing belt on the sprockets in the following sequence: crankshaft sprocket, tensioner sprocket, water pump sprocket and camshaft sprocket.

17. Check the timing marks to be sure that they did not move.

18. Loosen, then retighten the timing belt adjusting nut; this will apply the proper amount of tension to the timing belt.

19. Install the timing balancer belt drive sprocket and the lower timing belt cover.

20. Install the crankshaft pulley and bolt, tighten the bolt to 181 ft. lbs. (245 Nm). Rotate the crankshaft sprocket 5–6 turns to position the timing belt on the sprockets.

21. Set the No. 1 cylinder to TDC and loosen the timing belt adjusting nut 1 turn. Turn the crankshaft counterclockwise until the cam sprocket has moved 3 teeth; this creates tension on the timing belt.

22. Tighten the timing belt adjusting nut.

23. Set the crankshaft sprocket and the camshaft sprocket to TDC. If the sprockets do not align, remove the belt to realign the marks, then install the belt.

24. Remove the crankshaft pulley and the lower cover.

25. With the timing marks aligned, lock the timing belt adjuster in place with one of the lower cover mounting bolts.

26. Loosen the adjusting nut and ensure the timing balancer belt adjuster moves freely.

27. Align the rear timing balancer sprocket using a 6 x 100mm bolt or rod. Mark the bolt or rod at a point 2.9 in. (74mm) from the end. Remove the bolt from the maintenance hole on the side of the block; insert the bolt/rod into the hole and align the 2.9 in. (74mm) mark with the face of the hole. This will hold the shaft in place during installation.

28. Align the groove on the front balancer shaft sprocket with the pointer on the oil pump.

29. Install the balancer belt. Once the belts are in place, be sure that all the engine alignment marks are still correct. If not, remove the belts, realign the engine and reinstall the belts. Once the belts are properly installed, slowly loosen the adjusting nut, allowing the tensioner to move against the belt. Remove the bolt from the maintenance hole and reinstall the bolt and washer.

30. Install the crankshaft pulley, then turn the crankshaft sprocket 1 turn counterclockwise and tighten the timing belt adjusting nut to 33 ft. lbs. (45 Nm).

31. Remove the crankshaft pulley and the bolt locking the timing belt adjuster in place.

32. Install the lower and upper timing belt covers, and all applicable components. When installing the crankshaft pulley, coat the threads and seating face of the pulley bolt with engine oil, then install and tighten the bolt to 181 ft. lbs. (250 Nm).

33. Install the cylinder head cover gasket cover to the groove of the cylinder head cover. Before installing the gasket thoroughly clean the seal and the groove. Seat the recesses for the camshaft first, then work it into the groove around the outside edges. Be sure the gasket is seated securely in the corners of the recesses.

34. Apply liquid gasket to the 4 corners of the recesses of the cylinder head cover gasket. Do not install the parts if 5 minutes or more have elapsed since applying liquid gasket. After assembly, wait at least 20 minutes before filling the engine with oil.

35. Install the cylinder head (valve) cover and all other applicable components.

36. Connect the negative battery cable.

2.2L (H22A1) ENGINE

1. Disconnect the negative battery cable.

2. Turn the crankshaft so the No. 1 piston is at Top Dead Center (TDC) on the compression stroke. The No. 1 piston is TDC when the pointer on the block aligns with the white painted mark on the driveplate.

3. Remove all necessary components for access to the cylinder head and upper timing belt covers. Then, remove the covers.

4. Ensure the words **UP** embossed on the camshaft pulleys are aligned in the upward position.

5. Support the engine with a floor jack below the center of the center beam. Tension the jack so that it is just supporting the beam but not lifting it. Remove the 2 rear bolts from the center beam to allow the engine to drop down for clearance to remove the lower cover.

6. Remove and discard the rubber seal from the timing belt adjuster. Do not loosen the adjusting nut.

7. Remove the lock pin from the maintenance bolt.

8. Remove the lower timing belt cover.

9. There are 2 belts in this system; the one running to the

camshaft pulley is the timing belt. The other, shorter one drives the balance shafts and is referred to as the balancer belt or timing balancer belt.

10. Loosen the balancer shafts tensioner adjusting nut, do not loosen the nut more than 1 turn. Push the tensioner for the balancer belt away from the belt to relieve the tension. Hold the tensioner and tighten the adjusting nut to hold the tensioner in place.

11. Carefully remove the balancer belt by sliding it off of the pulleys. Do not crimp or bend the belt; protect it from contact with oil or coolant.

12. Remove the balancer belt drive sprocket from the crankshaft.

13. Remove the bolts attaching the Crankshaft Position/Top Dead Center (CKP/TDC) sensor and remove the sensor.

14. Remove the timing belt by sliding it off of the pulleys. Do not crimp or bend the belt; protect it from contact with oil or coolant.

15. If defective, remove the 2 bolts mounting the timing belt auto-tensioner and remove the tensioner from the vehicle.

16. If defective, remove the balancer belt tensioner by performing the following:

    a. Remove the spring from the balancer belt tensioner.

    b. Remove the adjusting nut.

    c. Remove the bolt from the balancer belt adjuster lever, then remove the lever and the tensioner pulley.

17. This is an excellent time to check or replace the water pump. Even if the timing belt is only being replaced as part of a good maintenance schedule, consider replacing the pump at the same time.

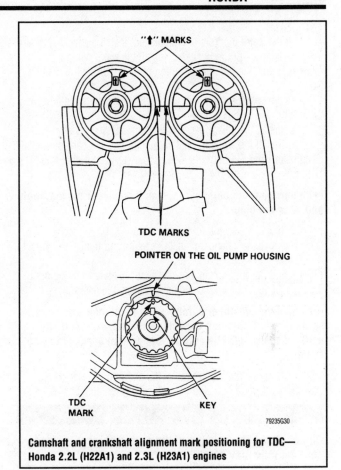

Camshaft and crankshaft alignment mark positioning for TDC—Honda 2.2L (H22A1) and 2.3L (H23A1) engines

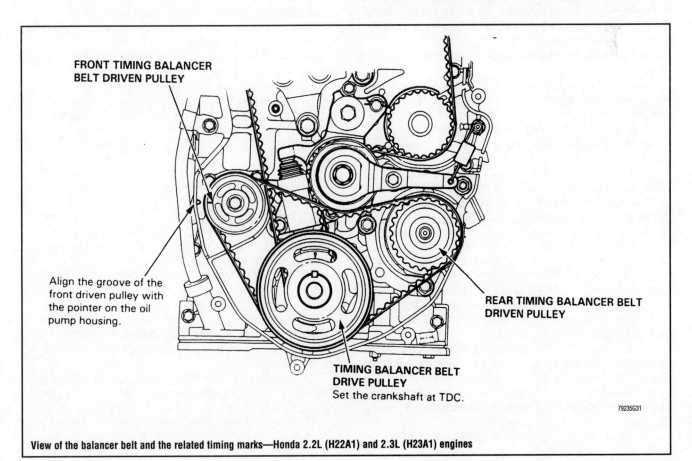

View of the balancer belt and the related timing marks—Honda 2.2L (H22A1) and 2.3L (H23A1) engines

To install:

18. If the water pump is to be replaced, install a new O-ring and make certain it is properly seated. Install the water pump and retaining bolts. Tighten the mounting bolts to 106 inch lbs. (12 Nm).

19. If the balancer tensioner was removed, perform the following to install it:

    a. Install the balancer belt pulley and adjuster lever.

    b. Install the adjusting nut and the bolt to the balancer belt adjuster lever. Do not tighten the adjuster nut or bolt at this time.

➡**Be sure that the balancer lever and tensioner moves freely and does not bind.**

    c. Install the spring to the tensioner.

    d. Move the balancer its full deflection and tighten the adjusting nut.

20. Hold the auto-tensioner with the maintenance bolt pointing up. Remove the maintenance bolt and discard the gasket.

➡**Handle the tensioner carefully so the oil inside does not spill or leak. Replenish the auto-tensioner with oil if any spills or leaks out. The auto-tensioner total capacity is ¼ oz. (8 ml).**

21. Clamp the mounting boss of the auto-tensioner in a vise. Use pieces of wood or a cloth to protect the mounting boss.

✳✳ **WARNING**

Do not clamp the housing of the auto-tensioner, component damage may occur.

22. Insert a flat-bladed prytool into the maintenance hole. Place the stopper (part No. 14540-P13-003) on the auto-tensioner while turning the prytool clockwise to compress the tensioner. Take care not to damage the threads or the gasket contact surface with the prytool.

23. Remove the prytool and install the maintenance bolt with a new gasket. Tighten the maintenance bolt to 71 inch lbs. (8 Nm).

24. Be sure no oil is leaking from the maintenance bolt and install the auto-tensioner to the engine. Tighten the auto-tensioner mounting bolts to 16 ft. lbs. (22 Nm).

25. The pointer on the crankshaft pulley should be aligned with the pointer on the oil pump. The camshaft pulley must be aligned so that the word **UP** is at the top of the pulley and the marks on the edge of the pulley are aligned with the surfaces of the head.

26. Install the timing belt.

27. Remove the stopper from the timing belt adjuster.

28. Install the CKP/TDC sensors and tighten the bolts to 106 inch lbs. (12 Nm). Connect the CKP/TDC sensor connectors.

29. Install the balancer belt drive sprocket to the crankshaft.

30. Align the groove on the front balancer shaft pulley with the pointer on the oil pump.

31. Align the rear timing balancer pulley using a 6 x 100mm bolt or rod. Mark the bolt or rod at a point 2.9 in. (74mm) from the end. Remove the bolt from the maintenance hole on the side of the block; insert the bolt/rod into the hole and align the 2.9 in. (74mm) mark with the face of the hole. This pin will hold the shaft in place during installation.

32. Ensure the timing balancer belt adjuster moves freely.

33. Install the balancer belt. Once the belts are in place, be sure that all the engine alignment marks are still correct. If not, remove the belts, realign the engine and reinstall the belts. Once the belts are properly installed, slowly loosen the adjusting nut, allowing the tensioner to move against the belt. Remove the pin from the maintenance hole and reinstall the bolt and washer.

34. Turn the crankshaft pulley 1 full turn, then tighten the adjusting nut to 33 ft. lbs. (45 Nm).

✳✳ **WARNING**

Do not apply extra pressure to the pulleys or tensioners while performing the adjustment.

35. Install the timing belt and cylinder head covers, and any other applicable components. When installing the side engine mount, tighten the bolt and nut attaching the mount to the engine to 40 ft. lbs. (55 Nm), and the through-bolt and nut to 47 ft. lbs. (65 Nm).

36. Connect the negative battery cable.

2.3L (H23A1) ENGINE

1. Disconnect the negative battery cable.

2. Turn the crankshaft so the No. 1 piston is at Top Dead Center (TDC) on the compression stroke.

➡**The No. 1 piston is at top dead center when the pointer on the block aligns with the white painted mark on the flywheel for manual transmission or driveplate for automatic transmission.**

3. Remove all necessary components for access to the cylinder head cover, then remove the cylinder head cover.

4. Ensure the words UP embossed on the camshaft pulleys are aligned in the upward position.

5. Insert a 5.0mm pin punch in each of the camshaft caps, nearest to the pulleys, through the holes provided.

6. Remove the upper and middle timing belt covers.

7. Support the engine with a floor jack below the center of the center beam. Tension the jack so that it is just supporting the beam but not lifting it. Remove the 2 rear bolts from the center beam to allow the engine to drop down for clearance to remove the lower cover.

8. Remove and discard the rubber seal from the timing belt adjuster. Do not loosen the adjusting nut.

9. Remove the lower timing belt cover.

10. There are 2 belts in this system; the one running to the camshaft pulley is the timing belt. The other, shorter one drives the balance shaft and is referred to as the balancer belt or timing balancer belt. Lock the timing belt adjuster in position, by installing one of the lower timing belt cover bolts to the adjuster arm.

11. Loosen the timing belt and balancer shaft tensioner adjuster nut(s), do not loosen the nut(s) more than 1 turn. Push the tensioner for the balancer belt away from the belt to relieve the tension. Hold the tensioner and tighten the adjusting nut to hold the tensioner in place.

12. Carefully remove the balancer belt by sliding it off of the pulleys. Do not crimp or bend the belt; protect it from contact with oil or coolant.

13. Remove the balancer belt drive sprocket from the crankshaft.

14. Loosen the lockbolt installed in the timing belt adjuster and

loosen the adjusting nut. Push the timing belt adjuster to remove the tension on the timing belt, then tighten the adjuster nut.

15. Remove the timing belt by sliding it off of the pulleys. Do not crimp or bend the belt; protect it from contact with oil or coolant.

16. If defective, remove the belt tensioners by performing the following:

    a. Remove the springs from the balancer belt and the timing belt tensioners.

    b. Remove the adjusting nut.

    c. Remove the bolt from the balancer belt adjuster lever, then remove the lever and the tensioner pulley.

    d. Remove the lockbolt from the timing belt tensioner lever, then remove the tensioner pulley and lever from the engine.

17. This is an excellent time to check or replace the water pump. Even if the timing belt is only being replaced as part of a good maintenance schedule, consider replacing the pump at the same time.

To install:

18. If the water pump is to be replaced, install a new O-ring and make certain it is properly seated. Install the water pump and retaining bolts. Tighten the mounting bolts to 106 inch lbs. (12 Nm).

19. If the tensioners were removed perform the following to install them:

    a. Install the timing belt tensioner lever and tensioner pulley.

➡**The tensioner lever must be properly positioned on its pivot pin located on the oil pump. Be sure that the timing belt lever and tensioner moves freely and does not bind.**

    b. Install the lockbolt to the timing belt tensioner, do not tighten the lockbolt at this time.

    c. Install the balancer belt pulley and adjuster lever.

    d. Install the adjusting nut and the bolt to the balancer belt adjuster lever. Do not tighten the adjuster nut or bolt at this time.

➡**Be sure that the balancer lever and tensioner moves freely and does not bind.**

    e. Install the springs to the tensioners.

    f. Move the timing belt tensioner its full deflection and tighten the lockbolt.

    g. Move the balancer its full deflection and tighten the adjusting nut.

20. The crankshaft timing pointer must be perfectly aligned with the white mark on the flywheel or flex-plate. The camshaft pulley must be aligned so that the word **UP** is at the top of the pulley and the marks on the edge of the pulley are aligned with the surfaces of the head.

21. Install the timing belt.

22. Install the balancer belt drive sprocket to the crankshaft.

23. Remove the 2, 5.0mm pin punches from the camshaft bearing caps.

24. Loosen the bolt used to lock the timing belt tensioner. Loosen, then tighten the timing belt adjuster nut.

25. Turn the crankshaft counterclockwise until the cam pulley has moved 3 teeth; this creates tension on the timing belt. Loosen, then tighten the adjusting nut and tighten it to 33 ft. lbs. (45 Nm). Tighten the bolt used to lock the timing belt tensioner.

26. Realign the timing belt timing marks.

27. Align the groove on the front balancer shaft pulley with the pointer on the oil pump.

28. Align the rear timing balancer pulley using a 6 x 100mm bolt or rod. Mark the bolt or rod at a point 2.913 in. (74mm) from the end. Remove the bolt from the maintenance hole on the side of the block; insert the bolt/rod into the hole and align the 74mm mark with the face of the hole. This pin will hold the shaft in place during installation.

29. Loosen the adjusting nut and ensure the timing balancer belt adjuster moves freely.

30. Install the balancer belt. Once the belts are in place, be sure that all the engine alignment marks are still correct. If not, remove the belts, realign the engine and reinstall the belts. Once the belts are properly installed, slowly loosen the adjusting nut, allowing the tensioner to move against the belt. Remove the pin from the maintenance hole and reinstall the bolt and washer.

31. Turn the crankshaft pulley 1 full turn and tighten the adjusting nut to 33 ft. lbs. (45 Nm).

➡**Both belt adjusters are spring loaded to properly tension the belts. Do not apply extra pressure to the pulleys or tensioners while performing the adjustment.**

32. Remove the 6 x 100mm bolt from the timing belt adjuster arm.

33. Install the timing belt and cylinder head covers. Reinstall all applicable components. When installing the crankshaft pulley, coat the threads and seating face of the pulley bolt with engine oil, then install and tighten the bolt to 181 ft. lbs. (250 Nm). When installing the side engine mount, tighten the bolt and nut attaching the mount to the engine to 40 ft. lbs. (55 Nm), and the through-bolt and nut to 47 ft. lbs. (65 Nm). Remove the jack from under the center beam.

34. Connect the negative battery cable.

2.7L & 3.0L ENGINES

1. Disconnect the negative battery cable.

2. Turn the engine to align the timing marks and set cylinder No. 1 to Top Dead Center (TDC) on the compression stroke. The white mark on the crankshaft pulley should align with the pointer on the timing belt cover. Remove the inspection caps on the upper timing belt covers to check the alignment of the timing marks. The pointers for the camshafts should align with the green marks on the camshaft sprockets.

3. Remove all necessary components for access to the timing belt covers, then remove the covers.

➡**Do not use the covers to store removed items.**

4. Loosen the timing belt adjuster bolt 180 degrees (½ turn). Push the tensioner to remove the tension from the timing belt, then retighten the adjusting bolt.

5. Remove the timing belt. Do not crimp or bend the belt; protect it from contact with oil or coolant. Slide the belt off the sprockets.

6. Remove the bolts attaching the camshaft sprockets to the camshafts, then remove the sprockets.

7. If the timing belt tensioner is defective, remove the spring from the timing belt tensioner. Remove the tensioner pulley adjusting bolt and the adjuster assembly from the engine.

➡**This is an excellent time to check or replace the water pump. Even if the timing belt is only being replaced as part of a good maintenance schedule, consider replacing the pump at the same time.**

To install:

8. If the water pump is to be replaced, install a new O-ring and make certain it is properly seated. Install the water pump and retaining bolts. Tighten the mounting bolts to 16 ft. lbs. (22 Nm).

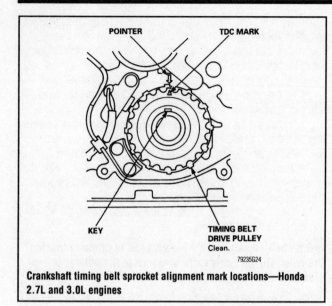

Crankshaft timing belt sprocket alignment mark locations—Honda 2.7L and 3.0L engines

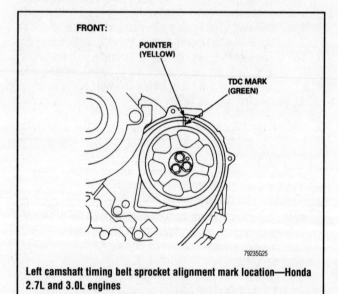

Left camshaft timing belt sprocket alignment mark location—Honda 2.7L and 3.0L engines

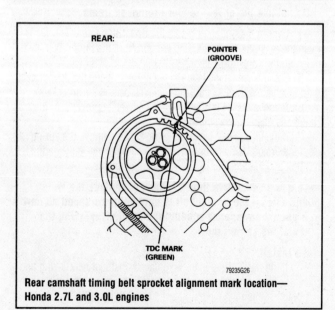

Rear camshaft timing belt sprocket alignment mark location—Honda 2.7L and 3.0L engines

9. If removed, install the tensioner pulley and the adjusting bolt, be sure the tensioner is properly positioned on its pivot pin. Install the spring to the tensioner, then push the tensioner to its full deflection and tighten the adjusting bolt.

10. Set the timing belt drive sprocket so that the No. 1 piston is at TDC. Align the TDC mark on the tooth of the timing belt drive sprocket with the pointer on the oil pump.

11. Set the camshaft sprockets so that the No. 1 piston is at TDC. Align the TDC marks (green mark) on the camshaft sprockets to the pointers on the back covers.

12. Install the timing belt onto the sprockets in the following sequence: crankshaft sprocket, tensioner pulley, front camshaft sprocket, water pump pulley and rear camshaft sprocket.

13. Loosen, then retighten the timing belt adjuster bolt to tension the timing belt.

14. Install the lower timing belt cover.

15. Install the crankshaft sprocket and the crankshaft pulley bolt. Tighten the bolt to 181 ft. lbs. (245 Nm) with the aid of the crank pulley holder.

16. Rotate the crankshaft 5–6 turns clockwise so that the timing belt positions on the sprockets.

17. Set cylinder No. 1 to TDC by aligning the timing marks. If the timing marks do not align, remove the timing belt, then adjust the components and reinstall the timing belt.

18. Loosen the timing belt adjusting bolt 180 degrees (½ turn) and retighten the adjusting bolt. Tighten the adjusting bolt to 31 ft. lbs. (42 Nm).

19. Install the upper timing belt cover and all other applicable components. When installing the side engine mount to the engine, use 3 new attaching bolts. Tighten the new bolts to 40 ft. lbs. (54 Nm).

20. Connect the negative battery cable.

Hyundai

✳✳ WARNING

Timing belt maintenance is extremely important. All Hyundai models use interference-type non-freewheeling engines. Should the timing belt break in these engines, the valves in the cylinder head will come in contact with the pistons, causing major engine damage. The recommended replacement interval for timing belts is 60,000 miles.

1.5L ENGINES

1986–00

1. Disconnect the negative battery cable.
2. Remove the engine undercover.
3. Using the proper equipment, slightly raise the engine to take the weight off the side engine mount.
4. Remove the engine mount bracket.
5. Remove the accessory drive belts.
6. Remove the water pump pulley and the alternator mounting bracket.
7. Remove the crankshaft pulley by removing the 4 retaining bolts.
8. Remove all attaching screws and remove the upper and lower timing belt covers.
9. Rotate the crankshaft clockwise and align the timing marks

so the No. 1 piston will be at Top Dead Center (TDC) of the compression stroke.

10. Loosen the tensioning bolt and the pivot bolt on the timing belt tensioner. Move the tensioner as far as it will go toward the water pump. Tighten the adjusting bolt.

11. If the timing belt is to be reused, mark the belt with an arrow showing direction of rotation.

12. Remove the timing belt.

13. Inspect the belt thoroughly. The back surface must be pliable and rough. If it is hard and glossy, the belt should be replaced. Any cracks in the belt backing or teeth or missing teeth mean the belt must be replaced. The canvas cover should be intact on all the teeth. If rubber is exposed anywhere, the belt should be replaced.

14. Inspect the tensioner for grease leaking from the grease seal and any roughness in rotation. Replace a tensioner for either defect.

15. The sprockets should be inspected and replaced, if there is any sign of damaged teeth or cracking. Do not immerse sprockets in solvent, as solvent that has soaked into the metal may cause deterioration of the timing belt later. Do not clean the tensioner in solvent either, as this may wash the grease out of the bearing.

To install:

16. Align the timing marks of the camshaft sprocket and check that the crankshaft timing marks are still in alignment.

17. If removed, install the timing belt tensioner, spring and spacer with the bottom end of the spring free.

18. Tighten the adjusting bolt slightly with the tensioner moved as far as possible away from the water pump.

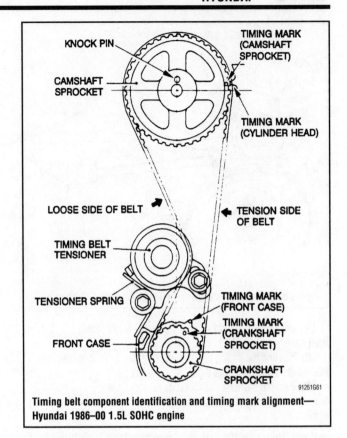

Timing belt component identification and timing mark alignment—Hyundai 1986–00 1.5L SOHC engine

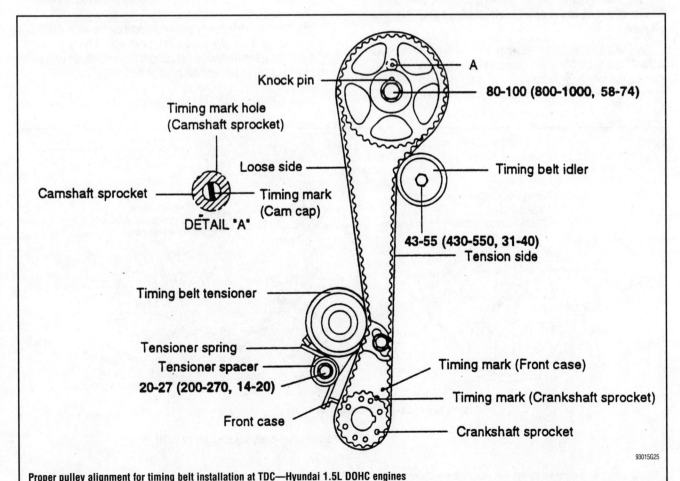

Proper pulley alignment for timing belt installation at TDC—Hyundai 1.5L DOHC engines

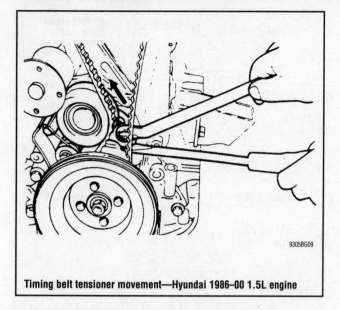

Timing belt tensioner movement—Hyundai 1986–00 1.5L engine

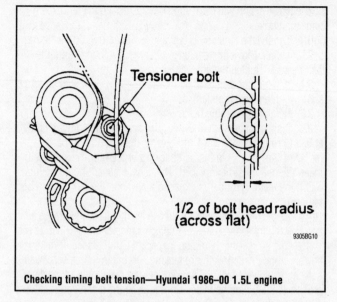

Checking timing belt tension—Hyundai 1986–00 1.5L engine

19. Install the free end of the spring into the locating tang on the front case.

20. Position the timing belt over the crankshaft sprocket, then over the camshaft sprocket. Slip the back of the belt over the tensioner pulley.

21. Turn the camshaft sprocket in the opposite direction of rotation until the straight side of the belt is tight and be sure the timing marks align.

➡**If the timing marks are not properly aligned, shift the belt 1 tooth at a time in the appropriate direction until they are aligned.**

22. Loosen the tensioner mounting bolts so the tensioner works, without the interference of any friction, under spring pressure. Be sure the belt follows the curve of the camshaft pulley so the teeth are engaged all the way around. Correct the path of the belt, if necessary.

23. Torque the tensioner adjusting bolt, then the tensioner pivot bolt to 15–18 ft. lbs. (20–26 Nm).

➡**Bolts must be tightened in the stated order or tension will not be correct.**

24. Turn the crankshaft 1 turn clockwise until timing marks again align to seat the belt.

25. Loosen both tensioner attaching bolts and let the tensioner position itself under spring tension. Retighten the bolts.

26. Check belt tension by putting a finger on the water pump side of the tensioner wheel and pull the belt toward the water pump. The belt should move toward the pump until the teeth are approximately ½ of the way across the head of the tensioner adjusting bolt. Re-tension the belt, if necessary.

27. Install the timing belt covers.

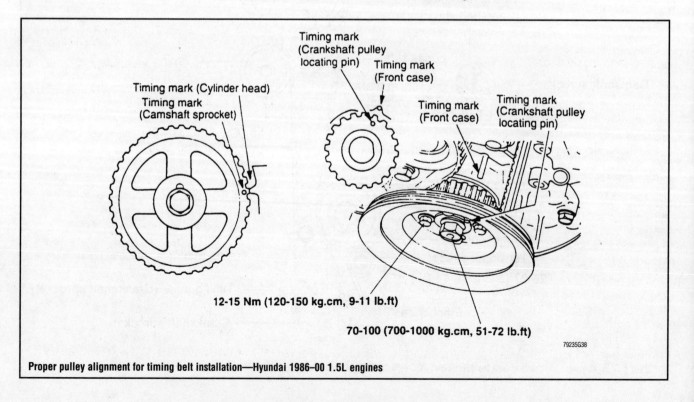

Proper pulley alignment for timing belt installation—Hyundai 1986–00 1.5L engines

28. Install the crankshaft pulley, making sure the pin on the crankshaft sprocket fits through the hole in the rear surface of the pulley. Install the bolts and torque to 90–102 inch lbs. (9–12 Nm).

29. Install water pump pulley and torque the bolts to 72–84 inch lbs. (8–10 Nm).

30. Install drive belts and adjust to proper tension.

31. Install engine bracket and lower the engine.

32. Install any previously removed covers.

33. Connect the negative battery cable.

34. Check and adjust ignition timing.

1.6L, 1.8L & 2.0L (VIN P) ENGINES

1992–95 1.6L, 1.8L and 1993–98 2.0L (VIN P) Engines

1. Disconnect the negative battery cable.

2. Remove the engine undercover.

3. Using the proper equipment, slightly raise the engine to take the weight off the side engine mount.

4. Remove the engine mount bracket.

5. Remove the accessory drive belts and any mounting brackets that may interfere with timing cover removal.

6. Remove the water pump pulley and the crankshaft pulley.

7. Remove all attaching screws and remove the upper and lower timing belt covers.

8. Rotate the crankshaft clockwise and align the timing marks so No. 1 piston will be at Top Dead Center (TDC) of the compression stroke. At this time the timing marks on the camshaft sprocket and the upper surface of the cylinder head should coincide, and the dowel pin of the camshaft sprocket should be at the upper side.

➡**Always rotate the crankshaft in a clockwise direction. If the timing belt is to be reused, make a mark on the back of the timing belt indicating the direction of rotation so it may be reassembled in the same direction.**

9. Remove the auto tensioner and remove the outermost timing belt.

10. Remove the timing belt tensioner pulley, tensioner arm, idler pulley, oil pump sprocket, special washer, flange and spacer.

11. Remove the silent shaft (inner) belt tensioner and remove the inner belt.

To install:

12. Align the timing marks on the crankshaft sprocket and the silent shaft sprocket. Fit the inner timing belt over the crankshaft and silent shaft sprocket. Ensure that there is no slack in the belt.

13. While holding the inner timing belt tensioner with your fingers, adjust the timing belt tension by applying a force towards the center of the belt, until the tension side of the belt is taut. Tighten the tensioner bolt.

➡**When tightening the bolt of the tensioner, ensure that the tensioner pulley shaft does not rotate with the bolt. Allowing it to rotate with the bolt can cause excessive tension on the belt.**

14. Check belt for proper tension by depressing the belt on its long side with your finger and noting the belt deflection. The desired reading is 0.20–0.28 in. (5–7mm). If tension is not correct, readjust and check belt deflection.

15. Install the flanged crankshaft sprocket. The flange on the crankshaft sprocket must be installed towards the inner timing belt sprocket.

16. Install the bolt and special washer and torque the bolt to 80–94 ft. lbs. (110–130 Nm).

17. To install the oil pump sprocket, insert a Phillips screwdriver with a shaft 0.31 in. (8mm) in diameter into the plug hole in the left side of the cylinder block to hold the left silent shaft. Tighten the nut to 36–43 ft. lbs. (50–60 Nm).

18. Carefully push in the auto tensioner rod until the set hole in the rod is aligned with the hole in the cylinder body. Place a wire into the hole to retain the rod.

19. Install the tensioner pulley onto the tensioner arm. Locate the pinhole in the tensioner pulley shaft to the left of the center bolt. Then, tighten the center bolt finger-tight.

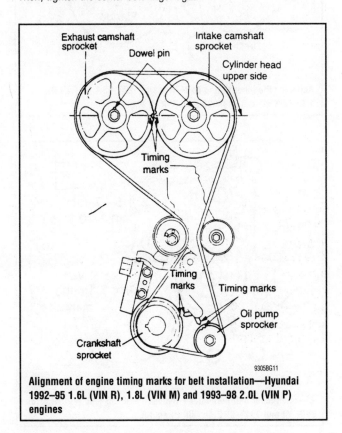

Alignment of engine timing marks for belt installation—Hyundai 1992–95 1.6L (VIN R), 1.8L (VIN M) and 1993–98 2.0L (VIN P) engines

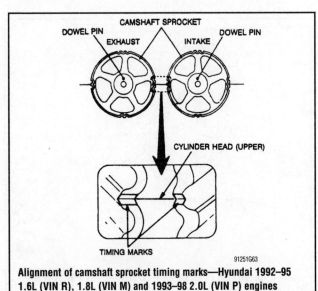

Alignment of camshaft sprocket timing marks—Hyundai 1992–95 1.6L (VIN R), 1.8L (VIN M) and 1993–98 2.0L (VIN P) engines

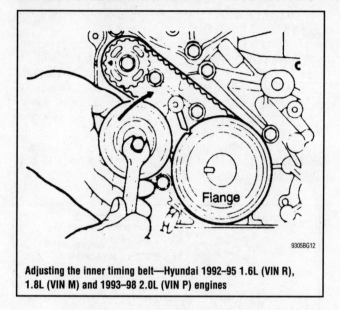

Adjusting the inner timing belt—Hyundai 1992–95 1.6L (VIN R), 1.8L (VIN M) and 1993–98 2.0L (VIN P) engines

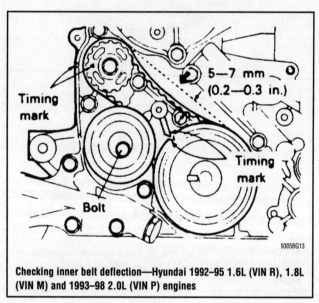

Checking inner belt deflection—Hyundai 1992–95 1.6L (VIN R), 1.8L (VIN M) and 1993–98 2.0L (VIN P) engines

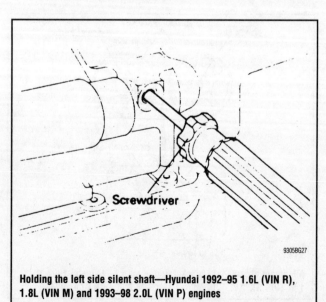

Holding the left side silent shaft—Hyundai 1992–95 1.6L (VIN R), 1.8L (VIN M) and 1993–98 2.0L (VIN P) engines

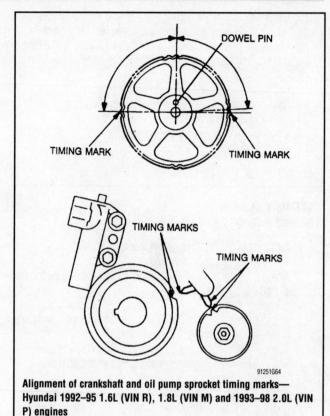

Alignment of crankshaft and oil pump sprocket timing marks—Hyundai 1992–95 1.6L (VIN R), 1.8L (VIN M) and 1993–98 2.0L (VIN P) engines

20. When installing the timing belt, turn the 2 camshaft sprockets so their dowel pins are located on top. Align the timing marks facing each other with the top surface of the cylinder head. When you let go of the exhaust camshaft sprocket, it will rotate 1 tooth in the counterclockwise direction. This should be taken into account when installing the timing belts on the sprocket.

➡ **Both camshaft sprockets are used for the intake and exhaust camshafts and are provided with 2 timing marks. When the sprocket is mounted on the exhaust camshaft, use the timing mark on the right with the dowel pinhole on top. For the intake camshaft sprocket, use the mark on the left with the dowel pinhole on top.**

21. Align the crankshaft sprocket and oil pump sprocket timing marks.

22. After alignment of the oil pump sprocket timing marks, remove the plug on the cylinder block and insert a Phillips screwdriver with a shaft diameter of 0.31 in. (8mm) through the hole. If the shaft can be inserted 2.4 in. (70mm) deep, the silent shaft is in the correct position. If the shaft of the tool can only be inserted 0.8—1.0 in. (20–25mm) deep, turn the oil pump sprocket 1 turn and realign the marks. Reinsert the tool making sure it is inserted 2.4 in. (70mm) deep. Keep the tool inserted in hole for the remainder of this procedure.

➡ **The above step assures that the oil pump socket is in correct orientation to the silent shafts. This step must not be skipped or a vibration may develop during engine operation.**

23. Install the timing belt as follows:
   a. Install the timing belt around the intake camshaft sprocket and retain it with 2 spring clips or binder clips.
   b. Install the timing belt around the exhaust sprocket, aligning

the timing marks with the cylinder head top surface using 2 wrenches. Retain the belt with 2 spring clips.

   c. Install the timing belt around the idler pulley, oil pump sprocket, crankshaft sprocket and the tensioner pulley. Remove the 2 spring clips.

   d. Lift upward on the tensioner pulley in a clockwise direction and tighten the center bolt. Make sure all timing marks are aligned.

   e. Rotate the crankshaft ¼ turn counterclockwise. Then, turn clockwise until the timing marks are aligned again.

24. To adjust the timing (outer) belt, turn the crankshaft ¼ turn counterclockwise, then turn it clockwise to move No. 1 cylinder to TDC.

25. Loosen the center bolt. Using tool MD998738 or equivalent and a torque wrench, apply a torque of 22–24 inch lbs. (2.6–2.8 Nm). Tighten the center bolt.

26. Screw the special tool into the engine left support bracket until its end makes contact with the tensioner arm. At this point, screw the special tool in some more and remove the set wire attached to the auto tensioner, if the wire was not previously removed. Then remove the special tool.

27. Rotate the crankshaft 2 complete turns clockwise and let it sit for approximately 15 minutes. Then, measure the auto tensioner protrusion (the distance between the tensioner arm and auto tensioner body) to ensure that it is within 0.15–0.18 in. (3.8–4.5mm). If out of specification, repeat Step 1–4 until the specified value is obtained.

28. If the timing belt tension adjustment is being performed with the engine mounted in the vehicle, and clearance between the tensioner arm and the auto tensioner body cannot be measured, the following alternative method can be used:

   a. Screw in special tool MD998738 or equivalent, until its end makes contact with the tensioner arm.

   b. After the special tool makes contact with the arm, screw it in some more to retract the auto tensioner pushrod while counting the number of turns the tool makes until the tensioner arm is brought into contact with the auto tensioner body. Make sure the number of turns the special tool makes conforms with the standard value of 2½–3 turns.

   c. Install the rubber plug to the timing belt rear cover.

29. Install the timing belt covers.

30. Install the water pump pulley and torque the bolts to 84 inch lbs. (10 Nm).

31. Install the crankshaft pulley and torque the bolts to 22 ft. lbs. (30 Nm).

32. Install drive belts and adjust to proper tension.

33. Install the engine mounting bracket and lower the engine.

34. Connect the negative battery cable.

1996–98 1.8L (VIN M), 1997–00 2.0L (VIN F) Engines

1. Disconnect the negative battery cable.
2. Remove the engine undercover.
3. Using the proper equipment, slightly raise the engine to take the weight off the side engine mount. Remove the engine mount bracket.
4. Remove the accessory drive belts, tension pulley brackets, water pump pulley and crankshaft pulley.
5. Remove all attaching screws and remove the upper and lower timing belt covers.

➡**Always rotate the crankshaft in a clockwise direction.**

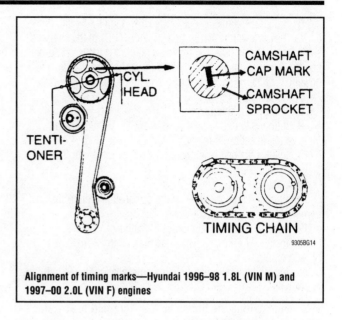

Alignment of timing marks—Hyundai 1996–98 1.8L (VIN M) and 1997–00 2.0L (VIN F) engines

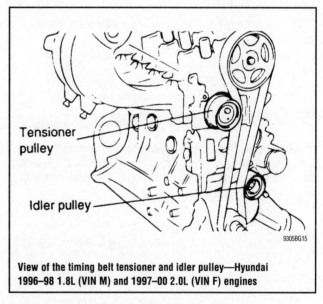

View of the timing belt tensioner and idler pulley—Hyundai 1996–98 1.8L (VIN M) and 1997–00 2.0L (VIN F) engines

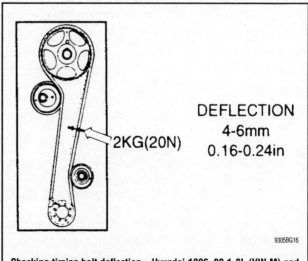

Checking timing belt deflection—Hyundai 1996–98 1.8L (VIN M) and 1997–00 2.0L (VIN F) engines

6. Rotate the crankshaft clockwise and align the timing marks so No. 1 piston will be at Top Dead Center (TDC) of the compression stroke. At this time the timing marks on the camshaft sprocket and the upper surface of the cylinder head should coincide, and the dowel pin of the camshaft sprocket should be at the upper side.

7. Remove the timing belt tensioner.

8. Remove the timing belt.

9. If timing belt is to be reused, mark the belt with an arrow to indicate the direction of rotation.

10. Inspect the belt tensioner and the idler pulley for free rotation and wear. Replace as necessary.

To install:

11. Align the timing marks on the camshaft and crankshaft sprockets.

12. Loosely install the timing belt tensioner.

13. Install the idler pulley. Torque the bolt to 32–41 ft. lbs. (43–55 Nm).

14. Install the timing belt onto the crankshaft and then onto the crankshaft.

➡**When installing the timing belt, ensure that the tension side has no slack.**

15. Rotate the belt tensioner toward the water pump to take up any slack. Tighten the tensioner bolt.

16. Turn the crankshaft 1 full turn in a clockwise rotation and realign the crankshaft timing mark.

17. Check the belt for proper tension by depressing the belt on its long side with your finger and noting the belt deflection. The desired deflection should be 0.16–0.24 in. (4–6mm).

18. Torque the tensioner bolt to 32–41 ft. lbs. (43–55 Nm).

➡**When tightening the bolt of the tensioner, ensure that the tensioner pulley shaft does not rotate with the bolt. Allowing it to rotate with the bolt can cause excessive tension on the belt.**

19. Ensure the timing marks are in alignment.

20. Install the timing belt covers and torque the bolts to 84 inch lbs. (10 Nm).

21. Install the crankshaft pulley and torque the bolt to 125–133 ft. lbs. (170–180 Nm).

22. Install the water pump pulley. Torque the bolts to 84 inch lbs. (10 Nm).

23. Install the drive belts and properly adjust tension.

24. Install the engine mount bracket.

25. Install the engine undercover.

26. Connect the negative battery cable.

2.4L ENGINE

1989–1991

➡**An 8mm diameter metal bar is needed for this procedure.**

1. Disconnect the negative battery cable.

2. Remove the water pump drive belt and pulley.

3. Remove the crank adapter and the crankshaft pulley.

4. Remove the upper and lower timing belt covers.

5. Move the tensioner fully in the direction of the water pump and temporarily secure it there.

6. If the timing belt is to be reused, make an arrow mark on the belt to indicate the direction of rotation.

7. Remove the timing belt.

8. Remove the crankshaft sprocket bolt and pull the crankshaft sprocket and flange from the crankshaft.

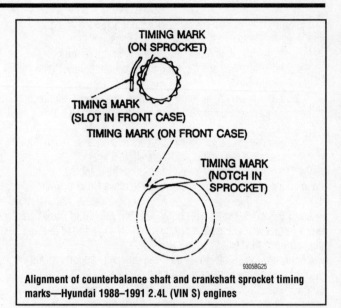

Alignment of counterbalance shaft and crankshaft sprocket timing marks—Hyundai 1988–1991 2.4L (VIN S) engines

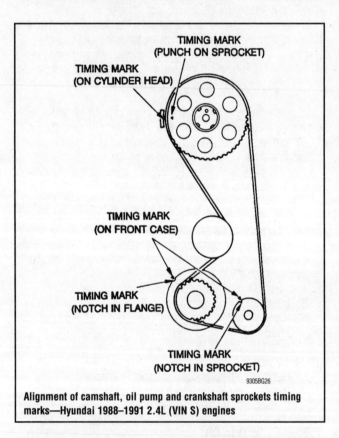

Alignment of camshaft, oil pump and crankshaft sprockets timing marks—Hyundai 1988–1991 2.4L (VIN S) engines

9. Remove the plug on the left side of the block and insert an 8mm diameter metal bar in the opening to keep the silent shaft in position.

10. Remove the oil pump sprocket retaining nut and remove the oil pump sprocket.

11. Loosen the right silent shaft mounting bolt until it can be turned by hand.

12. Remove the belt tensioner and remove the silent shaft belt.

➡**Do not attempt to turn the silent shaft sprocket or loosen its bolt while the belt is off.**

13. Remove the silent shaft belt sprocket from the crankshaft.

14. Check the belt for wear, damage or glossing. Replace it if any cracks, damage, brittleness or excessive wear is found.

15. Check the tensioners for a smooth rate of movement.

16. Replace any tensioner that shows grease leakage through the seal.

To install:

17. Install the silent shaft belt sprocket on the crankshaft, with the flat face toward the engine.

18. Apply a light coat of engine oil on the outer face of the spacer and install the spacer on the right silent shaft. The side with the rounded shoulder faces the engine.

19. Install the sprocket on the right silent shaft and install the bolt but do not tighten completely at this time.

➡**Align the silent shaft and oil pump sprockets using the timing marks. If the 8mm metal bar can not be inserted into the hole 2.36 inches (60mm) deep, the oil pump sprocket will have to be turned 1 full rotation until the bar can be to the full length. This step assures that the oil pump sprocket and the silent shafts are in correct orientation. This step must not be skipped or a vibration may develop during engine operation.**

20. Install the silent shaft belt and adjust the tension by moving the tensioner into contact with the belt. Adjust the tensioner enough to remove the slack from the belt. Torque the tensioner bolt to 21 ft. lbs. (28 Nm).

21. Torque the silent shaft sprocket bolt to 28 ft. lbs. (36 Nm).

22. Install the flange and crankshaft sprocket on the crankshaft. The flange conforms to the front of the silent shaft sprocket and the timing belt sprocket is installed with the flat face toward the engine.

➡**The flange must be installed correctly or a broken belt will result.**

23. Install the washer and bolt in the crankshaft and torque to 94 ft. lbs. (130 Nm).

24. Install the timing belt tensioner, spacer and spring.

25. Align the timing mark on each sprocket with the corresponding mark on the front case.

26. Install the timing belt on the sprockets and move the tensioner against the belt with sufficient force to allow a deflection of 5–7mm along its longest straight run.

27. Torque the tensioner bolt to 21 ft. lbs. (28 Nm).

28. Install the upper and lower timing belt covers.

29. Install the crankshaft pulley and the crank adapter. Torque the bolts to 21 ft. lbs. (28 Nm).

30. Remove the 8mm bar and install the plug.

31. Install the water pump pulley and the drive belt.

32. Connect the negative battery cable.

33. Start the engine and check for proper operation.

34. Check and adjust ignition timing.

1999–00

1. Disconnect the negative battery cable.

2. Remove any engine or side covers necessary to access the timing belt covers.

3. Using the proper equipment, slightly raise the engine to take the weight off the side engine mount. Remove the engine mount bracket.

4. Remove the accessory drive belts.

5. Remove the water pump pulley and crankshaft pulley.

6. Remove all attaching screws and remove the upper and lower timing belt covers.

7. Turn the crankshaft in a clockwise rotation to align the timing marks on the crankshaft and the camshafts.

8. Remove the auto tensioner.

➡**If the timing belt is to be reused, mark the belt with an arrow to indicate the direction of rotation.**

9. Remove the tensioner pulley and arm assembly.

10. Remove the timing belt.

11. Remove the balancer belt tensioner and remove the balancer belt.

12. Check camshaft and crankshaft sprockets, tensioner and idler pulleys for abnormal wear, cracks or damage. Replace as necessary.

13. Check the auto tensioner assembly for leaks and replace if necessary.

To install:

14. To install the balancer shaft belt, perform the following procedure:

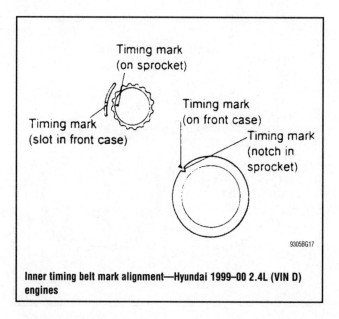

Inner timing belt mark alignment—Hyundai 1999–00 2.4L (VIN D) engines

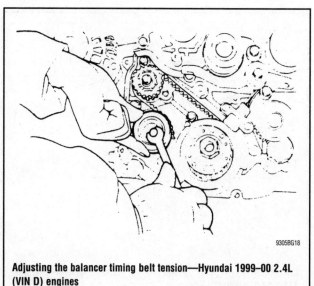

Adjusting the balancer timing belt tension—Hyundai 1999–00 2.4L (VIN D) engines

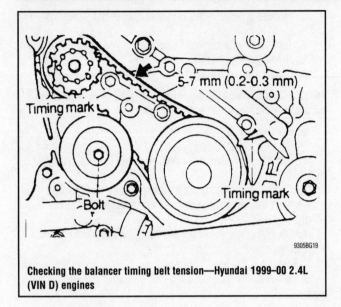

Checking the balancer timing belt tension—Hyundai 1999–00 2.4L (VIN D) engines

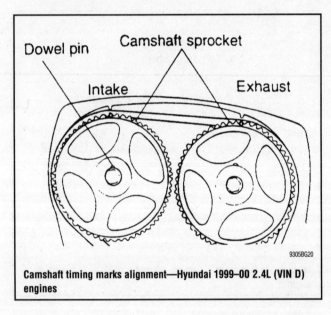

Camshaft timing marks alignment—Hyundai 1999–00 2.4L (VIN D) engines

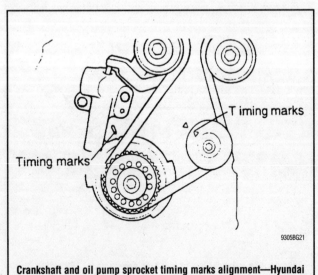

Crankshaft and oil pump sprocket timing marks alignment—Hyundai 1999–00 2.4L (VIN D) engines

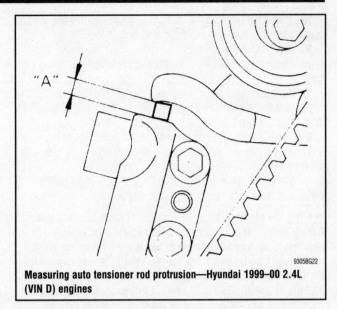

Measuring auto tensioner rod protrusion—Hyundai 1999–00 2.4L (VIN D) engines

a. Align the balancer timing belt marks with the corresponding marks on the front case.

b. Install the balancer belt onto the crankshaft and counter balance shaft sprockets.

c. Make sure the timing marks are properly aligned and that the tension side of the belt has no slack.

d. Install the balancer belt tensioner with center of pulley located on the left side of the mounting bolt and with the pulley flange toward the front of the engine.

e. Rotate the balancer belt tensioner in a clockwise direction so that the tension side of the belt will be pulled tight. Tighten the tensioner bolt.

➡ **When the tensioner bolt is tightened, use care not to move the tensioner pulley. The belt will be over tightened.**

f. Check that the balancer timing belt marks are still aligned.

g. Check the balancer belt tension. Deflection should be 0.2–0.3 in. (5–7mm).

15. Using a vise with soft jaw adapters, push the rod of the auto tensioner slowly into the tensioner body until the hole in the rod is aligned with the hole in the tensioner body. Install a set pin through the holes to hold the rod in place.

16. Install the auto tensioner assembly leaving the set pin in place.

17. Install the tensioner pulley and arm assembly.

18. Align the camshaft timing marks. The marks will be at the top of each sprocket aligning with the marks on the upper surface of the rocker cover.

19. Align the crankshaft and the oil pump sprocket marks.

20. Install the timing belt around the tensioner pulley and the crankshaft sprocket. Hold the belt onto the tensioner pulley with your left hand.

21. With your right hand, install the timing belt around the oil pump sprocket and the idler pulley.

22. Install the belt around the camshaft sprockets, making sure the timing marks are aligned.

23. Raise the tensioner pulley against the timing belt so that the belt does not sag. Temporarily tighten the tensioner pulley bolt.

24. Check to make sure that all timing marks are aligned.

25. Remove the set pin from the auto tensioner body to release the rod against the tensioner pulley arm.

26. Rotate the crankshaft 2 revolutions and check that the timing marks are correctly aligned.

27. Measure the auto tensioner rod protrusion between the tensioner body and the tensioner arm. The measurement should be 0.15–0.18 in. (6–8mm).

28. Install the lower and upper timing belt covers.

29. Install the water pump pulley and crankshaft pulley.

30. Install the accessory drive belts.

31. Install the front engine bracket and lower the engine.

32. Install any covers that were previously removed for access.

33. Connect the negative battery cable.

2.5L V6 ENGINE

1999–00

1. Disconnect the negative battery cable.

2. Remove the engine cover.

3. Using the proper equipment, slightly raise the engine to take the weight off the side engine mount. Remove the engine mount bracket.

4. Remove the accessory drive belt.

5. Remove the accessory belt tensioner and remove the crankshaft pulley.

6. Remove the upper and lower timing belt covers.

7. Turn the crankshaft in a clockwise direction to align the timing marks.

8. Remove the auto tensioner assembly.

9. Remove the timing belt.

10. Inspect the timing belt tensioner pulley, the idler pulley, crankshaft sprocket and the camshaft sprockets for cracks, abnormal wear or damage. Replace as necessary.

To install:

11. Using a vise with soft jaw adapters, push the rod of the auto tensioner slowly into the tensioner body until the hole in the rod is aligned with the hole in the tensioner body. Install a set pin through the holes to hold the rod in place.

12. Install the auto tensioner assembly leaving the set pin in place.

13. Install the tensioner pulley and arm assembly.

14. Make sure that all timing marks are aligned.

15. Install the timing belt in the following order:
   a. Crankshaft sprocket
   b. Idler pulley
   c. Intake cam sprocket (left side)
   d. Water pump pulley
   e. Intake cam sprocket (right side)
   f. Tensioner pulley

16. Raise the tensioner pulley against the timing belt so that the belt does not sag. Temporarily tighten the tensioner pulley bolt.

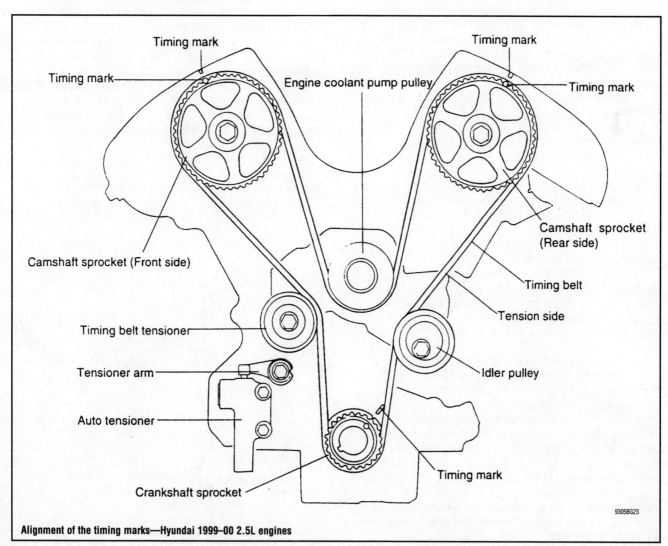

Alignment of the timing marks—Hyundai 1999–00 2.5L engines

9305BG23

17. Check to make sure that all timing marks are aligned.

18. Remove the set pin from the auto tensioner body to release the rod against the tensioner pulley arm.

19. Rotate the crankshaft 2 revolutions and check that the timing marks are correctly aligned.

20. Measure the auto tensioner rod protrusion between the tensioner body and the tensioner arm. The measurement should be 0.15–0.18 in. (6–8mm).

21. Install the upper and lower timing covers.

22. Install the accessory belt tensioner and install the crankshaft pulley.

23. Install the accessory drive belt.

24. Install the engine mount and bracket.

25. Install the engine cover.

26. Connect the negative battery cable.

3.0L V6 ENGINE

1990–94

1. Disconnect the negative battery cable.

2. To remove the air conditioning compressor belt, loosen the adjustment pulley locknut, turn the screw counterclockwise to reduce the drive belt tension and remove the belt.

3. To remove the serpentine drive belt, insert a ½ in. breaker bar in to the square hole of the tensioner pulley, rotate it counterclockwise to reduce the drive belt tension and remove the belt.

4. Remove the air conditioning compressor and the air compressor bracket, power steering pump and alternator from the mounts and support them to the side. Remove power steering pump/alternator automatic belt tensioner bolt and the tensioner.

5. Raise the vehicle and support safely. Remove the right inner fender splash shield.

6. Remove the crankshaft pulley bolt and the pulley/damper assembly from the crankshaft.

7. Lower the vehicle and place a floor jack under the engine to support it.

8. Separate the front engine mount insulator from the bracket. Raise the engine slightly and remove the mount bracket.

9. Remove the timing belt cover bolts and the upper and lower covers from the engine.

10. Turn the crankshaft until the timing marks on the camshaft sprocket and cylinder head are aligned.

11. Loosen the tensioning bolt, it runs in the slotted portion of the tensioner, and the pivot bolt on the timing belt tensioner.

12. Move the tensioner counterclockwise as far as it will go. Tighten the adjusting bolt.

13. Mark the timing belt with an arrow showing direction of rotation.

14. Remove the timing belt from the camshaft sprocket.

15. Remove the crankshaft pulley. Then, remove the timing belt. Remove the timing belt tensioner. Remove the retainer bolts from the timing sprockets and remove as required.

16. Inspect the belt thoroughly. The back surface must be pliable and rough. If it is hard and glossy, the belt should be replaced. Any cracks in the belt backing or teeth or missing teeth mean the belt must be replaced. The canvas cover should be intact on all the teeth. If rubber is exposed anywhere, the belt should be replaced.

17. Inspect the tensioner for grease leaking from the grease seal and any roughness in rotation. Replace a tensioner for either defect.

18. The sprockets should be inspected and replaced if there is any sign of damaged teeth or cracking anywhere.

19. Do not immerse sprockets in solvent, as solvent that has soaked into the metal may cause deterioration of the timing belt later.

20. Do not clean the tensioner in solvent either, as this may wash the grease out of the bearing.

To install:

21. Align the timing marks of the camshaft sprocket. Check that the crankshaft timing marks are still in alignment, the locating pin on the front of the crankshaft sprocket is aligned with a mark on the front case.

22. Mount the tensioner, spring and spacer with the bottom end of the spring free. Then, install the bolts and tighten the adjusting bolt slightly with the tensioner moved as far as possible away from the water pump. Install the free end of the spring into the locating tang on the front case. Position the belt over the crankshaft sprocket and then over the camshaft sprocket. Slip the back of the belt over the tensioner wheel. Turn the camshaft sprocket in the opposite of its normal direction of rotation until the straight side of the belt is tight and make sure the timing marks align. If not, shift the belt 1 tooth at a time in the appropriate direction until this occurs.

23. Loosen the tensioner mounting bolts so the tensioner works, without the interference of any friction, under spring pressure. Make sure the belt follows the curve of the camshaft pulley so the teeth are engaged all the way around.

24. Correct the path of the belt, if necessary. Torque the tensioner adjusting bolt to 16–21 ft. lbs. (22–29 Nm). Then, tighten the tensioner pivot bolt to the same figure. Bolts must be torqued in that order or tension won't be correct.

25. Turn the crankshaft 1 turn clockwise until timing marks again align to seat the belt. Then loosen both tensioner attaching bolts and let the tensioner position itself under spring tension as before. Finally, torque the bolts in the proper order exactly as before. Check belt tension by putting a finger on the water pump side of the tensioner wheel and pull the belt toward it. The belt should move toward the pump until the teeth are about ¼ of the way across the head of the tensioner adjusting bolt. Tension the belt, if necessary.

26. Install the timing belt covers.

27. Install the crankshaft pulley, making sure the pin on the crankshaft sprocket fits through the hole in the rear surface of the pulley. Install the retaining bolt and torque to 108–116 ft. lbs. (147–157 Nm).

28. Install the engine mount bracket and secure with the mounting hardware.

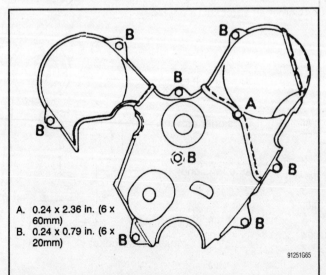

A. 0.24 x 2.36 in. (6 x 60mm)
B. 0.24 x 0.79 in. (6 x 20mm)

91251G65

Timing belt cover bolt locations—Hyundai 1990–94 3.0L engine

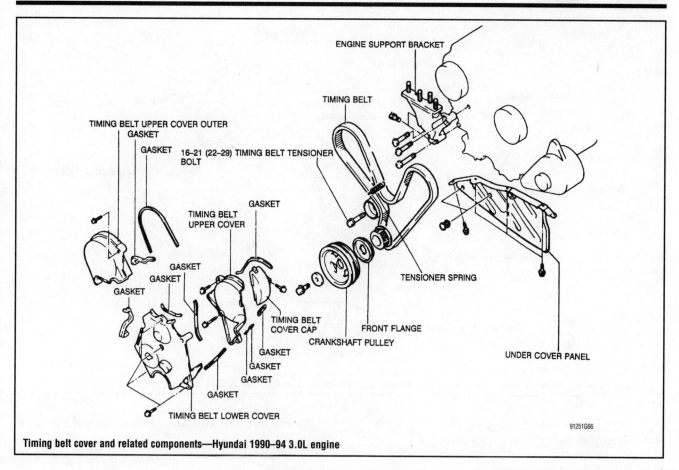

Timing belt cover and related components—Hyundai 1990–94 3.0L engine

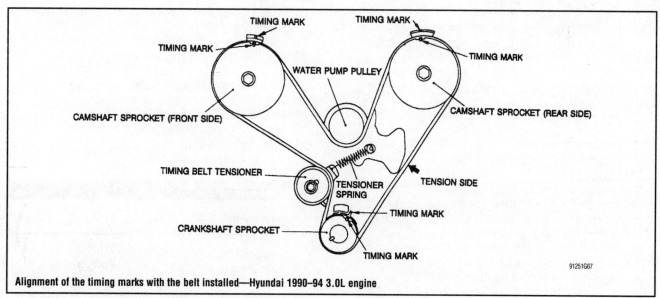

Alignment of the timing marks with the belt installed—Hyundai 1990–94 3.0L engine

29. Install the pulley damper assembly to the crankshaft. Torque the bolt to 110 ft. lbs. (149 Nm). Install the splash shield.

30. Install the power steering pump/alternator automatic belt tensioner.

31. Install the air conditioning compressor bracket, compressor, power steering pump and alternator.

32. Install the accessory drive belt.

33. Connect the negative battery cable and check all disturbed components for proper operation.

1995–98

1. Disconnect the negative battery cable.

2. Remove the engine undercover.

3. Remove the accessory drive belts.

4. Remove the air conditioner compressor tension pulley assembly.

5. Remove the tension pulley bracket.

6. Using the proper equipment, slightly raise the engine to take the weight off the side engine mount.

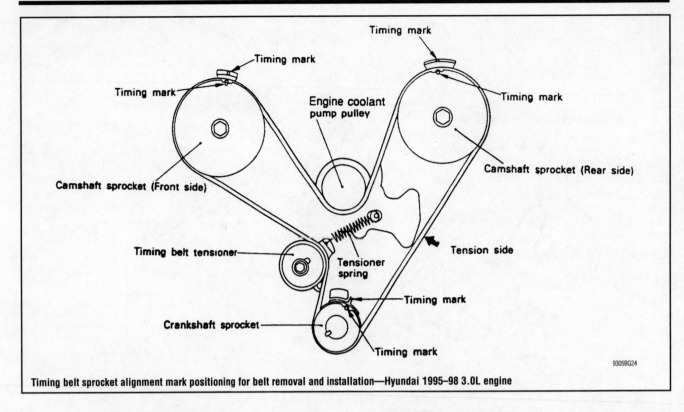

Timing belt sprocket alignment mark positioning for belt removal and installation—Hyundai 1995–98 3.0L engine

7. Disconnect the power steering pump pressure switch connector. Remove the power steering pump and wire aside.

8. Remove the engine support bracket.

9. Remove the crankshaft pulley.

10. Remove the timing belt cover cap.

11. Remove the timing belt upper and lower covers.

12. Turn the crankshaft until the timing marks on the camshaft sprocket and cylinder head are aligned.

13. Loosen the timing belt tensioner bolt and turn the tensioner counterclockwise as far as it will go. Tighten the adjusting bolt.

14. Mark the timing belt with an arrow showing direction of rotation.

15. Remove the timing belt.

16. If defective, remove the timing belt tensioner.

To install:

17. If necessary, install the timing belt tensioner.

18. Attach the top of the tensioner spring on the engine coolant pump pin. Ensure the hook on the pin is facing down and the hook on the tensioner is facing away from the engine

19. Rotate the timing belt tensioner to the extreme counterclockwise position. Temporarily lock the tensioner in place.

20. Align the timing marks of the camshaft and crankshaft sprockets.

21. Install the timing belt on the crankshaft sprocket, then onto the rear camshaft sprocket.

22. Route the belt to the coolant pump pulley, the front camshaft sprocket and the timing belt tensioner.

23. Apply force counterclockwise to the rear camshaft sprocket with tension on the tight side of the belt and check that timing marks are aligned.

24. Loosen the tensioner bolt 1–2 turns and tighten the timing belt to a tension of 57–84 lbs. (260–380 N).

25. Turn the crankshaft 2 turns clockwise.

26. Readjust the sprocket timing marks and tighten the tensioner bolts.

27. Install the timing covers. Make sure all pieces of packing are positioned in the inner grooves of the covers when installing.

28. Install the crankshaft pulley. Torque the bolt to 108–116 ft. lbs. (150–160 Nm).

29. Install the engine support bracket.

30. Install the power steering pump and reconnect wire harness at the power steering pump pressure switch.

31. Install the engine mounting bracket and remove the engine support fixture.

32. Install the tension pulleys and drive belts.

33. Install the cruise control actuator.

34. Install the engine undercover.

35. Connect the negative battery cable.

Infiniti

3.0L (VG30) ENGINE

1. Disconnect the negative battery cable.

2. Raise and support the front of the vehicle safely.

3. Remove the engine undercovers.

4. Drain the cooling system.

5. Remove the front right-side wheel.

6. Remove the engine side cover.

7. Remove the alternator, power steering and air conditioning compressor drive belts from the engine. When removing the power steering drive belt, loosen the idler pulley from the right-side wheel housing.

8. Remove the upper radiator and water inlet hoses; remove the water pump pulley.

9. Remove the idler bracket of the compressor drive belt.

10. Remove the crankshaft pulley with a suitable puller.

11. Remove the upper and lower timing belt covers and gaskets.

12. Rotate the engine with a socket wrench on the crankshaft pulley bolt to align the punch mark on the left hand camshaft pulley with the mark on the upper rear timing belt cover. Align the punch-mark on the crankshaft with the notch on the oil pump housing and temporarily install the crankshaft pulley bolt to allow for crankshaft rotation.

13. Use a hex wrench to turn the belt tensioner clockwise and tighten the tensioner locknut just enough to hold the tensioner in position. Then, remove the timing belt.

To install:

14. Before installing the timing belt, confirm that No. 1 cylinder is at Top Dead Center (TDC) of the compression stroke. Install tensioner and tensioner spring. If stud is removed, apply locking sealant to threads before installing.

15. Swing tensioner fully clockwise with hexagon wrench and temporarily tighten locknut.

16. Point the arrow on the timing belt toward the front belt cover. Align the white lines on the timing belt with the punch marks on all 3 pulleys.

➡There are 133 total timing belt teeth. If timing belt is installed correctly there will be 40 teeth between left hand and right hand camshaft sprocket timing marks. There will be 43 teeth between left hand camshaft sprocket and crank-shaft sprocket timing marks.

17. Loosen tensioner locknut, keeping tensioner steady with a hexagon wrench.

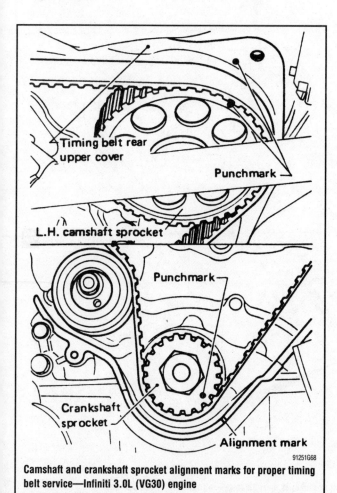

Camshaft and crankshaft sprocket alignment marks for proper timing belt service—Infiniti 3.0L (VG30) engine

18. Swing tensioner 70–80 degrees clockwise with hexagon wrench and temporarily tighten locknut.

19. Turn crankshaft clockwise 2–3 times, then slowly set No. 1 cylinder at TDC of the compression stroke.

20. Push middle of timing belt between right hand camshaft sprocket and tensioner pulley with a force of 22 lbs. (10 kg).

21. Loosen tensioner locknut, keeping tensioner steady with a hexagon wrench.

22. Insert a 0.138 in. (0.35mm) thick and 0.5 in. (12.7mm) wide feeler gauge between the bottom of tensioner pulley and timing belt. Turn crankshaft clockwise and position gauge completely between tensioner pulley and timing belt. The timing belt will move about 2.5 teeth.

23. Tighten tensioner locknut, keeping tensioner steady with a hexagon wrench.

24. Turn crankshaft clockwise or counterclockwise and remove the gauge.

25. Rotate the engine 3 times, then set No. 1 to Top Dead Center (TDC) of the compression stroke.

26. Install the upper and lower timing belt covers with new gaskets.

27. Install the crankshaft pulley. Tighten the pulley bolt to 90–98 ft. lbs. (123–132 Nm).

28. Install the compressor drive belt idler bracket.

29. Install the water pump pulley and tighten the nuts to 12–15 ft. lbs. (16–21 Nm). Install the upper radiator and water inlet hoses.

30. Install the drive belts.

31. Install the engine side cover.

32. Mount the front right wheel.

33. Install the engine undercovers.

34. Lower the vehicle.

35. Fill the cooling system and connect the negative battery cable.

3.0L (VG30DE) ENGINE

1993–94

1. Disconnect the negative battery cable.

2. Remove the engine undercover.

3. Drain the cooling system. Remove both cylinder block drain plug to drain coolant from the block.

4. Remove the air ducts.

5. Remove the radiator, drive belts, cooling fan and coupling.

6. Remove the crankshaft pulley bolt. Remove the crankshaft pulley using a puller.

7. Remove the water inlet and outlet. Remove the starter motor.

8. Remove the front timing covers.

9. Install a suitable stopper bolt into the tensioner arm so the auto-tensioner pusher does not spread out.

10. Set the No. 1 cylinder on Top Dead Center (TDC) of the compression stroke.

11. Remove the auto-tensioner and timing belt.

To install:

12. Confirm that the No. 1 cylinder is on TDC of the compression stroke.

13. Align matchmarks on camshaft and crankshaft sprockets with aligning marks on rear belt cover and oil pump housing.

14. Remove all spark plugs.

15. Align white lines on timing belt with matchmarks on camshaft sprocket and crankshaft sprocket. Point arrow on timing belt towards front and install.

16. Adjust the tensioner arm to give 0.16 in. (4mm) clearance

with pusher of auto-tensioner using a suitable vise, and then insert stopper bolt into tensioner arm so clearance does not change.

17. Install the auto-tensioner and tighten lower bolts hand tight. Push auto-tensioner toward belt until contact is made. Then push slightly more. Turn the crankshaft 10 degrees clockwise and tighten the tensioner nuts to 12–15 ft. lbs. (16–21 Nm).

18. Turn the crankshaft 120 degrees counterclockwise.

19. Loosen tensioner bolts ½ turn and move tensioner away from timing belt as far as it will move.

20. Turn the crankshaft clockwise and set No. 1 cylinder to TDC on its compression stroke.

21. Push the end of a pusher tool J-38387 with 13 lbs. (6 kg) of force and tighten the auto-tensioner bolts to 12–15 ft. lbs. (16–21 Nm).

22. Turn the crankshaft 120 degrees clockwise, then turn crankshaft 120 degrees counterclockwise. Set No. 1 cylinder to TDC on the compression stroke.

23. Prepare a steel plate measuring 0.12 x 0.39 in. (3 x 10mm). Set the plate on the timing belt and push it using a Pusher tool J-38387 or equivalent, with 11 lbs. (5 kg) of force at a point midway between the camshaft sprockets. Also use the tool between the camshaft sprockets and the idler/tensioner pulleys. Deflection should be 0.24–0.28 in. (6–7mm). If not, readjust the tensioner.

24. Confirm the auto-tensioner mounting nuts are tightened to 12–15 ft. lbs. (16–21 Nm). Remove the auto-tensioner stopper bolt.

25. After 5 minutes, check the clearance between the tensioner arm and the pusher stays at 0.138–0.205 in. (3.5–5.2mm).

26. Check for proper installation of the timing belt at all pulleys, then install the timing belt covers. Tighten bolts to 24–38 inch lbs. (3–5 Nm).

27. Install the water inlet and outlet.

28. Install the crankshaft pulley and bolt. Tighten the crankshaft pulley bolt to 159–174 ft. lbs. (216–235 Nm).

29. Install the cooling fan and fan coupling, drive belts and radiator.

30. Install the air ducts.

31. Fill the cooling system and install the engine undercover.

1995–00

1. Disconnect the negative battery cable.

2. Remove all necessary components for access to the front timing covers, then remove the covers.

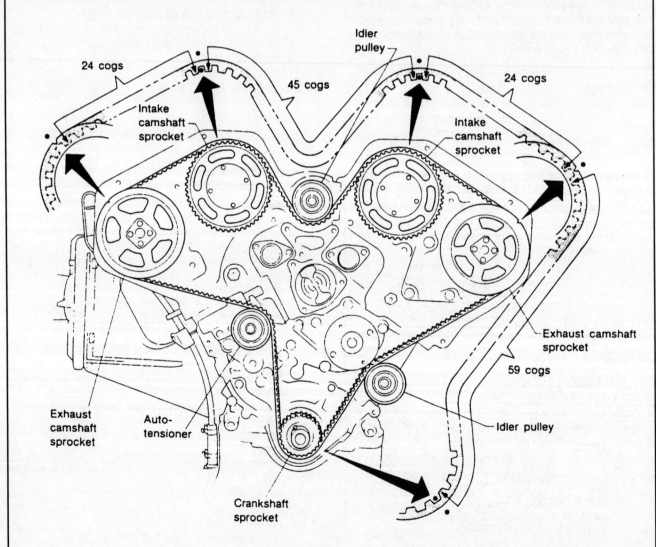

For proper timing belt positioning, ensure the number of teeth between each sprocket is as indicated—Infiniti 3.0L (VG30DE) engine

79235G44

3. Set the No. 1 cylinder on Top Dead Center (TDC) of the compression stroke.

4. The automatic belt tensioner is oil damped and spring operated. Install a 6mm bolt to hold the tensioner back against the spring and release tension on the belt.

5. Remove the auto-tensioner and timing belt.

❊❊ WARNING

Do not rotate the crankshaft or camshaft separately because the pistons will strike the valves causing engine damage.

To install:

6. Confirm that the No. 1 cylinder is at TDC of the compression stroke.

7. Align the marks on the camshaft and crankshaft sprockets with the marks on the rear belt cover and oil pump housing.

8. With the arrows on the timing belt pointing towards the front, align the white lines on the timing belt with the marks on the sprockets and install the belt.

9. To prepare the auto-tensioner for installation, perform the following:

a. Remove the bolt holding the tensioner in position.

b. Use a vise to adjust the gap between the tensioner arm and pusher body to 0.160 in. (4mm).

c. Install the bolt again to hold the arm in this position. Do not try to use the bolt to adjust the gap or the threads will be damaged.

10. Install the auto-tensioner, push it towards the belt to just take up the slack, then tighten the bolts finger-tight.

11. Before adjusting the timing belt tension, the slack must be properly distributed:

a. Turn the crankshaft 10 degrees clockwise and tighten the tensioner bolts and nut to 12–15 ft. lbs. (16–21 Nm). Do not push the auto-tensioner hard or the belt will be adjusted too tight.

b. Turn the crankshaft 120 degrees (⅓ turn) counterclockwise.

c. Loosen the tensioner bolts and nut ½ turn and move the tensioner body away from the timing belt as far as it will move.

d. Turn the crankshaft clockwise to TDC again.

e. Push the tensioner against the belt with a force of 13 lbs. (59 N) using a spring scale or similar tool and tighten the bolts again to 12–15 ft. lbs. (16–21 Nm). The pressure specification is important and a special spring scale tool, J-38387, is available to measure the tensioner force.

12. To check the timing belt tension:

a. Turn the crankshaft 120 degrees (⅓ turn) clockwise, then turn counterclockwise and return the engine to TDC.

b. Prepare a steel plate that is approximately ⅜ in. (10mm) wide and longer than the width of the belt.

c. Set the plate on the timing belt between 2 camshaft sprockets and push against the plate with a force of 11 lbs. (49 N). Note the belt deflection.

d. Repeat the procedure between the other camshaft sprockets and between the exhaust sprockets and idler/tensioner pulleys. There will be a total of 4 measurements.

e. Add the deflection measurements and divide by 4. The average deflection must be 0.240–0.280 in. (6–7mm). If belt tension is not correct, start the entire adjustment procedure again.

13. Confirm the auto-tensioner mounting nuts are tightened to 12–15 ft. lbs. (16–21 Nm) and remove the auto-tensioner stopper bolt.

14. After 5 minutes, measure the clearance between the tensioner arm and the pusher. It should be 0.138–0.205 in. (3.5–5.2mm).

15. Be sure all the sprocket timing marks are correctly aligned. Install the timing belt covers and tighten the bolts to 24–38 inch lbs. (3–5 Nm).

16. Install all applicable components.

Isuzu

1.8L (4FB1) DIESEL ENGINE

1. Disconnect the negative battery cable.
2. Remove the lower engine shrouds.
3. Remove the accessory drive belts.
4. Remove the cooling fan and shroud.
5. Remove the upper timing cover.
6. Drain the cooling system and remove the bypass hose.
7. Set the No. 1 cylinder at Top Dead Center (TDC) on the compression stroke.
8. Insure that the injection pump pulley setting mark is aligned with the front plate, then lock the pulley in place with an 8mm x 1.25 bolt.
9. Remove the valve cover and back off the valve adjustment until all rocker arms are loose.
10. Lock the camshaft in place by fitting a plate to the slot in the rear of the camshaft.
11. Remove the crankshaft pulley and the lower timing cover.
12. Remove the timing belt holder.
13. Remove the timing belt tensioner spring.
14. Loosen the tension pulley and plate bolts and remove the timing belt.

To install:

15. Loosen the camshaft pulley bolt so that the pulley turns freely.
16. Insure that the crankshaft does not turn and install the timing belt.
17. Gather any slack in the timing belt at the tensioner pulley and install the tensioner spring.

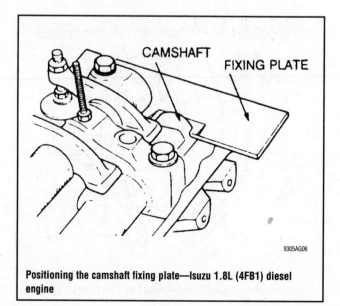

Positioning the camshaft fixing plate—Isuzu 1.8L (4FB1) diesel engine

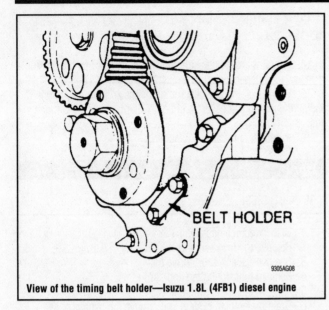

View of the timing belt holder—Isuzu 1.8L (4FB1) diesel engine

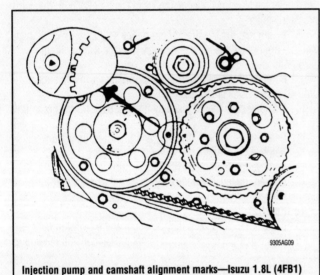

Injection pump and camshaft alignment marks—Isuzu 1.8L (4FB1) diesel engine

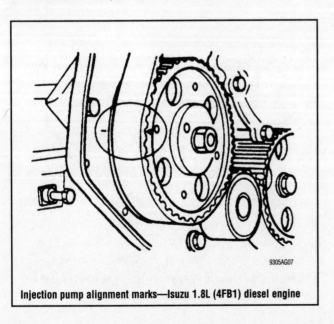

Injection pump alignment marks—Isuzu 1.8L (4FB1) diesel engine

18. Temporarily install the crankshaft pulley. Insure that the crankshaft is at TDC.

19. Tighten the tensioner pulley and plate bolts.

20. Tighten the camshaft pulley bolt.

21. Remove the injection pump lockbolt and the camshaft fixing plate.

22. Remove the crankshaft pulley.

23. Install the timing belt holder and install the lower timing cover. Reinstall the crankshaft pulley.

24. Adjust the valves and install the valve cover.

25. Install the bypass hose and the upper timing cover.

26. Install the cooling fan and shroud.

27. Install the accessory drive belts.

28. Install the lower engine shrouds.

29. Connect the negative battery cable and refill the cooling system.

1.5L (4XC1-U AND 4XC1-T) & 1.6L (4XE1) SOHC ENGINES

1. Disconnect the negative battery cable.

2. Set the crankshaft to Top Dead Center (TDC) on the compression stroke.

3. Support the engine with a jack and remove the right engine mount and brackets.

4. Remove the rear torque rod.

5. Remove the accessory drive belts.

6. Remove the crankshaft pulley.

7. Remove the timing cover.

8. Loosen the tensioner pulley and remove the timing belt.

To install:

9. Insure that the timing marks are aligned and install the timing belt.

10. Use an Allen wrench to rotate the tensioner eccentric and torque the tensioner bolt to 37 ft. lbs. (49 Nm).

11. Rotate the crankshaft 2 complete turns and insure that the timing marks align.

12. Use a belt tension gauge to check the timing belt tension. The belt tension should be 44 lbs. (20 kg).

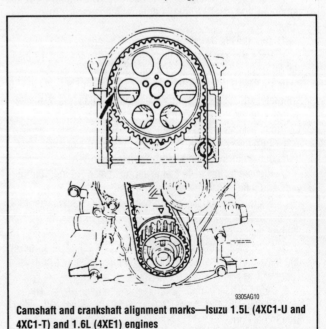

Camshaft and crankshaft alignment marks—Isuzu 1.5L (4XC1-U and 4XC1-T) and 1.6L (4XE1) engines

13. Install the timing cover and crankshaft pulley.
14. Install the accessory drive belts.
15. Install the rear torque rod and the right engine mount.
16. Remove the engine support and connect the negative battery cable.

1.6L (4XE1) & 1.8L (4XF1) DOHC ENGINES

1. Disconnect the negative battery cable.
2. Set the crankshaft to Top Dead Center (TDC) on the compression stroke.
3. Remove the accessory drive belts.
4. Remove the upper timing cover.
5. Support the engine with a jack and remove the right engine mount and brackets.
6. Remove the rear torque rod.
7. Remove the crankshaft pulley and the lower timing cover.
8. Align the timing marks and lock the camshafts in place by inserting bolts in the holes provided.
9. Loosen the tensioner pulley bolt and remove the timing belt.

To install:
10. Install the timing belt in the following order:
   a. Crankshaft timing pulley
   b. Water pump pulley
   c. Idler pulley
   d. Exhaust camshaft pulley
   e. Intake camshaft pulley
   f. Tensioner pulley
11. Use an Allen wrench to rotate the tensioner pulley eccentric and tighten the tensioner bolt.
12. Remove the camshaft pulley locking bolts.
13. Rotate the crankshaft 2 complete turns and insure that the timing marks align.

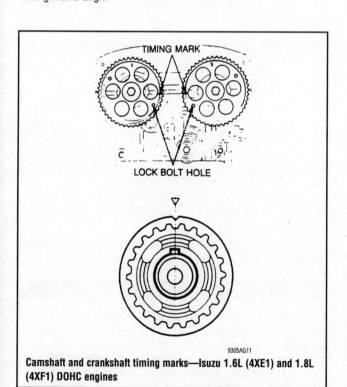

Camshaft and crankshaft timing marks—Isuzu 1.6L (4XE1) and 1.8L (4XF1) DOHC engines

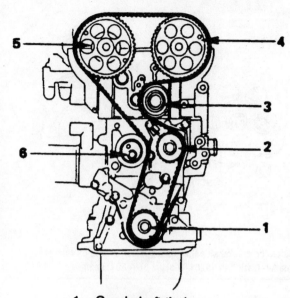

1. **Crankshaft timing sprocket**
2. **Water pump pulley**
3. **Idler pulley**
4. **Exhaust camshaft sprocket**
5. **Intake camshaft sprocket**
6. **Tensioner pulley.**

9305AG12

View of the timing belt assembly—Isuzu 1.6L (4XE1) and 1.8L (4XF1) DOHC engines

14. Install the lower timing cover and the crankshaft pulley. Tighten the crankshaft pulley bolt to 123 ft. lbs. (170 Nm).
15. Install the rear torque rod.
16. Install the right engine mount and brackets. Remove the engine support.
17. Install the accessory drive belts and connect the negative battery cable.

2.0L (4ZC1-T) & 2.3L (4ZD1) ENGINES

1. Disconnect the negative battery cable.
2. Loosen and remove the engine accessory drive belts.
3. Remove the cooling fan assembly and the water pump pulley.
4. Drain the fluid from the power steering reservoir.
5. Unbolt and remove the power steering pump. Unbolt the hydraulic line brackets from the upper timing cover and move the pump out of the work area without disconnecting the hydraulic lines.
6. Disconnect and remove the starter motor if a flywheel holder tool No. J-38674 or equivalent, is to be used.
7. Remove the upper timing belt cover.
8. Rotate the crankshaft to set the engine at Top Dead Center (TDC) of its compression for the No. 1 cylinder. The arrow mark on the camshaft sprocket will be aligned with the mark on the rear timing cover.

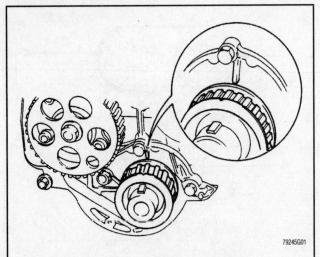

Align the crankshaft pulley timing mark the with oil retainer setting mark—Isuzu 2.0L (4ZC1-T) and 2.3L (4ZD1) engines

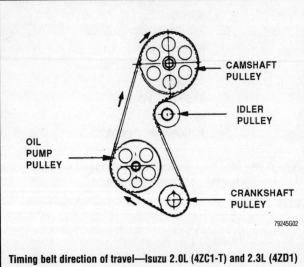

Timing belt direction of travel—Isuzu 2.0L (4ZC1-T) and 2.3L (4ZD1) engines

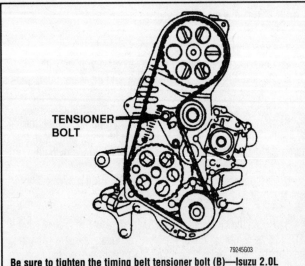

Be sure to tighten the timing belt tensioner bolt (B)—Isuzu 2.0L (4ZC1-T) and 2.3L (4ZD1) engines

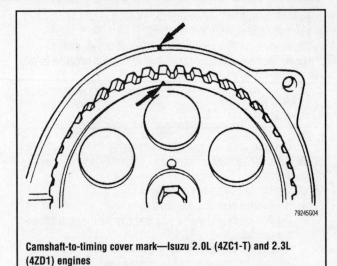

Camshaft-to-timing cover mark—Isuzu 2.0L (4ZC1-T) and 2.3L (4ZD1) engines

9. Remove the crankshaft pulley.

10. Remove the lower timing belt cover.

11. Verify that the engine is set at TDC/compression for the No. 1 cylinder. The notch on the crankshaft sprocket will be aligned with the pointer on the oil seal retainer.

12. Release and remove the tensioner spring to release the timing belt's tension.

13. Remove the timing belt.

14. Unbolt the tensioner pulley bracket from the engine's front cover.

15. If necessary, unbolt and remove the camshaft sprockets. Use a puller to remove the crankshaft pulley if necessary. Don't lose the crankshaft sprocket key.

To install:

16. If removed, install the camshaft and crankshaft sprockets. Align the camshaft and crankshaft timing marks and be sure to install any keys. Tighten the camshaft sprocket bolt to 43 ft. lbs. (59 Nm).

17. Install the tensioner assembly. Tighten the tensioner mounting bolt to 14 ft. lbs. (19 Nm) and the cap bolt to 108 inch lbs. (13 Nm).

18. Be sure the crankshaft and the camshaft sprockets are aligned with their timing marks. Install the timing belt onto the sprockets using the following sequence: first around the crankshaft sprocket; second around the oil pump sprocket; third around the camshaft sprocket.

19. Loosen the tensioner mounting bolt. This will allow the tensioner spring to apply pressure to the timing belt.

20. After the spring has pulled the timing belt as far as possible, temporarily tighten the tensioner mounting bolt to 14 ft. lbs. (19 Nm).

➡ Remove the flywheel holder before rotating the crankshaft. Reinstall the holder to tighten the crankshaft pulley bolt.

❄ WARNING

If any binding is felt when adjusting the timing belt tension by turning the crankshaft, STOP turning the engine, because the pistons may be hitting the valves.

21. Rotate the crankshaft counterclockwise 2 complete revolutions to check the rotation of the belt and the alignment of the tim-

ing marks. Listen for any rubbing noises which may mean the belt is binding.

22. Loosen the tensioner pulley bolt to allow the spring to adjust the correct tension. Then, retighten the tensioner pulley bolt to 14 ft. lbs. (19 Nm).

23. Install the lower timing cover and the crankshaft pulley.

24. Tighten the crankshaft pulley bolt to 87 ft. lbs. (118 Nm). Tighten the small pulley bolts to 72 inch lbs. (8 Nm).

25. Install the upper timing cover.

26. If removed, install the starter and tighten the bolts to 30 ft. lbs. (40 Nm).

27. Install the power steering pump. If the hydraulic lines were disconnected, refill and bleed the power steering system.

28. Install the water pump pulley and tighten its nut to 20 ft. lbs. (26 Nm).

29. Install the cooling fan assembly.

30. Install and adjust the accessory drive belts.

31. Connect the negative battery cable.

KIA

SINGLE OVERHEAD CAM ENGINE

1994–95

1. Disconnect the negative battery cable.
2. Raise and support the vehicle.
3. Remove the accessory drive belts.
4. Remove the water pump pulley.
5. Remove the 4 crankshaft pulley bolts and remove the crankshaft pulley and pulley plate.
6. Remove the crankshaft pulley center bolt and remove the crankshaft pulley boss.
7. Remove the upper and lower timing belt covers.
8. Rotate the crankshaft to align the crankshaft and camshaft timing marks.
9. Loosen the tensioner pulley lockbolt.
10. Pry the tensioner pulley back to remove the tension on the timing belt, and tighten the lockbolt.
11. Remove the timing belt.

To install:

12. Insure that the crankshaft and camshaft timing marks are aligned and install the timing belt.

13. Loosen the tensioner lockbolt and allow the tensioner spring to apply tension to the timing belt. Do not apply additional tension. Tighten the lockbolt to 14–19 ft. lbs. (19–25 Nm).

14. Rotate the crankshaft 2 full turns clockwise and check that the timing marks on the crankshaft and camshafts align with the marks on the engine.

15. Check timing belt deflection by applying 22 lbs. (10 kg) pressure to the belt midway between the crankshaft and camshaft pulleys. Belt deflection should be 0.43–0.51 inches. (11–13 Nm).

16. Install the upper and lower timing covers.

17. Install the crankshaft pulley boss and the center bolt. Hold the crankshaft from turning and tighten the center bolt to 116–122 ft. lbs. (157–166 Nm).

18. Install the water pump pulley.

19. Install the crankshaft pulley and pulley plate.

20. Install the accessory drive belts.

21. Lower the vehicle.

22. Connect the negative battery cable.

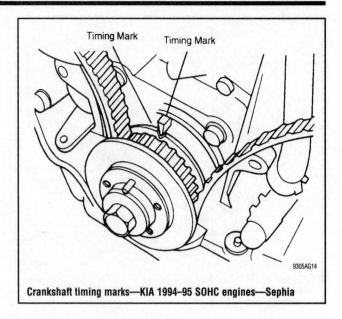

Crankshaft timing marks—KIA 1994–95 SOHC engines—Sephia

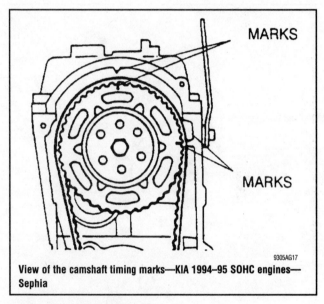

View of the camshaft timing marks—KIA 1994–95 SOHC engines—Sephia

DOUBLE OVERHEAD CAM ENGINES

1996–97

1. Disconnect the negative battery cable.
2. Raise and support the vehicle.
3. Remove the accessory drive belts.
4. Remove the water pump pulley.
5. Remove the 4 crankshaft pulley bolts and remove the crankshaft pulley and pulley plate.
6. Remove the spark plug wires and the spark plugs.
7. Remove the valve cover.
8. Remove the upper, middle, and lower timing belt covers.
9. Turn the crankshaft so that the timing mark on the crankshaft timing pulley aligns with the timing mark on the engine, and the camshaft pulley timing marks align with the timing marks on the seal plate.
10. Loosen the tensioner pulley lockbolt.
11. Pry the tensioner pulley back to remove the tension on the timing belt, and tighten the lockbolt.

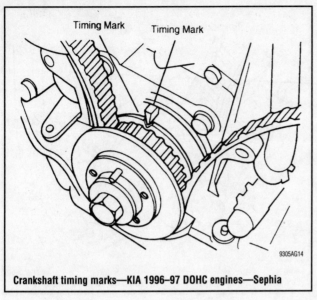

Crankshaft timing marks—KIA 1996–97 DOHC engines—Sephia

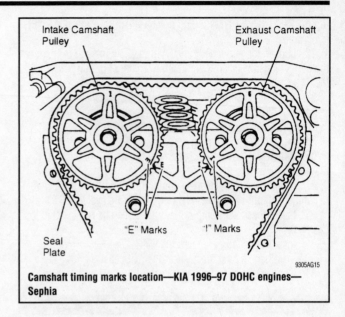

Camshaft timing marks location—KIA 1996–97 DOHC engines—Sephia

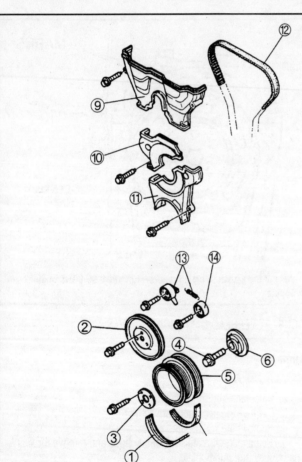

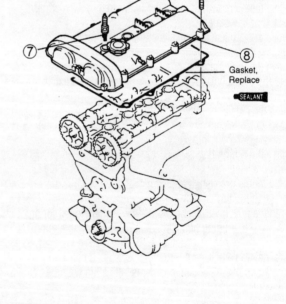

1. Drive Belts
2. Water Pump Pulley
3. Crankshaft Pulley Plate
4. Crankshaft Pulley Lock Bolt
5. Crankshaft Pulley
6. Crankshaft Pulley Boss
7. Spark Plug
8. Cylinder Head Cover
9. Upper Timing Belt Cover
10. Middle Timing Belt Cover
11. Lower Timing Belt Cover
12. Timing Belt
13. Tensioner Pulley, Tensioner Spring
14. Idler Pulley

Exploded view of the timing belt assembly—KIA 1996–97 DOHC engines—Sephia

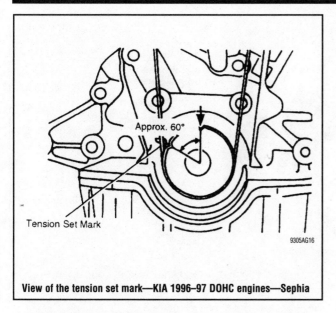

View of the tension set mark—KIA 1996–97 DOHC engines—Sephia

12. Remove the crankshaft pulley center bolt and remove the crankshaft pulley boss.

13. Remove the timing belt.

To install:

14. Insure that the crankshaft and camshaft timing marks are aligned and install the timing belt.

15. Install the crankshaft pulley boss and the center bolt. Hold the crankshaft from turning and tighten the center bolt to 116–122 ft. lbs. (157–166 Nm).

16. Loosen the tensioner lockbolt and allow the tensioner spring to apply tension to the timing belt. Do not apply additional tension. Tighten the lockbolt to 28–38 ft. lbs. (38–51 Nm).

17. Rotate the crankshaft 2 full turns clockwise and check that the timing marks on the crankshaft and camshafts align with the marks on the engine and seal plate.

➡**Never rotate the crankshaft counterclockwise.**

18. Rotate the crankshaft 1⅚ turns and align the crankshaft timing mark with the tension set mark.

19. Check the timing belt tension by applying 22 lbs. (10 kg) of force to the belt between the camshaft pulleys. Timing belt deflection should be 0.36–0.45 inches. (10–11 Nm).

20. Install the lower, middle, and upper timing covers.

21. Install the valve cover, spark plugs, and spark plug wires.

22. Install the water pump pulley.

23. Install the crankshaft pulley and pulley plate.

24. Install the accessory drive belts.

25. Lower the vehicle.

26. Connect the negative battery cable.

1998–00

1. Disconnect the negative battery cable.

2. Remove the accessory drive belts.

3. Remove the generator.

4. Remove the water pump and crankshaft pulleys.

5. Remove the timing belt guide plate.

6. Remove the upper and lower timing belt covers.

7. Position the crankshaft so that the timing mark is aligned with the timing mark on the engine.

8. Verify that the **I** and **E** mark on the camshaft pulley align with the mark on the cylinder head.

➡**Do not move the crankshaft or camshaft once the timing marks have been correctly positioned.**

9. If the timing belt is to be reused, mark the direction of rotation on the timing belt.

10. Remove the timing belt tensioner pulley.

11. Remove the timing belt.

To install:

12. Install the timing belt tensioner pulley, move the tensioner to its furthest point and tighten the lockbolt.

13. Install the timing belt onto the pulleys, as follows, crankshaft pulley first, then the idler pulley, exhaust camshaft pulley, intake camshaft pulley and the tensioner pulley.

14. Loosen the tensioner pulley and allow the tensioner spring to apply tension on the belt, then tighten the lockbolt to 28–38 ft. lbs. (38–51 Nm).

15. Rotate the crankshaft clockwise 2 turns and be sure all marks are still correctly aligned.

16. Install the remaining components in the revere order of the removal noting the following torque specifications:

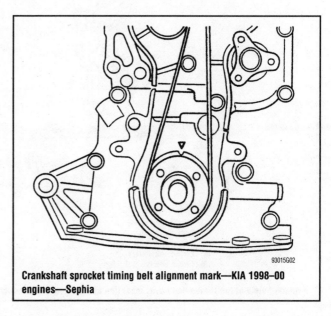

Crankshaft sprocket timing belt alignment mark—KIA 1998–00 engines—Sephia

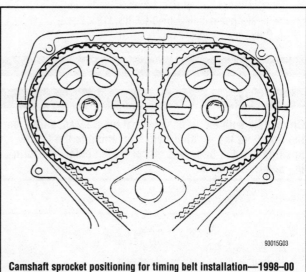

Camshaft sprocket positioning for timing belt installation—1998–00 KIA Sephia

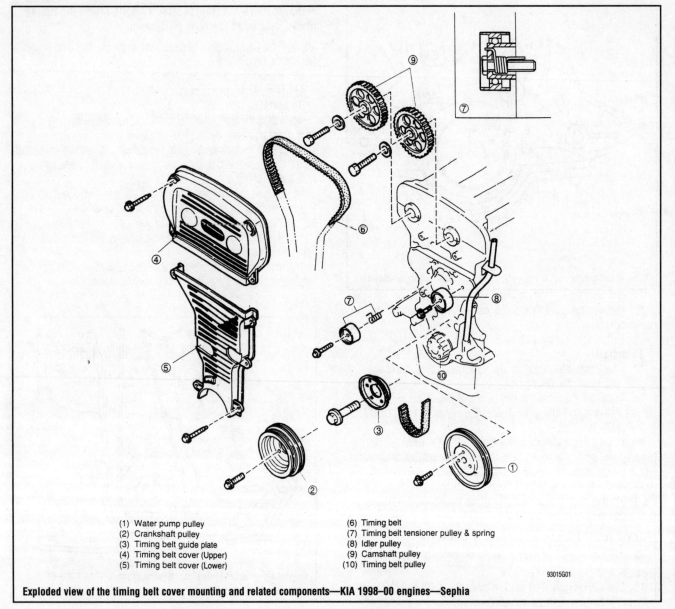

(1) Water pump pulley
(2) Crankshaft pulley
(3) Timing belt guide plate
(4) Timing belt cover (Upper)
(5) Timing belt cover (Lower)

(6) Timing belt
(7) Timing belt tensioner pulley & spring
(8) Idler pulley
(9) Camshaft pulley
(10) Timing belt pulley

93015G01

Exploded view of the timing belt cover mounting and related components—KIA 1998–00 engines—Sephia

- Crankshaft pulley: 10–13 ft. lbs. (13–17 Nm)
- Water pump pulley: 10–13 ft. lbs. (13–17 Nm)
17. Connect the negative battery cable.

Lexus

❄ CAUTION

On models with an air bag, wait at least 90 seconds from the time that the ignition switch is turned to the LOCK position and the battery is disconnected before performing any further work.

2.5L (2VZ-FE) ENGINE

1. Disconnect the negative battery cable.
2. Remove the strut tower brace.
3. Remove the cruise control actuator.
4. Remove the power steering reservoir without disconnecting the hoses.

5. Raise and support the vehicle.
6. Remove the right front wheel and inner splash shield.
7. Remove the accessory drive belts.
8. Support the engine with a jack and remove the right engine mount stays and the engine mount.
9. Remove the No. 2 (upper) timing cover.
10. Remove the engine mount bracket.
11. Turn the crankshaft to align the camshaft timing marks with the timing marks on the No. 3 (rear) timing cover. Align the crankshaft pulley timing mark with the **0** mark on the No. 1 (lower) timing cover.
12. Remove the crankshaft pulley bolt and press the pulley off the crankshaft with special tool 09213-60017 or equivalent.
13. Remove the No. 1 (lower) timing cover.
14. Remove the timing belt guide.
15. Remove the timing belt tensioner and remove the timing belt.
To install:
16. Install the timing belt to the crankshaft timing pulley and install the timing belt guide.
17. Install the No. 1 timing cover and install the crankshaft pulley.

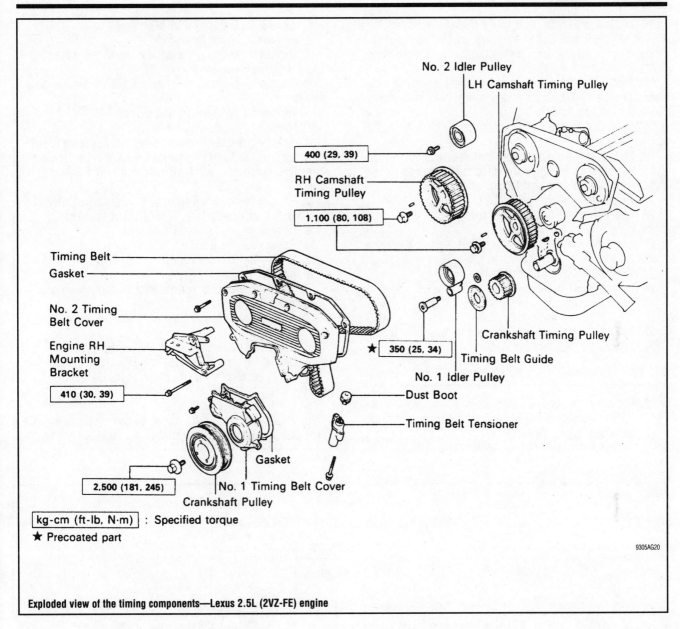

No. 2 Idler Pulley
LH Camshaft Timing Pulley
400 (29, 39)
RH Camshaft Timing Pulley
1,100 (80, 108)
Timing Belt
Gasket
No. 2 Timing Belt Cover
Engine RH Mounting Bracket
410 (30, 39)
★ 350 (25, 34)
Crankshaft Timing Pulley
Timing Belt Guide
No. 1 Idler Pulley
Dust Boot
Timing Belt Tensioner
Gasket
2,500 (181, 245)
No. 1 Timing Belt Cover
Crankshaft Pulley
kg-cm (ft-lb, N·m) : Specified torque
★ Precoated part
9305AG20

Exploded view of the timing components—Lexus 2.5L (2VZ-FE) engine

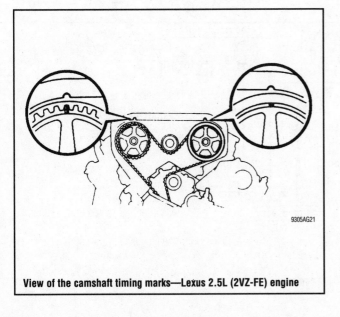

View of the camshaft timing marks—Lexus 2.5L (2VZ-FE) engine

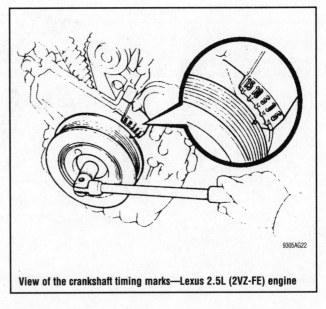

View of the crankshaft timing marks—Lexus 2.5L (2VZ-FE) engine

18. Insure that the timing mark on the crankshaft pulley is aligned with the **0** mark on the No. 1 cover, and that the camshaft timing marks are aligned with the marks on the No. 3 timing cover.

19. Install the timing belt.

20. Prepare the timing belt tensioner by compressing it in a vise and installing a small Allen wrench through the holes in the tensioner body and pushrod.

21. Install the timing belt tensioner and torque the bolts to 20 ft. lbs. (26 Nm).

22. Remove the Allen wrench from the tensioner.

23. Rotate the crankshaft 2 complete turns clockwise and insure that the crankshaft and camshaft timing marks align.

24. Tighten the crankshaft pulley bolt to 181 ft. lbs. (245 Nm).

25. Install the engine mount bracket and install the No. 2 timing cover.

26. Install the right engine mount and stays. Remove the engine support jack.

27. Install the accessory drive belts.

28. Install the splash shield and the right front wheel.

29. Lower the vehicle.

30. Install the power steering reservoir.

31. Install the cruise control actuator and the strut tower brace.

32. Connect the negative battery cable.

3.0L (3VZ-FE) ENGINE

1. Disconnect the negative battery cable.
2. Remove the washer fluid tank and the coolant reservoir tank.
3. Raise and support the vehicle.
4. Remove the right front wheel and inner splash shield.

5. Remove the accessory drive belts.

6. Remove the engine torque rod.

7. Support the engine with a jack and remove the right engine mount stays and the engine mount.

8. Remove the power steering reservoir without disconnecting the hoses.

9. Remove the No. 2 (upper) timing cover.

10. Remove the engine mount bracket.

11. Turn the crankshaft to align the camshaft timing marks with the timing marks on the No. 3 (rear) timing cover. Align the crankshaft pulley timing mark with the **0** mark on the No. 1 (lower) timing cover.

12. Remove the crankshaft pulley bolt and press the pulley off the crankshaft with special tool 09213-60017 or equivalent.

13. Remove the No. 1 (lower) timing cover.

14. Remove the timing belt guide.

15. Remove the timing belt tensioner and remove the timing belt.

To install:

16. Install the timing belt to the crankshaft timing pulley and install the timing belt guide.

17. Install the No. 1 timing cover and install the crankshaft pulley.

18. Insure that the timing mark on the crankshaft pulley is aligned with the **0** mark on the No. 1 cover, and that the camshaft timing marks are aligned with the marks on the No. 3 timing cover.

19. Install the timing belt.

20. Prepare the timing belt tensioner by compressing it in a vise and installing a small Allen wrench through the holes in the tensioner body and pushrod.

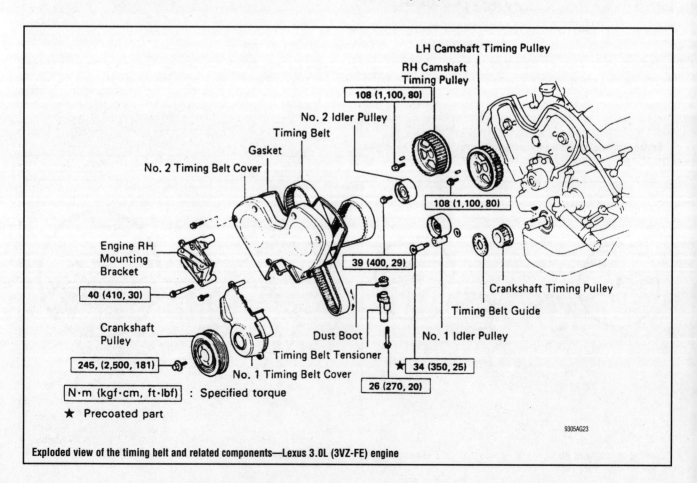

Exploded view of the timing belt and related components—Lexus 3.0L (3VZ-FE) engine

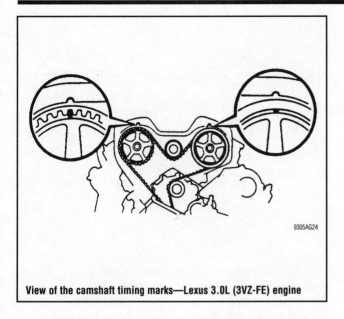

View of the camshaft timing marks—Lexus 3.0L (3VZ-FE) engine

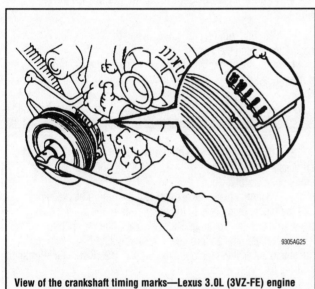

View of the crankshaft timing marks—Lexus 3.0L (3VZ-FE) engine

21. Install the timing belt tensioner and torque the bolts to 20 ft. lbs. (26 Nm).
22. Remove the Allen wrench from the tensioner.
23. Rotate the crankshaft 2 complete turns clockwise and insure that the crankshaft and camshaft timing marks align.
24. Tighten the crankshaft pulley bolt to 181 ft. lbs. (245 Nm).
25. Install the engine mount bracket and install the No. 2 timing cover.
26. Install the power steering reservoir.
27. Install the right engine mount and stays.
28. Install the engine torque rod. Remove the engine support jack.
29. Install the accessory drive belts.
30. Install the splash shield and the right front wheel.
31. Lower the vehicle.
32. Install the coolant reservoir tank and the washer fluid tank.
33. Connect the negative battery cable.

3.0L (1MZ-FE) ENGINE

1. Disconnect the negative battery cable.
2. Remove all necessary components for access to the upper timing belt cover. Remove the 8 bolts and lift off the upper (No. 2) cover.
3. Paint matchmarks on the timing belt at all points where it meshes with the pulleys and the lower timing cover.
4. Set the No. 1 cylinder to Top Dead Center (TDC) of the compression stroke and check that the timing marks on the camshaft timing pulleys are aligned with those on the No. 3 timing cover. If not, turn the engine 1 complete revolution (360 degrees) and check again.
5. Remove the timing belt tensioner and the dust boot.
6. Turn the right camshaft pulley clockwise slightly to release tension, then remove the timing belt from the pulleys.
7. Remove the upper (No. 3) and lower (No. 1) timing belt covers.
8. Remove the timing belt guide.
9. Remove the timing belt from the engine.

➡If the timing belt is to be reused, draw a directional arrow on the timing belt in the direction of engine rotation (clockwise) and place matchmarks on the timing belt and crankshaft gear to match the drilled mark on the pulley.

To install:

➡If the old timing belt is being reinstalled, be sure the directional arrow is facing in the original direction and that the belt and crankshaft gear matchmarks are properly aligned.

10. Install the lower (No. 1) timing cover and tighten the bolts.
11. Set the No. 1 cylinder to TDC again. Turn the right camshaft

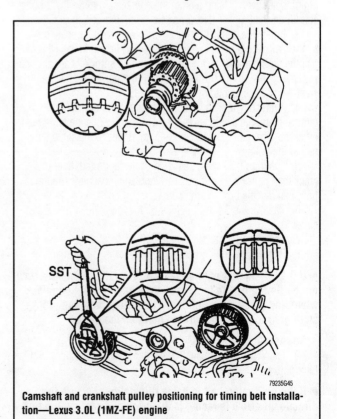

Camshaft and crankshaft pulley positioning for timing belt installation—Lexus 3.0L (1MZ-FE) engine

until the knock pin hole is aligned with the timing mark on the No. 3 belt cover. Turn the left pulley until the marks on the pulley are aligned with the mark on the No. 3 timing cover.

12. Check that the mark on the belt matches with the edge of the lower cover. If not, shift it on the crank pulley until it does. Turn the left pulley clockwise a bit and align the mark on the timing belt with the timing mark on the pulley. Slide the belt over the left pulley. Now move the pulley until the marks on it align with the one on the No. 3 cover. There should be tension on the belt between the crankshaft pulley and the left camshaft pulley.

13. Align the installation mark on the timing belt with the mark on the right side camshaft pulley. Hang the belt over the pulley with the flange facing inward. Align the timing marks on the right pulley with the one on the No. 3 cover and slide the pulley onto the end of the camshaft. Move the pulley until the camshaft knock pin hole is aligned with the groove in the pulley, then install the knock pin. Tighten the bolt to 55 ft. lbs. (75 Nm).

14. Position a plate washer between the timing belt tensioner and the block, then press in the pushrod until the holes are aligned between it and the housing. Slide a 0.05 in. Allen wrench through the hole to keep the pushrod set. Install the dust boot, then install the tensioner. Tighten the bolts to 20 ft. lbs. (26 Nm). Don't forget to pull out the Allen wrench.

15. Turn the crankshaft clockwise 2 complete revolutions and check that all marks are still in alignment. If they aren't, remove the timing belt and start over again.

16. Install the remaining components.

17. Connect the negative battery cable.

3.0L (2JZ-GE) ENGINE

1. Disconnect the negative battery cable.

2. Remove all necessary components for access to the upper timing belt covers. Using a 5mm Allen wrench, remove the 9 bolts and lift off the 2 upper (No. 2 and No. 3) timing belt covers.

3. Rotate the crankshaft pulley clockwise so its groove is aligned with the **0** mark in the No. 1 (lower) timing cover. Check that the timing marks on the camshaft timing sprockets are aligned with the marks on the No. 4 (inner) cover. If not, rotate the crankshaft 1 complete revolution (360 degrees).

4. Alternately loosen the 2 tensioner mounting bolts and remove them, the tensioner and the dust boot. Slide the timing belt off of the 2 camshaft sprockets. It's a good idea to matchmark the belt to the pulleys.

5. Ensuring the timing belt is securely supported, hold the crankshaft pulley with a spanner wrench and loosen the mounting bolt. Remove the bolt and the pulley.

6. Remove the 5 bolts, then lift off the lower No. 1 timing belt cover.

7. Remove the timing belt guide.

8. Remove the timing belt.

➡️ **If the timing belt is to be reused, draw a directional arrow on the timing belt in the direction of engine rotation (clockwise) and place matchmarks on the timing belt and crankshaft gear to match the drilled mark on the pulley.**

To install:

9. Install the timing belt on the crankshaft timing pulley and the idler pulleys.

➡️ **If the old timing belt is being reinstalled, be sure the directional arrow is facing in the original direction and that**

79235G46

Set the engine to TDC by aligning the marks before removing the lower timing cover—Lexus 3.0L (2JZ-GE) engine

the belt and crankshaft gear matchmarks are properly aligned.

10. Install the timing belt guide. Install the lower (No. 1) timing cover and tighten the bolts.

11. Align the crankshaft pulley set key with the key groove on the pulley and slide the pulley on. Tighten the bolt to 239 ft. lbs. (324 Nm).

12. Set the No. 1 cylinder to TDC again. Turn the camshaft until the sprocket timing marks are aligned with the timing marks on the No. 4 belt cover.

13. Check that the marks on the belt matches with those on the sprockets, then slide it over the sprockets. If not, shift it on the crank pulley until it does.

14. Position a plate washer between the timing belt tensioner and the block, then press in the pushrod until the holes are aligned between it and the housing. Slide a 1.5mm Allen wrench through the hole to keep the pushrod set. Install the dust boot, then install the tensioner. Tighten the bolts to 20 ft. lbs. (26 Nm). Don't forget to pull out the Allen wrench.

15. Turn the crankshaft clockwise 2 complete revolutions and check that all marks are still in alignment. If they aren't, remove the timing belt and start over again.

16. Position new gaskets, then install the upper (No. 2 and No. 3) timing covers.

17. Connect the negative battery cable.

4.0L (1UZ-FE) ENGINE

1. Disconnect the negative battery cable.

2. Remove all necessary components for access to the right-hand side No. 3 and No. 2, and left-hand side No. 2 timing belt covers, then remove the covers.

3. Turn the crankshaft pulley and align it's groove with the timing mark **0** of the No. 1 timing cover. Check that the timing marks of the camshaft timing pulleys and timing belt rear plates are aligned. If not, turn the crankshaft 1 full revolution (360 degrees).

4. Remove the timing belt tensioner. Using the proper tool, loosen the tension between the left side and right side timing pulleys by slightly turning the left side camshaft clockwise.

5. Disconnect the timing belt from the camshaft timing pulleys. Using the proper tool, remove the bolt and the timing pulleys.

6. Remove the bolt and the crankshaft pulley with the proper tool. Remove the fan bracket. On the SC400, remove the hydraulic pump.

7. Remove the mounting bolts and the No. 1 timing belt cover.

8. Remove the 2 upper and lower timing belt covers.

9. Remove the timing belt guide (No. 1 crank position sensor plate).

10. Remove the timing belt.

➡**If the timing belt is to be reused, draw a directional arrow on the timing belt in the direction of engine rotation (clockwise) and place matchmarks on the timing belt and crankshaft gear to match the drilled mark on the pulley.**

To install:

11. Align the installation mark on the timing belt with the drilled mark of the crankshaft timing pulley. Install the timing belt on the crankshaft timing pulley, No. 1 idler pulley and the No. 2 idler pulley.

➡**If the old timing belt is being reinstalled, be sure the directional arrow is facing in the original direction and that the belt and crankshaft gear matchmarks are properly aligned.**

12. Install the timing belt guide (No. 1 crank angle sensor plate) with the cup side facing forward. Replace the timing belt cover spacer.

13. Install the No. 1 timing belt cover and tighten the mounting bolts. On the SC400, install the hydraulic pump. Install the fan bracket.

14. Align the pulley set key on the crankshaft with the key groove of the pulley. Install the pulley, using the proper tool to tap in the pulley. Tighten the pulley bolt to 181 ft. lbs. (245 Nm).

15. Align the knock pin on the right side camshaft with the knock pin of the timing pulley. Slide on the timing pulley with the right side mark facing forward. Tighten the bolt to 80 ft. lbs. (108 Nm).

16. Align the knock pin on the left side camshaft with the knock pin of the timing pulley. Slide on the timing pulley with the left side mark facing forward. Tighten the bolt to 80 ft. lbs. (108 Nm).

17. Turn the crankshaft pulley and align its groove with the **0** timing mark on the No. 1 timing belt cover. Using the proper tool, turn the crankshaft timing pulley and align the timing marks of the camshaft timing pulley and the timing belt rear plate.

18. Install the timing belt to the left side camshaft timing pulley by:

   a. Using the proper tool, slightly turn the left side timing pulley clockwise. Align the installation mark of the timing belt with the timing mark of the camshaft timing pulley and hang the timing belt on the left side camshaft pulley.

   b. Using the proper tool, align the timing marks of the left side camshaft pulley and the timing belt rear plate.

   c. Check that the timing belt has tension between crankshaft timing pulley and the left side camshaft pulley.

19. Install the timing belt to the right side camshaft timing pulley by:

   a. Using the proper tool, slightly turn the right side timing pulley clockwise. Align the installation mark of the timing belt with the timing mark of the camshaft timing pulley and hang the timing belt on the right side camshaft pulley.

   b. Using the proper tool, align the timing marks of the right side camshaft pulley and the timing belt rear plate.

   c. Check that the timing belt has tension between the crankshaft timing pulley and the right side camshaft pulley.

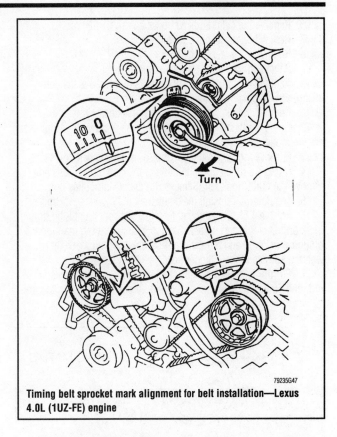

Timing belt sprocket mark alignment for belt installation—Lexus 4.0L (1UZ-FE) engine

79235G47

20. The timing belt tensioner must be set prior to installation. The tensioner can be set as follows:

   a. Place a plate washer between the tensioner and a block. Using a suitable press, press in the pushrod using 220–2205 lbs. (100–1000kg) of pressure.

   b. Align the holes of the pushrod and housing, pass the proper tool (0.05 in. Allen wrench) through the holes to keep the setting position of the pushrod.

   c. Release the press and install the dust boot on the tensioner.

21. Install the tensioner and tighten the bolts to 20 ft. lbs. (26 Nm). Remove the tool from the tensioner.

22. Turn the crankshaft pulley 2 complete revolutions from TDC-to-TDC. Always turn the crankshaft clockwise. Check that each pulley aligns with the timing marks.

23. Install all remaining components in the reverse order of removal.

24. Connect the negative battery cable.

Mazda

1.5L (Z5), 1.6L (B6) &
1.8L (BP) DOHC ENGINES

1. Disconnect the negative battery cable.

2. Raise and safely support the vehicle.

3. Remove the right front wheel and the right side splash shield.

4. Lower the vehicle.

5. Label and disconnect the spark plug wires. Remove the spark plugs.

➡**Spark plugs are removed to make it easier to rotate the engine.**

6. Remove the cylinder head cover.

7. Remove the accessory drive belts.

8. Remove the water pump pulley.

9. Raise and safely support the vehicle.

10. Remove the crankshaft pulley bolts and remove the crankshaft pulley and baffle plate.

11. Remove the oil dipstick and tube, if necessary.

12. Using a suitable engine holding brace, remove the front engine mount bracket.

13. Remove the upper, middle and lower timing belt covers.

14. Using a suitable tool, hold the crankshaft pulley boss and remove the pulley lockbolt. Remove the crankshaft pulley boss.

15. Temporarily reinstall the crankshaft pulley boss lockbolt.

16. For the 1.5L (ZD) engine, turn the crankshaft until the timing mark on the crankshaft sprocket aligns with the timing mark on the oil pump housing and the camshaft sprocket timing marks align with the surface of the cylinder head.

17. For the 1.6L (B6) or 1.8L (BP) engine, turn the crankshaft until the timing mark on the crankshaft sprocket aligns with the tim-

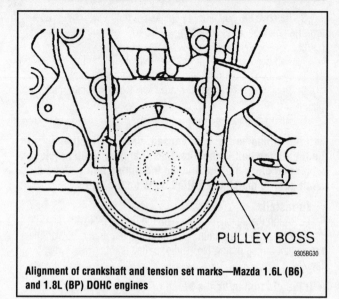

Alignment of crankshaft and tension set marks—Mazda 1.6L (B6) and 1.8L (BP) DOHC engines

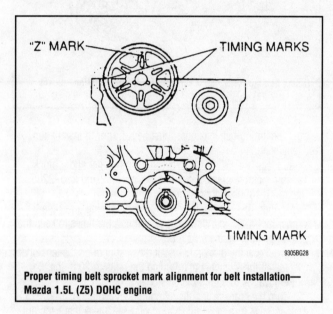

Proper timing belt sprocket mark alignment for belt installation—Mazda 1.5L (Z5) DOHC engine

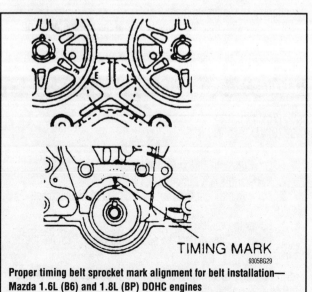

Proper timing belt sprocket mark alignment for belt installation—Mazda 1.6L (B6) and 1.8L (BP) DOHC engines

ing mark on the oil pump housing and the camshaft sprocket timing marks align with the marks on the seal plate.

18. Loosen the tensioner pulley lockbolt and pull the tensioner pulley away from the center of the engine to reduce the tension on the timing belt.

19. If the timing belt is to be reused, mark the direction of rotation on the timing belt.

20. Remove the timing belt.

21. Remove the crankshaft pulley boss lockbolt.

22. On the 1.6L (B6) or 1.8L (BP) engine, insert a camshaft sprocket holding tool between the camshaft sprockets.

23. Remove the tensioner and idler pulleys and check for free movement and abnormal noises. Replace as needed.

To install:

24. Be sure the timing marks on the camshaft and crankshaft sprockets are still aligned.

25. Install the idler pulleys and torque the bolts to 28–38 ft. lbs. (38–51 Nm).

26. Install the tensioner pulley and position the tensioner with the spring fully extended and temporarily torque the lockbolt.

27. On 1.5L (ZD) engine, install the timing belt in the following order:

    a. Crankshaft pulley

    b. Lower idler pulley

    c. Upper idler pulley

    d. Camshaft pulley

    e. Tensioner

28. On the 1.6L (B6) or 1.8L (BP) engine, install the timing belt in the following order:

    a. Crankshaft pulley

    b. Idler pulley

    c. Left side camshaft pulley

    d. Right side camshaft pulley

    e. Tensioner

29. Install the crankshaft pulley boss and lockbolt.

30. Remove the holding tool from between the camshaft sprockets.

31. Rotate the crankshaft in the direction of rotation 1⅚ turns to align the crankshaft timing mark with the tension set mark.

32. Loosen the tensioner lockbolt to apply tension to the timing belt. Torque the mounting bolt to 28–38 ft. lbs. (38–51 Nm).

33. Rotate the crankshaft in the direction of rotation 2⅛ turns and be sure all marks are correctly aligned.

34. Torque the crankshaft pulley boss lockbolt to 116–122 ft. lbs. (157–166 Nm).

35. Install the upper, middle and lower timing belt covers.

36. Install the front engine mount bracket and remove the engine brace.

37. Install the oil dipstick and tube, if removed.

38. Install the crankshaft pulley, baffle plate and bolts. Torque the bolts to 12 ft. lbs. (17 Nm).

39. Lower the vehicle.

40. Install the water pump pulley.

41. Install the accessory drive belts.

42. Install the cylinder head cover.

43. Install the spark plugs and plug wires.

44. Raise and safely support the vehicle.

45. Install the right front wheel and the right side splash shield.

46. Connect the negative battery cable.

1.6L (B6E) & 1.8L (BPE) SOHC ENGINES

1. Disconnect the negative battery cable.
2. Raise and safely support the vehicle.
3. Remove the right front wheel and right side splash shield.
4. Remove the accessory drive belts.
5. Remove the water pump pulley.
6. Remove the crankshaft pulley bolts and remove the crankshaft pulley and baffle plate.
7. Using a suitable tool to hold the crankshaft pulley boss, remove the pulley boss lockbolt. Remove the crankshaft pulley boss.
8. Remove the upper and lower timing belt covers.
9. Label and disconnect the spark plug wires. Remove the spark plugs.

➡**Spark plugs are removed to make it easier to rotate the engine.**

10. Temporarily reinstall the crankshaft pulley boss and lockbolt.

11. Turn the crankshaft until the camshaft sprocket and crankshaft sprocket timing marks are aligned.

12. If the timing belt is to be reused, mark the direction of rotation on the timing belt.

13. Loosen the timing belt tensioner lockbolt. Move the tensioner away from the belt to relieve tension. Torque the lockbolt.

14. Remove the timing belt.

➡**Do not rotate the engine after the timing belt has been removed.**

15. Inspect the belt for wear, peeling, cracking, hardening or signs of oil contamination. Inspect the tensioner for free and smooth rotation. Check the tensioner spring free length; it should not exceed 2.52 in. (64mm) from end to end. Inspect the sprocket teeth for wear or damage. Replace parts, as necessary.

To install:

16. Make sure the timing marks on the camshaft and crankshaft sprockets are properly aligned.

17. Install the timing belt so there is no slack on the tension side. If reusing the old timing belt, make sure it is reinstalled in the same direction of rotation as marked.

18. Loosen the tensioner lockbolt to set tension against the belt, then torque the bolt to 19 ft. lbs. (25 Nm).

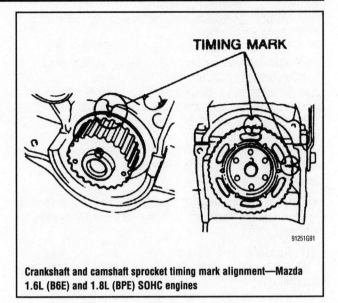

Crankshaft and camshaft sprocket timing mark alignment—Mazda 1.6L (B6E) and 1.8L (BPE) SOHC engines

19. Turn the crankshaft 2 turns in the direction of rotation and check the timing mark alignment.

20. Apply approximately 22 lbs. (10 kg) pressure to the timing belt on the side opposite the tensioner, at a point midway between the sprockets. The belt should deflect 0.43–0.51 in. (11–13mm). If the tension is not as specified, replace the tensioner spring.

21. Install the upper and lower timing belt covers. Torque the bolts to 95 inch lbs. (11 Nm).

22. Using a suitable tool, install the crankshaft pulley boss and torque the lockbolt to 123 ft. lbs. (167 Nm).

23. Install the crankshaft pulley and baffle plate.

24. Install the water pump pulley.

25. Install the accessory drive belts.

26. Install the right front wheel and right side splash shield.

27. Lower the vehicle.

28. Install the spark plugs and connect the spark plug wires.

29. Connect the negative battery cable.

30. Start the engine and check for proper operation.

31. Check the ignition timing.

1.6L (B6ZE) DOHC ENGINE

Miata

1. Disconnect the negative battery cable.
2. Drain the cooling system.
3. Remove the air intake pipe.
4. Remove the upper radiator hose and disconnect the coolant hoses at the thermostat housing.
5. Remove the accessory drive belts.
6. Remove the water pump pulley.
7. Remove the crankshaft pulley bolts and the crankshaft pulley. On 1992–94 vehicles, hold the pulley boss with a suitable tool and remove the pulley lockbolt. Remove the pulley boss.
8. On 1990–91 vehicles, remove the outer and inner timing belt guide plates.
9. Label and disconnect the spark plug wires from the spark plugs. Remove the ignition coil and plug wires assembly. Remove the spark plugs.
10. Remove the cylinder head cover.
11. Remove the upper, middle and lower timing belt covers.

12. On 1992–94 vehicles, temporarily reinstall the pulley boss and lockbolt.

13. Turn the crankshaft until the crankshaft and camshaft sprocket timing marks are aligned. On 1992–94 vehicles, the pin on the pulley boss must face upward.

14. On 1992–94 vehicles, remove the pulley boss and lockbolt, being careful not to disturb the crankshaft.

15. If the timing belt is to be reused, mark the direction of rotation on the timing belt.

16. Loosen the tensioner lockbolt and pry the tensioner outward away from the belt. Temporarily torque the lockbolt with the tensioner spring fully extended.

➡**Protect the tensioner with a shop towel before prying on it. Do not rotate the crankshaft after the timing belt has been removed.**

17. Remove the timing belt.

18. Remove the tensioner and spring. If necessary, remove the idler pulley.

19. Inspect the belt for wear, peeling, cracking, hardening or signs of oil contamination. Inspect the tensioner for free and smooth rotation. Check the tensioner spring free length; it should not exceed 2.315 in. (58.8mm) from end to end. Inspect the sprocket teeth for wear or damage. Replace parts, as necessary.

To install:

20. If removed, install the idler pulley and torque the bolt to 38 ft. lbs. (52 Nm).

21. Install the tensioner and tensioner spring. Pry the tensioner outward and temporarily torque the tensioner lockbolt with the tensioner spring fully extended.

22. Make sure the crankshaft sprocket timing mark is aligned with the mark on the oil pump housing and the camshaft sprocket timing marks are aligned with the marks on the seal plate.

23. Install the timing belt so there is no looseness at the idler

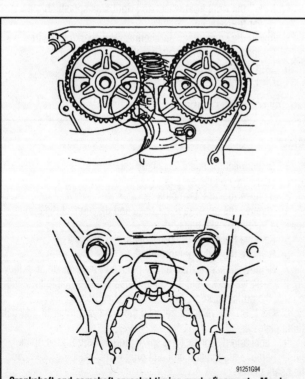

Crankshaft and camshaft sprocket timing mark alignment—Mazda 1.6L (B6ZE) DOHC engine—Miata

pulley side or between the camshaft sprockets. If reusing the old belt, make sure it is installed in the same direction of rotation as marked.

24. On 1992–94 vehicles, temporarily install the pulley boss and lockbolt.

25. Turn the crankshaft 2 turns in the direction of rotation and align the crankshaft sprocket timing mark. On 1992–94 vehicles, face the pin on the pulley boss upright. Make sure the camshaft sprocket timing marks are aligned.

26. Turn the crankshaft 1⅙ turns clockwise and align the crankshaft sprocket timing mark with the tension set mark for proper belt tension adjustment. On 1992–94 vehicles, remove the lockbolt and pulley boss.

27. Make sure the crankshaft sprocket timing mark is aligned with the tension set mark. Loosen the tensioner lockbolt and allow the spring to apply tension to the belt. Torque the tensioner lockbolt to 38 ft. lbs. (52 Nm).

28. On 1992–94 vehicles, install the pulley boss and lockbolt.

29. Turn the crankshaft 2⅙ turns clockwise and make sure the timing marks are correctly aligned.

30. Apply approximately 22 lbs. (10 kg) pressure to the timing belt at a point midway between the camshaft sprockets. The belt should deflect 0.35–0.45 in. (9.0–11.5mm).

31. Install the timing belt covers and torque the bolts to 95 inch lbs. (11 Nm).

32. Apply silicone sealer to the cylinder head in the area adjacent to the front and rear camshaft caps.

33. Install the cylinder head cover and torque the bolts to 78 inch lbs. (8.8 Nm).

34. Install the spark plugs. Install the ignition coil and torque the bolts to 19 ft. lbs. (25 Nm). Connect the spark plug wires.

35. On 1990–91 vehicles, install the inner timing belt guide plate with the dished side facing away from the engine. Install the outer guide plate.

36. On 1992–94 vehicles, hold the pulley boss with a suitable tool and torque the lockbolt to 123 ft. lbs. (167 Nm).

37. Install the crankshaft pulley. Torque the bolts to 13 ft. lbs. (17 Nm).

38. Install the water pump pulley.

39. Install the accessory drive belts and adjust the belt tension.

40. Connect the coolant hoses to the thermostat housing and install the upper radiator hose.

41. Install the air intake pipe.

42. Connect the negative battery cable.

43. Fill and bleed the cooling system.

44. Start the engine and bring to normal operating temperature.

45. Check for leaks and proper operation.

46. Check the ignition timing.

1.6L (ZM) ENGINE

1. Disconnect the negative battery cable.

2. Raise and safely support the vehicle.

3. Remove the right front wheel.

4. Remove the right side splash shield.

5. Lower the vehicle.

6. Remove the camshaft position sensor.

7. Remove the ignition coils.

8. Remove the accessory drive belt.

9. Remove the crankshaft pulley bolts and remove the crankshaft pulley and crankshaft position sensor plate.

10. Remove the water pump pulley.

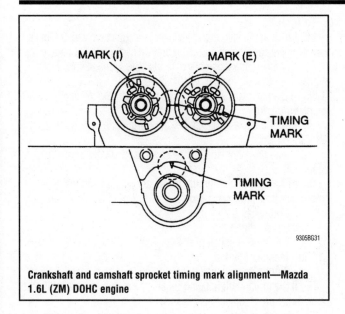

Crankshaft and camshaft sprocket timing mark alignment—Mazda 1.6L (ZM) DOHC engine

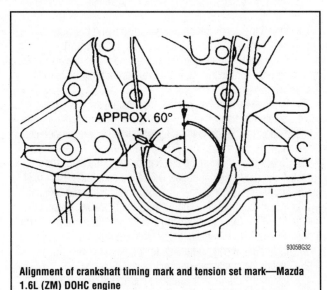

Alignment of crankshaft timing mark and tension set mark—Mazda 1.6L (ZM) DOHC engine

11. Remove the cylinder head cover.

12. Using special tool 49G0175A0 or equivalent, support the engine and remove the front engine mount and brace.

13. Remove the timing belt covers.

14. Rotate the engine in the direction of rotation and align the crankshaft and camshaft timing marks.

15. Using a suitable tool, hold the crankshaft pulley boss and remove the pulley bolt. Remove the pulley boss.

16. Loosen the tensioner lockbolt and move the tensioner away from the timing belt. Temporarily torque the lockbolt to hold the tensioner in place.

17. Remove the timing belt.

18. If the timing belt is to be reused, mark the direction of rotation on the belt.

19. Remove the tensioner and check the tensioner pulley for ease of movement and abnormal noises. Check the tensioner spring free length. The free length should measure 2.43 in. (61.8mm). If out of specification, replace the spring.

To install:

20. Install the tensioner and spring assembly. Move the ten-

sioner fully to the left until the spring is in its fully extended position. Temporarily torque the lockbolt.

21. Be sure that the timing marks on the camshaft sprockets and the crankshaft sprocket are aligned.

22. Install the timing belt in the following order:
   a. Crankshaft sprocket
   b. Idler pulley
   c. Left camshaft sprocket
   d. Right camshaft sprocket
   e. Tensioner pulley

23. If reusing the timing belt, be sure that it is installed in the direction of rotation as marked.

24. Install the crankshaft pulley boss and lockbolt.

25. Loosen the tensioner lockbolt to apply tension to the timing belt.

26. Rotate the crankshaft 1⅚ turns in the direction of rotation and align the crankshaft sprocket timing mark with the tension set mark.

27. Remove the crankshaft pulley boss and lockbolt.

28. Check that the crankshaft timing mark is aligned with the tension set mark.

29. Again, loosen the tensioner lockbolt to apply tension to the timing belt. Torque the lockbolt to 38 ft. lbs. (51 Nm).

30. Install the crankshaft pulley boss and lockbolt. Torque the bolt to 122 ft. lbs. (166 Nm).

31. Turn the crankshaft 2⅙ turns clockwise and make sure the timing marks are correctly aligned.

32. Apply approximately 22 lbs. (10 kg) pressure to the timing belt at a point midway between the camshaft sprockets. The belt should deflect 0.24–0.29 in. (6.0–7.5mm).

33. Install the timing belt covers.

34. Install the front engine mount and brace and remove the engine support tool.

35. Install the cylinder head cover.

36. Install the water pump pulley and torque the bolts to 95inch lbs. (11 Nm).

37. Install the crankshaft position sensor plate, crankshaft pulley and crankshaft pulley bolts. Torque the bolts to 151 inch lbs. (18 Nm).

38. Install the accessory drive belt.

39. Install the ignition coils.

40. Install the camshaft position sensor.

41. Raise the vehicle.

42. Install the right side splash shield.

43. Install the right front wheel.

44. Lower the vehicle.

45. Connect the negative battery cable.

1.8L V6 (K8) ENGINE

1. Disconnect the negative battery cable.

2. Remove the accessory drive belts.

3. Remove the water pump pulley.

4. Remove the crankshaft pulley bolt and remove the crankshaft pulley.

5. Remove the oil dipstick and tube assembly.

6. Remove the crankshaft position sensor.

7. Remove left and right side timing belt covers.

8. Support the engine using a suitable engine brace. Remove the nuts and through-bolt from the right-side engine mount sub bracket. Remove the sub bracket.

9. Temporarily install the crankshaft pulley bolt.

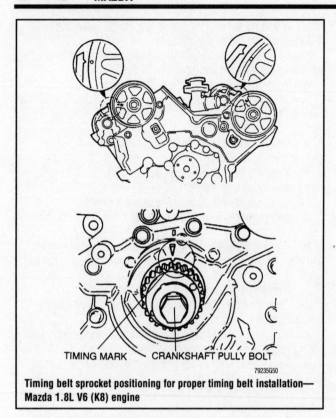

Timing belt sprocket positioning for proper timing belt installation—Mazda 1.8L V6 (K8) engine

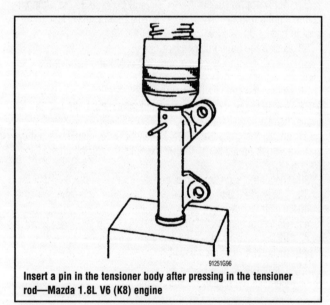

Insert a pin in the tensioner body after pressing in the tensioner rod—Mazda 1.8L V6 (K8) engine

10. Turn the crankshaft until the timing mark on the crankshaft sprocket aligns with the timing mark on the oil pump and the camshaft sprocket timing marks align with the marks on the cylinder head. The No. 1 piston should be at Top Dead Center (TDC) of the compression stroke.

11. Remove the 2 bolts from the automatic tensioner, removing the lower one first. Keep the bolt holes aligned by holding the tensioner to reduce the chance of stripping the threads on the bolts.

12. If the timing belt is to be reused, mark the direction of rotation on the timing belt.

13. Remove the upper and lower idler pulleys and remove the timing belt.

To install:

14. Position the automatic tensioner in a suitable press. Set a flat washer under the tensioner body to prevent damage to the body plug.

15. Compress the tensioner until the hole in the piston is aligned with the 2nd hole in the tensioner case. Insert a 0.060 in. (1.6mm) diameter wire or pin through the 2nd hole to keep the piston compressed.

16. Install the lower idler pulley and torque the bolt to 28–38 ft. lbs. (38–51 Nm).

17. Install the automatic belt tensioner and snugly torque the upper bolt.

18. Make sure that all timing marks are aligned and install the timing belt in the following order:
    a. Crankshaft pulley
    b. Lower idler pulley
    c. Left side camshaft pulley
    d. Tensioner pulley
    e. Right side camshaft pulley

19. If the timing belt is being reused, be sure it is installed in the same direction of rotation as marked.

20. Make sure that there is no slack on the tension side of the belt.

21. Install the upper idler pulley while applying pressure on the timing belt. Torque the bolt to 28–38 ft. lbs. (38–51 Nm).

22. Push the tensioner body to the left and install the lower mounting bolt. Torque the upper and lower mounting bolts to 18 ft. lbs. (25 Nm).

23. Remove the wire or pin from the tensioner.

24. Rotate the crankshaft 2 turns in the normal direction of rotation and align the timing marks. Be sure all marks are aligned correctly aligned.

25. Check the timing belt deflection between the crankshaft sprocket and the tensioner pulley. Deflection should be 0.24–0.31 in. (6–8mm). If it is out of specification, replace the auto-tensioner.

26. Install the right-side engine mount sub bracket. Torque the nuts to 55–77 ft. lbs. (75–104 Nm) and the through-bolt to 63–86 ft. lbs. (86–116 Nm). Remove the engine support.

27. Install the left and right side timing belt covers.

28. Install the crankshaft position sensor.

29. Install the oil dipstick and tube assembly.

30. Install the water pump pulley.

31. Install the crankshaft pulley and bolt. Torque the bolt to 116–122 ft. lbs. (157–166 Nm).

32. Install the accessory drive belts and adjust tension.

33. Connect the negative battery cable.

1.8L (FP) ENGINE

1. Disconnect the negative battery cable.
2. Raise and safely support the vehicle.
3. Remove the right front wheel.
4. Remove the right side splash shield.
5. Lower the vehicle.
6. Remove the camshaft position sensor and the crankshaft position sensor.
7. Label and disconnect the spark plug wires.
8. Remove the spark plugs.
9. Remove the cylinder head cover.
10. Remove the accessory drive belt.
11. Remove the water pump pulley.
12. Using a suitable tool, hold the crankshaft pulley and remove the pulley bolt. Remove the crankshaft pulley and guide plate.

13. Remove the oil dipstick and tube.

14. Using special tool 49G0175AO or equivalent, Support the engine and remove the front engine mount and bracket.

15. Remove the timing belt covers.

16. Reinstall the crankshaft pulley bolt. Rotate the engine in the direction of rotation to align the timing marks.

17. Using an Allen wrench, rotate the tensioner clockwise and remove the tensioner spring.

18. If timing belt is to be reused, mark the direction of rotation on the timing belt.

19. Remove the timing belt.

20. Remove the tensioner and check for ease of movement and abnormal noises. Check the tensioner spring free length. The measurement should be 1.44 in. (36.6mm). If the out of specification, replace the spring.

To install:

21. Install the tensioner and lockbolt. Torque the lockbolt to 38 ft. lbs. (51 Nm).

22. Be sure that the camshaft and crankshaft timing marks are aligned.

23. Install the timing belt so that there is no slack on the idler side of the belt.

24. Turn the tensioner clockwise and install the tensioner spring.

25. Rotate the crankshaft 2 full turns in the direction of rotation. Check that the timing marks are properly aligned.

26. Install the timing belt covers.

27. Install the front engine mount and bracket and remove the engine support.

28. Install the oil dipstick and tube.

29. Install the crankshaft pulley guide plate and crankshaft pulley. Using a suitable tool, hold the crankshaft pulley, install the pulley bolt and torque to 122 ft. lbs. (166 Nm).

30. Install the water pump pulley and torque the bolts to 95 inch lbs. (11 Nm).

31. Install the accessory drive belt.

32. Remove the cylinder head cover.

33. Install the cylinder head cover.

34. Install the spark plugs.

35. Connect the spark plug wires.

36. Install the camshaft position sensor and the crankshaft position sensor.

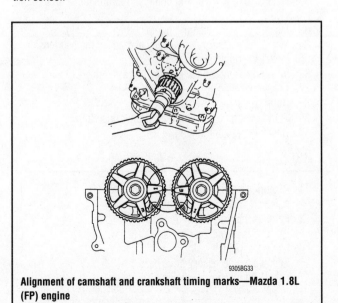

9305BG33

Alignment of camshaft and crankshaft timing marks—Mazda 1.8L (FP) engine

37. Raise the vehicle.

38. Install the right side splash shield.

39. Install the right front wheel.

40. Lower the vehicle.

41. Connect the negative battery cable.

2.0L (FE) & 2.0L DIESEL (RF) ENGINES

1. Disconnect the negative battery cable.

2. Remove any side or undercovers necessary for access to timing belt cover.

3. Remove accessory drive belts.

4. Remove crankshaft pulley and baffle plate.

5. Remove water pump pulley.

6. Remove upper and lower timing belt covers.

7. Turn crankshaft in the direction of rotation to align timing marks on the camshaft and crankshaft sprockets.

8. If timing belt is to be reused, mark the belt with an arrow as to indicate direction of rotation.

9. Remove the tensioner pulley lockbolt and remove the tensioner and spring assembly.

10. Remove the timing belt.

11. Inspect the timing belt for wear, damage or cracks. If the belt is coated with oil or grease, it must be replaced.

12. Check belt tensioner and idler pulley for smooth movement and abnormal noises. Replace as necessary.

To install:

13. Make sure that the timing mark on the camshaft sprocket is aligned with the mark on the upper seal plate and that the mark on the crankshaft sprocket is aligned with the mark on the oil pump housing.

14. Install the belt tensioner and spring assembly and install the lockbolt.

15. Position the tensioner so that the spring is in its fully extended position. Temporarily torque the lockbolt.

16. Install the timing belt onto the crankshaft sprocket and then onto the camshaft sprocket from the tension side.

17. If reusing the timing belt. Make sure the belt is installed in the marked direction.

18. Loosen the tensioner lockbolt to apply pressure to the belt. Make sure that there is equal pressure on both sides of the belt and that the timing marks are still aligned.

19. Turn the crankshaft 2 full revolutions in the direction of rotation. Make sure the timing marks are correctly aligned.

20. Loosen the tensioner lockbolt to allow the tensioner spring to apply pressure to the timing belt.
Torque the lockbolt and torque to 38 ft. lbs. (52 Nm).

21. Check for proper belt tension by measuring the deflection midway between the camshaft sprocket and the crankshaft sprocket on the tension side. Deflection should measure 0.43–0.51 in. (11–13mm).

22. Install the upper and lower timing belt covers.

23. Install the water pump pulley.

24. Install the crankshaft pulley and the baffle plate.

25. Install the drive belts and adjust to proper tension.

26. Install side covers and undercovers.

27. Connect the negative battery cable.

2.0L (FS) ENGINE

1. Disconnect the negative battery cable.

2. Raise and safely support the vehicle.

3. Remove the right front wheel and tire assembly.

4. Remove the right side splash shield.

5. Remove the accessory drive belts and the water pump pulley.

6. Remove the power steering pump pulley shield.

7. Remove the power steering pump and position aside, leaving the hoses connected.

8. Hold the crankshaft pulley using a suitable tool and remove the pulley bolt. Remove the crankshaft pulley and the guide plate.

9. Disconnect the spark plug wires and remove the spark plugs.

10. Remove the cylinder head cover.

11. Remove the engine oil dipstick and dipstick tube.

12. Remove the upper and lower timing belt covers.

13. Support the engine using engine support tool 49 G017 5A0. Remove the right-side engine mount bracket.

14. Turn the crankshaft, in the normal direction of rotation, until the crankshaft and camshaft sprocket timing marks are aligned.

15. If the timing belt is to be reused, mark the direction of rotation on the belt.

16. Turn the belt tensioner clockwise and disconnect the tensioner spring from the hook pin. Remove the timing belt.

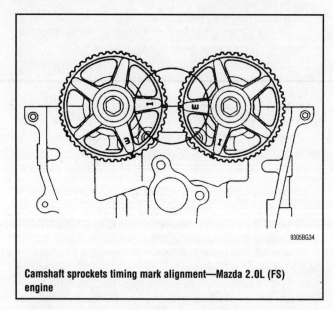

Camshaft sprockets timing mark alignment—Mazda 2.0L (FS) engine

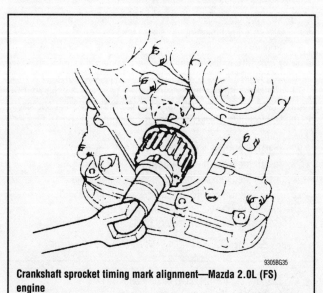

Crankshaft sprocket timing mark alignment—Mazda 2.0L (FS) engine

➡**Do not rotate the engine after the timing belt has been removed.**

17. Inspect the belt for wear, peeling, cracking, hardening or signs of oil contamination. Inspect the tensioner pulley for free and smooth rotation and for oil leaks. Check the spring bracket and grommet for looseness or damage. Measure the tensioner spring free length; it should not exceed 1.441 in. (36.6mm) from end to end. Check the sprockets for worn teeth or other damage.

To install:

18. Make sure the crankshaft and camshaft timing marks are aligned.

19. Install the timing belt so there is no looseness at the idler side or between the camshaft sprockets. If reusing the old timing belt, make sure it is installed in the same direction of rotation.

20. Turn the crankshaft 2 turns in the direction of rotation and make sure the timing marks are correctly aligned.

21. Turn the tensioner clockwise and connect the tensioner spring to the hook pin. Make sure tension is applied to the timing belt.

22. Turn the crankshaft 2 turns again in the direction of rotation and make sure the timing marks are correctly aligned.

23. Install the right-side engine mount bracket. Torque the mount-to-engine nuts to 76 ft. lbs. (102 Nm) and the mount through bolt to 86 ft. lbs. (116 Nm). Install the ground harness and torque the nut to 65 ft. lbs. (89 Nm).

24. Remove the engine support tool.

25. Install the timing belt covers and torque the bolts to 95 inch lbs. (10.7 Nm).

26. Install the dipstick tube and dipstick.

27. Apply silicone sealant to the contact surfaces of the cylinder head cover. Also apply sealant to the cylinder head surface in the area adjacent to the front camshaft caps.

28. Install the cylinder head cover and torque the bolts in 2–3 steps to 69 inch lbs. (7.8 Nm), in the proper sequence.

29. Install the spark plugs and connect the spark plug wires.

30. Install the guide plate and the crankshaft pulley. Hold the pulley with a suitable tool and torque the lockbolt to 122 ft. lbs. (166 Nm).

31. Install the water pump pulley and torque the bolts to 95 inch lbs. (10 Nm).

32. Install the power steering pump and torque the bolts to 33 ft. lbs. (46 Nm).

33. Install the power steering pump and pulley shield.

34. Install the accessory drive belts and adjust the belt tension.

35. Install the right side splash shield.

36. Install the right front wheel and tire assembly.

37. Lower the vehicle.

38. Connect the negative battery cable.

39. Start the engine and check for proper operation.

40. Check the ignition timing.

2.2L (F2) ENGINE

1. Disconnect the negative battery cable.

2. Raise and safely support the vehicle.

3. Remove the right front wheel.

4. Remove the right side splash shield.

5. Lower the vehicle.

6. Remove the accessory drive belts.

7. Label and disconnect the spark plug wires and remove the spark plugs.

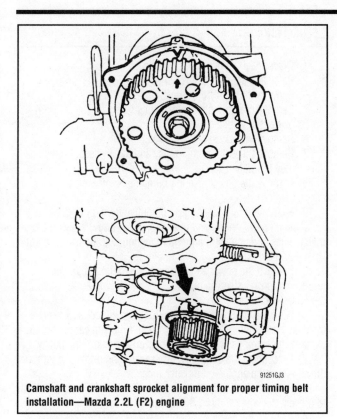

Camshaft and crankshaft sprocket alignment for proper timing belt installation—Mazda 2.2L (F2) engine

8. Remove the retaining bolts and remove the crankshaft pulley.

9. Remove the water pump pulley.

10. Remove the upper and lower timing belt covers and remove the baffle plate from in front of the crankshaft sprocket.

11. Turn the crankshaft clockwise until the "arrow" **1** mark on the camshaft is aligned with the mark on top of the seal plate.

12. Loosen the tensioner lockbolt. Position the tensioner with the spring in a fully extended position. Torque the lockbolt.

13. Remove the timing belt.

14. If the timing belt is to be reused, mark the direction of rotation.

➡**Do not rotate the engine after the timing belt has been removed.**

15. Inspect the belt for wear, peeling, cracking, hardening or signs of oil contamination. Inspect the tensioner pulley for free and smooth rotation. Measure the tensioner spring free length; it should not exceed 2.480 in. (63mm) from end to end. Check the camshaft and crankshaft sprockets for worn teeth or other damage.

To install:

16. Make sure the crankshaft and camshaft timing marks are aligned.

17. Install the timing belt tensioner and spring. Move the tensioner until the spring is fully extended and temporarily torque the tensioner bolt to hold it in place.

18. Install the timing belt. Make sure there is no slack at the side of the water pump and idler pulleys. If reusing the old belt, be sure it is installed in the original direction of rotation as marked.

19. Turn the crankshaft 2 turns in the direction of rotation and make sure the timing marks are aligned.

20. Loosen the tensioner lockbolt to apply tension to the belt. Torque the tensioner bolt to 38 ft. lbs. (52 Nm).

21. Turn the crankshaft again 2 turns in the direction of rotation and make sure the timing marks are aligned.

22. Apply approximately 22 lbs. (98 N) pressure to the timing belt at a point midway between the idler pulley and camshaft sprocket. A new belt should deflect 0.31–0.35 in. (8–9mm). A used belt should deflect 0.35–0.39 in. (9–10mm). If the deflection is not as specified, replace the tensioner spring.

23. Install the baffle plate with the dished side facing away from the engine.

24. Install the timing belt covers and torque the bolts to 87 inch lbs. (10 Nm).

25. Install the crankshaft pulley and torque the bolts to 13 ft. lbs. (17 Nm).

26. Install the water pump pulley.

27. Install the accessory drive belts and adjust the belt tension.

28. Install the right side splash shield.

29. Install the right side wheel.

30. Lower the vehicle.

31. Install the spark plugs and connect the spark plug wires.

32. Connect the negative battery cable.

33. Start the engine and check for proper operation.

34. Check the ignition timing.

2.3L (KJ) ENGINE

1. Disconnect the negative battery cable.

2. Raise and safely support the vehicle.

3. Remove the right front wheel.

4. Remove the right side splash shield.

5. Lower the vehicle.

6. Remove the accessory drive belt.

7. Remove the water pump pulley.

8. Remove the oil dipstick and tube assembly.

9. Remove the crankshaft pulley bolts and remove the crankshaft pulley.

10. Remove the crankshaft position sensor.

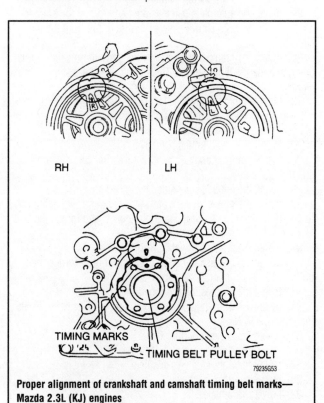

Proper alignment of crankshaft and camshaft timing belt marks—Mazda 2.3L (KJ) engines

11. Remove the power steering auto-tensioner and pulley.

12. Remove the left and right side timing belt covers.

13. Support the engine with a suitable engine brace. Remove the front engine mount bracket.

14. Remove the upper timing belt cover.

15. Turn the crankshaft until the timing mark on the crankshaft sprocket aligns with the timing mark on the oil pump housing and the camshaft sprocket timing marks align with the marks on the cylinder heads. The No. 1 piston should be at Top Dead Center (TDC) of the compression stroke.

16. Remove the 2 bolts from the automatic tensioner, removing the lower one first. Keep the bolt holes aligned by holding the tensioner to reduce the chance of stripping the threads on the bolts.

17. If the timing belt is to be reused, mark the direction of rotation on the timing belt.

18. Remove the timing belt.

To install:

19. Position the automatic tensioner in a press. Set a flat washer under the tensioner body to prevent damage to the body plug.

20. Compress the tensioner until the hole in the piston is aligned with the 2nd hole in the tensioner case. Insert a 0.063 in. (1.6mm) diameter wire or pin through the 2nd hole to keep the piston compressed.

21. Be sure the camshaft and crankshaft sprocket timing marks are aligned.

22. Install the timing belt in the following order:
    a. Crankshaft sprocket
    b. Right side lower idler pulley
    c. Right side camshaft sprocket
    d. Left and right upper idler pulleys
    e. Left side camshaft sprocket
    f. Tensioner pulley

23. If the original belt is being reused, be sure it is installed in the same direction of rotation as marked.

24. Install the automatic belt tensioner and torque the bolts to 14–18 ft. lbs. (19–25 Nm). Remove the wire or pin from the tensioner.

25. Rotate the crankshaft 2 turns in the normal direction of rotation and align the timing marks. Be sure all marks are still correctly aligned.

26. Inspect timing belt deflection, 0.24–0.31 in. (6–8mm), between the crankshaft sprocket and the tensioner pulley. If it is out of specification, replace the auto-tensioner.

27. Install the upper timing belt cover.

28. Install the front engine mount bracket. Remove the engine support brace.

29. Install the left and right side timing belt covers.

30. Install the power steering auto-tensioner and torque the bolts to 14–18 ft. lbs. (19–25 Nm). Install the pulley, and torque the bolt to 29–34 ft. lbs. (40–47 Nm).

31. Install the crankshaft position sensor.

32. Install the crankshaft pulley and pulley bolts. Torque the bolts to 19–22 ft. lbs. (26–30 Nm).

33. Install the oil dipstick and tube assembly.

34. Install the water pump pulley and torque the bolts to 70–95 inch lbs. (7–10 Nm).

35. Install the accessory drive belt.

36. Raise the vehicle.

37. Install the right side splash shield.

38. Install the right front wheel.

39. Lower the vehicle.

40. Connect the negative battery cable.

2.5L (KL) ENGINE

1. Disconnect the negative battery cable.

2. Raise and safely support the vehicle.

3. Remove the right front wheel.

4. Remove the right side splash shield.

5. Lower the vehicle.

6. Remove the accessory drive belts.

7. Remove the water pump pulley and the accessory drive belt idler pulley bracket.

8. Remove the power steering pump reservoir bolts and secure the reservoir aside.

9. Keep the power steering pump pulley from turning, by installing a socket on the end of a breaker bar through one of the pulley holes and engaging a pump mounting bolt. Remove the pulley nut and the pulley.

10. Remove the power steering pump mounting bolts and remove the power steering pump. Secure the pump aside, leaving the hoses connected.

11. Hold the crankshaft pulley with a suitable tool and remove the pulley bolt. Remove the crankshaft pulley, being careful not to damage the crank position sensor rotor on the rear of the pulley.

12. Disconnect the crank position sensor connector and remove the clip from the engine oil dipstick tube.

13. Remove the oil dipstick and tube. Plug the hole after removal to prevent the entry of dirt or foreign material.

14. Remove the crank position sensor harness bracket and wiring harness bracket from the timing belt cover.

15. Support the engine with engine support tool 49 G017 5A0 or equivalent.

16. Remove the right-side engine mount.

17. Remove the right and left timing belt covers.

18. Install the crankshaft pulley bolt and turn the crankshaft until the No. 1 piston is at Top Dead Center (TDC) of the compression stroke.

19. If the timing belt is to be reused, mark the direction of rotation on the timing belt.

20. Loosen the automatic tensioner bolts and remove the lower bolt. Hold the tensioner so the bolt threads are not damaged during removal.

21. Hold the upper idler pulley to reduce the belt resistance and remove the pulley bolts and pulley. Remove the timing belt.

22. Remove the automatic tensioner and the remaining idler pulley.

➡**Do not rotate the engine after the timing belt has been removed.**

23. Inspect the belt for wear, peeling, cracking, hardening or signs of oil contamination. Inspect the tensioner pulley for free and smooth rotation. Check the automatic tensioner for oil leakage. Check the tensioner rod projection (free length); it should be 0.55–0.63 in. (14–16mm). Inspect the sprocket teeth for wear or damage. Replace parts, as necessary.

To install:

24. Position the automatic tensioner in a suitable press. Place a flat washer under the tensioner body to prevent damage to the body plug.

25. Slowly press in the tensioner rod, but do not exceed 2200 lbs. (998 kg) force. Insert a pin through the hole in the tensioner body to hold the rod in place.

26. Install the tensioner and loosely torque the upper bolt so the tensioner can move.

➡**This is done to reduce the timing belt resistance when the upper idler pulley is installed.**

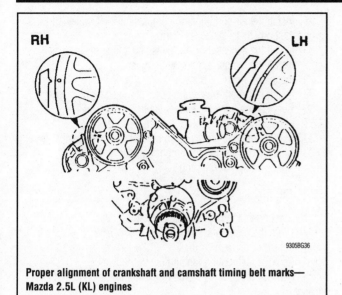

Proper alignment of crankshaft and camshaft timing belt marks—Mazda 2.5L (KL) engines

27. If removed, install the lower idler pulley and torque the bolt to 38 ft. lbs. (52 Nm).

28. Make sure the crankshaft and camshaft sprocket timing marks are aligned.

29. Install the timing belt in the following order:
   a. Crankshaft sprocket
   b. Lower idler pulley
   c. Left camshaft sprocket
   d. Tensioner pulley
   e. Right camshaft sprocket
   f. Make sure the belt has no looseness at the tension side.

30. If reusing the old timing belt, make sure it is installed in the same direction of rotation as marked.

31. Install the upper idler pulley while applying pressure on the timing belt. Be careful not to damage the pulley bolt threads when installing. Torque the upper idler pulley bolt to 34 ft. lbs. (46 Nm).

32. Push the bottom of the automatic tensioner away from the belt and torque the mounting bolts to 19 ft. lbs. (25 Nm). Remove the pin from the tensioner, applying tension to the belt.

33. Turn the crankshaft 2 full turns in the normal direction of rotation and make sure the timing marks are aligned.

34. Apply approximately 22 lbs. (98 N) pressure to the timing belt at a point midway between the automatic tensioner and the crankshaft sprocket. The belt should deflect 0.24–0.31 in. (6–8mm). If the deflection is not as specified, replace the automatic tensioner.

35. Install the right and left timing belt covers. Torque the bolts to 95 inch lbs. (11 Nm).

36. Install the crank position sensor and harness brackets and torque the bolts to 95 inch lbs. (11 Nm).

37. Install the right-side engine mount. Torque the mount-to-engine nuts to 76 ft. lbs. (103 Nm) and the mount through bolt to 86 ft. lbs. (116 Nm).

38. Remove the engine support tool.

39. Apply clean engine oil to a new O-ring and install on the dipstick tube. Remove the plug and install the dipstick tube and dipstick. Torque the tube bracket bolt to 95 inch lbs. (11 Nm).

40. Install the crank angle sensor harness and clip to the dipstick tube. Connect the electrical connector.

41. Remove the crankshaft pulley bolt and install the crankshaft pulley. Reinstall the bolt and hold the pulley with a suitable tool. Torque the bolt to 122 ft. lbs. (166 Nm).

42. Install the power steering pump. Torque the mounting bolts to 34 ft. lbs. (46 Nm) except the bolt to the right of the idler pulley. Torque that bolt to 19 ft. lbs. (25 Nm).

43. Install the power steering pump pulley and loosely torque the nut. Hold the pulley with the socket and breaker bar and torque the nut to 69 ft. lbs. (93 Nm).

44. Install the power steering fluid reservoir and engine ground. Torque to 87 inch lbs. (9.8 Nm).

45. Install the water pump pulley and loosely torque the bolts. Install the accessory drive belts and adjust the belt tension.

46. Torque the water pump pulley bolts to 95 inch lbs. (11 Nm).

47. Raise the vehicle.

48. Install the right side splash shield.

49. Install the right front wheel.

50. Lower the vehicle.

51. Connect the negative battery cable.

52. Start the engine and check for proper operation.

53. Check the ignition timing.

3.0L (JE) ENGINE

SOHC Engine

1. Disconnect the negative battery cable.

2. Drain the cooling system.

3. Label and disconnect the spark plug wires. Remove the spark plugs.

4. Remove the fresh air duct. Remove the cooling fan and the fan shroud.

5. Remove the accessory drive belts and the air conditioning compressor idler pulley.

6. Remove the crankshaft pulley and baffle plate.

7. Remove the coolant bypass hose and the upper radiator hose.

8. Remove the timing belt covers and gaskets.

9. Turn the crankshaft, in the normal direction of rotation and align the crankshaft and camshaft sprocket timing marks. Mark the direction of rotation on the timing belt.

10. Remove the upper idler pulley and remove the timing belt. Remove the automatic tensioner.

➡**Do not rotate the engine after the timing belt has been removed.**

11. Inspect the belt for wear, peeling, cracking, hardening or signs of oil contamination. Inspect the tensioner pulley for free and smooth rotation. Check the automatic tensioner for oil leakage. Check the tensioner rod projection (free length); it should be 0.47–0.55 in. (12–14mm). Inspect the sprocket teeth for wear or damage. Replace parts, as necessary.

To install:

12. Position the automatic tensioner in a suitable press. Place a flat washer under the tensioner body to prevent damage to the body plug.

13. Slowly press in the tensioner rod, but do not exceed 2200 lbs. (998 kg) force. Insert a pin into the tensioner body to hold the rod in place.

14. Install the tensioner and torque the bolts to 19 ft. lbs. (25 Nm).

15. Make sure the crankshaft and camshaft sprocket timing marks are aligned.

16. Install the timing belt in the following order:
   a. Crankshaft sprocket
   b. Lower idler pulley
   c. Left camshaft sprocket

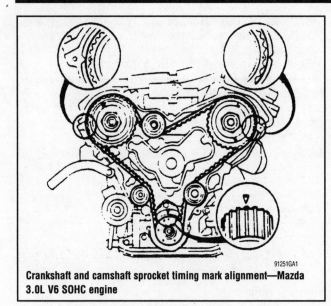

Crankshaft and camshaft sprocket timing mark alignment—Mazda 3.0L V6 SOHC engine

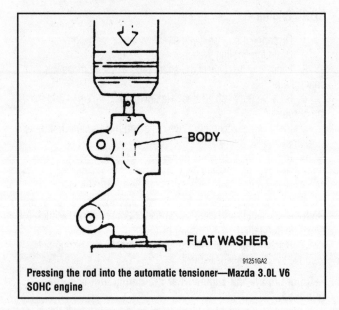

Pressing the rod into the automatic tensioner—Mazda 3.0L V6 SOHC engine

d. Right camshaft sprocket

e. Tensioner pulley

17. If reusing the old timing belt, make sure it is installed in the same direction of rotation as marked.

18. Install the upper idler pulley and torque the bolt to 38 ft. lbs. (52 Nm).

19. Turn the crankshaft 2 turns, in the normal direction of rotation, and align the timing marks.

20. Remove the pin from the automatic tensioner. Turn the crankshaft 2 turns, in the normal direction of rotation, and make sure the timing marks are aligned.

21. Apply approximately 22 lbs. (98 N) pressure to the timing belt at a point midway between the right camshaft sprocket and tensioner pulley. The belt should deflect 0.20–0.28 in. (5–7mm). If the deflection is not as specified, replace the automatic tensioner.

22. Install the timing belt covers with new gaskets. Torque the bolts to 95 inch lbs. (11 Nm).

23. Install the upper radiator hose and coolant bypass hose.

24. Install the baffle plate and crankshaft pulley. Torque the bolts to 130 inch lbs. (15 Nm).

25. Install the air conditioning compressor idler pulley and torque the bolts to 19 ft. lbs. (25 Nm).

26. Install the accessory drive belts and adjust the tension.

27. Install the fan shroud and cooling fan.

28. Install the fresh air duct.

29. Install the spark plugs and connect the spark plug wires.

30. Connect the negative battery cable.

31. Fill and bleed the cooling system.

32. Start the engine and bring to normal operating temperature.

33. Check for leaks and for proper operation.

34. Check the ignition timing.

DOHC Engine

1. Disconnect the negative battery cable.

2. Drain the cooling system.

3. Remove the fresh air duct.

4. Remove the cooling fan and fan shroud.

5. Remove the air intake pipe from the throttle body and air cleaner.

6. Disconnect the spark plug wires and remove the spark plugs.

7. Remove the idler pulleys and the accessory drive belts.

8. Remove the coolant bypass hose and the upper radiator hose.

9. Disconnect the electrical connector and remove the distributor.

10. Hold the crankshaft pulley with a suitable tool and remove the bolt. Remove the crankshaft pulley. On 1992–94 vehicles, be careful not to damage the sensor rotor.

11. Remove the timing belt covers.

12. Rotate the crankshaft until the camshaft and crankshaft sprocket timing marks are aligned.

13. Remove the upper idler pulley.

14. Remove the automatic tensioner and pulley.

15. If timing belt is to be reused, mark the direction of rotation on the timing belt.

16. Remove the timing belt.

➡**Do not rotate the engine after the timing belt has been removed.**

17. Inspect the belt for wear, peeling, cracking, hardening or signs of oil contamination. Inspect the tensioner pulley for free and smooth rotation. Check the automatic tensioner for oil leakage. Check the tensioner rod projection (free length); it should be 0.47–0.55 in. (12–14mm). Inspect the sprocket teeth for wear or damage. Replace parts, as necessary.

To install:

18. Position the automatic tensioner in a suitable press. Place a flat washer under the tensioner body to prevent damage to the body plug.

19. Slowly press in the tensioner rod, but do not exceed 2200 lbs. (998 kg) force. Insert a pin into the tensioner body to hold the rod in place.

20. Install the tensioner and torque the bolts to 19 ft. lbs. (25 Nm).

21. Make sure the crankshaft and camshaft sprocket timing marks are aligned.

22. Install the timing belt in the following order:

a. Crankshaft sprocket

b. Lower idler pulley

c. Left exhaust camshaft sprocket and left intake camshaft sprocket

d. Tensioner pulley

e. Right exhaust camshaft sprocket and right intake camshaft sprocket

23. Push the belt down and install the upper idler pulley. Torque the bolt to 38 ft. lbs. (52 Nm). Make sure the timing marks are still aligned after installing the upper idler pulley.

24. Turn the crankshaft 2 revolutions in the direction of normal rotation and realign the timing marks.

25. Remove the pin from the automatic tensioner. Again rotate the crankshaft 2 turns in the direction of rotation and make sure the timing marks are aligned.

26. Apply approximately 22 lbs. (98 N) pressure to the timing belt at a point midway between the right exhaust camshaft sprocket and the tensioner pulley. The belt should deflect 0.20–0.28 in. (5–7mm). If the deflection is not as specified, replace the automatic tensioner.

27. Install the timing belt cover and torque the bolts to 95 inch lbs. (11 Nm).

28. Install the crankshaft pulley. Hold the pulley with a suitable tool and torque the lockbolt to 123 ft. lbs. (167 Nm). On 1992–99 vehicles, be careful not to damage the sensor rotor.

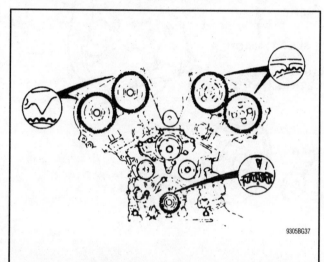

Crankshaft and camshaft sprocket timing mark alignment—Mazda 3.0L DOHC engine

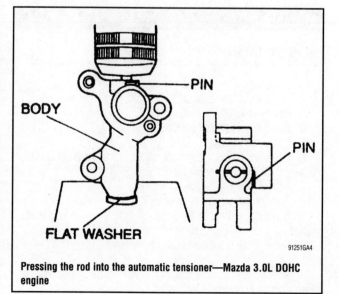

Pressing the rod into the automatic tensioner—Mazda 3.0L DOHC engine

29. Install the distributor and connect the electrical connector.

30. Install the upper radiator hose and coolant bypass hose.

31. Install the idler pulleys and the accessory drive belts. Adjust the belt tension.

32. Install the spark plugs and connect the spark plug wires.

33. Install the air intake pipe to the throttle body and air cleaner.

34. Install the cooling fan and radiator shroud.

35. Install the fresh air duct.

36. Connect the negative battery cable.

37. Fill and bleed the cooling system.

38. Start the engine and bring to normal operating temperature.

39. Check for leaks and for proper operation.

40. Check the ignition timing.

Merkur

2.3L (VIN W) ENGINE

1. Disconnect the negative battery cable.

2. Matchmark the distributor body to indicate the location of the No. 1 spark plug wire tower, and remove the distributor cap.

3. Rotate the crankshaft clockwise so that the timing mark is at Top Dead Center (TDC) of the compression stroke and the distributor rotor is pointing to the No. 1 spark plug wire tower mark.

4. Remove the accessory drive belts.

5. Remove the cooling fan and shroud.

6. Remove the water pump pulley.

7. Remove the front timing cover.

8. Check that the camshaft timing mark is aligned with the timing mark on the rear timing cover.

9. Hold the crankshaft from turning and remove the crankshaft pulley bolt.

10. Using crankshaft puller took T74P-6312-A, or equivalent, remove the crankshaft pulley.

11. Loosen the tensioner adjustment and pivot bolts. Using the tensioner tool T74P-6254-A, or equivalent, lever the tensioner away from the timing belt and tighten the adjustment bolt.

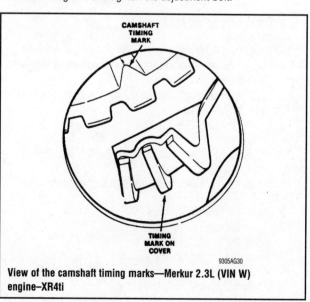

View of the camshaft timing marks—Merkur 2.3L (VIN W) engine-XR4ti

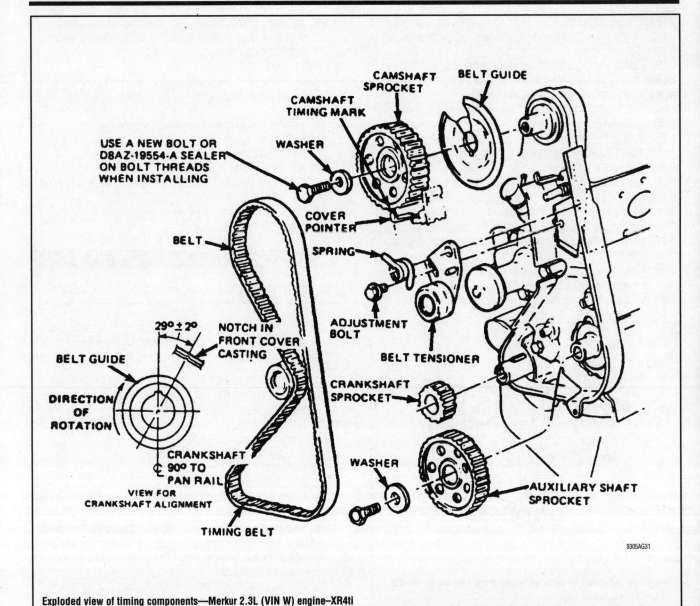

USE A NEW BOLT OR
D8AZ-19554-A SEALER
ON BOLT THREADS
WHEN INSTALLING

CAMSHAFT
TIMING MARK

CAMSHAFT
SPROCKET

BELT GUIDE

WASHER

COVER
POINTER

BELT

SPRING

ADJUSTMENT
BOLT

BELT TENSIONER

29° ± 2°

NOTCH IN
FRONT COVER
CASTING

BELT GUIDE

DIRECTION
OF
ROTATION

CRANKSHAFT
SPROCKET

CRANKSHAFT
¢ 90° TO
PAN RAIL

WASHER

AUXILIARY SHAFT
SPROCKET

VIEW FOR
CRANKSHAFT ALIGNMENT

TIMING BELT

9305AG31

Exploded view of timing components—Merkur 2.3L (VIN W) engine–XR4ti

12. Remove the timing belt.
To install:
13. Install the timing belt.
14. Loosen the tensioner adjustment bolt, allowing it to spring back against the belt.
15. Rotate the crankshaft clockwise 2 complete turns and ensure that the timing marks align.
16. Tighten the tensioner adjustment bolt to 14–21 ft. lbs. (19–28 Nm).
17. Tighten the tensioner pivot bolt to 28–40 ft. lbs. (38–54 Nm).
18. Install the crankshaft pulley and tighten the center bolt to 100–120 ft. lbs. (136–163 Nm).
19. Install the front timing cover.
20. Install the water pump pulley.
21. Install the accessory drive belts.
22. Install the cooling fan and shroud.
23. Install the distributor cap.
24. Connect the negative battery cable.

Mitsubishi

1.5L (G15B/4G15) ENGINE

1. Disconnect the negative battery cable.
2. Raise and safely support the vehicle.
3. Remove the left front wheel.
4. Remove the left side splash shield.
5. Lower the vehicle.
6. Using the proper equipment, slightly raise the engine to take the weight off the side engine mount.
7. Remove the engine mount bracket.
8. Remove the accessory drive belts.
9. Remove the water pump pulley and the alternator mounting bracket.
10. Remove the crankshaft pulley by removing the 4 retaining bolts.
11. Remove all attaching screws and remove the upper and lower timing belt covers.

12. Rotate the crankshaft in the direction of rotation and align the timing marks so that the No. 1 piston is at Top Dead Center (TDC) of the compression stroke.

13. Loosen the tensioning bolt and the pivot bolt on the timing belt tensioner. Move the tensioner as far as it will go toward the water pump. Tighten the adjusting bolt.

14. If the timing belt is to be reused, mark the belt with an arrow showing direction of rotation.

15. Remove the timing belt.

16. Inspect the belt thoroughly. The back surface must be pliable and rough. If it is hard and glossy, the belt should be replaced. Any cracks or missing teeth mean the belt must be replaced.

17. Inspect the tensioner for grease leaking from the grease seal and any roughness in rotation. Replace as necessary.

18. The sprockets should be inspected for damaged teeth and

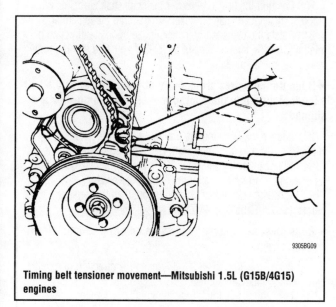

Timing belt tensioner movement—Mitsubishi 1.5L (G15B/4G15) engines

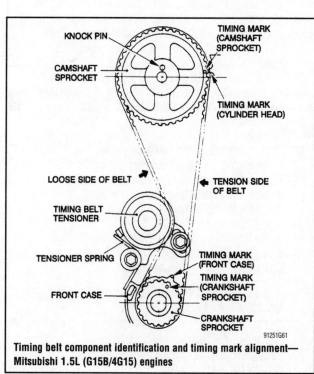

Timing belt component identification and timing mark alignment—Mitsubishi 1.5L (G15B/4G15) engines

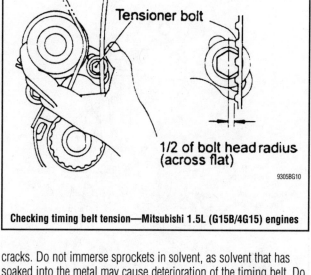

Checking timing belt tension—Mitsubishi 1.5L (G15B/4G15) engines

cracks. Do not immerse sprockets in solvent, as solvent that has soaked into the metal may cause deterioration of the timing belt. Do not clean the tensioner in solvent, as this may wash the grease out of the bearing.

To install:

19. Make sure the timing marks on the camshaft sprocket and crankshaft sprocket are in alignment.

20. Position the timing belt over the crankshaft sprocket, then over the camshaft sprocket. Slip the back of the belt over the tensioner pulley.

21. Rotate the camshaft sprockets in the opposite direction of rotation until the tension side of the belt is tight and that the timing marks are still aligned.

➡**If the timing marks are not properly aligned, shift the belt 1 tooth at a time in the appropriate direction until they are aligned.**

22. Loosen the tensioner mounting bolts so the tensioner applies pressure to the timing belt. Be sure the belt follows the curve of the camshaft pulley and that the teeth are engaged all the way around. Correct the path of the belt, if necessary.

23. Torque the tensioner adjusting bolt, then the tensioner pivot bolt to 15–18 ft. lbs. (20–26 Nm).

➡**Bolts must be tightened in the stated order or tension will not be correct.**

24. Rotate the crankshaft 2 full turns in the direction of rotation until timing marks are aligned.

25. Loosen both tensioner attaching bolts and let the tensioner position itself under spring tension. Tighten the bolts to the specified torque.

26. Check belt tension by putting a finger on the water pump side of the tensioner wheel and pull the belt toward the water pump. The belt should move toward the pump until the teeth are approximately ½ of the way across the head of the tensioner adjusting bolt.

27. Install the timing belt covers.

28. Install the crankshaft pulley, making sure the pin on the crankshaft sprocket fits through the hole in the rear surface of the pulley. Install the bolts and torque to 90–102 inch lbs. (10–11 Nm).

29. Install water pump pulley and torque the bolts to 72–84 inch lbs. (8–10 Nm).

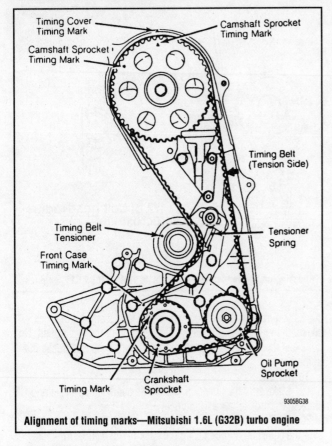

Alignment of timing marks—Mitsubishi 1.6L (G32B) turbo engine

Labels on diagram:
- Timing Cover Timing Mark
- Camshaft Sprocket Timing Mark
- Camshaft Sprocket Timing Mark
- Timing Belt (Tension Side)
- Timing Belt Tensioner
- Tensioner Spring
- Front Case Timing Mark
- Timing Mark
- Crankshaft Sprocket
- Oil Pump Sprocket
- 9305BG38

30. Install drive belts and adjust to proper tension.
31. Install engine mounting brackets and lower the engine.
32. Raise and safely support the vehicle.
33. Install the left side splash shield.
34. Install the left front wheel.
35. Lower the vehicle.
36. Connect the negative battery cable.
37. Start the engine and check for proper operation.
38. Check and adjust ignition timing.

1.6L (G32B) TURBO ENGINE

1. Disconnect the negative battery cable.
2. Raise and safely support the vehicle.
3. Remove the left front wheel.
4. Remove the left side splash shield.
5. Lower the vehicle.
6. Using the proper equipment, slightly raise the engine to take the weight off the side engine mount.
7. Remove the engine mount bracket.
8. Remove the accessory drive belts.
9. Remove the water pump pulley and the alternator mounting bracket.
10. Remove the crankshaft pulley by removing the 4 retaining bolts.
11. Remove all attaching screws and remove the upper and lower timing belt covers.
12. Rotate the crankshaft in the direction of rotation and align the timing marks so that the No. 1 piston is at Top Dead Center (TDC) of the compression stroke.
13. Loosen the tensioning bolt and the pivot bolt on the timing belt tensioner. Move the tensioner as far as it will go toward the water pump. Tighten the adjusting bolt.

14. If the timing belt is to be reused, mark the belt with an arrow showing direction of rotation.
15. Remove the timing belt.
16. Inspect the belt thoroughly. The back surface must be pliable and rough. If it is hard and glossy, the belt should be replaced. Any cracks or missing teeth mean the belt must be replaced.
17. Inspect the tensioner for grease leaking from the grease seal and any roughness in rotation. Replace as necessary.
18. The sprockets should be inspected for damaged teeth and cracks. Do not immerse sprockets in solvent, as solvent that has soaked into the metal may cause deterioration of the timing belt. Do not clean the tensioner in solvent, as this may wash the grease out of the bearing.
   To install:
19. Align the timing marks of the camshaft sprocket, oil pump sprocket and crankshaft sprocket.
20. Position the timing belt over the crankshaft sprocket, oil pump sprocket, camshaft sprocket and the onto the tensioner pulley.
21. Rotate the camshaft sprockets in the opposite direction of rotation until the tension side of the belt is tight and that the timing marks are still aligned.

➡**If the timing marks are not properly aligned, shift the belt 1 tooth at a time in the appropriate direction until they are aligned.**

22. Loosen the tensioner mounting bolts so the tensioner applies pressure to the timing belt. Be sure the belt follows the curve of the camshaft pulley and that the teeth are engaged all the way around. Correct the path of the belt, if necessary.
23. Torque the tensioner adjusting bolt, then the tensioner pivot bolt to 15–18 ft. lbs. (20–26 Nm).

➡**Bolts must be tightened in the stated order or tension will not be correct.**

24. Rotate the crankshaft 2 full turns in the direction of rotation until timing marks are aligned.
25. Loosen the tensioner mounting bolts so the tensioner applies pressure to the timing belt. Be sure the belt follows the curve of the camshaft pulley and that the teeth are engaged all the way around. Tighten the bolts to the specified torque.
26. Make sure the timing marks are correctly aligned.
27. Install the timing belt covers.
28. Install the crankshaft pulley, making sure the pin on the crankshaft sprocket fits through the hole in the rear surface of the pulley. Install the bolts and torque to 90–102 inch lbs. (9–12 Nm).
29. Install water pump pulley and torque the bolts to 72–84 inch lbs. (8–10 Nm).
30. Install drive belts and adjust to proper tension.
31. Install engine mounting bracket and lower the engine.
32. Raise and safely support the vehicle.
33. Install the left side splash shield.
34. Install the left front wheel.
35. Lower the vehicle.
36. Connect the negative battery cable.
37. Start the engine and check for proper operation.
38. Check and adjust ignition timing.

1989–1992 1.6L (4G61) AND 1989–95 2.0L (4G63) DOHC ENGINES

1. Disconnect the negative battery cable.
2. Raise and safely support the vehicle.
3. Remove the left front wheel.

4. Remove the left front splash shield.

5. Using the proper equipment, slightly raise the engine to take the weight off the side engine mount.

6. Remove the engine mount bracket.

7. Remove the accessory drive belts and any mounting brackets that may interfere with timing cover removal.

8. Remove the water pump pulley.

9. Remove the crankshaft pulley bolts and remove the crankshaft pulley.

10. Remove all attaching screws and remove the upper and lower timing belt covers.

11. Label and remove the spark plug wires.

12. Remove the spark plugs.

➥**Removal of the spark plugs is so that the engine can be rotated easily.**

13. Rotate the crankshaft in the direction of rotation and align the timing marks. The No. 1 piston will be at Top Dead Center (TDC) of the compression stroke. At this time the timing marks on the camshaft sprocket and the upper surface of the cylinder head should coincide, and the dowel pin of the camshaft sprocket should be at the upper side.

➥**Always rotate the crankshaft in the normal direction of rotation.**

14. While holding the auto tensioner, remove the tensioner mounting bolts. This will keep from stripping the threads on the mounting bolts. Remove the auto tensioner.

15. Remove the timing belt.

16. If the timing belt is to be reused, mark the belt with an arrow indicating the direction of rotation.

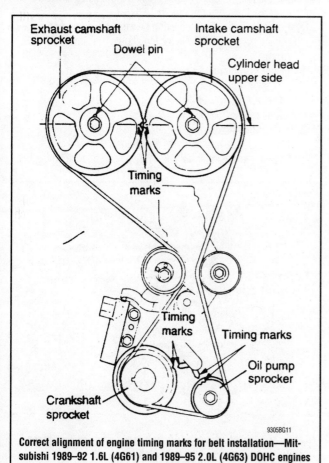

Correct alignment of engine timing marks for belt installation—Mitsubishi 1989–92 1.6L (4G61) and 1989–95 2.0L (4G63) DOHC engines

9305BG11

17. Check the tensioner and idler pulleys for ease of movement and abnormal noises. Replace as necessary.

18. Check crankshaft and camshaft sprockets for cracks and excessive wear. Replace as necessary.

19. Check the auto tensioner for leaks and damage. Measure the rod protrusion from the tensioner body and the tip of the rod. Measurement should be 0.47 in. (12mm). If out of specification, the tensioner must be replaced.

To install:

20. Carefully push in the auto tensioner rod until the set hole in the rod is aligned with the hole in the cylinder body. Place a wire pin into the hole to retain the rod.

21. Install the auto tensioner and torque the bolts to 14–20 ft. lbs. (20–27 Nm).

22. When installing the timing belt, turn the 2 camshaft sprockets so their dowel pins are located on top. Align the timing marks facing each other with the top surface of the cylinder head. When you let go of the exhaust camshaft sprocket, it will rotate 1 tooth in the counterclockwise direction. This should be taken into account when installing the timing belts on the sprocket.

➥**Both camshaft sprockets are used for the intake and exhaust camshafts and are provided with 2 timing marks. When the sprocket is mounted on the exhaust camshaft, use the timing mark on the right with the dowel pinhole on top. For the intake camshaft sprocket, use the mark on the left with the dowel pinhole on top.**

23. Align the crankshaft sprocket and oil pump sprocket timing marks.

24. Install the timing belt as follows:

   a. Install the timing belt around the tensioner pulley and the crankshaft sprocket while holding the timing belt onto the tensioner pulley with your left hand.

   b. Pull the belt with your right hand and install it around the oil pump sprocket.

   c. Install the belt around the idler pulley and the intake camshaft sprocket.

   d. Rotate the exhaust camshaft sprocket, 1 tooth, in the direction of rotation to align the timing mark with the intake camshaft timing mark.

   e. Install the belt around the exhaust camshaft sprocket.

   f. Raise the tensioner pulley clockwise against the timing belt and temporarily tighten the bolt.

   g. Rotate the crankshaft ¼ turn counterclockwise. Then, turn clockwise until the timing marks are aligned again.

25. To adjust the timing belt, turn the crankshaft ¼ turn opposite direction of rotation, then turn it in the direction of rotation to move No. 1 cylinder to TDC.

26. Loosen the tensioner center bolt. Using tool MD998738 or equivalent and a torque wrench, apply a torque of 22.5–24.3 inch lbs. (2.6–2.8 Nm) to the tensioner pulley. Torque the center bolt to 31–40 ft. lbs. (43–55 Nm).

27. Remove the rubber plug from the rear timing belt cover and screw special tool MD99873801 or equivalent, into the engine left support bracket until its end makes contact with the tensioner arm. At this point, screw the special tool in some more and remove the set wire attached to the auto tensioner. Then remove the special tool.

28. Rotate the crankshaft 2 complete turns clockwise and let it sit for approximately 15 minutes. Then, measure the auto tensioner protrusion (the distance between the tensioner arm and auto tensioner body) to ensure that it is within 0.15–0.18 in. (3.8–4.5mm).

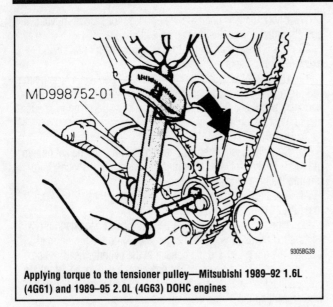

Applying torque to the tensioner pulley—Mitsubishi 1989–92 1.6L (4G61) and 1989–95 2.0L (4G63) DOHC engines

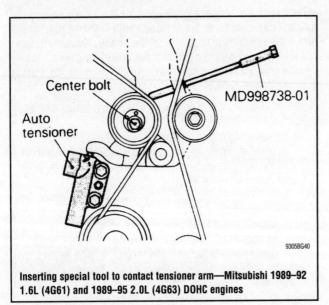

Inserting special tool to contact tensioner arm—Mitsubishi 1989–92 1.6L (4G61) and 1989–95 2.0L (4G63) DOHC engines

29. Install the rubber plug to the timing belt rear cover.
30. Install the upper and lower timing belt covers and torque the bolts to 84–108 inch lbs. (9–12 Nm).
31. Install the water pump pulley and torque the bolts to 84 inch lbs. (10 Nm).
32. Install the crankshaft pulley and torque the bolts to 22 ft. lbs. (30 Nm).
33. Install the mounting brackets and the accessory drive belts.
34. Install the engine mount bracket and engine mount. Lower the engine.
35. Install the spark plugs.
36. Install the spark plug wires.
37. Raise and safely support the vehicle.
38. Install the left front splash shield.
39. Install the left front wheel.
40. Lower the vehicle
41. Connect the negative battery cable.
42. Start the engine and check for proper operation.
43. Check ignition timing.

1983–00 1.8L (G62B), 2.0L (G63B/4G63) & 2.4L 8-VALVE (G64B/4G64) SOHC ENGINES

1. Disconnect the negative battery cable.
2. Remove the accessory drive belts.
3. Remove the water pump pulley.
4. Remove the crankshaft pulley.
5. Remove the power steering pump pulley and all accessories that may interfere with the removal of the timing belt covers.
6. Remove the timing belt covers.
7. Rotate the crankshaft in the direction of rotation to align the camshaft timing mark with the timing mark on the cylinder head.
8. Loosen the tensioner bolts and move the tensioner towards the water pump to relieve pressure on the timing belt. Tighten the bolts.
9. If timing belt is to be reused, mark the belt with an arrow to indicate the direction of rotation.
10. Remove the timing belt.
11. Loosen the lockbolt for the inner timing belt. Move the tensioner away from the belt and remove the inner belt.
12. Inspect the tensioner pulley for ease of movement and abnormal noises. Replace as necessary.
13. Inspect the crankshaft, camshaft and oil pump sprockets for cracks and excessive wear. Replace as necessary.
To install:
14. Make sure that all timing marks are aligned.
15. Install the inner timing belt over the crankshaft inner sprocket and the silent shaft sprocket.
16. Move the inner belt tensioner against the belt to remove any slack from the tension side. Make sure that the timing marks are aligned. Torque the tensioner bolt to 14 ft. lbs. (19 Nm).
17. Check the inner belt deflection between the crankshaft pulley and the silent shaft sprocket. Deflection should measure 0.20–0.30 in. (5–7mm).
18. Install the outer timing belt in the following order:
    a. Crankshaft sprocket
    b. Oil pump sprocket
    c. Camshaft sprocket
    d. Tensioner pulley
19. Make sure the timing marks are aligned.
20. Loosen the tensioner bolts to apply pressure to the timing belt.

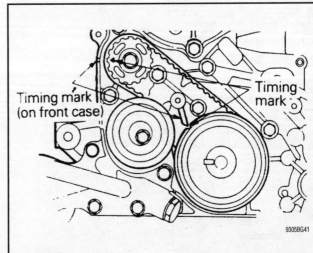

Inner timing belt mark alignment—Mitsubishi 1983–00 1.8L (G62B), 2.0L (G63B) and 2.4L 8-valve (G64B) SOHC engines

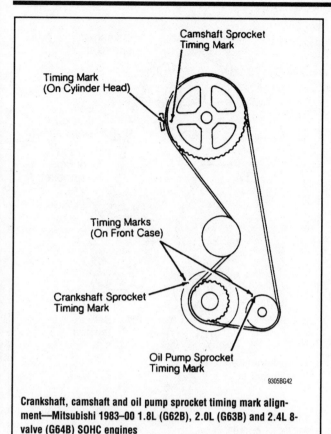

Crankshaft, camshaft and oil pump sprocket timing mark alignment—Mitsubishi 1983–00 1.8L (G62B), 2.0L (G63B) and 2.4L 8-valve (G64B) SOHC engines

21. Rotate the crankshaft 2 full turns in the direction of rotation and align the timing marks. Make sure that there is no slack on the tension side of the belt.

22. Loosen the tensioner bolts again to apply pressure to the timing belt. Torque the bolts to 35 ft. lbs. (49 Nm).

23. Install the timing belt covers.

24. Install all accessories that were previously removed to access the timing belt covers. Install the power steering pump pulley.

25. Install the crankshaft pulley and torque the bolts to 15–21 ft. lbs. (20–29 Nm).

26. Install the water pump pulley and torque the bolts to 84 inch lbs. (9 Nm).

27. Install the accessory drive belts and adjust to proper tension.

28. Connect the negative battery cable.

29. Start the engine and check for proper operation.

30. Check ignition timing and adjust if necessary.

1990–1994 1.8L (4G37) ENGINE

1. Disconnect the negative battery cable.
2. Raise and safely support the vehicle.
3. Remove the left front wheel.
4. Remove the left side splash shield.
5. Lower the vehicle.
6. Using a suitable engine brace, remove the front engine mount and bracket.
7. Remove the accessory drive belts.
8. Remove the accessory belt tensioner pulley and bracket.
9. Remove the water pump and power steering pump pulleys.
10. Remove the damper pulley and adapter.

11. Remove the crankshaft pulley.
12. Remove the upper and lower timing belt covers.
13. Rotate the crankshaft in the direction of rotation to align the timing marks.
14. Loosen the tensioner bolts and move the tensioner towards the water pump to relieve pressure on the timing belt. Tighten the tensioner bolts.
15. If the timing belt is to be reused, mark the belt with an arrow indicating the direction of rotation.
16. Remove the timing belt.
17. Loosen the tensioner lockbolt for the inner timing belt. Move the tensioner away from the belt and remove the inner belt.
18. Inspect the tensioner pulley for ease of movement and abnormal noises. Replace as necessary.
19. Inspect the crankshaft, camshaft and oil pump sprockets for cracks and excessive wear. Replace as necessary.

To install:

20. Make sure that all timing marks are aligned.
21. Install the inner timing belt over the crankshaft inner sprocket and the silent shaft sprocket.

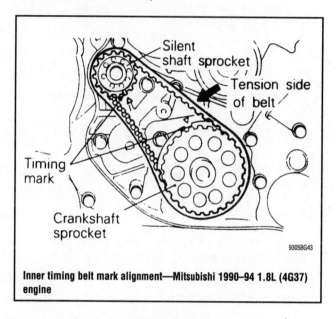

Inner timing belt mark alignment—Mitsubishi 1990–94 1.8L (4G37) engine

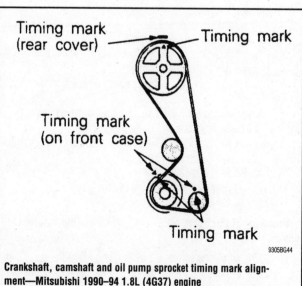

Crankshaft, camshaft and oil pump sprocket timing mark alignment—Mitsubishi 1990–94 1.8L (4G37) engine

22. Move the inner belt tensioner clockwise against the belt to remove any slack from the tension side. Make sure that the timing marks are aligned. Torque the tensioner bolt to 14 ft. lbs. (19 Nm).

23. Check the inner belt deflection between the crankshaft pulley and the silent shaft sprocket. Deflection should measure 0.20–0.28 in. (5–7mm).

24. Install the outer timing belt in the following order:
    a. Crankshaft sprocket
    b. Oil pump sprocket
    c. Camshaft sprocket
    d. Tensioner pulley

25. Make sure the timing marks are aligned.

26. Loosen the tensioner bolts to apply pressure to the timing belt.

27. Rotate the crankshaft 2 full turns in the direction of rotation and align the timing marks. Make sure that there is no slack on the tension side of the belt.

28. Loosen the tensioner bolts again to apply pressure to the timing belt. Torque the bolts to 35 ft. lbs. (49 Nm).

29. Install the upper and lower timing belt covers.

30. Install the crankshaft pulley and damper adapter. Torque the bolts to 11–13 ft. lbs. (15–18 Nm).

31. Install the damper pulley and torque the bolts to 11–13 ft. lbs. (15–18 Nm).

32. Install the water pump and power steering pump pulleys and torque the bolts to 72–84 ft. lbs. (8–10 Nm).

33. Install the accessory belt tensioner pulley and bracket. Torque the bolts to 17–20 ft. lbs. (23–27 Nm).

34. Install the front engine mount and bracket. Remove the engine brace.

35. Install the accessory drive belts and adjust to proper tension.

36. Raise and safely support the vehicle.

37. Install the left side splash shield.

38. Install the left front wheel.

39. Lower the vehicle.

40. Connect the negative battery cable.

41. Start the engine and check for proper operation.

42. Check ignition timing and adjust if necessary.

1992–00 1.8L (4G93) ENGINE

1. Disconnect the negative battery cable

2. Raise and safely support the vehicle.

3. Remove the left front wheel.

4. Remove the left side splash shield.

5. Lower the vehicle.

6. Remove the accessory drive belts.

7. Remove the upper alternator bracket and position the alternator aside.

8. Using a suitable engine support brace, remove the front engine mount and bracket.

9. Remove the crankshaft pulley bolt and remove the crankshaft pulley.

10. Remove the timing belt covers.

11. Tag and remove the spark plug wires.

12. Remove the spark plugs.

➥**Removal of the spark plugs is to make engine rotation easier.**

13. Reinstall the crankshaft pulley bolt and rotate the engine in the direction of rotation to align the timing marks on the camshaft sprocket and the crankshaft sprocket.

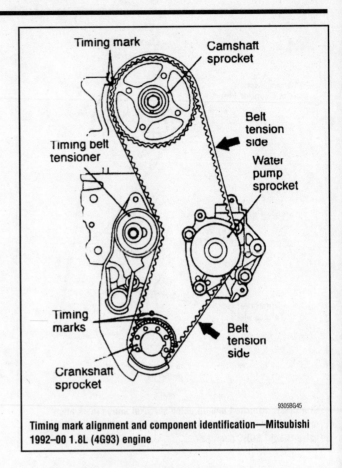

Timing mark alignment and component identification—Mitsubishi 1992–00 1.8L (4G93) engine

14. Loosen the tensioner lockbolt and move the tensioner counterclockwise to relieve pressure on the timing belt.

15. Temporarily tighten the tensioner lockbolt.

16. If the timing belt is to be reused, mark the belt with an arrow to indicate the direction of rotation.

17. Remove the timing belt.

18. Inspect the tensioner pulley for ease of movement and abnormal noises. Replace as necessary.

19. Inspect the crankshaft and camshaft sprockets for cracks and excessive wear. Replace as necessary.

To install:

20. Make sure that the timing marks are correctly aligned.

21. Install the timing belt in the following order:
    a. Crankshaft sprocket
    b. Water pump sprocket
    c. Camshaft sprocket
    d. Tensioner pulley

22. Make sure that here is no slack on the tension side of the belt.

23. Loosen the tensioner lockbolt to apply tension to the timing belt.

24. Temporally tighten the tensioner lockbolt.

25. Rotate the crankshaft 2 full turns in the direction of rotation and align the timing marks.

26. Loosen the tensioner lockbolt and make sure that there is no slack in the timing belt and that the timing marks are aligned.

27. Torque the tensioner lockbolt to 17 ft. lbs. (24 Nm).

28. Install the upper and lower timing belt covers.

29. Install the front engine mount and bracket and lower the engine. Remove the engine brace.

30. Install the crankshaft pulley and torque the pulley bolt to 130–137 ft. lbs. (177–186 Nm).

31. Install the upper alternator bracket and mount the alternator.
32. Install the accessory drive belts and adjust tension.
33. Install the spark plugs and plug wires.
34. Raise and safely support the vehicle.
35. Install the left side splash shield.
36. Install the left front wheel.
37. Lower the vehicle.
38. Connect the negative battery cable.
39. Start the engine and check for proper operation.
40. Check and adjust ignition timing if necessary.

1995–00 2.0L (4G63) DOHC TURBO ENGINE

1. Disconnect the negative battery cable.
2. Remove the engine undercover.
3. Remove the engine mount bracket.
4. Remove the drive belts.
5. Remove the belt tensioner pulley.
6. Remove the water pump pulleys.
7. Remove the crankshaft pulley.
8. Using a suitable engine support brace, remove the stud bolt from the engine support bracket and remove the bracket and mount.
9. Remove the timing belt covers.
10. Rotate the crankshaft in the direction of rotation to align the camshaft timing marks. Always turn the crankshaft in the normal direction of rotation only.
11. Loosen the tension pulley center bolt.

➡ **If the timing belt is to be reused, mark the direction of rotation on the flat side of the belt with an arrow.**

12. Move the tension pulley towards the water pump and remove the timing belt.
13. Remove the crankshaft sprocket center bolt using special tool MB990767, or equivalent, to hold the crankshaft sprocket while removing the center bolt. Then, use MB998778 or equivalent puller to remove the sprocket.
14. Mark the direction of rotation on the timing belt **B** with an arrow.
15. Loosen the center bolt on the tensioner and remove the belt.

❊❊ WARNING

Do not rotate the camshafts or the crankshaft while the timing belt is removed.

To install:

16. Place the crankshaft sprocket on the crankshaft. Use tool MB990767, or equivalent, to hold the crankshaft sprocket while tightening the center bolt. Torque the center bolt to 80–94 ft. lbs. (108–127 Nm).
17. Align the timing marks on the crankshaft sprocket **B** and the balance shaft.
18. Install timing belt **B** on the sprockets. Position the center of the tensioner pulley to the left and above the center of the mounting bolt.
19. Push the pulley clockwise toward the crankshaft to apply tension to the belt and torque the mounting bolt to 14 ft. lbs. (19 Nm). Do not let the pulley turn when tightening the bolt because it will cause excessive tension on the belt. The belt should deflect 0.20–0.28 in. (5–7mm) when finger pressure is applied between the pulleys.
20. Install the crankshaft sensing blade and the crankshaft sprocket. Apply engine oil to the mounting bolt and torque the bolt to 80–94 ft. lbs. (108–127 Nm).

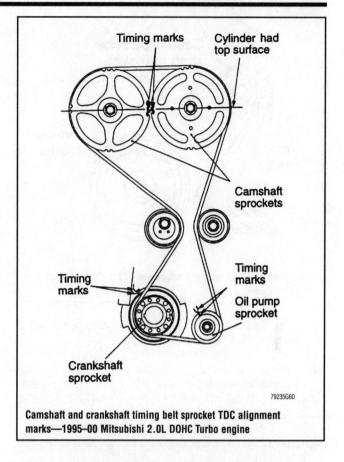

Camshaft and crankshaft timing belt sprocket TDC alignment marks—1995–00 Mitsubishi 2.0L DOHC Turbo engine

21. Use a press or vise to compress the auto-tensioner pushrod. Insert a set pin when the holes in the rod and the tensioner body are aligned.

❊❊ WARNING

Do not compress the pushrod too quickly, damage to the pushrod can occur.

22. Install the auto-tensioner on the engine.
23. Align the timing marks on the camshaft sprocket, crankshaft sprocket and the oil pump sprocket.
24. After aligning the mark on the oil pump sprocket, remove the cylinder block plug and insert a prytool in the hole to check the position of the counterbalance shaft. The prytool should go in at least 2.36 in. (60mm) or more, if not, rotate the oil pump sprocket once and realign the timing mark so the prytool goes in. Do not remove the prytool until the timing belt is installed.
25. Install the timing belt on the intake camshaft and secure it with a clip.
26. Install the timing belt on the exhaust camshaft. Align the timing marks with the cylinder head top surface using 2 wrenches. Secure the belt with another clip.
27. Install the belt around the idler pulley, oil pump sprocket, crankshaft sprocket and the tensioner pulley.
28. Turn the tensioner pulley so the pinholes are at the bottom. Press the pulley lightly against the timing belt.
29. Screw the special tool into the left engine support bracket until it contacts the tensioner arm, then screw the tool in a little more and remove the pushrod pin from the auto-tensioner. Remove the special tool and torque the center bolt to 35 ft. lbs. (48 Nm).

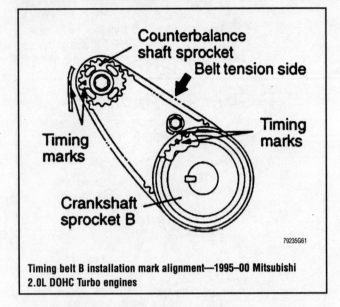

Timing belt B installation mark alignment—1995–00 Mitsubishi 2.0L DOHC Turbo engines

30. Turn the crankshaft ¼ turn counterclockwise, then clockwise until the timing marks are aligned.

31. Loosen the center bolt. Install Mitsubishi Special Tool MD998767 or equivalent, on the tensioner pulley. Turn the tensioner pulley counterclockwise with a torque of 31 inch lbs. (3.5 Nm) and torque the center bolt to 35 ft. lbs. (48 Nm). Do not let the tensioner pulley turn when tightening the bolt.

32. Turn the crankshaft clockwise 2 revolutions and align the timing marks. After 15 minutes, measure the protrusion of the pushrod on the auto-tensioner. The standard measurement is 0.150–0.177 in. (3.8–4.5mm). If the protrusion is out of specification, loosen the tensioner pulley, apply the proper torque to the belt and retighten the center bolt.

33. Install the timing belt covers and all applicable components.

1993–00 2.4L 16-VALVE (G64B) ENGINE

1. Disconnect the negative battery cable.
2. Remove the splash shield under the engine.
3. Safely support the weight of the engine and remove the engine mount and bracket assembly.
4. Remove the drive belts.
5. Remove the water pump pulley and the crankshaft pulley.
6. Remove the timing belt covers.
7. Position the engine so that the No. 1 piston is at Top Dead Center (TDC) of the compression stroke.

➡If the timing belts are going to be reused, mark the direction of rotation on the belt. This will ensure the belt is reinstalled in same direction, extending belt life.

8. To loosen the timing belt tensioner, install Mitsubishi Special tool MD998738 or equivalent, into the slot on the rear timing belt cover and screw inward to move the tensioner toward the water pump. Once the tension has been relieved, remove the outer timing belt.

9. Align the pin hole in the tensioner rod to the hole in the tensioner cylinder. Insert a 0.055 in. (1.4mm) wire in the hole and remove the special tool from the slot. With the cylinder tension relieved, remove the auto-tensioner cylinder assembly 2 mounting bolts.

10. Remove the outer crankshaft sprocket and flange.

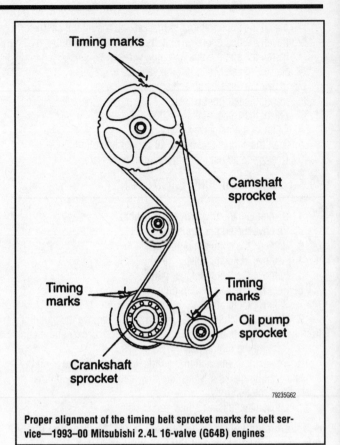

Proper alignment of the timing belt sprocket marks for belt service—1993–00 Mitsubishi 2.4L 16-valve (G64B) engines

11. Loosen the silent shaft (inner) belt tensioner and remove the belt.

To install:

✳✳ WARNING

Do not spray or immerse the sprockets or tensioners in cleaning solvent. The sprocket may absorb the solvent and transfer it to the belt. The tensioners are internally lubricated and the solvent will dilute or dissolve the lubricant.

12. Align the timing marks of the silent shaft sprockets and the crankshaft sprocket with the timing marks on the front case. Route the timing belt around the sprockets so there is no slack on the tension side of the belt and the timing marks are still aligned.

13. Install the tensioner pulley and move the pulley by hand so that the tension side of the belt deflects approximately ¼ in. (6mm).

14. Hold the pulley tightly so the pulley cannot rotate when the bolt is tightened. Torque the bolt to 14 ft. lbs. (19 Nm) and recheck the deflection.

15. Align the timing marks of the camshaft, crankshaft and oil pump sprockets with their corresponding marks on the front case or rear cover.

➡There is a possibility to align all timing marks and have the oil pump sprocket and silent shaft out of time, causing an engine vibration during operation. If the following step is not followed exactly, there is a 50 percent chance that the silent shaft alignment will be 180 degrees (½ turn) off.

16. Before installing the timing belt, ensure that the left-side (rear) silent shaft (oil pump sprocket) is in the correct position as follows:

a. Remove the plug from the rear side of the block and insert a tool with shaft diameter of 0.31 in. (8mm) into the hole.

b. With the timing marks still aligned, the shaft of the tool must be able to go in at least 2 ½ in. (63.5mm). If the tool can only go in approximately 1 in. (25mm), the shaft is not in the correct orientation and will cause a vibration during engine operation. Remove the tool from the hole and turn the oil pump sprocket 1 complete revolution. Realign the timing marks and insert the tool. The shaft of the tool must go in at least 2¼ in. (63.5mm).

c. Recheck and realign the timing marks.

d. Leave the tool in place to hold the silent shaft while continuing.

17. If the camshaft belt tensioner was removed, use a vise to carefully push the auto-tensioner rod in until the set hole in the rod is aligned with the hole in the cylinder. Place a wire into the hole to retain the rod. Mount the tensioner to the engine block and torque the mounting bolt to 17 ft. lbs. (23 Nm).

18. Install the belt to the crankshaft sprocket, oil pump sprocket, then camshaft sprocket, in that order. While doing so, be sure there is no slack between the sprocket except where the tensioner is installed.

19. To adjust the timing (outer) belt perform the following steps:

a. Turn the crankshaft ¼ turn counterclockwise, then turn it clockwise to move No. 1 cylinder to TDC.

b. Loosen the tensioner roller center bolt. Using tool MD998752 or equivalent, and a torque wrench, apply a torque of 31 inch lbs. (3.6 Nm) to the tensioner. Tighten the center bolt.

c. Screw the special tool into the engine left support bracket until its end makes contact with the tensioner arm. At this point, screw the special tool in some more and remove the set wire attached to the auto-tensioner. Then, remove the special tool.

d. Rotate the crankshaft 2 complete turns in the direction of rotation and let it sit for approximately 15 minutes. Then, measure the auto-tensioner protrusion (the distance between the tensioner arm and auto-tensioner body) to ensure that it is within 0.15–0.18 in. (3.8–4.5mm). If out of specification, the auto-tensioner should be replaced.

20. Install the upper and lower timing belt covers.

21. Install the water pump pulley and the crankshaft pulley.

22. Install the drive belts.

23. Install the engine mount and bracket assembly.

24. Install the splash shield under the engine.

25. Connect the negative battery cable.

26. Start the engine and check for proper operation.

3.0L (6G72) SOHC V6 ENGINE

1988–94

1. Disconnect the negative battery cable.
2. Remove the engine undercover.
3. Remove the cruise control actuator.
4. Remove the accessory drive belts.
5. Remove the air conditioner compressor tension pulley assembly.
6. Remove the tension pulley bracket.
7. Using the proper equipment, slightly raise the engine to take the weight off the side engine mount. Remove the engine mounting bracket.
8. Disconnect the power steering pump pressure switch connector. Remove the power steering pump and wire aside.
9. Remove the engine support bracket.

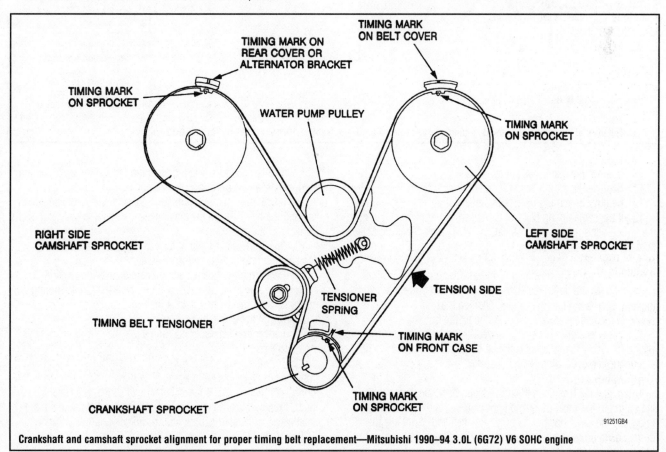

Crankshaft and camshaft sprocket alignment for proper timing belt replacement—Mitsubishi 1990–94 3.0L (6G72) V6 SOHC engine

1. Engine support bracket
2. Bolt
3. Washer
4. Crankshaft pulley
5. Access cover
6. Right side upper front cover
7. Cap
8. Left side upper front cover
9. Front lower cover
10. Flange

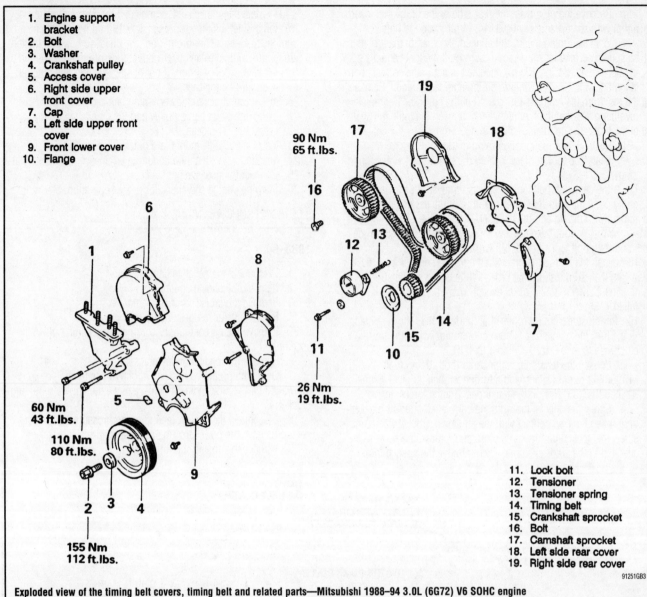

90 Nm
65 ft.lbs.

60 Nm
43 ft.lbs.

110 Nm
80 ft.lbs.

155 Nm
112 ft.lbs.

26 Nm
19 ft.lbs.

11. Lock bolt
12. Tensioner
13. Tensioner spring
14. Timing belt
15. Crankshaft sprocket
16. Bolt
17. Camshaft sprocket
18. Left side rear cover
19. Right side rear cover

91251GB3

Exploded view of the timing belt covers, timing belt and related parts—Mitsubishi 1988–94 3.0L (6G72) V6 SOHC engine

10. Remove the crankshaft pulley.

11. Remove the timing belt cover cap.

12. Remove the timing belt upper and lower covers.

13. If the same timing belt will be reused, mark the direction of the timing belt's rotation for installation in the same direction. Make sure the engine is positioned so the No. 1 cylinder is at the Top Dead Center (TDC) of the compression stroke and the sprockets' timing marks are aligned with the engine's timing mark indicators.

14. Loosen the timing belt tensioner bolt and remove the belt. If the tensioner is not being removed, position it as far away from the center of the engine as possible and tighten the bolt.

15. If the tensioner is being removed, paint the outside of the spring to ensure that it is not installed backwards. Unbolt the tensioner and remove it along with the spring.

To install:

16. Install the tensioner, if removed, and hook the upper end of the spring to the water pump pin and the lower end to the tensioner in exactly the same position as originally installed. If not already done, position both camshafts so the marks align with those on the

rear. Rotate the crankshaft in the direction of rotation so the timing mark aligns with the mark on the oil pump.

17. Install the timing belt on the crankshaft sprocket and while keeping the belt tight on the tension side, install the belt on the front camshaft sprocket.

18. Install the belt on the water pump pulley, then the rear camshaft sprocket and then the tensioner.

19. Rotate the front camshaft counterclockwise to tension the belt between the front camshaft and the crankshaft. If the timing marks became misaligned, repeat the procedure.

20. Install the crankshaft sprocket flange.

21. Loosen the tensioner bolt and allow the spring to apply tension to the belt.

22. Turn the crankshaft 2 full turns in the clockwise direction until the timing marks align again. Now that the belt is properly tensioned, torque the tensioner lockbolt to 21 ft. lbs. (29 Nm). Measure the belt tension between the rear camshaft sprocket and the crankshaft with belt tension gauge. The specification is 46–68 lbs. (210–310 N).

23. Install the timing covers. Make sure all pieces of packing are positioned in the inner grooves of the covers when installing.

24. Install the crankshaft pulley. Torque the bolt to 108–116 ft. lbs. (150–160 Nm).

25. Install the engine support bracket.

26. Install the power steering pump and reconnect wire harness at the power steering pump pressure switch.

27. Install the engine mounting bracket and remove the engine support fixture.

28. Install the tension pulleys and drive belts.

29. Install the cruise control actuator.

30. Install the engine undercover.

31. Connect the negative battery cable.

32. Start the engine and check for proper operation.

1995–00

1. Disconnect the negative battery cable.

⁂ CAUTION

Wait at least 90 seconds after the negative battery cable is disconnected to prevent possible deployment of the air bag.

2. Remove the engine undercover.

3. Remove the front undercover panel.

4. Remove the cruise control pump and the link assembly.

5. Remove the accessory drive belts.

6. Remove the alternator.

7. Raise and suspend the engine so that force is not applied to the engine mount.

8. Remove the engine mount and bracket.

9. Position the engine so the No. 1 cylinder is at Top Dead Center (TDC) of the compression stroke.

10. Remove the timing covers from the engine.

11. If the same timing belt will be reused, mark the direction of the timing belt's rotation for installation in the same direction. Make sure the engine is positioned so the No. 1 cylinder is at the TDC of its compression stroke and the timing marks are aligned with the engine's timing mark indicators on the valve covers or head.

12. Loosen the center bolt of tensioner pulley and unbolt the tensioner assembly. Remove the timing belt.

To install:

13. Install the tensioner, if removed, and hook the upper end of the spring to the water pump pin and the lower end to the tensioner in exactly the same position as originally installed.

14. Ensure both camshafts are still positioned so the timing marks align with those on the rear timing covers. Rotate the crankshaft so the timing mark aligns with the mark on the front cover.

15. Install the timing belt on the crankshaft sprocket and while

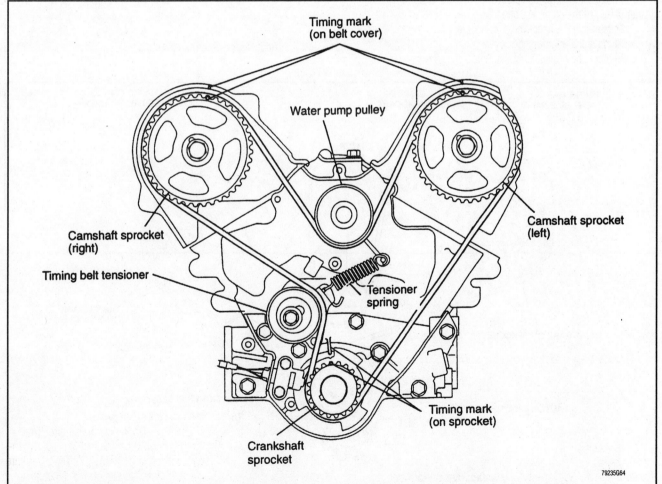

Align the sprockets properly before removing or installing the timing belt—1995–00 Mitsubishi 3000 GT and Diamante with the 3.0L (6G72) V6 SOHC engine

keeping the belt tight on the tension side, install the belt on the front (left) camshaft sprocket.

16. Install the belt on the water pump pulley, then the rear (right) camshaft sprocket and the tensioner.

17. Loosen the bolt that secures the adjustment of the tensioner and lightly press the tensioner against the timing belt.

18. Check that the timing marks are in alignment.

19. Rotate the crankshaft 2 full turns in the direction of rotation, then realign the timing marks.

20. Torque the bolt that secures the tensioner to 19 ft. lbs. (26 Nm).

21. Install the lower and the upper timing belt covers.

22. Raise and suspend the engine so that force is not applied to the engine mount.

23. Install the engine mount and bracket.

24. Install the alternator.

25. Install the accessory drive belts.

26. Install the cruise control pump and the link assembly.

27. Install the front undercover panel.

28. Connect the negative battery cable.

29. Start the engine and check for proper operation.

3.0L (6G72) DOHC V6 ENGINE

1990–00

1. Position the engine so the No. 1 cylinder is at Top Dead Center (TDC) of the compression stroke.

2. Disconnect the negative battery cable.

3. Remove the engine undercover.

4. Remove the cruise control actuator.

5. Remove the alternator.

6. Remove the air hose and pipe.

7. Remove the belt tensioner assembly and the power steering belt.

8. Remove the crankshaft pulley.

9. Disconnect the brake fluid level sensor.

10. Remove the timing belt upper cover.

11. Using the proper equipment, slightly raise the engine to take the weight off the side engine mount. Remove the engine mount bracket.

12. Remove the alternator/air conditioner idler pulley.

13. Remove the engine support bracket. The mounting bolts are different lengths; mark them for proper installation.

14. Remove the timing belt covers. Timing bolt cover mounting bolts are different in length, note their position during removal.

15. If the same timing belt will be reused, mark the direction of the timing belt's rotation for installation in the same direction. Make sure the engine is positioned so the No. 1 cylinder is at the Top Dead Center (TDC) of its compression stroke and the sprockets' timing marks are aligned with the engine's timing mark indicators on the valve covers or head.

16. Loosen the timing belt tensioner bolt and remove the belt.

17. Remove the tensioner assembly.

To install:

18. If the auto tensioner rod is fully extended, reset it as follows:

   a. Clamp the tensioner in a soft-jaw vise in level position.

   b. Slowly push the rod in with the vise until the set hole in the rod is aligned with the hole in the cylinder.

   c. Insert a stiff wire into the set holes to retain the position.

   d. Remove the assembly from the vise.

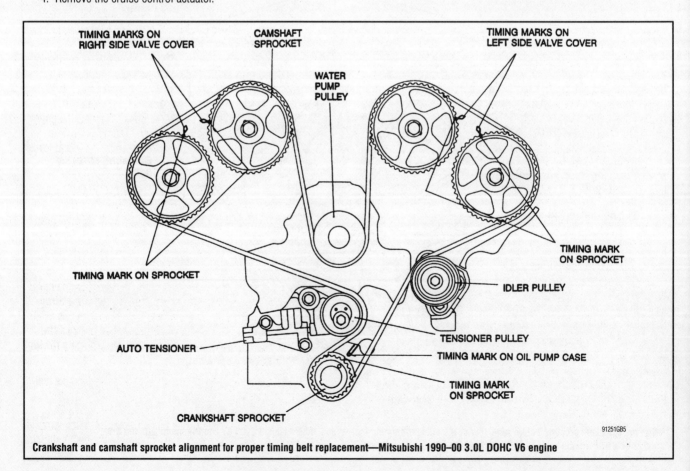

Crankshaft and camshaft sprocket alignment for proper timing belt replacement—Mitsubishi 1990–00 3.0L DOHC V6 engine

19. Leave the retaining wire in the tensioner and install to the engine.

→On 1991 DOHC 3.0L engines, clean and inspect both auto tensioner mounting bolts. Coat the threads of the old bolts with thread sealer. If new bolts are installed, inspect the heads of the new bolts. If there is white paint on the bolt head, no sealer is required. If there is no paint on the head of the bolt, apply a coat of thread sealer to the bolt. Install both bolts and torque to 17 ft. lbs. (24 Nm).

20. If the timing marks of the camshaft sprockets and crankshaft sprocket are not aligned at this point, proceed as follows:

a. Align the mark on the crankshaft sprocket with the mark on the front case. Then move the sprocket 2 teeth clockwise to lower the piston so the valve can't touch the piston when the camshafts are being moved.

b. Turn each camshaft sprocket 1 at a time to align the timing marks with the mark on the valve cover or head. If the intake and exhaust valves of the same cylinder are opened simultaneously, they could interfere with each other. Therefore, if any resistance is felt, turn the other camshaft to move the valve.

c. Align the timing mark of the crankshaft sprocket, then continue 1 tooth farther in the counterclockwise direction to facilitate belt installation.

21. Using 4 spring loaded paper clips to hold the belt on the cam sprockets, install the belt to the sprockets in the following order:
- **1st:** exhaust camshaft sprocket for the front head
- **2nd:** intake camshaft sprocket for the front head
- **3rd:** water pump pulley
- **4th:** intake camshaft sprocket for the rear head
- **5th:** exhaust camshaft sprocket for the rear head
- **6th:** idler pulley
- **7th:** crankshaft sprocket
- **8th:** tensioner pulley

22. Turn the tensioner pulley so its pin holes are located above the center bolt. Then press the tensioner pulley against the timing belt and simultaneously tighten the center bolt.

23. Make certain that all timing marks are still aligned. If so, remove the 4 clips.

24. Turn the crankshaft ¼ turn counterclockwise, then turn it clockwise until all timing marks are aligned.

25. Loosen the center bolt on the tensioner pulley. Using tool MD998767 or equivalent and a torque wrench, apply a torque of 86 inch lbs. (10 Nm) to the tensioner roller. Tighten the tensioner bolt making sure the tensioner doesn't rotate with the bolt.

26. Remove the set wire attached to the auto tensioner.

27. Rotate the crankshaft 2 complete turns clockwise and let it sit for approximately 5 minutes. Then, make sure the set pin can easily be inserted and removed from the hole in the tensioner.

28. Measure the auto tensioner protrusion (the distance between the tensioner arm and auto tensioner body) to ensure that it is within 0.15–0.18 in. (3.8–4.5mm). If out of specification, repeat Steps 1–4 until the specified value is obtained.

29. Make sure all pieces of packing are positioned in the inner grooves of the lower cover, position cover on engine and install mounting bolts in their original location.

30. Install the engine support bracket and secure using mounting bolts in their original location. Lubricate the reaming area of the reamer bolt and tighten slowly.

31. Install the idler pulley.

32. Install the engine mount bracket. Remove the engine support fixture.

33. Make sure all pieces of packing are positioned in the inner grooves of the upper cover and install.

34. Connect the brake fluid level sensor.

35. Install the crankshaft pulley. Torque the bolt to 130–137 ft. lbs. (180–190 Nm).

36. Install the belt tensioner assembly and the power steering belt.

37. Install the air hose and pipe.

38. Install the alternator.

39. Install the cruise control actuator.

40. Install the engine undercover.

41. Connect the negative battery cable.

3.5L (6G74) ENGINE

1998–00

1. Disconnect the negative battery cable.
2. Drain the cooling system.
3. Remove the drive belts.

❊❊ CAUTION

Never open, service or drain the radiator or cooling system when hot; serious burns can occur from the steam and hot coolant. Always drain coolant into a sealable container. Coolant should be reused unless it is contaminated or is several years old.

4. Remove the upper radiator shroud.
5. Remove the accessory drive belts.
6. Remove the fan and fan pulley.
7. Without disconnecting the lines, remove the power steering pump from its bracket and position aside. Remove the pump brackets.
8. Remove the belt tensioner pulley bracket.
9. Without releasing the refrigerant, remove the air conditioning compressor from its bracket and position aside. Remove the bracket.
10. Remove the cooling fan bracket.
11. On some vehicles it may be necessary to remove the pulley from the crankshaft to access the lower cover bolts.
12. Remove the timing belt cover bolts and the upper and lower covers from the engine.
13. Remove the crankshaft position sensor connector.
14. Using SST MB990767-01 and MD998754 or their equivalents, remove the crankshaft pulley from the crankshaft.
15. Use a shop rag to clean the timing marks to assist in properly aligning them.
16. Loosen the center bolt on the tension pulley and remove the timing belt.

→If the same timing belt will be reused, mark the direction of timing belt's rotation, for installation in the same direction. Be sure engine is positioned so No. 1 cylinder is at the Top Dead Center (TDC) of the compression stroke and the sprockets timing marks are aligned with the engine's timing mark indicators.

17. Remove the auto-tensioner, the tension pulley and the tension arm assembly.

To install:

18. Install the crankshaft pulley and turn the crankshaft sprocket timing mark forward (clockwise) 3 teeth to move the piston slightly past No. 1 cylinder top dead center.

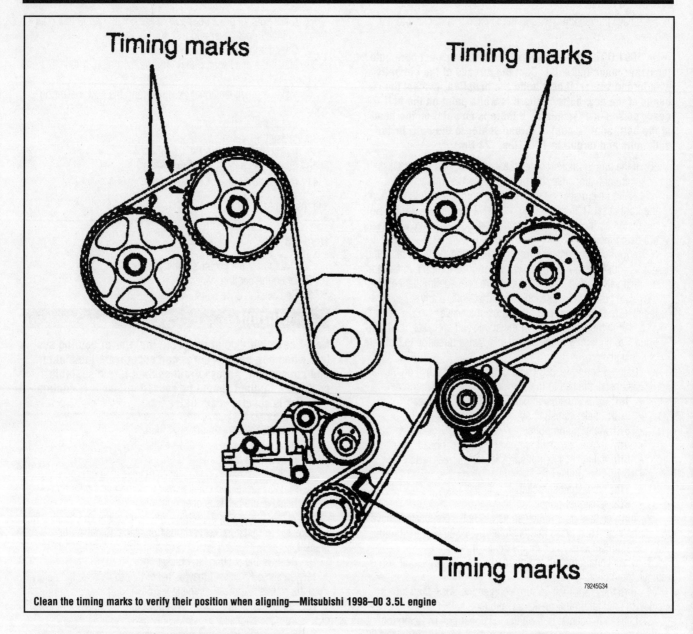

Timing marks

Timing marks

Timing marks

79245G34

Clean the timing marks to verify their position when aligning—Mitsubishi 1998–00 3.5L engine

19. If removed, install the camshaft sprockets and torque the bolts to 64 ft. lbs. (88 Nm).

20. Align the timing mark of the left bank side camshaft sprocket.

21. Align the timing mark of the right bank side camshaft sprocket, and hold the sprocket with a wrench so that it doesn't turn.

22. Set the timing belt onto the water pump pulley.

23. Check that the camshaft sprocket timing mark of the left bank side is aligned and clamp the timing belt with double clips.

24. Set the timing belt onto the idler pulley.

❈❈ WARNING

If any binding is felt when adjusting the timing belt tension by turning the crankshaft, STOP turning the engine, because the pistons may be hitting the valves.

25. Turn the crankshaft 1 turn counterclockwise and set the timing belt onto the crankshaft sprocket.

26. Set the timing belt on the tension pulley.

27. Place the tension pulley pin hole so that it is towards the top. Press the tension pulley onto the timing belt, and then provisionally tighten the adjusting bolt. Torque the bolt to 35 ft. lbs. (48 Nm).

28. Slowly turn the crankshaft 2 full turns in the direction of rotation until the timing marks align. Remove the 4 double clips.

29. Install the crankshaft position sensor connector.

30. Install the upper and lower covers on the engine and secure them with the retaining screws. Be sure the packing is properly positioned in the inner grooves of the covers when installing.

31. Install the crankshaft pulley if it was removed. Torque the bolt to 110 ft. lbs. (150 Nm).

32. Install the air conditioning bracket and compressor on the engine. Install the belt tensioner.

33. Install the power steering pump into position.

34. Install the fan pulley and fan.

35. Install the accessory drive belts.

36. Install the fan shroud on the radiator.

37. Refill the cooling system.
38. Connect the negative battery cable.
39. Start the engine and check for fluid leaks.

Nissan/Datsun

1.5L (E15, E15ET) & 1.6L (E16, E16S) ENGINES

1. Disconnect the negative battery cable.
2. Remove the accessory drive belts.
3. Support the engine with a jack and remove the right motor mount.
4. Remove the right engine splash guards.
5. Remove the water pump pulley.
6. Remove the crankshaft pulley.
7. Remove the timing covers.
8. Rotate the crankshaft so that the timing marks on the camshaft and crankshaft pulleys align with the timing marks on the rear timing cover.
9. Loosen the timing belt tensioner locknut.
10. Turn the tensioner away from the timing belt and tighten the locknut.
11. Remove the timing belt.
To install:
12. Ensure that the timing marks are aligned and install the timing belt.
13. Loosen the tensioner locknut.
14. Carefully turn the camshaft pulley clockwise 2 teeth and tighten the tensioner locknut.
15. Rotate the crankshaft 2 complete turns and check that the timing marks align.
16. Install the timing covers.
17. Install the crankshaft pulley and tighten the bolt to 83–108 ft. lbs. (113–147 Nm).
18. Install the water pump pulley.
19. Install the accessory drive belts.
20. Install the right motor mount and the splash guards. Remove the engine support jack.
21. Connect the negative battery cable.

1.7L (CD17) DIESEL ENGINE

Valve Timing Belt

1. Disconnect the negative battery cable.
2. Remove the accessory drive belts.
3. Support the engine with a jack and remove the right motor mount.
4. Remove the right engine splash guards.
5. Remove the crankshaft damper pulley.
6. Remove the timing covers.
7. Rotate the crankshaft so that the timing marks on the camshaft pulley and the crankshaft timing pulley align with the marks on the inner timing cover.
8. Loosen the tensioner locknut and set the tensioner in the free position. Tighten the locknut.
9. Remove the crankshaft timing pulley and the timing belt.
To install:
10. Install the crankshaft timing pulley and timing belt.
11. Ensure that the timing marks align and loosen the tensioner locknut.
12. Carefully turn the camshaft pulley 2 teeth and tighten the tensioner locknut.
13. Rotate the crankshaft 2 complete turns and check that the timing marks align.
14. Install the timing covers.
15. Install the crankshaft damper pulley and tighten the bolt to 90–98 ft. lbs. (122–133 Nm).
16. Install the accessory drive belts.
17. Install the right splash guards and motor mount. Remove the engine support jack.
18. Connect the negative battery cable.

Injection Pump Timing Belt

1. Disconnect the negative battery cable.
2. Rotate the crankshaft to set the No. 1 cylinder at Top Dead Center (TDC) of the compression stroke. The crankshaft damper pulley mark should align with the front cover pointer.
3. Remove the rear outer timing cover.
4. Loosen the belt tensioner locknut and set the tensioner to the free position. Tighten the locknut.

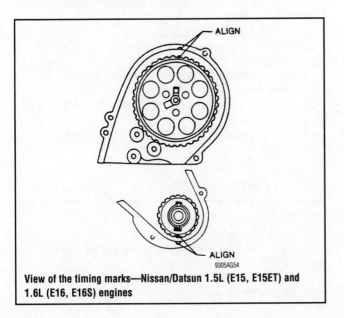

View of the timing marks—Nissan/Datsun 1.5L (E15, E15ET) and 1.6L (E16, E16S) engines

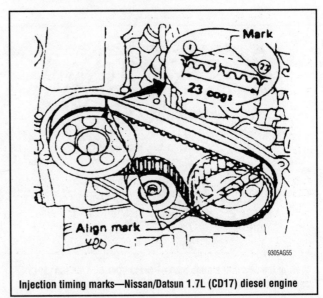

Injection timing marks—Nissan/Datsun 1.7L (CD17) diesel engine

5. Remove the injection pump timing belt.

To install:

6. Prepare the belt for installation by placing matchmarks 23 cogs apart. Align the matchmarks with the timing marks on the camshaft and injection pump pulleys and install the belt.

7. Rotate the crankshaft 2 complete turns and tighten the injection pump timing belt tensioner locknut.

8. Install the rear outer timing cover and connect the negative battery cable.

1.6L (CA16DE) & 1.8L (CA18DE) ENGINES

1. Disconnect the negative battery cable.
2. Drain the cooling system.
3. Remove the upper radiator hose.
4. Remove the right engine undercover.
5. Remove the right front inner fender access cover.
6. Remove the accessory drive belts.
7. Remove the water pump pulley.
8. Rotate the crankshaft to set the No. 1 piston at Top Dead Center (TDC) of the compression stroke.
9. Matchmark the Camshaft Position (CMP) sensor to the upper timing cover and remove the sensor.
10. Support the engine with a jack and remove the right engine mount and bracket.
11. Remove the upper timing cover. Check that the camshaft timing marks are aligned.
12. Remove the crankshaft pulley.
13. Remove the lower timing cover.
14. Loosen the tensioner pulley locknut. Rotate the tensioner clockwise to loosen the timing belt. Tighten the tensioner locknut
15. Remove the timing belt.

To install:

16. Ensure that the timing marks are aligned and install the timing belt. When properly installed, there will be 39 cogs between the intake and exhaust camshaft pulley timing marks, and 48 cogs between the exhaust camshaft pulley timing mark and the crankshaft timing mark.

17. Loosen the tensioner locknut.

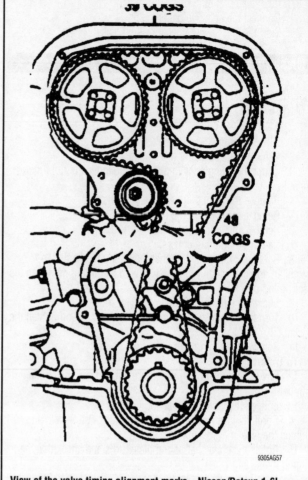

View of the valve timing alignment marks—Nissan/Datsun 1.6L (CA16DE) and 1.8L (CA18DE) engines

9305AG57

18. Rotate the crankshaft 2 complete turns and check that the timing marks align.
19. Tighten the tensioner locknut to 16–22 ft. lbs. (22–30 Nm).
20. Install the upper and lower timing covers.
21. Install the crankshaft pulley and tighten the bolt to 105–112 ft. lbs. (143–152 Nm).
22. Install the water pump pulley.
23. Install the engine mount and bracket. Remove the engine support jack.
24. Align the matchmarks and install the CMP sensor.
25. Install the accessory drive belts.
26. Install the right engine under cover and inner fender access cover.
27. Install the upper radiator hose and refill the cooling system.
28. Connect the negative battery cable.

1.8L (CA18ET) & 2.0L (CA20, CA20E) ENGINES

1. Disconnect the negative battery cable.
2. Remove the accessory drive belts.
3. Support the engine with a jack and remove the right motor mount.
4. Remove the right engine splash guards.
5. Remove the water pump pulley.

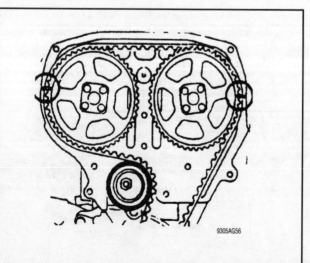

9305AG56

View of the camshaft timing marks—Nissan/Datsun 1.6L (CA16DE) and 1.8L (CA18DE) engines

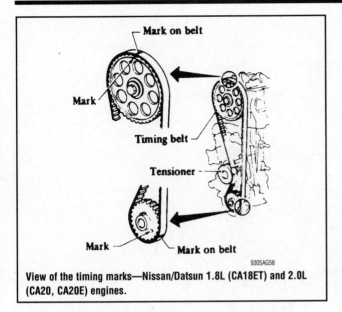

View of the timing marks—Nissan/Datsun 1.8L (CA18ET) and 2.0L (CA20, CA20E) engines.

6. Remove the crankshaft pulley.
7. Remove the timing covers.
8. Rotate the crankshaft so that the timing marks on the camshaft and crankshaft pulleys align with the timing marks on the rear timing cover.
9. Loosen the timing belt tensioner locknut.
10. Turn the tensioner away from the timing belt and tighten the locknut.
11. Remove the timing belt.
To install:
12. Ensure that the timing marks are aligned and install the timing belt.
13. Loosen the tensioner locknut.
14. Rotate the crankshaft 2 complete turns and check that the timing marks align. Tighten the tensioner locknut.
15. Install the timing covers.
16. Install the crankshaft pulley and tighten the bolt to 83–108 ft. lbs. (113–147 Nm).
17. Install the water pump pulley.
18. Install the accessory drive belts.
19. Install the right motor mount and the splash guards. Remove the engine support jack.
20. Connect the negative battery cable.

3.0L (VG30E, VG30ET) ENGINES

1. Disconnect the negative battery cable.
2. Drain the cooling system.
3. For Maxima models, perform the following procedure:
   a. Raise and safely support the vehicle.
   b. Remove the right front wheel.
   c. Remove the right front inner fender access cover.
   d. Remove the right engine under cover.
4. For 300 ZX models, perform the following procedure:
   a. Remove the cooling fan and shroud.
   b. Remove the radiator.
5. Remove the accessory drive belts.
6. Remove the intake air hose.
7. Remove the upper radiator hose.
8. Remove the air conditioning compressor belt tensioner pulley and bracket.

9. Remove the water pump pulley.
10. Remove the crankshaft pulley.
11. Remove the upper and lower timing belt covers.
12. Rotate the crankshaft so that the camshaft and crankshaft pulley timing marks align with the marks on the rear timing cover and oil pump.
13. Loosen the timing belt tensioner locknut. Rotate the tensioner clockwise to loosen the timing belt. Tighten the tensioner locknut.
14. Remove the timing belt.
To install:
15. Ensure that the timing marks align and install the timing belt.

➡**When the belt is properly installed, there will be 40 cogs between the camshaft pulley timing marks, and 43 cogs between the left camshaft timing mark and the crankshaft timing mark.**

16. To tension the timing belt, perform the following:
   a. Loosen the tensioner lockbolt.
   b. Pivot the tensioner through an arc of 70–80 degrees with an Allen wrench and release the tensioner. Tighten the locknut.

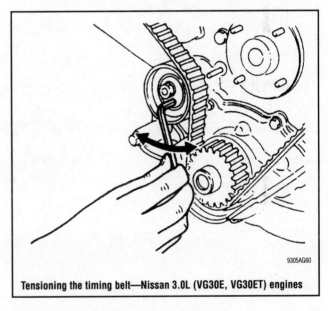

Tensioning the timing belt—Nissan 3.0L (VG30E, VG30ET) engines

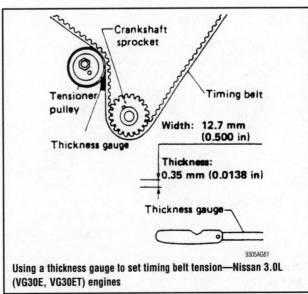

Using a thickness gauge to set timing belt tension—Nissan 3.0L (VG30E, VG30ET) engines

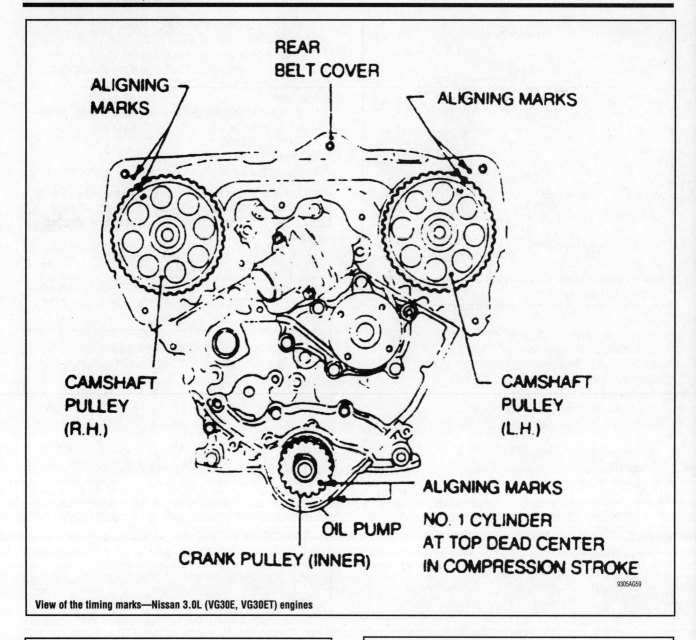

View of the timing marks—Nissan 3.0L (VG30E, VG30ET) engines

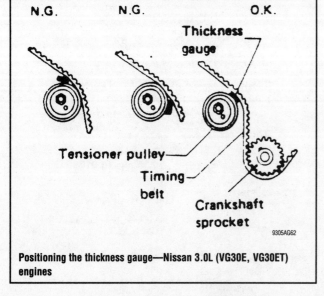

Positioning the thickness gauge—Nissan 3.0L (VG30E, VG30ET) engines

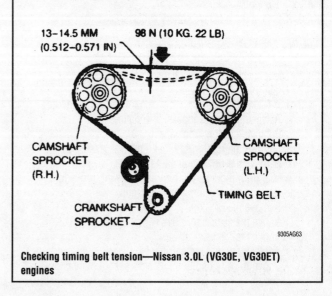

Checking timing belt tension—Nissan 3.0L (VG30E, VG30ET) engines

c. Rotate the crankshaft 2 complete turns and check that the timing marks align.

d. Apply 22 lbs. (10 kg) of force to the timing belt between the camshaft pulleys and loosen the tensioner locknut.

e. Place a 0.0138 inch x ½ inch thickness gauge between the tensioner pulley and the timing belt.

f. Rotate the crankshaft slowly clockwise until the gauge is drawn into position.

g. Hold the tensioner from turning with an Allen wrench and tighten the locknut.

h. Rotate the crankshaft 2 complete turns and check that the timing marks align.

i. Apply 22 lbs. (10 kg) of force to the timing belt between the camshaft pulleys and measure belt deflection. Deflection should be 0.512–0.571 inches.

17. Install the timing covers.

18. Install the crankshaft pulley and tighten the bolt to 90–98 ft. lbs. (122–136 Nm).

19. Install the water pump pulley.

20. Install the air conditioning compressor belt tensioner pulley and bracket.

21. For Maxima models, perform the following procedure:

a. Install the accessory drive belts.

b. Install the right inner fender access cover and the right engine under cover.

c. Install the right front wheel.

d. Lower the vehicle.

22. For 300 ZX models, perform the following procedure:

a. Install the radiator.

b. Install the cooling fan and shroud.

c. Install the accessory drive belts.

23. Install the upper radiator hose and the air intake hose.

24. Refill the cooling system

25. Connect the negative battery cable.

3.0L (VG30DE, VG30DETT) ENGINES

1. Remove the spark plugs and position the engine so that No. 1 piston is at Top Dead Center (TDC) of the compression stroke.

2. Remove all necessary components for access to the timing belt covers, then remove the covers and gaskets.

❊❊ WARNING

After the timing belt has been removed, DO NOT rotate the camshafts or the crankshaft. Severe internal engine damage will result from piston and valve contact.

3. Install a suitable 6mm stopper bolt in the tensioner arm of the auto-tensioner so the length of the pusher does not change.

4. Remove the automatic tensioner and the timing belt.

5. Check the automatic tensioner for oil leaks in the pusher rod and diaphragm. If oil is evident, replace the automatic tensioner assembly.

6. Inspect the timing gear teeth for wear.

To install:

7. Verify that the No. 1 piston is at TDC of the compression stroke.

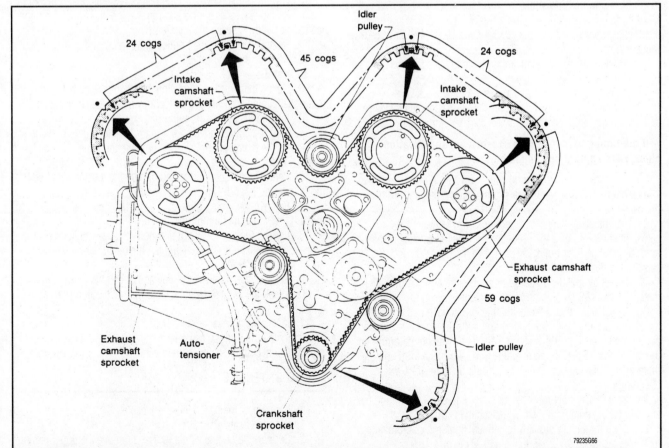

To ensure proper installation of the timing belt, be sure the proper number of belt teeth are between each sprocket, as indicated—Nissan 3.0L (VG30DE, VG30DETT) engines

8. Align the timing marks on the camshaft and crankshaft sprockets with the timing marks on the rear timing belt cover and the oil pump housing.

9. With a feeler gauge, check the clearance between the tensioner arm and the pusher of the automatic tensioner. The clearance should be 0.16 in. (4mm) with a slight drag on the feeler gauge. If the clearance is not as specified, mount the tensioner in a vise and adjust the clearance. When the clearance is set, insert the stopper bolt into the tensioner arm to retain the adjustment.

➡ **When adjusting the clearance, do not push the tensioner arm with the stopper bolt fitted, because damage to the threaded portion of the bolt will result.**

10. Mount the automatic tensioner and tighten the nuts and bolts by hand.

11. Install the timing belt. Ensure the timing sprockets are free of oil and water. Do not bend or twist the timing belt. Align the white lines on the belt with the timing marks on the camshaft and crankshaft sprockets. Point the arrow on the belt towards the front.

12. Push the automatic tensioner slightly towards the timing belt to prevent the belt from slipping. At the same time, turn the crankshaft 10 degrees clockwise and tighten the tensioner fasteners to 12–15 ft. lbs. (16–21 Nm).

➡ **Do not push the tensioner too hard because it will create excessive tension on the belt.**

13. Turn the crankshaft 120 degrees (⅓ turn) counterclockwise.

14. Back off on the automatic tensioner fasteners ½ turn.

15. Turn the crankshaft clockwise and set the No. 1 piston at TDC of the compression stroke.

16. Using Push-Pull gauge No. EG1486000 (J-38387) or equivalent, apply approximately 15.2–18.3 lbs. (67.7–81.4 N) of force to the tensioner.

17. Tighten the tensioner mounting bolts to 12–15 ft. lbs. (17–21 Nm).

18. Turn the crankshaft 120 degrees (⅓ turn) clockwise.

19. Turn the crankshaft 120 degrees (⅓ turn) counterclockwise and set the No. 1 piston at TDC of the compression stroke.

➡ **If the timing belt deflection exceeds the specification, change the applied pushing force.**

20. Fabricate a 0.35 in. (9mm) wide x 0.10 in. (2mm) steel plate. The length of the plate should be slightly longer than the width of the belt.

21. Set the steel plate at positions mid-way between the camshaft sprockets on each head, between the left exhaust camshaft sprocket and the idler pulley, and between the right exhaust camshaft sprocket and the tensioner.

22. Using the push-pull gauge or equivalent, apply approximately 11 lbs. (49 N) of force to the tensioner.

23. Check and record the belt deflection at each position with the steel plate in place. The timing belt deflection at each position should be 0.24–0.28 in. (6–7mm). Another means of determining the belt deflection is to add all deflection readings and divide them by 4. This average deflection should be 0.24–0.28 in. (6–7mm).

24. If the belt deflection is not as specified, repeat the timing belt adjusting procedure until the belt deflection is correct.

25. Once the belt is properly tensioned, tighten the automatic tensioner fasteners to 12–15 ft. lbs. (16–21 Nm).

26. Remove the stopper bolt from the tensioner and wait 5 minutes. After 5 minutes, check the clearance between the tensioner

arm and the pusher of the automatic tensioner. The clearance should remain at 0.138–0.205 in. (3.5–5.2mm).

27. Be sure the belt is installed and aligned properly on each pulley and the timing sprocket. There must be no slippage or misalignment.

28. Install the timing belt covers with new gaskets. Tighten the covers bolts to 24–48 inch lbs. (3–5 Nm).

29. Install the remaining components in the reverse order of removal.

Peugeot

1.9L (XU9J2) ENGINE

1. Disconnect the negative battery cable.
2. Remove the right splash guard.
3. Remove the accessory drive belts.
4. Remove the crankshaft pulley.
5. Remove the timing covers.
6. Temporarily install the crankshaft pulley.
7. Rotate the crankshaft clockwise so that Timing Pin No. (-) 01663 can be installed through the crankshaft pulley into the timing hole in the engine block, and Timing Pin No. (-) 0163 can be installed through the camshaft pulley and into the timing hole in the cylinder head.
8. Loosen the tensioner nut and bolts.
9. If reusing the timing belt, mark the direction of rotation on the back side of the belt.
10. Turn the square drive on the rear of the tensioner housing to compress the tensioner spring and remove the timing belt from the camshaft pulley. Release the square drive.
11. Remove the crankshaft pulley and timing pin.
12. Remove the timing belt.

To install:

13. Noting the direction of rotation, install the timing belt to the crankshaft timing sprocket and temporarily install the crankshaft pulley.
14. Install the crankshaft timing pin.
15. Turn the square drive on the rear of the tensioner housing to compress the tensioner spring and complete the timing belt installation.
16. Release the square drive and tighten the tensioner nut and bolts to 11 ft. lbs. (15 Nm).
17. Remove the timing pins and rotate the crankshaft 2 complete turns. Check that the timing holes align.
18. Loosen the tensioner nut and bolts to allow the tensioner to take up any remaining slack. Retighten the tensioner nut and bolts to 11 ft. lbs. (15 Nm).
19. Remove the crankshaft pulley and install the timing covers.
20. Install the crankshaft pulley and tighten the bolt to 81 ft. lbs. (110 Nm).
21. Install the accessory drive belts.
22. Install the right splash guard.
23. Connect the negative battery cable.

1.9L (XU9J4) ENGINE

1. Disconnect the negative battery cable.
2. Raise and safely support the vehicle.
3. Support the engine with a jack and remove the upper engine mount and bracket.
4. Remove the right splash guard.

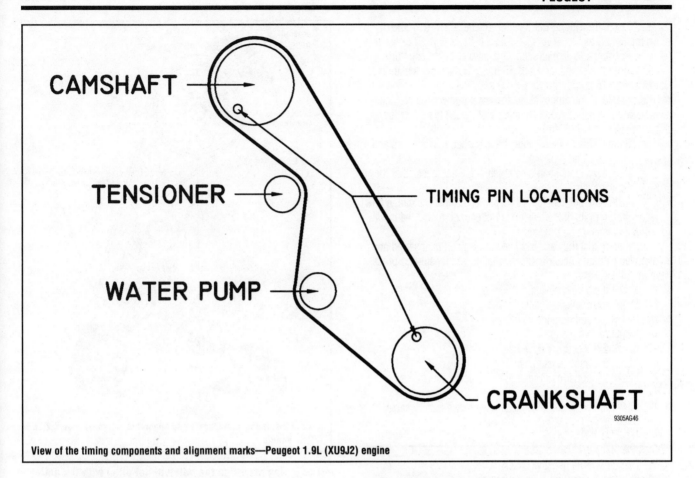

View of the timing components and alignment marks—Peugeot 1.9L (XU9J2) engine

5. Remove the accessory drive belts.
6. Remove the alternator.
7. Remove the upper timing cover.
8. Rotate the crankshaft clockwise so that Timing Pin No. (-) 0153G can be installed through the crankshaft pulley into the timing hole in the engine block, and Timing Pins No. (-) 0153M can be installed through the camshaft pulleys and into the timing holes in the cylinder head.
9. Remove the bell housing cover and install the Flywheel Locking tool No. 91.71.

10. Remove the crankshaft timing pin and the crankshaft pulley.
11. Remove the lower timing cover.
12. If reusing the timing belt, mark the direction of rotation on the back of the belt.
13. Loosen both tensioner lockbolts and remove the timing belt.

To install:
14. Noting the direction of rotation, install the timing belt.
15. Adjust the timing belt tension by performing the following procedure:

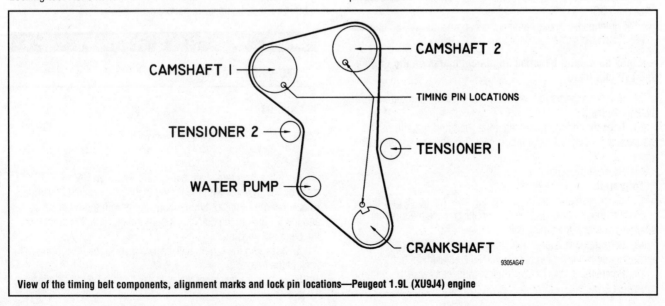

View of the timing belt components, alignment marks and lock pin locations—Peugeot 1.9L (XU9J4) engine

a. Install the Timing Belt Tension Gauge SEEM C.Tronic 87 on the timing belt.

b. Rotate the tensioner opposite the water pump counterclockwise until the gauge reads 15 units. Tighten the tensioner lockbolt to 15 ft. lbs. (20 Nm).

c. Rotate the tensioner above the water pump counterclockwise until the gauge reads 25 units. Tighten the tensioner lockbolt to 15 ft. lbs. (20 Nm).

d. Remove the tension gauge, the camshaft locking pins and the flywheel locking tool.

e. Rotate the crankshaft 2 complete turns and check that the timing holes align.

f. Install the belt tension gauge and check that the gauge now reads 40–50 units. If the tension is not correct, repeat the installation and tensioning procedure.

g. Ensure that the camshaft locking pins can be inserted and removed without force. Repeat installation and tensioning procedure, if necessary.

16. Install the timing covers.

17. Install the crankshaft pulley.

18. Install the bell housing cover.

19. Install the alternator.

20. Install the accessory drive belts.

21. Install the right splash guard.

22. Install the upper motor mount and bracket. Remove the engine support jack.

23. Lower the vehicle and connect the negative battery cable.

2.2L (ZDJL) ENGINE

1. Disconnect the negative battery cable.
2. Remove the cooling fan and shroud.
3. Remove the accessory drive belts.
4. Install the Flywheel Locking tool No. (-) 0144B.
5. Remove the crankshaft pulley.
6. Remove the flywheel locking tool.
7. Remove the timing cover.
8. Rotate the crankshaft so that the timing marks are aligned as follows:

   • Camshaft pulley timing mark approximately 45 degrees before Top Dead Center (TDC)
   • Crankshaft pulley timing mark approximately 90 degrees before TDC
   • Intermediate pulley timing mark approximately 120 degrees after TDC

➡ It may be helpful to scribe alignment marks on the engine and cylinder head.

9. If reusing the timing belt, mark the direction of rotation on the back of the belt.

10. Loosen the timing belt tensioner bolt and nut and push the tensioner away from the timing belt. Tighten the adjustment nut.

11. Remove the timing belt.

To install:

12. Align the marks on the timing belt with the marks on the crankshaft, intermediate and camshaft pulleys. Note the direction of rotation arrows on the timing belt.

13. Loosen the tensioner adjustment nut and allow tensioner to apply tension to the timing belt. Tighten the adjustment bolt.

14. Rotate the crankshaft 2 complete turns clockwise.

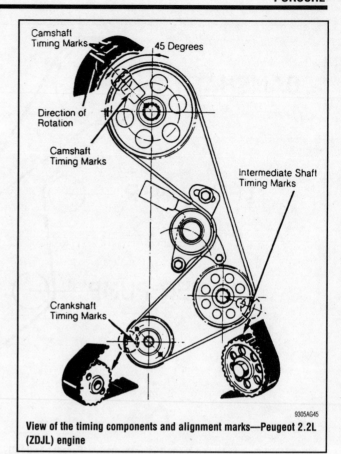

View of the timing components and alignment marks—Peugeot 2.2L (ZDJL) engine

➡ The timing marks on the belt will not align with the pulley marks after the crankshaft has been rotated.

15. Loosen the tensioner adjustment nut and allow the tensioner to take up any remaining slack in the timing belt. Tighten the nut and bolt to 18 ft. lbs. (25 Nm).

16. Install the timing cover.

17. Install the crankshaft pulley. Install the flywheel locking tool and tighten the crankshaft pulley bolt to 96 ft. lbs. (130 Nm). Remove the flywheel locking tool.

18. Install the accessory drive belts.

19. Install the cooling fan and shroud.

20. Connect the negative battery cable.

Porsche

2.0L (VC & M31) ENGINES

924 and 944

1. Disconnect the negative battery cable.
2. Remove the accessory drive belts.
3. Remove the timing cover.
4. Rotate the crankshaft to align the Top Dead Center (TDC) mark on the flywheel with the timing mark on the clutch housing. Check that the camshaft timing mark aligns with the valve cover, and the **V** notch in the crankshaft pulley aligns with the pointer on the oil pump housing.
5. If reusing the timing belt, mark the rotational direction on the back of the belt.
6. Loosen the tensioner pulley locknut and remove the timing belt.

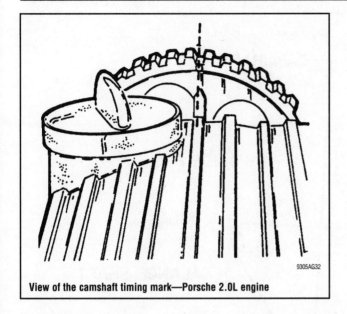

View of the camshaft timing mark—Porsche 2.0L engine

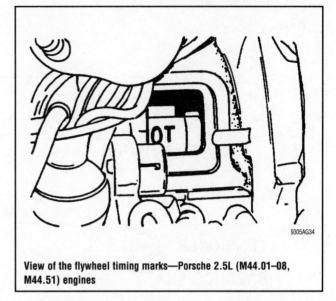

View of the flywheel timing marks—Porsche 2.5L (M44.01–08, M44.51) engines

3. Remove the air cleaner and Mass Air Flow (MAF) sensor.
4. Remove the accessory drive belts.
5. Remove the distributor cap.
6. Remove the rotor arm.
7. Remove the plastic cap and remove the distributor cap mount.
8. Remove the crankshaft pulley.
9. Remove the timing covers.
10. Rotate the crankshaft to align the flywheel timing mark with the mark on the clutch housing.
11. Check that the camshaft and balance shaft timing marks are aligned with the marks on the rear timing cover.
12. Loosen the balance shaft belt guide pulley.
13. Loosen the balance shaft belt tensioner pulley and remove the balance shaft belt.
14. If reusing the timing belt, mark the rotational direction on the back of the belt.
15. Loosen the timing belt tensioner pulley and remove the timing belt.

To install:
16. Ensure that the camshaft and flywheel timing marks are aligned. Noting the direction of rotation, install the timing belt.
17. Turn the timing belt tensioner pulley counterclockwise to tension the timing belt and tighten the tensioner pulley bolt.
18. Turn the crankshaft 10 degrees counterclockwise and install the Tension Gauge No. 9201 on the timing belt between the crankshaft and camshaft pulleys.
19. Timing belt tension should be 3.7–4.3 gauge units for a new belt or 2.4–3.0 gauge units for a used belt.
20. If the belt tension is not correct, loosen the belt tensioner pulley lockbolt and rotate the tensioner pulley clockwise to decrease tension and counterclockwise to increase tension. Tighten the tensioner pulley lockbolt.
21. Rotate the crankshaft clockwise to Top Dead Center (TDC) and check the alignment of the flywheel and camshaft timing marks. Rotate the crankshaft counterclockwise 10 degrees and repeat the belt tension measurement.
22. When the timing belt tension is correct, tighten the tensioner pulley to 33 ft. lbs. (45 Nm). Rotate the crankshaft clockwise to TDC.

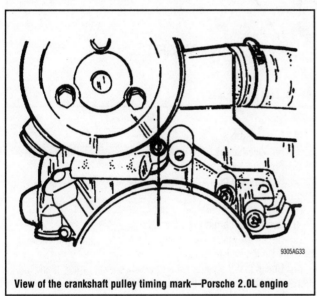

View of the crankshaft pulley timing mark—Porsche 2.0L engine

To install:
7. Ensure that the crankshaft and camshaft timing marks are aligned. Noting the direction of rotation, install the timing belt.
8. Rotate the tensioner pulley counterclockwise to tension the timing belt. Timing belt tension is correct when the belt can be twisted no more than 90 degrees midway between the crankshaft and camshaft pulleys.
9. Tighten the tensioner locknut to 30 ft. lbs. (41 Nm).
10. Rotate the crankshaft clockwise 2 complete turns and check that the timing marks align.
11. Install the timing cover.
12. Install the accessory drive belts.
13. Connect the negative battery cable.

2.5L (M44.01–08, M44.51) ENGINES

924 and 944

1. Disconnect the negative battery cable.
2. Remove the engine splash guard.

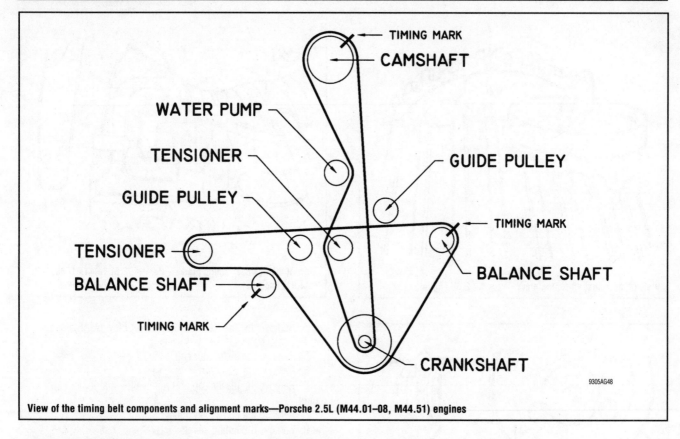

View of the timing belt components and alignment marks—Porsche 2.5L (M44.01–08, M44.51) engines

➡ **The balance shaft belt has teeth on both sides. Install the belt with the color coded tooth facing out.**

23. Ensure that the balance shaft timing marks are aligned and install the balance shaft belt.

24. Install the Belt Tension Gauge No. 9201 on the balance shaft belt between the tensioner pulley and the left balance shaft pulley.

25. Turn the balance shaft belt tensioner pulley clockwise until the gauge reads 2.4–3.0 units. Tighten the belt tensioner pulley lockbolt to 33 ft. lbs. (45 Nm).

26. Rotate the crankshaft 2 complete turns and align all timing marks. Recheck the balance shaft belt tension.

27. When the balance shaft belt tension is correct, adjust the guide pulley so that there is 0.5mm clearance between the guide pulley and the balance shaft belt. Tighten the guide pulley lockbolt to 33 ft. lbs. (45 Nm).

28. Install the timing covers.
29. Install the crankshaft pulley.
30. Install the distributor cap mount.
31. Install the plastic cap and the rotor arm.
32. Install the distributor cap.
33. Install the accessory drive belts.
34. Install the MAF sensor and the air filter.
35. Install the engine splash guard.
36. Connect the negative battery cable.

2.5L, 2.7L, 3.0L (M44.05–12, M44.40–41, M44.52) ENGINES

924 and 944

1. Disconnect the negative battery cable.
2. Remove the engine splash guard.
3. Remove the air cleaner and Mass Air Flow (MAF) sensor.

4. Remove the accessory drive belts.
5. Remove the distributor cap.
6. Remove the rotor arm.
7. Remove the plastic cap and remove the distributor cap mount.
8. Remove the crankshaft pulley.
9. Remove the timing covers.
10. Rotate the crankshaft to align the flywheel timing mark with the mark on the clutch housing.
11. Check that the camshaft and balance shaft timing marks are aligned with the marks on the rear timing cover.
12. Loosen the balance shaft belt guide pulley.
13. Loosen the balance shaft belt tensioner pulley and remove the balance shaft belt.
14. Remove the guide plate and guide pulley.
15. If reusing the timing belt, mark the rotational direction on the back of the belt.
16. Loosen the timing belt tensioner pulley lockbolt and pivot bolt.
17. Using tool No. 9200, pivot the belt tensioner away from the timing belt and tighten the lockbolt.
18. Remove the timing belt tensioner assembly and remove the timing belt.

To install:

19. Ensure that the flywheel and camshaft timing marks are aligned. Noting the direction of rotation, install the timing belt.
20. Install the timing belt tensioner assembly.
21. Loosen the timing belt tensioner lockbolt to tension the timing belt.
22. Tighten the tensioner lockbolt.
23. Rotate the crankshaft 2 complete turns and check that the flywheel and camshaft marks align.
24. Loosen the tensioner lockbolt and allow the tensioner to take

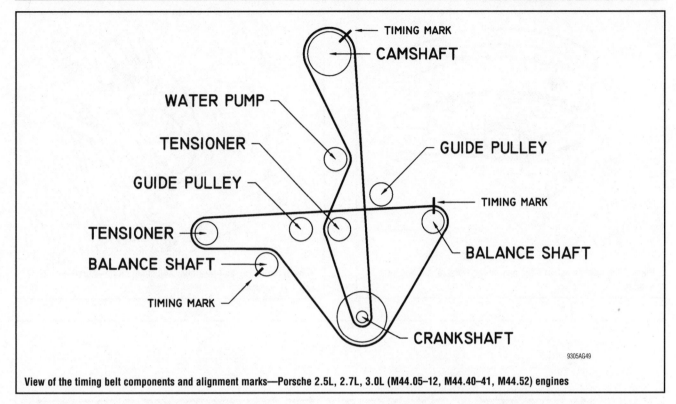

View of the timing belt components and alignment marks—Porsche 2.5L, 2.7L, 3.0L (M44.05–12, M44.40–41, M44.52) engines

up any remaining slack. Tighten the lockbolt and pivot bolt to 15 ft. lbs. (20 Nm).

➡️**The balance shaft belt has teeth on both sides. Install the belt with the color coded tooth facing out.**

25. Ensure that the balance shaft timing marks are aligned and install the balance shaft belt.

26. Install the Belt Tension Gauge No. 9201 on the balance shaft belt between the tensioner pulley and the left balance shaft pulley.

27. Turn the balance shaft belt tensioner pulley clockwise until the gauge reads 2.4–3.0 units. Tighten the belt tensioner pulley lockbolt to 33 ft. lbs. (45 Nm).

28. Rotate the crankshaft 2 complete turns and align all timing marks. Recheck the balance shaft belt tension.

29. When the balance shaft belt tension is correct, adjust the guide pulley so that there is 0.5mm clearance between the guide pulley and the balance shaft belt. Tighten the guide pulley lockbolt to 33 ft. lbs. (45 Nm).

30. Install the timing covers.
31. Install the crankshaft pulley.
32. Install the distributor cap mount.
33. Install the plastic cap and the rotor arm.
34. Install the distributor cap.
35. Install the accessory drive belts.
36. Install the MAF sensor and the air filter.
37. Install the engine splash guard.
38. Connect the negative battery cable.

4.5L & 4.7L (M28) ENGINES

928

1. Disconnect the negative battery cable.
2. Remove the air intake hoses.

3. Remove the oil dipstick and fill tubes.
4. Disconnect the throttle cable, cruise control cable and transmission cable, if equipped with automatic transmission.
5. Disconnect the cable retainers from the fan shroud.
6. Remove the fan shroud.
7. Remove the accessory drive belts.
8. Remove the cooling fan, bracket and pulley.

❄️ WARNING

Do not lay the fan assembly flat. The silicone oil filler will leak out and the fan will become inoperative.

9. Remove the upper timing covers.
10. Remove the power steering pump and position aside.

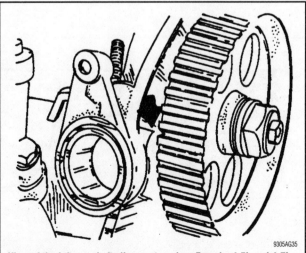

View of the left camshaft alignment marks—Porsche 4.5L and 4.7L engines

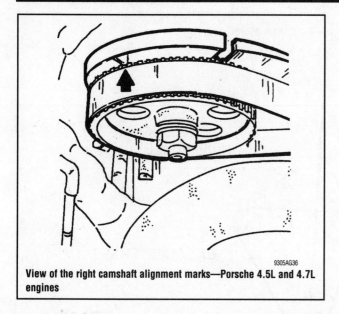

View of the right camshaft alignment marks—Porsche 4.5L and 4.7L engines

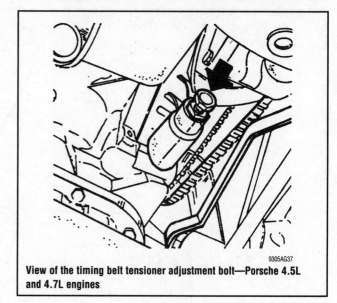

View of the timing belt tensioner adjustment bolt—Porsche 4.5L and 4.7L engines

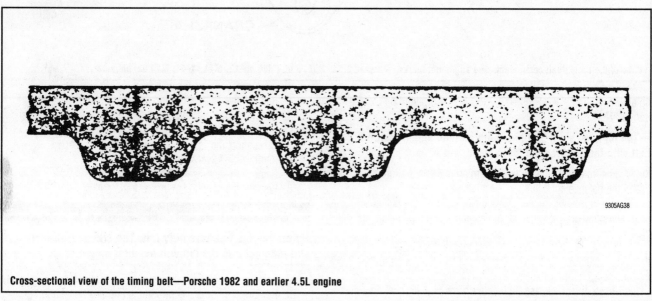

Cross-sectional view of the timing belt—Porsche 1982 and earlier 4.5L engine

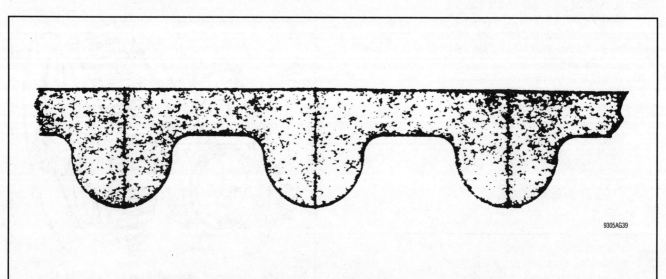

Cross-sectional view of the timing belt with High Torque Drive (HTD) teeth—Porsche 1983 and later 4.7L engine

❄❄ WARNING

Never rotate the crankshaft counterclockwise.

11. Rotate the crankshaft clockwise to align the timing marks on the camshaft pulleys with the marks on the rear timing cover, and the Top Dead Center (TDC) mark on the crankshaft damper with the red pointer on the lower timing cover. Check that all marks align.

12. If equipped with a manual transmission, remove the clutch slave cylinder. If equipped with an automatic transmission, remove the slave cylinder hole cover.

13. Install the Crankshaft Locking tool No. 9161.

14. Remove the crankshaft pulley and the lower timing cover.

15. If reusing the timing belt, mark the rotational direction on the back of the belt.

16. Loosen the timing belt tensioner bolt and remove the timing belt.

➡**Be sure to use the correct replacement timing belt.**

To install:

17. Ensure that the camshaft timing marks are aligned. Noting the direction of rotation, install the timing belt.

18. Install the lower timing cover and the crankshaft pulley. Ensure that the crankshaft timing mark aligns with the pointer on the lower cover.

19. If all timing marks align, tighten the timing belt tensioner adjustment bolt until correct belt tension is achieved.

20. For belts with square teeth, correct tension is achieved when the belt can be twisted 90 degrees between the tensioner pulley and the right bank camshaft pulley.

21. For belts with round teeth, use Tension tool No. 9201 and adjust the tensioner until the gauge reads 4.5 units.

22. Remove the crankshaft locking tool and rotate the crankshaft 2 complete turns. Check that the timing marks align and recheck the timing belt tension. If the belt tension is correct, tighten the tensioner locknut.

23. Install the crankshaft locking tool again and tighten the crankshaft pulley bolt to 181 ft. lbs. (246 Nm). Remove the crankshaft locking tool.

24. Install the clutch slave cylinder or cover.

25. Install the upper timing covers.

26. Install the power steering pump.

27. Install the cooling fan, bracket and pulley.

28. Install the accessory drive belts.

29. Install the fan shroud.

30. Connect the throttle cable, the cruise control cable and the transmission cable, if equipped with automatic transmission.

31. Connect the cable retainers.

32. Install the oil dipstick and fill tubes.

33. Install the air intake hoses.

34. Connect the negative battery cable.

5.0L (M28) ENGINE

928

1. Disconnect the negative battery cable.

2. Remove the air intake hoses.

3. Remove the oil dipstick and fill tubes.

4. Disconnect the throttle cable, cruise control cable and transmission cable, if equipped with automatic transmission.

5. Disconnect the cable retainers from the fan shroud.

6. Remove the air intake guide.

7. Remove the fan shroud.

8. Remove the electric cooling fans.

9. Remove the accessory drive belts.

10. Remove the distributor caps and rotor arms.

11. Remove the upper timing covers and the power steering pump.

❄❄ CAUTION

Never rotate the crankshaft counterclockwise.

12. Rotate the crankshaft to align the pointer on the lower timing cover with the **45** degrees Before Top Dead Center (BTDC) mark on the crankshaft damper.

13. If equipped with a manual transmission, remove the clutch slave cylinder. If equipped with an automatic transmission, remove the slave cylinder hole cover.

14. Install the Crankshaft Locking tool No. 9161/1.

15. Remove the crankshaft pulley and the lower timing cover.

16. If reusing the timing belt, mark the rotational direction on the back of the belt.

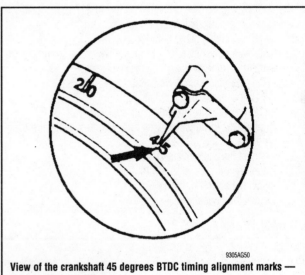

9305AG50

View of the crankshaft 45 degrees BTDC timing alignment marks — Porsche 5.0L engine

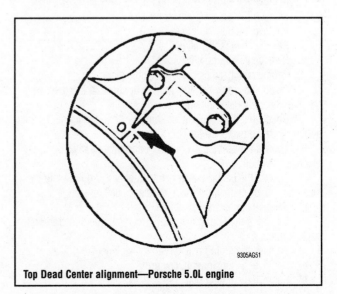

9305AG51

Top Dead Center alignment—Porsche 5.0L engine

17. Loosen the timing belt tensioner bolt and remove the timing belt.

To install:

18. Temporarily install the lower timing cover and crankshaft pulley.

19. Rotate the camshafts to align the pulley timing marks with the marks on the rear timing cover.

20. Hold the camshafts in place (may require an assistant).

21. Remove the crankshaft locking tool and carefully rotate the crankshaft clockwise to align the TDC mark with the pointer on the timing cover. Reinstall the crankshaft locking tool.

22. Remove the crankshaft pulley and lower timing cover.

23. Noting the direction of rotation, install the timing belt.

24. Install the lower timing cover and crankshaft pulley.

25. Install the Belt Tension tool No. 9201 and tighten the timing belt tensioner adjustment bolt until correct belt tension is achieved.

26. Correct belt tension is 4.7–5.3 gauge units.

27. Remove the crankshaft locking tool and rotate the crankshaft 2 complete turns. Check that all timing marks align.

28. Repeat the belt tension test.

29. If the timing belt tension is correct, reinstall the crankshaft locking tool and tighten the crankshaft pulley to 217 ft. lbs. (295 Nm). Remove the crankshaft locking tool.

30. Install the clutch slave cylinder or cover.

31. Install the power steering pump and the upper timing covers.

32. Install the rotor arms and the distributor caps.

33. Install the accessory drive belts.

34. Install the electric cooling fans.

35. Install the fan shroud and the air intake guide.

36. Connect the throttle cable, the cruise control cable and the transmission cable, if equipped with automatic transmission.

37. Connect the cable retainers.

38. Install the oil dipstick and fill tubes.

39. Install the air intake hoses.

40. Connect the negative battery cable.

3.0L (M44.43, M44.44) ENGINES

968

1. Disconnect the negative battery cable.
2. Remove the engine splash guard.
3. Remove the accessory drive belts.
4. Remove the distributor cap.
5. Remove the crankshaft pulley.
6. Remove the timing covers.
7. Remove the balance shaft belt guide plate.
8. Rotate the crankshaft to align the flywheel timing mark with the mark on the clutch housing.
9. Check that the camshaft and balance shaft timing marks are aligned with the marks on the rear timing cover.
10. Loosen the guide pulley locknut and rotate away from the balance shaft belt.
11. Loosen the balance shaft belt tensioner locknut. Rotate the tensioner pulley counterclockwise to release belt tension.
12. Remove the balance shaft belt.
13. If reusing the timing belt, mark the rotational direction on the back of the belt.
14. Remove the timing belt hydraulic tensioner.
15. Remove the tensioner pulley arm.
16. Remove the timing belt.

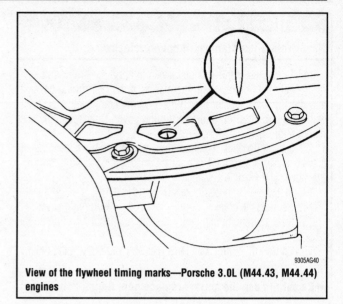

View of the flywheel timing marks—Porsche 3.0L (M44.43, M44.44) engines

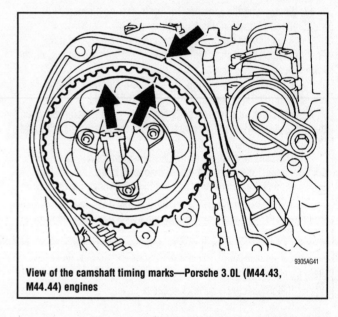

View of the camshaft timing marks—Porsche 3.0L (M44.43, M44.44) engines

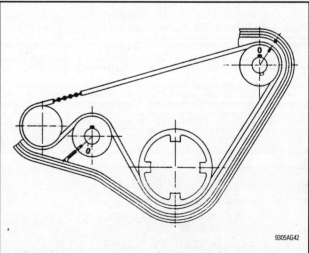

View of the balance shaft timing marks—Porsche 3.0L (M44.43, M44.44) engines

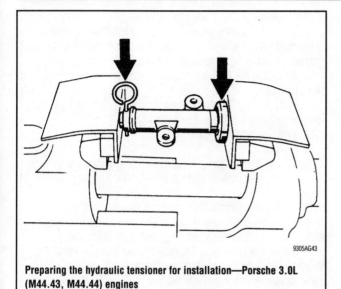

Preparing the hydraulic tensioner for installation—Porsche 3.0L (M44.43, M44.44) engines

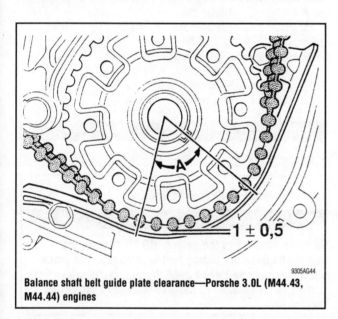

Balance shaft belt guide plate clearance—Porsche 3.0L (M44.43, M44.44) engines

To install:

➡**Use a large washer to protect the bottom of the tensioner while compressing it.**

17. Compress the hydraulic tensioner slowly in a vise and install a retainer pin.

18. Ensure that the camshaft and flywheel timing marks are aligned. Noting the direction of rotation, install the timing belt.

19. Install the tensioner pulley arm.

20. Install the hydraulic tensioner and remove the retainer pin.

21. Rotate the crankshaft 2 complete turns and check that the timing marks align.

➡**The balance shaft belt has teeth on both sides. Install the belt with the color coded tooth facing out.**

22. Align the balance shaft timing marks and install the balance shaft belt.

23. Turn the balance shaft belt tensioner pulley clockwise until the gauge reads 2.4–3.0 units. Tighten the belt tensioner pulley lockbolt to 33 ft. lbs. (45 Nm).

24. Rotate the crankshaft 2 complete turns and align all timing marks. Recheck the balance shaft belt tension.

25. When the balance shaft belt tension is correct, adjust the guide pulley so that there is 0.5mm clearance between the guide pulley and the balance shaft belt. Tighten the guide pulley lockbolt to 33 ft. lbs. (45 Nm).

26. Install the balance shaft belt guide plate so that the clearance measured over 7 teeth is 0.5–1.5 mm.

27. Install the timing covers.

28. Install the crankshaft pulley and tighten the bolt to 155 ft. lbs. (210 Nm).

29. Install the distributor cap and the accessory drive belts.

30. Install the engine splash guard.

31. Connect the negative battery cable.

Renault

➡**On Renault engines, the No. 1 cylinder is closest to the flywheel.**

1.7L (F3N) ENGINE

1. Disconnect the negative battery cable.

2. Remove the accessory drive belts.

3. Remove the crankshaft pulley and remove the timing cover.

4. Matchmark the distributor to indicate the location of the No. 1 spark plug wire tower.

5. Remove the distributor cap.

6. Rotate the crankshaft so that the timing mark on the camshaft pulley aligns with the mark on the rear timing cover and the distributor rotor points to the No. 1 spark plug wire tower matchmark.

7. Remove the access cover on the engine block behind the alternator and insert the Top Dead Center (TDC) locating tool Mot. 861 into the alignment hole in the crankshaft.

8. Loosen the tensioner lockbolt and pry the tensioner pulley away from the timing belt. Tighten the tensioner lockbolt.

9. If reusing the timing belt, mark the rotational direction on the back of the belt.

10. Remove the timing belt.

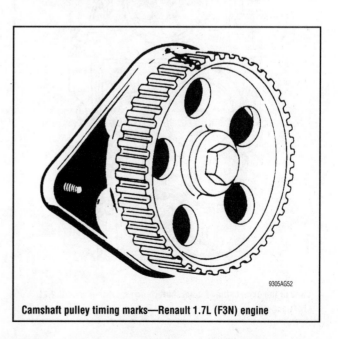

Camshaft pulley timing marks—Renault 1.7L (F3N) engine

To install:

11. Ensure that the camshaft timing marks are aligned and that the TDC tool is in place and the distributor rotor aligns with the No. 1 spark plug wire tower matchmark.

12. Noting the direction of rotation, install the timing belt.

13. Loosen the tensioner lockbolt and allow the tensioner to apply tension to the timing belt. Tighten the tensioner lockbolt.

14. Remove the TDC locating tool and rotate the crankshaft 2 complete turns. Install the TDC locating tool and check that the timing marks align.

15. Loosen the tensioner lockbolt and allow the tensioner to take up any remaining slack in the timing belt. Tighten the tensioner lockbolt.

16. Remove the TDC locating tool and install the access cover.

17. Install the timing cover.

18. Install the crankshaft pulley and tighten the bolt to 70 ft. lbs. (95 Nm).

19. Install the distributor cap.

20. Install the accessory drive belts.

21. Connect the negative battery cable.

2.2L (J7T) ENGINE

1. Disconnect the negative battery cable.

2. Remove the accessory drive belts.

3. Remove the alternator.

4. Remove the AIR pump.

5. Unbolt the air conditioning compressor and position it out of the work area.

6. Remove the crankshaft pulley and remove the timing cover.

7. Matchmark the distributor to indicate the location of the No. 1 spark plug wire tower.

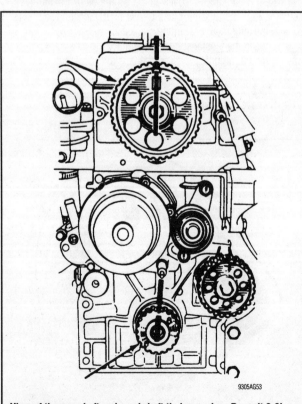

View of the camshaft and crankshaft timing marks—Renault 2.2L (J7T) engine

8. Remove the distributor cap.

9. Rotate the crankshaft so that the crankshaft and camshaft timing marks align with the marks on the engine and the distributor rotor points to the No. 1 spark plug wire tower matchmark.

10. If reusing the timing belt, mark the rotational direction on the back of the belt.

11. Loosen the timing belt tensioner and pry the tensioner away from the belt. Tighten the tensioner lockbolt.

12. Remove the timing belt.

To install:

13. Ensure that the camshaft and crankshaft timing marks align, and the distributor rotor is aligned with the No. 1 spark plug wire tower matchmark.

14. Noting the direction of rotation, install the timing belt.

15. Loosen the tensioner lockbolt and allow the tensioner to apply tension to the timing belt. Tighten the tensioner lockbolt.

16. Rotate the crankshaft 2 complete turns and check that the timing marks align. Loosen the tensioner lockbolt and allow the tensioner to take up any remaining slack in the timing belt. Tighten the lockbolt and pivot bolt.

17. Install the distributor cap.

18. Install the timing cover and the crankshaft pulley. Tighten the pulley bolt to 96 ft. lbs. (130 Nm).

19. Install the air conditioning compressor.

20. Install the alternator and the AIR pump.

21. Install the accessory drive belts.

22. Connect the negative battery cable.

Saab

2.5L (B258L) ENGINE

1. Disconnect the negative battery cable.

✳✳ WARNING

To avoid damage to the valves, DO NOT rotate the camshafts once the timing belt is removed. The crankshaft may only be turned 0–60 degrees BTDC when the camshafts are locked in position with the appropriate locking tool.

2. Remove the necessary components for access to the timing cover, then remove the cover.

3. Remove the right front wheel and the cover in the wheel well.

4. Remove the 6 bolts holding the crankshaft pulley and remove the pulley. Do not remove the center bolt.

5. Put the No. 1 cylinder in the Top Dead Center (TDC) on the compression stroke.

6. The timing marks on the crankshaft and camshafts should be in alignment with their respective marks on the engine. Insert Camshaft Locking tool 83-94-926 and install Crankshaft Locking tool 83-94-868 or their equivalents.

7. If reusing the timing belt, mark the direction of its rotation. To help with refitting, the belt can be marked at the camshaft timing marks and also at the crankshaft timing mark.

8. Remove the tensioning roller, both adjusting rollers and the timing belt.

9. Release tension from and remove the timing belt. Loosen the timing belt adjuster bolts.

10. Rotate the crankshaft back to 60 degrees BTDC, to prevent damage to the valves.

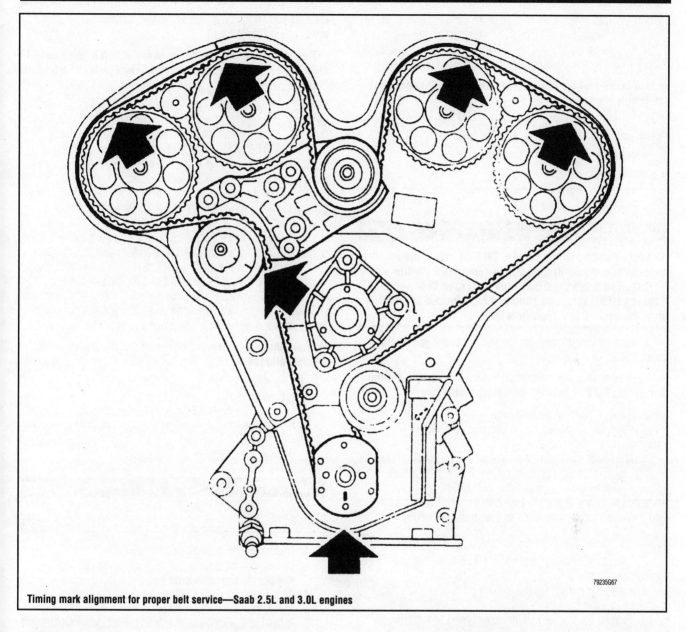

79235G67

Timing mark alignment for proper belt service—Saab 2.5L and 3.0L engines

11. Remove the bracket with the upper timing belt adjuster and tensioner rollers.

To install:

12. Install the bracket with the upper timing belt adjuster and tensioner pulleys.

13. Install and locking tools 83-94-926 or equivalent, between the camshaft sprockets to lock the camshafts of both heads in position.

14. Rotate the crankshaft forward to just before 0 degrees TDC and install the crankshaft locking tool on the crankshaft. Carefully rotate the engine until the arm of the tool is against the water pump flange. Be sure the crankshaft is at 0 degrees TDC and all timing marks are aligned. Remove the locking tool.

15. Install the timing belt, noting the direction of rotation marked at disassembly.

   a. Adjust the tensioner lightly by hand counterclockwise to keep the timing belt from falling off. Be sure the crankshaft is a 0 degrees TDC and all timing marks are aligned.

   b. Install the camshaft locking tool.

   c. Install tool 83-93-985 or equivalent, with a cut piece from an old timing belt, to measure belt tension.

   d. Snug the center bolts of the adjusting rollers. Turn the lower adjusting roller counterclockwise, until a belt tension of 202–221 ft. lbs. (275–300 Nm) is registered on the torque wrench.

   e. Tighten the adjusting roller center bolts to 30 ft. lbs. (40 Nm).

➡**This adjustment of the timing belt is only preparatory and should, therefore, not be used as a final check.**

16. Adjust the tensioner pulley until the marks are aligned. Tighten the tensioner pulley to 15 ft. lbs. (20 Nm). Remove the camshaft locking tool on camshaft sprockets No. 1 and 2. Adjust the upper adjusting roller until sprocket No. 2 moves 0.04–0.08 in. (1–2mm) clockwise. Tighten the upper adjusting roller to 30 ft. lbs. (40 Nm) and remove the upper locking tool.

17. Rotate the engine 2 complete revolutions to just before 0 degrees TDC and install the locking tool on the crankshaft. Carefully turn the crankshaft until the arm of the locking tool is against the water pump flange and tighten the locking tool. Set the camshaft

locking tool into position on the front of the camshaft sprockets. Be sure that the timing marks on the camshaft sprockets are aligned with the marks on the tool and that the edge of the timing belt is flush with the edge of the camshaft sprockets.

➡**Also check that the alignment marks on the tensioner pulley is still aligned.**

18. Install the timing belt cover. Tighten the bolts to 72 inch lbs. (8 Nm). Install all of the remaining components.
19. Connect the negative battery cable.

3.0L (B308L) ENGINE

1. Disconnect the negative battery cable.

❋❋ WARNING

To avoid damage to the valves, DO NOT rotate the camshafts once the timing belt is removed from the engine. The crankshaft may only be turned between 0–60 degrees BTDC when the camshafts are locked in position with the appropriate locking tool.

2. Remove all necessary components for access to the timing cover, then remove the cover.

➡**When removing the crankshaft pulley, remove the 6 outer bolts only, DO NOT remove the center bolt.**

3. Remove the crankshaft pulley.
4. Rotate the crankshaft to Top Dead Center (TDC) of No. 1 cylinder.
5. The timing marks on the crankshaft and camshafts should be in alignment with their respective marks on the engine. Install camshaft locking tools (such as Saab tools KM-800-1 for camshaft sprockets No. 1 and 2 and KM-800-2 for sprockets No. 3 and 4) and a flywheel locking tool (such as Saab tool 83-94-868).
6. Mark the direction of rotation of the timing belt for reassembly.
7. Release tension from and remove the timing belt. Loosen the timing belt adjuster bolts.
8. Rotate the crankshaft back to 60 degrees BTDC, to prevent damage to the valves.
9. Remove the bracket with the upper timing belt adjuster and tensioner rollers.

To install:
10. Remove the flywheel locking tool and install the flywheel inspection cover.
11. Install the bracket with the upper timing belt adjuster and tensioner pulleys.
12. Install both camshaft locking tools.
13. Rotate the crankshaft forward to just before 0 degrees TDC and install the crankshaft locking tool on the crankshaft. Carefully rotate the engine until the arm of the tool is against the water pump flange. Be sure the crankshaft is at 0 degrees TDC and all timing marks are aligned. Remove the locking tool.
14. If reusing the belt, fit the timing belt according to its marked direction of rotation and timing marks. Adjust the tensioning roller loosely by hand to prevent the belt from slipping out of the cogs. Always adjust counterclockwise.
15. Measure the belt tension with Saab tool 83-93-985 or equivalent.
16. Snug the center bolts of the adjusting rollers. Turn the lower adjusting roller counterclockwise, until a belt tension of 202–220 ft.

lbs. (275–300 Nm) is reached. Tighten the adjusting roller center bolts to 30 ft. lbs. (40 Nm).

➡**This is a preliminary adjustment of the belt tension and must not be used as a check when the belt is finally adjusted.**

17. Continue to carry out the adjustment by means of the tensioning roller, mark against mark. Remove the locking tool for camshaft sprockets No. 1 and 2. Carry out the final adjustment with the upper center adjusting roller until camshaft sprocket No. 2 moves 0.04–0.08 in. (1–2mm) forward.
18. Remove the locking tool for camshaft sprockets No. 3 and 4 and also remove the crankshaft locking tool.
19. Tighten the tensioning roller to 15 ft. lbs. (20 Nm). Tighten the upper adjusting roller to 30 ft. lbs. (40 Nm) and tighten the lower adjusting roller to 15 ft. lbs. (20 Nm).
20. Rotate the engine 2 complete revolutions to just before 0 degrees TDC and install the locking tool on the crankshaft. Carefully turn the crankshaft until the arm of the locking tool is against the water pump flange and tighten the locking tool. Set Saab tool KM-800-20 or equivalent, into position. Be sure that the timing marks on the camshaft sprockets are aligned with the marks on the tool and that the edge of the timing belt is flush with the edge of the camshaft sprockets.

➡**Check that the alignment marks on the tensioner pulley are still aligned.**

21. If necessary, install the crankshaft pulley and tighten the retaining bolts to 15 ft. lbs. (20 Nm).
22. Install the timing belt cover, and tighten the bolts to 72 inch lbs. (8 Nm). Install all of the remaining components in the reverse order of the removal procedure.
23. Connect the negative battery cable.

Subaru

1.2L ENGINE

1. Disconnect the negative battery cable.
2. Remove the air cleaner.
3. Remove the accessory drive belt.
4. Remove the crankshaft pulley.
5. Remove the front timing cover.
6. Rotate the crankshaft so that the camshaft pulley timing mark aligns with the timing mark on the back cover, and the crankshaft pulley timing mark aligns with the timing mark on the engine.
7. Loosen the timing belt tensioner lockbolt and pivot bolt. Push the tensioner away from the timing belt and tighten the tensioner lockbolt.
8. Remove the timing belt.
To install:
9. Remove the valve cover and loosen the valve adjustment so that the camshaft can rotate freely. This step is necessary for proper timing belt tension.
10. Ensure that the camshaft and crankshaft timing marks align and install the timing belt.
11. Loosen the tensioner lockbolt and allow the tensioner to operate.
12. Tighten the tensioner pivot bolt, then the lockbolt.
13. Rotate the crankshaft 2 complete turns and check that the camshaft and crankshaft timing marks align.
14. Adjust the valves and install the valve cover.
15. Install the front timing cover.

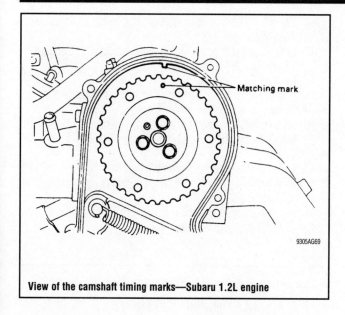

View of the camshaft timing marks—Subaru 1.2L engine

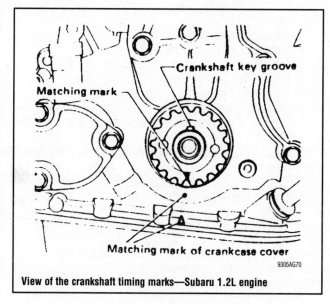

View of the crankshaft timing marks—Subaru 1.2L engine

16. Install the crankshaft pulley and tighten the bolt to 58–72 ft. lbs. (79–98 Nm).

17. Install the accessory drive belt and the air cleaner.

18. Connect the negative battery cable.

1.8L ENGINE

1980–1993

➡This engine is equipped with 2 timing belts.

1. Disconnect the negative battery cable.

2. Remove all necessary components for access to the timing belt covers, then remove the covers.

3. Loosen the timing belt tensioner mounting bolts ½ turn and slacken the timing belt. Tighten the mounting bolts.

4. Mark the rotating direction of the No. 1 timing belt, then remove the belt.

5. Perform the same procedure for the No. 2 timing belt. Remove the crankshaft sprockets.

6. Remove both tensioners together with the tensioner springs.

7. Remove the belt idler. Remove the camshaft sprockets.

8. Remove the No. 2 belt covers.

To install:

9. Inspect the timing belt for breaks, cracks and wear. Replace as required.

10. Check the belt tensioner and idler for smooth rotation. Replace if noisy or excessive play is noticed.

11. Install the left side belt cover seal No. 3 to the cylinder block.

12. Install the left side belt cover seal, left side belt cover seal No. 4, and belt cover mount to the right rear belt cover, then install the assembly on the cylinder block. Tighten to 34 ft. lbs. (45 Nm).

13. Install the left side belt cover seal No. 2 and belt cover mounts to left side belt cover No. 2, then install to the cylinder head and camshaft case. Tighten to 34 ft. lbs. (45 Nm).

14. Install the right side belt cover seal, belt cover seal No. 2 and belt cover mounts to the right side belt cover No. 2, then install to the cylinder head and camshaft case. Tighten to 34 ft. lbs. (45 Nm).

15. Install the camshaft sprockets to the right and left camshafts. Tighten the bolts gradually in 2–3 steps to 67 ft. lbs. (91 Nm).

16. Attach the tensioner spring to the tensioner, then install to the right side of the cylinder block. Tighten the bolts temporarily by hand.

17. Attach the tensioner spring to the bolt, tighten the right side bolt, then loosen it ½ turn.

18. Push down the tensioner until it stops, then temporarily tighten the left bolt.

19. Install the left side tensioner in the same manner.

20. Install the belt idler to the cylinder block using care not to turn the seal. Tighten to 29–35 ft. lbs. (39–47 Nm).

21. Install the sprockets on the crankshaft. Install the crankshaft pulley and tighten the bolt temporarily.

22. Align the center of the 3 lines scribed on the flywheel with the timing mark on the flywheel housing.

23. Align the timing mark on the left side camshaft sprocket with the notch on the belt cover.

24. Attach timing belt No. 2 to the crankshaft sprocket No. 2, oil pump sprocket, belt idler and camshaft sprocket in that order. Avoid downward slackening of the belt.

25. Loosen tensioner No. 2 lower bolt ½ turn to apply tension. Push timing belt by hand to ensure smooth movement of tensioner.

26. Apply 25 ft. lbs. (new belt) or 18 ft. lbs. (used belt) torque to the camshaft sprocket in counterclockwise direction. While applying torque tighten tensioner No. 2 lower bolt temporarily, then tighten upper bolt temporarily.

27. Tighten the lower bolt, then the upper bolt to 13–15 ft. lbs. (17–20 Nm) in that order.

28. Check that the flywheel timing mark and drivers side camshaft sprocket marks are in their proper positions.

29. Turn the crankshaft 1 turn clockwise from the position where timing belt No. 2 was installed and align the center of the 3 lines on the flywheel with the timing mark on the flywheel housing.

30. Align the timing mark on the right side camshaft sprocket with the notch in the belt cover.

31. Attach the timing belt to the crankshaft sprocket and camshaft sprocket, avoiding slackening of the belt on the upper side.

32. Loosen the tensioner ½ turn to apply tension to the belt. Push the belt by hand to ensure smooth operation.

33. Apply 25 ft. lbs. (34 Nm) for new belts or 18 ft. lbs. (24 Nm) for used belts, tighten to the camshaft sprocket in the counterclockwise direction. While applying torque, tighten the tensioner left bolt temporarily, then tighten right bolt temporarily.

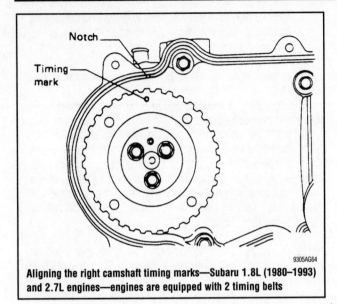

Aligning the right camshaft timing marks—Subaru 1.8L (1980–1993) and 2.7L engines—engines are equipped with 2 timing belts

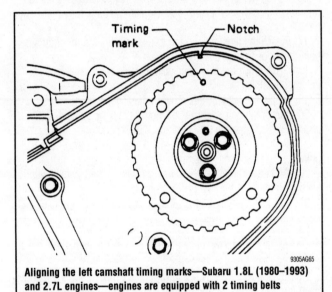

Aligning the left camshaft timing marks—Subaru 1.8L (1980–1993) and 2.7L engines—engines are equipped with 2 timing belts

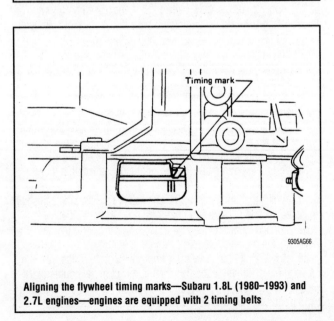

Aligning the flywheel timing marks—Subaru 1.8L (1980–1993) and 2.7L engines—engines are equipped with 2 timing belts

34. Tighten the left bolt, then the right bolt to 13–15 ft. lbs. (17–20 Nm) in that order.

35. Check that the flywheel timing mark and left side camshaft sprocket marks are in their proper positions.

36. Remove the crankshaft pulley.

37. Install the right front belt cover seals and belt cover plug. Install the belt covers to the cylinder block.

38. On turbocharged engines, install the belt cover plate.

39. Install the crankshaft pulley and tighten to 66–79 ft. lbs. (89–107 Nm).

40. Install the water pump pulley and tighten to 67 ft. lbs. (91 Nm). Install the pulley cover, oil level guide and gauge and oil pressure switch connector.

41. Install and properly tension the accessory drive belt.

42. Connect the negative battery cable.

2.7L ENGINE

➡**This engine is equipped with 2 timing belts.**

1. Disconnect the negative battery cable.
2. Remove the engine cooling fans.
3. Remove the accessory drive belts.
4. Remove the water pump pulley.
5. Remove the oil dipstick and tube.
6. Disconnect the oil pressure switch connector.
7. Remove the crankshaft pulley.
8. Remove the left, right and center timing belt covers.
9. Loosen the right timing belt tensioner lockbolt and pivot bolt.
10. Pivot the tensioner away from the timing belt and tighten the lockbolt.
11. Remove the right timing belt. Remove the right timing belt crankshaft sprocket.
12. Remove the plug rubber below the left timing belt tensioner.
13. Remove the plug screw from the bottom of the left timing belt tensioner.
14. Using a standard screwdriver, turn the adjustment screw clockwise to relieve timing belt tension.
15. Install a Belt Adjuster Stopper tool No. 13082AA000.
16. Remove the left belt tensioner pulley.
17. Remove the left timing belt.
18. Remove the left timing belt tensioner.

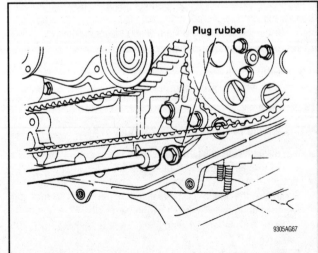

View of the rubber plug—Subaru 2.7L engine—engine is equipped with 2 timing belts

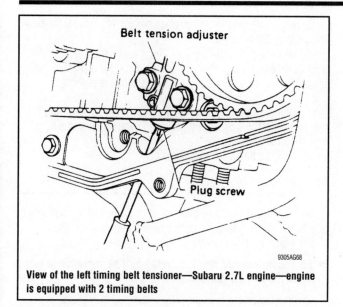

View of the left timing belt tensioner—Subaru 2.7L engine—engine is equipped with 2 timing belts

To install:

19. Fill the left timing belt tensioner with clean engine oil and replace the plug screw.

20. Install the left timing belt tensioner.

21. Rotate the crankshaft to align the middle timing belt alignment mark on the flywheel with the bell housing pointer.

22. Rotate the left camshaft to align the camshaft pulley mark with the rear cover timing mark.

23. Install the left timing belt.

24. Install the left timing belt tensioner pulley and remove the Belt Adjuster Stopper.

25. Rotate the crankshaft 2 complete turns and check that the flywheel and left camshaft timing marks align.

26. Install the plug rubber below the left timing belt tensioner.

27. Install the right timing belt crankshaft pulley.

28. Rotate the crankshaft 1 full turn and align the middle timing belt alignment mark on the flywheel with the bell housing pointer.

29. Rotate the right camshaft to align the camshaft pulley mark with the rear cover timing mark.

30. Install the right timing belt.

31. Loosen the right timing belt tensioner lockbolt.

32. Using a Belt Tension Wrench No. 499437100, apply 17–19 ft. lbs. (24–25 Nm) of torque to the right camshaft pulley in the counterclockwise direction. Tighten the right timing belt tensioner lockbolt and pivot bolt.

33. Install the center, left and right timing covers.

34. Install the crankshaft pulley and tighten the bolt to 66–79 ft. lbs. (89–107 Nm).

35. Connect the oil pressure switch.

36. Install the oil dipstick and tube.

37. Install the water pump pulley.

38. Install the accessory drive belts.

39. Install the engine cooling fans.

40. Connect the negative battery cable.

1.8L (1994–00) & 2.2L ENGINES

The engines use a single cam belt drive system with a serpentine type belt. The left side of the engine uses a hydraulic cam belt tensioner, which is self-adjusting.

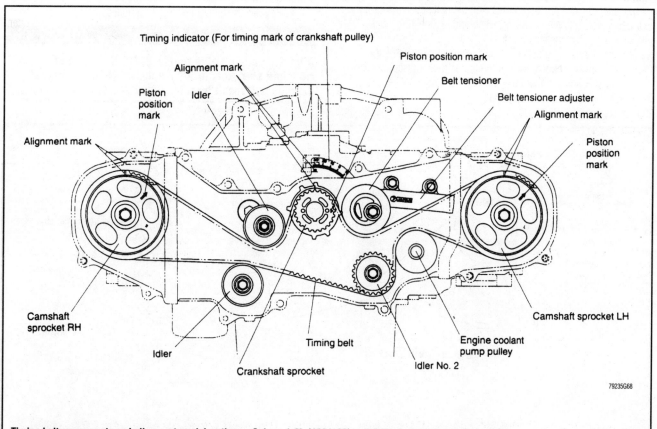

Timing belt components and alignment mark locations—Subaru 1.8L (1994–00) and 2.2L single timing belt engines

➥**It is recommended that the timing belt be replaced at least every 60,000 miles (96,618 km).**

1. Disconnect the negative battery cable.
2. Position the No. 1 piston to Top Dead Center (TDC) of its compression stroke.
3. Remove the engine drive belts.
4. Remove the timing belt covers.
5. Align the camshaft sprockets so each sprocket notch aligns with the cam cover notches. Align the crankshaft sprocket top tooth notch, located at the rear of the tooth, with the notch on the crank angle sensor boss. Mark the 3 alignment points as well as the direction of cam belt rotation.
6. Loosen the tensioner adjusting bolts. Remove the bottom 3 idlers, the cam belt and the cam belt tensioner. The cam sprockets can, then be removed with a modified camshaft sprocket wrench tool.
7. Remove the sprockets, if necessary. Note the reference sensor at the rear of the left cam sprocket.

To install:

8. Install the sprockets, if removed and tighten the retaining bolts to 47–54 ft. lbs. (64–74 Nm).
9. Install the crankshaft sprocket and the non-adjustable right side idler. Do not install the tensioner idler at this time.
10. Compress the hydraulic tensioner in a vise slowly and temporarily secure the plunger with a pin or suitable Allen wrench. Install the tensioner and the pulley with the adjustable idler pulley. Temporarily tighten the tensioner while the tensioner is pushed to the right.
11. Align the crankshaft sprocket notch on the rear sprocket tooth with the crank angle sensor boss. This places the sprocket notch in the 12 o'clock position.
12. Align the camshaft sprockets with the notches in the cam rear belt cover. This places the sprocket notch in the 12 o'clock position for each camshaft.
13. Install the timing belt with the directional mark and alignment marks properly positioned (if the belt is to be reused).
14. Loosen the tensioner retaining bolts and slide the tensioner to the left. Tighten the mounting bolts.

15. After verifying the timing marks are correct, remove the stopper pin from the tensioner.
16. Verify the correctness of the timing by noting that the notches on the 2 cam pulleys and the notch on the crankshaft pulley all point to the 12 o'clock position when the belt is properly installed.
17. Complete the engine component assembly by installing the cam belt covers, the crankshaft pulley bolt and pulley and the remaining components.
18. Connect the negative battery cable.

2.5L ENGINE

The engine uses a single cam belt drive system with a serpentine type belt. The left side of the engine uses a hydraulic cam belt tensioner, which is self-adjusting.

➥**It is recommended that the timing belt be replaced at least every 60,000 miles (96,618 km).**

1. Disconnect the negative battery cable.
2. Remove all necessary components for access to the left, right and center timing belt covers, then remove the covers.
3. Align the camshaft sprockets so each sprocket notch aligns with the rear cover notches. Align the crankshaft sprocket top tooth notch, located at the rear of the tooth with the notch. The crankshaft notch will be at 12 o'clock and the keyway will be at 6 o'clock.

➥**Mark the sprocket alignment points as well as the direction of cam belt rotation for reinstallation purposes if the belt is to be reused.**

4. Loosen the tensioner adjusting bolts.
5. Remove the lower timing belt idler.
6. Remove the timing belt from the pulleys.

✴✴ WARNING

After the timing belt is removed, DO NOT rotate the camshaft sprockets or the crankshaft. Severe internal damage will result from the valve and/or piston contact.

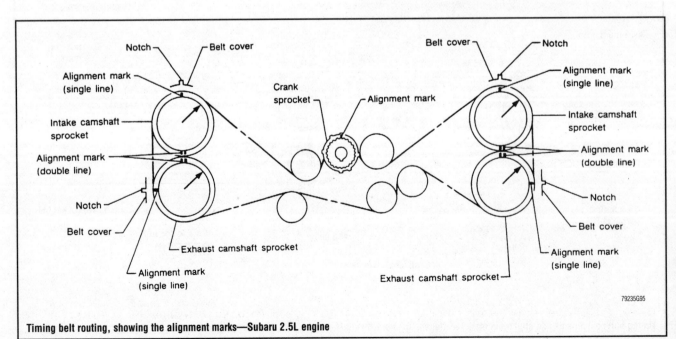

Timing belt routing, showing the alignment marks—Subaru 2.5L engine

79235G95

7. Remove the timing belt tensioner and the timing belt tension adjuster.

To install:

→**Inspect the timing belt and tensioner for wear or damage and replace as necessary.**

8. Inspect the timing belt tensioner as follows:

   a. When compressing the pushrod of the tensioner with a force of 33–110 lbs. (147–490 N), the tensioner should not sink within 8.5 seconds.

   b. Measure the extension of the rod beyond the body of the tensioner for a length of 0.606–0.646 in. (15.4–16.4mm). If not within specifications, replace the tensioner.

→**Check the idler sprockets for smooth operation. Replace as necessary.**

9. Using a press, compress the tensioner gradually, taking 3 minutes or more and insert a 0.059 in. (1.5mm) pin to secure the rod.

10. Install the tensioner and the pulley with the adjustable idler pulley. Temporarily tighten the tensioner while the tensioner is pushed to the right.

11. Align the crankshaft sprocket notch on the rear sprocket tooth with the crank angle sensor boss. This places the sprocket notch in the 12 o'clock position and the keyway at the 6 o'clock position.

12. Align the camshaft sprockets with the notches in the cam rear belt cover. This places the sprocket notch in the 12 o'clock position for each camshaft.

13. Install the timing belt in a clockwise direction starting at the crankshaft with the directional mark and alignment marks properly positioned (if the belt is to be reused).

14. Install the lower timing belt idler and tighten the mounting bolt to 29 ft. lbs. (39 Nm).

✳✳ WARNING

Be sure all the timing marks are properly aligned.

15. Loosen the tensioner retaining bolts and slide the tensioner to the left. Tighten the mounting bolts to 18 ft. lbs. (25 Nm).

16. After verifying that the timing marks are correct, remove the stopper pin from the tensioner and recheck the timing marks.

17. Install the center, right, then the center timing belt covers. Tighten the bolts to 44 inch lbs. (5 Nm).

18. Install the remaining components in the reverse order of the removal procedure. When installing the crankshaft sprocket, be sure to tighten the mounting bolt to 94 ft. lbs. (127 Nm).

19. Connect the negative battery cable.

3.3L ENGINE

1. Disconnect the negative battery cable.

2. Remove the timing belt covers.

3. Matchmark the timing belt to the sprocket, cover and block marks as follows:

   a. Turn the crankshaft to align the timing marks on the crankshaft sprocket with the mark on the block.

   b. With the crankshaft marks aligned be sure the left and right camshaft sprocket marks are aligned with marks on the timing covers.

   c. If all the marks align, use white paint to mark the direction

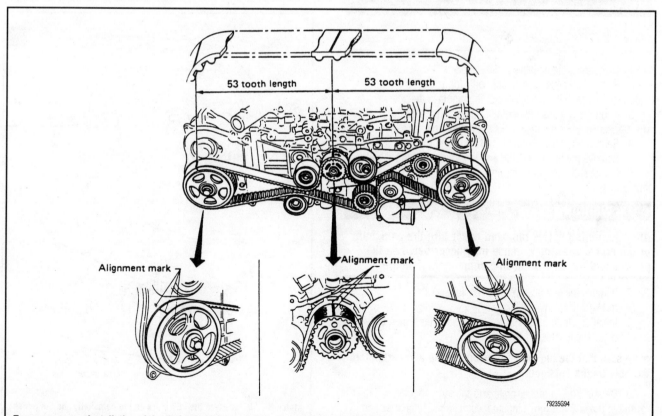

To ensure proper installation of the timing belt, be sure the proper number of belt teeth are between each sprockets, as indicated—Subaru 3.3L engine

of rotation of the belt as well as mark the spots on the belt where it crosses over the timing marks on the pulleys.

4. Loosen the belt tensioner bolts.
5. Remove belt idler pulley No. 1.
6. Remove belt idler pulley No. 2.
7. Remove the timing belt.
8. Remove the tensioner pulley bolt and remove the tensioner pulley.
9. Remove the 2 bolts and the tensioner assembly.

To install:

10. Insert a 0.059 in. (1.5mm) diameter stopper pin into place while pushing the tension adjuster rod into the tensioner body.

11. Install the tensioner and tighten the bolts to 18 ft. lbs. (24 Nm), while the tensioner is pushed all the way to the right.

12. Install the tensioner pulley and mounting bolt. DO NOT tighten the idler pulley bolt completely.

13. Be sure the crankshaft and both camshaft sprockets are still aligned with their respective timing marks.

14. Install the timing belt onto the sprockets with the direction of rotation arrow in the correct direction and the timing marks on the belt align with the marks on the sprockets.

15. Install No. 1 and 2 idler pulleys and tighten the mounting bolts to 29 ft. lbs. (39 Nm).

16. Loosen the tensioner pulley bolt and the tensioner assembly mounting bolts. Slide the tensioner assembly all the way to the left and tighten the bolts to 18 ft. lbs. (24 Nm).

17. Check again that all the timing marks are still in alignment. If they are remove the stopper pin from the tensioner assembly.

18. Install the timing belt covers.
19. Connect the negative battery cable.

Suzuki

1.0L & 1.3L SOHC ENGINES

1. Disconnect the negative battery cable.
2. Remove all necessary components for access to the upper and lower timing belt outside covers, then remove the covers.
3. Align the camshaft timing belt pulley with its timing marks. The crankshaft and camshaft marks are straight up.
4. Remove the resonator and the timing belt outside cover.
5. Remove the tensioner stud and loosen the tensioner bolt.
6. Remove the tensioner spring and damper, then remove the timing belt.

❊❊ WARNING

After the timing belt is removed never turn the camshaft or the crankshaft. Interference may occur between the pistons and the valves causing component damage.

7. Remove the tensioner and the tensioner plate.

To install:

8. Install the timing belt tensioner plate and tensioner. Only hand-tighten the tensioner bolt.

➡**Be sure that the lug on the tensioner plate is inserted into the hole on the tensioner.**

9. Be sure the tensioner plate and the tensioner move uniformly. If they do not move together remove the tensioner and the tensioner plate and reinsert the plate lug into the tensioner hole.

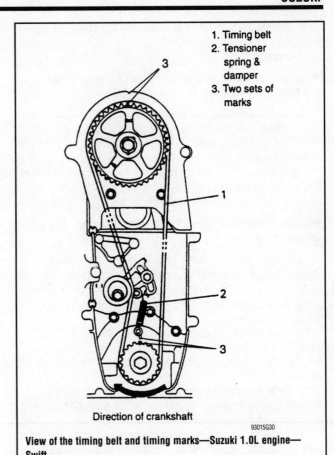

1. Timing belt
2. Tensioner spring & damper
3. Two sets of marks

Direction of crankshaft

93015G30

View of the timing belt and timing marks—Suzuki 1.0L engine—Swift

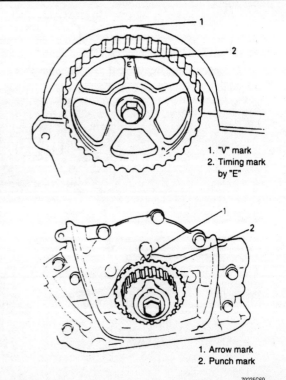

1. "V" mark
2. Timing mark by "E"

1. Arrow mark
2. Punch mark

79235G69

Match the "V" notch to the "E" mark on the camshaft, and the punch and arrow on the crankshaft to properly position the engine for belt service—Suzuki 1.3L (SOHC) and 1.6L engines—Esteem and Swift

10. Check the camshaft sprocket to verify that it has not moved.

11. Check the crankshaft alignment by verifying that the punch mark on the timing belt pulley is aligned with the arrow on the oil pump case.

12. Remove the cylinder head cover.

➡**This is to permit the free rotation of the camshaft. When installing the timing belt on the pulleys, the tensioner spring force should correctly tension the belt. If the camshaft does not rotate freely, the belt will not be correctly tensioned.**

13. With the timing marks aligned, hold the tensioner plate up by hand and install the timing belt on the pulleys so there is no slack on the drive side of the belt.

14. Turn the crankshaft 2 rotations clockwise. Confirm that the timing marks are still properly aligned.

15. If the belt is free of slack and the alignment marks are correct tighten the tensioner stud to 84–96 inch lbs. (9–12 Nm). Tighten the tensioner bolt to 17–21 ft. lbs. (24–30 Nm).

16. Install the timing belt upper and lower outside covers. Tighten the timing cover bolts to 84–96 inch lbs. (9–12 Nm).

17. Install all remaining components in the reverse order of the removal procedure.

18. Connect the negative battery cable.

1.3L DOHC ENGINE

1. Disconnect the negative battery cable.

2. Disconnect and remove the Mass Air Flow (MAF) meter and the air cleaner assembly.

3. Remove the air cleaner bracket.

4. Raise and safely support the vehicle.

5. Remove the right fender apron extension.

6. Remove the accessory drive belt.

7. Remove the water pump pulley.

8. Remove the crankshaft pulley.

➡**It may be necessary to remove the crankshaft timing pulley bolt for clearance.**

9. Remove the upper and lower timing covers.

10. Rotate the crankshaft so that the camshaft pulley timing marks align with the timing marks on the rear timing cover, and the crankshaft timing mark aligns with the mark on the oil pump.

11. Loosen the timing belt tensioner locknut and pivot bolt. Pry the tensioner pulley away from the timing belt and tighten the locknut.

12. Remove the timing belt.

To install:

13. Ensure that the camshaft and crankshaft timing marks are aligned and install the timing belt.

14. Loosen the tensioner locknut.

15. Rotate the crankshaft 2 complete turns and check that the timing marks align.

16. Tighten the tensioner locknut to 84–102 inch lbs. (9–12 Nm) and the pivot bolt to 17–22 ft. lbs. (24–30 Nm).

17. Install the upper and lower timing covers.

18. Install the crankshaft pulley and tighten the bolts to 84–102 inch lbs. (9–12 Nm). If removed, tighten the crankshaft timing pulley bolt to 76–83 ft. lbs. (105–115 Nm).

19. Install the water pump pulley and the accessory drive belt.

20. Install the fender apron extension.

21. Lower the vehicle.

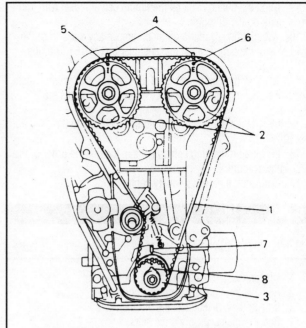

1. Timing belt
2. Camshft timing belt pulleys
3. Crankshaft timing belt pulley
4. Marks on cylinder head cover
5. Punch mark by "I" (intake side)
6. Punch mark by "E" (exhaust side)
7. Mark on oil pump case
8. Key on crankshaft

9305AG73

Alignment of the timing marks—Suzuki 1.3L DOHC engine—Swift

22. Install the air cleaner bracket, the air cleaner assembly and the MAF meter.

23. Connect the negative battery cable.

1.6L ENGINE

1. Disconnect the negative battery cable.

2. Remove all necessary components for access to the timing belt covers, then remove the covers.

3. Loosen but do not remove the tensioner bolt.

❊❊ WARNING

After the timing belt is removed, never turn the camshaft and crankshaft independently. This engine is an interference engine and if the camshaft or crankshaft is turned beyond a certain point, damage to the valves could occur.

4. Loosen the timing belt tensioner adjusting bolt and pivot nut. Apply pressure to the tensioner to loosen the timing belt, and remove the timing belt from the camshaft and crankshaft sprockets.

5. Remove the timing belt tensioner, tensioner plate and tensioner spring.

To install:

6. Install the timing belt tensioner, plate and spring. Hand-tighten the tensioner bolt and stud only at this time.

7. Turn the camshaft sprocket clockwise and align the timing marks.

8. Turn the crankshaft clockwise, using a 17mm wrench to crank the timing belt sprocket bolt.

9. Align the punch mark on the timing belt sprocket with the arrow mark on the oil pump.

10. With the timing marks aligned, remove any slack from the drive side of the belt. Tighten the tensioner bolt to 16–20 ft. lbs. (22–28 Nm).

11. To allow the belt to be free of any slack, turn the crankshaft clockwise 2 full rotations. Confirm that the timing marks are aligned.

12. Install the timing cover and tighten the bolts to 84–96 inch lbs. (9–12 Nm).

13. Install all remaining components in the reverse order of the removal procedure.

14. Connect the negative battery cable.

Toyota

1.5L (3E, 3E-E) ENGINES

1. Disconnect the negative battery cable.
2. Remove the right engine under cover.
3. If necessary, disconnect the throttle cable.
4. Remove the accessory drive belts.
5. Remove the alternator and bracket.
6. If necessary, remove the air cleaner and air intake collector.
7. Support the engine with a jack and remove the right engine mount.
8. Remove the upper timing cover.
9. Rotate the crankshaft so that the crankshaft timing mark aligns with the **0** mark on the timing cover and the timing hole in the camshaft pulley aligns with the timing mark on the oil seal retainer.
10. Remove the crankshaft pulley.
11. Remove the lower timing cover.
12. Loosen the timing belt tensioner lockbolt.
13. Pry the tensioner pulley away from the timing belt and tighten the lockbolt.
14. Remove the timing belt.

To install:

15. Ensure that the camshaft and crankshaft timing pulley marks align and install the timing belt.

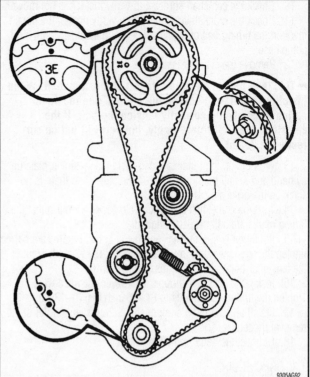

View of the timing belt alignment marks—Toyota 1.5L (3E, 3E-E) engines

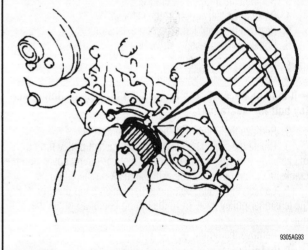

Crankshaft timing sprocket alignment—Toyota 1.5L (3E, 3E-E) engines

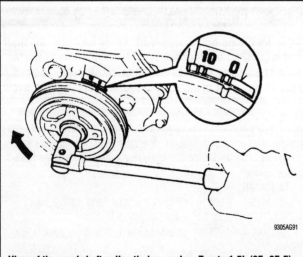

View of the crankshaft pulley timing marks—Toyota 1.5L (3E, 3E-E) engines

16. Loosen the tensioner lockbolt.
17. Rotate the crankshaft 2 complete turns and check that the timing marks align.
18. Tighten the tensioner lockbolt to 13 ft. lbs. (17 Nm).
19. Install the timing covers.
20. Install the crankshaft pulley and tighten the bolt to 112 ft. lbs. (152 Nm).
21. Install the right motor mount and remove the support jack.
22. Install the air cleaner and air intake collector, if removed.
23. Install the alternator and bracket.
24. Install the accessory drive belts.

25. Connect the throttle cable, if removed.
26. Install the engine under cover.
27. Connect the negative battery cable.

1.5L (5E-FE) ENGINE

1. Remove all necessary components for access to the timing belt covers.

✳✳ CAUTION

If equipped with an air bag, be sure to disconnect the negative battery cable and wait at least 90 seconds before proceeding.

2. Remove the No. 2 timing belt cover.
3. Rotate the engine clockwise until the crankshaft pulley is aligned with the 0 mark on the No. 1 timing belt cover. Verify that the hole in the camshaft timing pulley is aligned with the timing mark on the No. 1 bearing cap. If not as specified, rotate the crankshaft an additional 360 degrees.
4. If equipped with air conditioning and/or power steering, remove the 4 bolts to the No. 2 crankshaft pulley. Then, remove the No. 2 crankshaft pulley.
5. Using Toyota tools SST 09213-14010 and SST 09330-00021, or their equivalents, remove the No. 1 crankshaft pulley bolt.
6. Using a crankshaft pulley/damper puller (such as Toyota tool SST 09950-50010), remove the No. 1 crankshaft pulley from the crankshaft.
7. Remove the No. 3 timing belt cover (plug).
8. Remove the No. 1 timing belt cover and timing belt guide.
9. Place matchmarks on the timing belt on both sides of the cam and crankshaft gear timing marks. Also, place an arrow on the top surface of the belt to indicate the direction of travel.
10. Using pliers, remove the tension spring.
11. Loosen the No. 1 idler pulley bolt and push the pulley to the left as far as it will go, then temporarily tighten the bolt.
12. Remove the belt.
To install:
13. If equipped with a distributor (distributor ignition), use the crankshaft bolt to turn the crankshaft until the timing marks on the sprocket and oil pump body align. This is method is used to set the piston at Top Dead Center (TDC) before the marks on the belt cover can be seen.
14. If equipped with a crankshaft position sensor (distributorless ignition), use the crankshaft bolt to turn the crankshaft until the rotor side of the crankshaft position sensor faces inward.
15. Turn the camshaft and align the hole of the camshaft timing pulley with the timing mark of the bearing cap. The matchmarks, if using the old belt should align. Place the belt over the crankshaft pulley and the idler pulleys.
16. Install the belt on the crankshaft gear (using the matchmarks if reinstalling the old belt) and install the timing belt guide with flange out.
17. Loosen the No. 1 idler pulley bolt until the pulley is moved slightly by the spring tension.
18. Turn the crankshaft pulley 2 revolutions from TDC-to-TDC.

➡**Always rotate the crankshaft clockwise.**

19. Check that the pulleys align with the reference marks. If not, reinstall the belt.
20. When the timing is verified, tighten the adjuster pulley (No. 1 idler pulley) to 13 ft. lbs. (18 Nm).

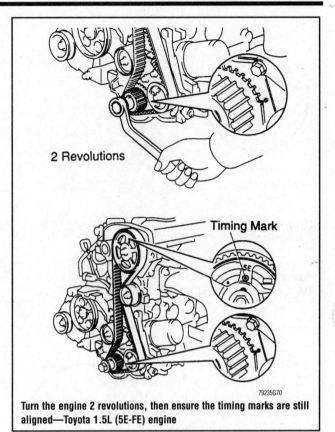

2 Revolutions

Timing Mark

79235G70

Turn the engine 2 revolutions, then ensure the timing marks are still aligned—Toyota 1.5L (5E-FE) engine

21. Install No. 1 and No. 3 lower timing belt covers.
22. Install the No. 1 crankshaft pulley and tighten the pulley bolt to 112 ft. lbs. (152 Nm).
23. Tighten the 4 No. 2 crankshaft pulley bolts to 14 ft. lbs. (19 Nm).
24. Install the No. 2 timing belt cover with the 4 bolts.
25. Install the remaining components in the reverse order of the removal procedure. When installing the right-hand engine mounting insulator, tighten the bracket bolt to 47 ft. lbs. (64 Nm) and the through-bolt to 54 ft. lbs. (73 Nm).

1.5L (1A-C, 3A, 3A-C), 1.6L (4A-C, 4A-CL, 4A-F, 4A-FE) & 1.8L (7A-FE) ENGINES

1. Disconnect the negative battery cable.
2. Remove the windshield washer fluid reservoir.
3. If equipped, remove the cruise control actuator.
4. Remove the accessory drive belts.
5. Raise and safely support the vehicle.
6. Remove the right front wheel.
7. Remove the right engine under cover.
8. If equipped with air conditioning, unbolt the air conditioning compressor and set aside. Remove the compressor bracket.
9. Support the engine with a jack and remove the right engine mount.
10. Remove the water pump pulley.
11. Remove the valve cover on 4A-FE and 7A-FE engines. Remove the upper timing cover on all others.
12. Rotate the crankshaft so that the camshaft pulley timing hole aligns with the timing mark on the bearing cap and the crankshaft pulley timing mark aligns with the **0** mark on the lower timing cover.
13. Remove the crankshaft pulley.

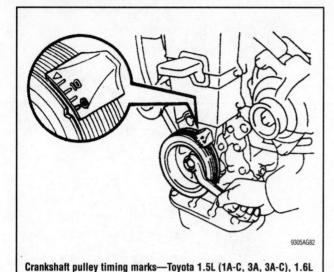

Crankshaft pulley timing marks—Toyota 1.5L (1A-C, 3A, 3A-C), 1.6L (4A-C, 4A-CL, 4A-F, 4A-FE) and 1.8L (7A-FE) engines

9305AG82

Camshaft timing marks—Toyota 1.5L (1A-C, 3A, 3A-C), 1.6L (4A-C, 4A-CL, 4A-F, 4A-FE) and 1.8L (7A-FE) engines

9305AG83

14. Remove the timing covers.
15. Loosen the tensioner pulley lockbolt.
16. Push the tensioner pulley away from the timing belt and tighten the lockbolt.
17. Remove the timing belt.

To install:

18. Install the timing belt to the crankshaft timing pulley.
19. Install the lower timing cover and the crankshaft pulley.
20. Ensure that the camshaft and crankshaft timing marks align and install the timing belt to the camshaft pulley.
21. If necessary for access to the tensioner lockbolt, remove the grommet in the lower timing cover.
22. Loosen the tensioner lockbolt.
23. Rotate the crankshaft 2 complete turns and check that the timing marks align.
24. Tighten the tensioner lockbolt to 27 ft. lbs. (37 Nm).
25. Replace the grommet, if removed.
26. Tighten the crankshaft pulley bolt to 87 ft. lbs. (118 Nm).
27. Install the remaining timing covers and the valve cover, if removed.

28. Install the water pump pulley.
29. Install the right engine mount and remove the support jack.
30. Install the air conditioning compressor bracket and compressor, if removed.
31. Install the right engine under cover.
32. Install the right front wheel and lower the vehicle.
33. Install the accessory drive belts.
34. Install the cruise control actuator, if removed.
35. Install the windshield washer fluid reservoir.
36. Connect the negative battery cable.

1.6L (4A-GE) ENGINE

1. Disconnect the negative battery cable.
2. Raise and safely support the vehicle.
3. For Corolla models, drain the cooling system and remove the following items:
   - Right front wheel
   - Right engine under cover
   - Cruise control actuator, if equipped
   - Windshield washer fluid reservoir
   - Radiator hose
4. For MR2 models, remove the following items:
   - Right rear wheel
   - Engine splash guard
5. Remove the accessory drive belts.
6. Support the engine with a jack and remove the right engine mount.
7. Remove the upper timing cover.
8. Remove the water pump pulley.
9. Remove the middle timing cover.
10. Rotate the crankshaft so that the camshaft pulley timing marks align with the timing marks on the rear timing cover and the crankshaft pulley timing mark aligns with the **0** mark on the lower timing cover.
11. Remove the crankshaft pulley.
12. Remove the lower timing cover.
13. Loosen the tensioner pulley lockbolt.
14. Push the tensioner pulley away from the timing belt and tighten the lockbolt.
15. Remove the timing belt.

To install:

16. Install the timing belt to the crankshaft timing pulley.
17. Install the lower timing cover and the crankshaft pulley. Tighten the crankshaft pulley bolt to 101 ft. lbs. (137 Nm).
18. Ensure that the camshaft and crankshaft timing marks align and install the timing belt to the camshaft pulleys.
19. Rotate the crankshaft 2 complete turns and check that the timing marks align.
20. Tighten the tensioner lockbolt to 27 ft. lbs. (37 Nm).
21. Install the middle timing cover and the water pump pulley.
22. Install the upper timing cover.
23. Install the right motor mount and remove the engine support jack.
24. Install the accessory drive belts.
25. For Corolla models, install the following:
   - Radiator hose
   - Windshield washer fluid reservoir
   - Cruise control actuator, if removed
   - Right engine under cover
   - Right front wheel
26. For MR2 models, install the following:

- Engine splash guard
- Right rear wheel
27. Lower the vehicle.
28. Connect the negative battery cable and fill the cooling system, if drained.

1.8L (1C-L, 1C-TL) & 2.0L (2C-T) ENGINES

1. Disconnect the negative battery cable.
2. Raise and safely support the vehicle.
3. Remove the right front wheel.
4. Remove the right front inner fender liner.
5. Remove the windshield washer fluid reservoir.
6. Remove the engine coolant overflow tank.
7. If equipped, remove the cruise control actuator.
8. Remove the accessory drive belts.
9. Unbolt the power steering pump and position aside.
10. Remove the alternator and bracket.
11. Remove the upper timing cover.
12. Rotate the crankshaft clockwise to align the camshaft pulley timing mark with the top edge of the cylinder head.
13. Remove the crankshaft pulley.
14. Remove the lower timing cover.
15. Support the engine with a jack and remove the right engine mount.
16. Loosen the timing belt tensioner lockbolt.
17. Pry the tensioner pulley away from the timing belt and tighten the lockbolt.
18. Remove the timing belt.
To install:
19. Ensure that the timing marks on the camshaft pulley, crankshaft timing pulley and injection pump pulley are aligned.

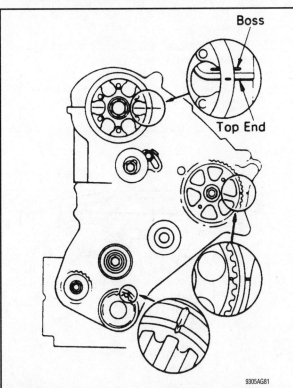

Timing mark alignment—Toyota 1.8L (1C, 1C-TL) and 2.0L (2C-T) engines

20. Install the timing belt.
21. Loosen the tensioner lockbolt.
22. Rotate the crankshaft 2 complete turns and check that the timing marks align.
23. Tighten the tensioner lockbolt to 27 ft. lbs. (37 Nm).
24. Install the right engine mount. Remove the engine support jack.
25. Install the upper and lower timing covers.
26. Install the crankshaft pulley and tighten the bolt to 72 ft. lbs. (98 Nm).
27. Install the alternator and bracket.
28. Install the power steering pump.
29. Install the accessory drive belts.
30. If removed, install the cruise control actuator.
31. Install the coolant overflow tank.
32. Install the windshield washer fluid reservoir.
33. Install the right inner fender liner.
34. Install the right front wheel and lower the vehicle.
35. Connect the negative battery cable.

2.0L (2S-E, 3S-FE) ENGINES

1. Disconnect the negative battery cable.
2. Raise and safely support the vehicle.
3. Remove the right front wheel.
4. Remove the right front inner fender liner.
5. Remove the windshield washer fluid reservoir.
6. Remove the engine coolant overflow tank.
7. If equipped, remove the cruise control actuator.
8. Remove the accessory drive belts.
9. Unbolt the power steering pump and position aside.
10. Remove the alternator and bracket.
11. Remove the upper timing cover.
12. Rotate the crankshaft so that the crankshaft timing mark aligns with the **0** mark on the timing cover and the timing hole in the camshaft pulley aligns with the timing mark on the oil seal retainer.

➡ **There is an "E" cast into the camshaft pulley. On USA models, this "E" will be at 12 o'clock when the timing is correct. On Canadian models, the "E" will be at 6 o'clock.**

13. Remove the crankshaft pulley.
14. Remove the lower timing cover.
15. Loosen the timing belt tensioner lockbolt.
16. Pry the tensioner pulley away from the timing belt and tighten the lockbolt.
17. Remove the timing belt.
To install:
18. Install the timing belt to the crankshaft timing pulley and install the lower cover.
19. Install the crankshaft pulley.
20. Ensure that the crankshaft and camshaft timing marks are aligned and install the timing belt to the camshaft pulley.
21. Loosen the tensioner lockbolt.
22. Rotate the crankshaft 2 complete turns and check that the timing marks align.
23. Tighten the tensioner lockbolt to 31 ft. lbs. (42 Nm).
24. Install the upper timing cover.
25. Tighten the crankshaft pulley bolt to 80 ft. lbs. (108 Nm).
26. Install the alternator and bracket.
27. Install the power steering pump.
28. Install the accessory drive belts.

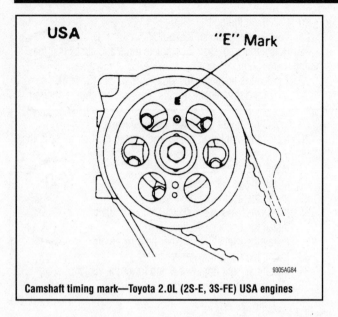

Camshaft timing mark—Toyota 2.0L (2S-E, 3S-FE) USA engines

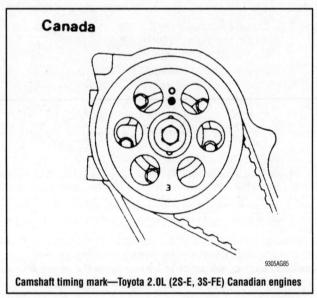

Camshaft timing mark—Toyota 2.0L (2S-E, 3S-FE) Canadian engines

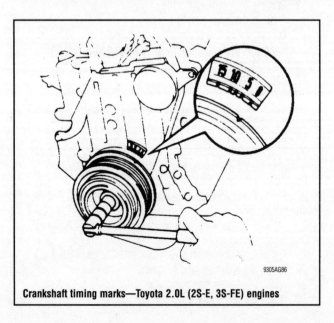

Crankshaft timing marks—Toyota 2.0L (2S-E, 3S-FE) engines

29. If removed, install the cruise control actuator.
30. Install the coolant overflow tank.
31. Install the windshield washer fluid reservoir.
32. Install the right inner fender liner.
33. Install the right front wheel and lower the vehicle.
34. Connect the negative battery cable.

2.0L (3S-GE) ENGINE

1. Disconnect the negative battery cable.
2. Raise and safely support the vehicle.
3. Remove the right front wheel.
4. Remove the right front inner fender liner.
5. Remove the windshield washer fluid reservoir.
6. Remove the engine coolant overflow tank.
7. If equipped, remove the cruise control actuator.
8. Remove the accessory drive belts.
9. Unbolt the power steering pump reservoir and position aside.
10. Remove the alternator and bracket.
11. Support the engine with a jack and remove the right engine mount.
12. Remove the upper timing cover.
13. Rotate the crankshaft so that the crankshaft timing mark aligns with the **0** mark on the timing cover and the timing holes in the camshaft pulleys align with the timing marks on the rear timing cover.
14. Remove the crankshaft pulley.
15. Remove the lower timing cover.
16. Loosen the timing belt tensioner lockbolt.
17. Pry the tensioner pulley away from the timing belt and tighten the lockbolt.
18. Remove the timing belt.
To install:
19. Install the timing belt to the crankshaft timing pulley and install the lower cover.
20. Install the crankshaft pulley.
21. Ensure that the crankshaft and camshaft timing marks are aligned and install the timing belt to the camshaft pulley.
22. Loosen the tensioner lockbolt.
23. Rotate the crankshaft 2 complete turns and check that the timing marks align.

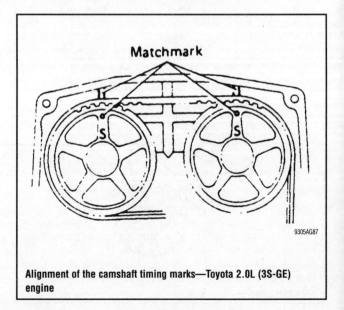

Alignment of the camshaft timing marks—Toyota 2.0L (3S-GE) engine

24. Tighten the tensioner lockbolt to 32 ft. lbs. (43 Nm).
25. Install the right engine mount. Remove the engine support jack.
26. Install the upper timing cover.
27. Tighten the crankshaft pulley bolt to 80 ft. lbs. (108 Nm).
28. Install the alternator and bracket.
29. Install the power steering pump reservoir.
30. Install the accessory drive belts.
31. If removed, install the cruise control actuator.
32. Install the coolant overflow tank.
33. Install the windshield washer fluid reservoir.
34. Install the right inner fender liner.
35. Install the right front wheel and lower the vehicle.
36. Connect the negative battery cable.

2.0L (3S-GTE) ENGINE

Celica

1. Disconnect the negative battery cable.
2. Raise and safely support the vehicle.
3. Remove the right front wheel.
4. Remove the right front fender liner.
5. Remove the windshield washer fluid reservoir.
6. Remove the engine coolant overflow tank.
7. If equipped, remove the cruise control actuator.
8. Remove the intercooler and throttle body.
9. Remove the Exhaust Gas Recirculation (EGR) vacuum modulator and vacuum solenoid valve.
10. Remove the accessory drive belts.
11. Unbolt the power steering pump and position aside.
12. Remove the alternator and bracket.
13. Support the engine with a jack and remove the right engine mount.
14. Remove the upper timing cover.
15. Rotate the crankshaft so that the crankshaft timing mark aligns with the **0** mark on the timing cover and the timing marks on the camshaft pulleys align with the timing marks on the rear timing cover.
16. Remove the crankshaft pulley.
17. Remove the lower timing cover.
18. Remove the timing belt tensioner.
19. Remove the timing belt.

To install:

20. Install the timing belt to the crankshaft timing pulley and install the lower cover.
21. Install the crankshaft pulley. Tighten the crankshaft pulley bolt to 80 ft. lbs. (108 Nm).
22. Ensure that the crankshaft and camshaft timing marks are aligned and install the timing belt to the camshaft pulleys.
23. Prepare the tensioner for installation by compressing it slowly in a vise.
24. Align the hole in the plunger with the holes in the tensioner body and insert a 3/64 (1.27mm) Allen wrench to retain the plunger.
25. To tension the timing belt, perform the following:
   a. Use a torque wrench and turn the tensioner pulley counterclockwise. Apply 13 ft. lbs. (18 Nm) torque to the tensioner pulley and install the tensioner.
   b. Slowly rotate the crankshaft pulley clockwise 5/6 of a turn and align the crankshaft timing mark at 60 degrees Before Top Dead Center (BTDC).

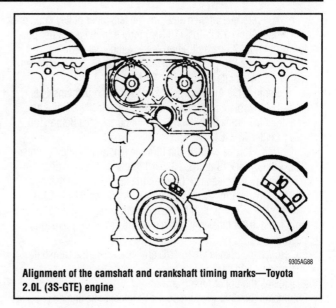

Alignment of the camshaft and crankshaft timing marks—Toyota 2.0L (3S-GTE) engine

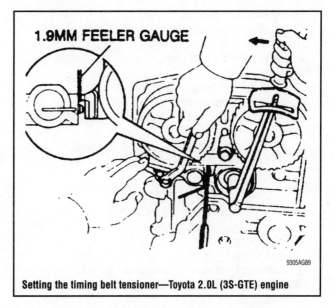

Setting the timing belt tensioner—Toyota 2.0L (3S-GTE) engine

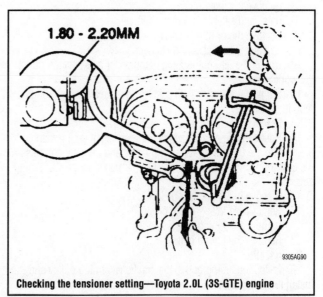

Checking the tensioner setting—Toyota 2.0L (3S-GTE) engine

c. Use a torque wrench and turn the tensioner pulley counterclockwise. Apply 13 ft. lbs. (18 Nm) torque to the tensioner pulley. Insert a 0.075 inch (1.9mm) feeler gauge between the tensioner pulley stopper and the tensioner body. Tighten the tensioner retainer bolts to 15 ft. lbs. 21 Nm).

d. Remove the Allen wrench from the timing belt tensioner.

e. Slowly rotate the crankshaft pulley clockwise ⅝ of a turn and align the crankshaft timing mark at 60 degrees Before Top Dead Center (BTDC).

f. Use a torque wrench and turn the tensioner pulley counterclockwise. Apply 13 ft. lbs. (18 Nm) torque to the tensioner pulley. Measure the clearance between the tensioner pulley stopper and the tensioner body. The clearance should be 0.071–0.087 inches (1.8–2.2mm). If this measurement is not within specification, remove the tensioner and repeat the timing belt tension procedure.

26. Rotate the crankshaft to TDC and check that the camshaft timing marks align.

27. Install the upper timing cover.

28. Install the right motor mount and remove the engine support jack.

29. Install the alternator and bracket.

30. Install the power steering pump.

31. Install the accessory drive belts.

32. Install the EGR vacuum modulator and vacuum solenoid valve.

33. Install the throttle body and intercooler.

34. Install the cruise control actuator, if removed.

35. Install the coolant overflow tank and the windshield washer fluid reservoir.

36. Install the right front fender liner and the wheel.

37. Lower the vehicle and connect the negative battery cable.

MR2

1. Disconnect the negative battery cable.
2. Remove the right engine hood side panel.
3. Remove the upper suspension brace.
4. Remove the No. 1 and No. 2 air intake connectors.
5. If equipped, remove the cruise control actuator.
6. Disconnect the accelerator cable.
7. Remove the electric cooling fan.
8. Remove the intercooler.
9. Raise and safely support the vehicle.
10. Remove the right rear wheel.
11. Remove the accessory drive belts.
12. Unbolt the air conditioning compressor and position out of the work area.
13. Support the engine with a jack and remove the right engine mount.
14. Remove the upper timing cover.
15. Rotate the crankshaft so that the crankshaft timing mark aligns with the **0** mark on the timing cover and the timing marks on the camshaft pulleys align with the timing marks on the rear timing cover.
16. Remove the crankshaft pulley.
17. Remove the lower timing cover.
18. Remove the timing belt tensioner.
19. Remove the timing belt.

To install:

20. Install the timing belt to the crankshaft timing pulley and install the lower cover.

21. Install the crankshaft pulley. Tighten the crankshaft pulley bolt to 80 ft. lbs. (108 Nm).

22. Ensure that the crankshaft and camshaft timing marks are aligned and install the timing belt to the camshaft pulleys.

23. Prepare the tensioner for installation by compressing it slowly in a vise.

24. Align the hole in the plunger with the holes in the tensioner body and insert a ³⁄₆₄ (1.27mm) Allen wrench to retain the plunger.

25. To tension the timing belt, perform the following:

a. Use a torque wrench and turn the tensioner pulley counterclockwise. Apply 13 ft. lbs. (18 Nm) torque to the tensioner pulley and install the tensioner.

b. Slowly rotate the crankshaft pulley clockwise ⅝ of a turn and align the crankshaft timing mark at 60 degrees Before Top Dead Center (BTDC).

c. Use a torque wrench and turn the tensioner pulley counterclockwise. Apply 13 ft. lbs. (18 Nm) torque to the tensioner pulley. Insert a 0.075 inch (1.9mm) feeler gauge between the tensioner pulley stopper and the tensioner body. Tighten the tensioner retainer bolts to 15 ft. lbs. 21 Nm).

d. Remove the Allen wrench from the timing belt tensioner.

e. Slowly rotate the crankshaft pulley clockwise ⅝ of a turn and align the crankshaft timing mark at 60 degrees Before Top Dead Center (BTDC).

f. Use a torque wrench and turn the tensioner pulley counterclockwise. Apply 13 ft. lbs. (18 Nm) torque to the tensioner pulley. Measure the clearance between the tensioner pulley stopper and the tensioner body. The clearance should be 0.071–0.087 inches (1.8–2.2mm). If this measurement is not within specification, remove the tensioner and repeat the timing belt tension procedure.

26. Rotate the crankshaft to TDC and check that the camshaft timing marks align.

27. Install the upper timing cover.

28. Install the right motor mount and remove the engine support jack.

29. Install the air conditioning compressor.

30. Install the accessory drive belts.

31. Install the right rear wheel and lower the vehicle.

32. Install the intercooler.

33. Install the electric cooling fan.

34. Connect the accelerator cable and install the cruise control actuator.

35. Connect the air intake connectors.

36. Install the upper suspension brace.

37. Install the right engine hood side panel.

38. Connect the negative battery cable.

2.2L (5S-FE) ENGINES

1. Disconnect the negative battery cable.

2. Remove all necessary components for access to the timing belt covers.

※ CAUTION

If equipped with an air bag, be sure to disconnect the negative battery cable and wait at least 90 seconds before proceeding.

3. Remove the No. 2 timing cover.

4. Position the No. 1 cylinder to Top Dead Center (TDC) on the

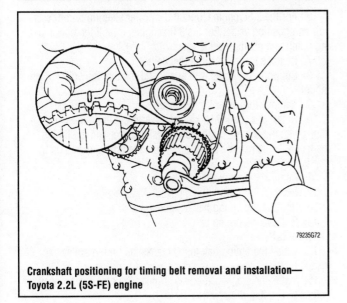

Crankshaft positioning for timing belt removal and installation—Toyota 2.2L (5S-FE) engine

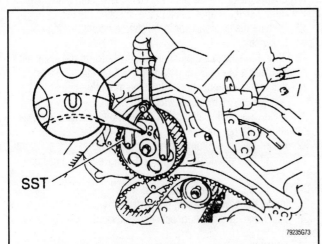

Using a spanner wrench, turn the camshaft into position so that the alignment mark is visible through the hole in the sprocket—Toyota 2.2L (5S-FE) engine

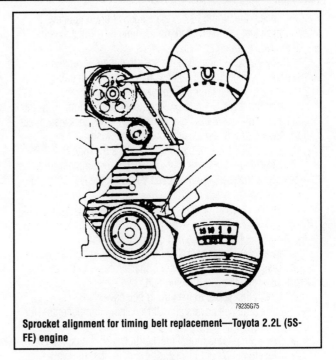

Sprocket alignment for timing belt replacement—Toyota 2.2L (5S-FE) engine

compression stroke by turning the crankshaft pulley and aligning its groove with the timing mark 0 of the No. 1 timing belt cover. Check that the hole of the camshaft timing pulley is aligned with the alignment mark of the bearing cap. If not, turn the crankshaft 1 revolution (360 degrees).

5. Remove the timing belt from the camshaft timing pulley, as follows:

a. If reusing the belt, place matchmarks on the timing belt and the camshaft pulley. Loosen the mount bolt of the No. 1 idler pulley and position the pulley toward the left as far as it will go. Tighten the bolt. Remove the belt from the camshaft pulley.

6. Remove the camshaft timing pulley as follows:

a. Using Toyota tools Nos. 09249-63010 and 09960-10010 or their equivalents, remove the bolt and the camshaft pulley.

7. Remove the crankshaft pulley as follows:

a. Using Toyota tools Nos. 09213-54015 and 09330-00021 or their equivalents, to hold the crankshaft pulley. Remove the pulley set bolt and remove the pulley using a puller.

8. Remove the No. 1 timing belt cover.

9. Remove the timing belt and the belt guide. If reusing the belt mark the belt and the crankshaft pulley in the direction of engine rotation and matchmark for correct installation.

To install:

10. Install the crankshaft timing pulley, as follows:

a. Align the timing pulley set key with the key groove of the pulley.

b. Slide on the timing pulley with the flange side facing inward.

11. Install the No. 2 idler pulley and tighten the bolt to 31 ft. lbs. (42 Nm). Be sure that the pulley moves freely.

12. Temporarily install the No. 1 idler pulley and tension spring. Pry the pulley toward the left as far as it will go. Tighten the bolt. Be sure that the pulley rotates freely.

13. Temporarily install the timing belt, as follows:

a. Using the crankshaft pulley bolt, turn the crankshaft and align the timing marks of the crankshaft timing pulley and the oil pump body.

b. If reusing the old belt, align the marks made during removal and install the belt with the arrow pointing in the direction of the engine revolution.

14. Install the timing belt guide with the cup side facing outward.

15. Install the No. 1 timing belt cover.

16. Install the crankshaft pulley. Align the pulley set key with the key groove of the pulley and slide on the pulley. Tighten the bolt to 80 ft. lbs. (108 Nm).

17. Install the camshaft timing pulley as follows:

a. Align the camshaft knock pin with the knock pin groove of the pulley and slide on the timing pulley. Tighten the bolt to 40 ft. lbs. (54 Nm).

18. With the No. 1 cylinder set at TDC on the compression stroke, install the timing belt (all timing marks aligned). If reusing the belt, align with the marks made during the removal procedure:

a. Turn the crankshaft pulley and align its groove with the timing mark 0 of the No. 1 timing belt cover. Be sure the camshaft sprocket hole is aligned with the mark on the bearing cap.

19. Connect the timing belt to the camshaft timing pulley.

20. Check that the matchmark on the timing belt matches the end of the No. 1 timing belt cover.

21. Once the belt is installed, be sure that there is tension between the crankshaft timing pulley and the camshaft pulley.

22. Check the valve timing as follows:

    a. Loosen the No. 1 idler pulley mount bolt ½ turn. Turn the crankshaft pulley 2 revolutions from TDC in the clockwise direction. Always turn the crankshaft pulley clockwise.

    b. Be sure that the all the timing marks are aligned.

    c. Slowly turn the crankshaft pulley 1⅞ revolutions. Align its groove with the mark at 45 degrees BTDC on the No. 1 timing belt cover for the No. 1 cylinder.

    d. Tighten the No. 1 idler pulley mount bolt to 31 ft. lbs. (42 Nm).

23. Install the No. 2 timing belt cover as follows:

    a. Install the upper gasket to the No. 1 timing belt cover.

    b. Disconnect the engine wire protector between the cylinder head cover and the No. 3 timing belt cover.

    c. Install the gasket to the timing belt cover.

    d. Install the belt covers and all remaining components. During assembly, tighten:

- The right engine mount bracket bolts: 38 ft. lbs. (52 Nm)
- The engine mount insulator bolt: 47 ft. lbs. (64 Nm)
- The through-bolt to 54 ft. lbs. (78 Nm)
- The power steering reservoir bracket bolt: 21 ft. lbs. (28 Nm)
- The power steering reservoir-to-bracket bolt to 27 ft. lbs. (37 Nm)
- The power steering reservoir-to-bracket nut: 38 ft. lbs. (52 Nm).

24. Connect the negative battery cable.

2.5L (2VZ-FE) & 3.0L (3VZ-FE) ENGINES

1. Disconnect the negative battery cable.
2. Raise and safely support the vehicle.
3. Remove the right front wheel.
4. Remove the right front fender liner.
5. If equipped, remove the cruise control actuator and the vacuum pump.

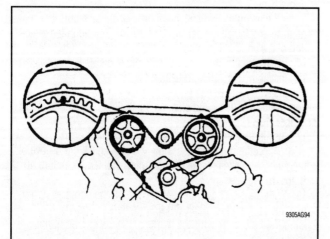

View of the camshaft timing marks—Toyota 2.5L (2VZ-FE) and 3.0L (3VZ-FE) engines

6. For the 2.5L engine, unbolt the power steering pump reservoir and position aside. For the 3.0L engine, unbolt the power steering pump and position aside.

7. Remove the accessory drive belts.

8. Support the engine with a jack and remove the right engine mount.

9. Remove the upper timing cover.

10. Rotate the crankshaft so that the crankshaft timing mark aligns with the **0** mark on the lower timing cover and the timing marks on the camshaft pulleys align with the timing marks on the rear timing cover.

11. Remove the crankshaft pulley.

12. Remove the lower timing cover.

13. Remove the timing belt tensioner.

14. Remove the timing belt.

To install:

15. Install the timing belt to the crankshaft timing pulley and install the lower cover.

16. Install the crankshaft pulley. Tighten the crankshaft pulley bolt to 181 ft. lbs. (245 Nm).

17. Ensure that the crankshaft and camshaft timing marks are aligned and install the timing belt to the camshaft pulleys.

18. Prepare the tensioner for installation by compressing it slowly in a vise.

19. Align the hole in the plunger with the holes in the tensioner body and insert a ³⁄₆₄ (1.27mm) Allen wrench to retain the plunger.

20. Install the tensioner and tighten the mounting bolts to 20 ft. lbs. (27 Nm). Remove the Allen wrench.

21. Rotate the crankshaft clockwise 2 complete turns and check that the timing marks align.

22. Install the upper timing cover.

23. Install the right engine mount and remove the engine support jack.

24. Install the accessory drive belts.

25. Install the power steering pump or reservoir.

26. Install the cruise control actuator and vacuum pump, if removed.

27. Install the right front fender liner and wheel.

28. Lower the vehicle and connect the negative battery cable.

2.8L (5M-GE) & 3.0L (7M-GE, 7M-GTE) ENGINES

1. Disconnect the negative battery cable.
2. Drain the cooling system and remove the upper radiator hose.
3. Remove the upper fan shroud.
4. For 2.8L engines, remove the air intake pipe.
5. For 3.0L engines, remove the water outlet.
6. Remove the accessory drive belts.
7. Remove the cooling fan and the water pump pulley.
8. Remove the air conditioning compressor and idler pulley, if equipped.
9. Remove the upper timing cover.
10. Rotate the crankshaft so that the crankshaft timing mark aligns with the **0** mark on the lower timing cover and the timing marks on the camshaft pulleys align with the timing marks on the rear timing cover.
11. Remove the crankshaft pulley.
12. Remove the lower timing cover.
13. Loosen the timing belt tensioner locknut.
14. Pry the tensioner away from the timing belt and tighten the locknut.

15. Remove the timing belt.
To install:
16. Install the timing belt to the crankshaft timing pulley and install the lower cover.
17. Install the crankshaft pulley. Tighten the crankshaft pulley bolt to 160 ft. lbs. (218 Nm) on 2.8L engines or 195 ft. lbs. (265 Nm) on 3.0L engines.
18. Ensure that the crankshaft and camshaft timing marks are aligned and install the timing belt to the camshaft pulleys.
19. Loosen the tensioner locknut.
20. Rotate the crankshaft 2 complete turns and check that the timing marks align.
21. Tighten the tensioner locknut to 36 ft. lbs. (49 Nm).
22. Install the upper timing cover.
23. Install the air conditioning compressor and idler pulley, if removed.
24. Install the water pump pulley and the cooling fan.
25. Install the accessory drive belts.
26. Install the water outlet, if removed.
27. Install the air intake pipe, if removed.

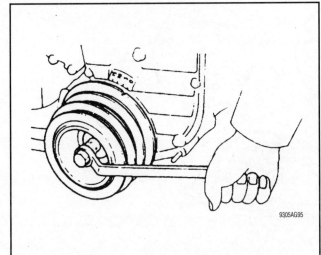

View of the crankshaft timing marks—Toyota 2.8L (5M-GE) and 3.0L (7M-GE, 7M-GTE) engines

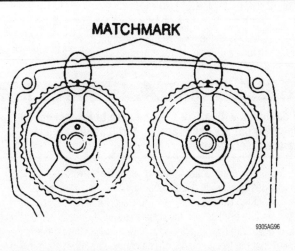

View of the camshaft timing marks—Toyota 2.8L (5M-GE) and 3.0L (7M-GE, 7M-GTE) engines

28. Install the upper fan shroud.
29. Install the upper radiator hose.
30. Connect the negative battery cable and fill the cooling system.

3.0L (1MZ-FE) ENGINE

1. Disconnect the negative battery cable.
2. Remove all necessary components for access to the timing belt covers.

✳✳ CAUTION

If equipped with an air bag, be sure to disconnect the negative battery cable and wait at least 90 seconds before proceeding.

3. Remove the lower timing belt cover by removing the 4 bolts.
4. Remove the No. 2 timing belt cover as follows:
   a. Remove the bolt and disconnect the engine wire protector from the No. 3 (rear) timing belt cover.
   b. Disconnect the engine wire protector clamp from the No. 3 timing belt cover.
   c. Remove the 5 bolts from the No. 2 timing belt cover.
   d. Remove the No. 2 cover from the engine.
5. Remove the right engine mounting bracket by removing the nut and 2 bolts.
6. Remove the crankshaft timing belt guide.
7. Temporarily install the crankshaft pulley bolt.
8. Turn the crankshaft and align the crankshaft timing pulley groove with the oil pump alignment mark. Always turn the engine clockwise.
9. Ensure the timing mark of the camshaft timing pulleys and rear timing belt covers are aligned. If not, turn the engine over an additional 360 degrees (one revolution).
10. Remove the crankshaft pulley bolt.

➡ If the belt is to be reused, align the installation marks on the belt to the marks on the pulleys. If the marks have worn off, make new ones.

11. Alternately loosen the 2 timing belt tensioner bolts. Remove the tensioner and dust boot.
12. Remove the timing belt.
To install:
13. Remove any oil or water from the pulleys.
14. Align the front mark of the timing belt with the dot mark of the crankshaft timing pulley.
15. Align the installation marks on the timing belt with the timing marks of the camshaft pulleys.
16. Install the timing belt in the following order:
   a. Crankshaft pulley.
   b. Water pump pulley.
   c. Left camshaft pulley.
   d. No. 2 idler pulley.
   e. Right camshaft pulley.
   f. No. 1 idler pulley.
17. Using a press, slowly press the timing belt tensioner until the holes of the pushrod and housing align. Insert a 0.05 in. (1.27mm) hexagonal Allen wrench through the holes to preserve the setting position.
18. Install the dust boot to the tensioner.
19. Install the tensioner with the 2 bolts. Alternately tighten the bolts to 20 ft. lbs. (27 Nm). Remove the Allen wrench.

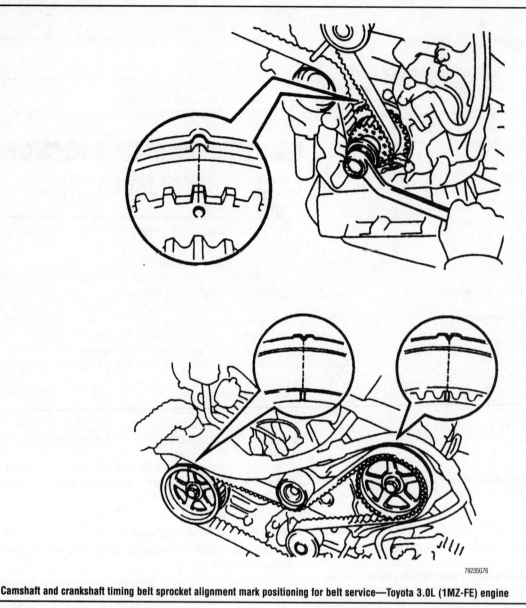

79235G76

Camshaft and crankshaft timing belt sprocket alignment mark positioning for belt service—Toyota 3.0L (1MZ-FE) engine

20. Turn the crankshaft clockwise and align the crankshaft timing pulley groove with the oil pump alignment mark.

21. Ensure the camshaft timing marks align with the timing marks on the rear timing belt cover.

22. Install the timing belt guide.

23. Install the right engine mounting bracket and tighten the bolts to 21 ft. lbs. (28 Nm).

24. Install the upper timing belt cover with the 5 bolts. Tighten the bolts to 74 inch lbs. (8 Nm).

25. Install the engine wire protector clamp to the No. 3 timing belt cover.

26. Install the engine wire protector to the No. 3 timing belt cover with the bolt.

27. Install the lower timing belt cover by installing the 4 bolts. Tighten the bolts to 74 inch lbs. (8 Nm).

28. Install the remaining components. During installation be sure to tighten the crankshaft pulley bolt to 159 ft. lbs. (215 Nm) and the No. 2 alternator bracket nut to 21 ft. lbs. (28 Nm).

29. Connect the negative battery cable.

3.0L (2JZ-GTE & 2JZ-GE) ENGINES

1. Disconnect the negative battery cable.

2. Remove all necessary components for access to the timing belt covers.

✷✷ CAUTION

If equipped with an air bag, be sure to disconnect the negative battery cable and wait at least 90 seconds before proceeding.

3. Remove the upper 2 timing belt covers (Nos. 2 and 3).

4. Remove the drive belt tensioner.

5. Set the No. 1 cylinder to Top Dead Center (TDC) on the compression stroke. Turn the crankshaft pulley clockwise to align the groove with the 0 mark on the lower (No. 1) timing belt cover. Check that the timing marks on the camshaft pulleys are aligned with the marks on the rear belt cover. If the camshaft marks do not align, turn the crankshaft another 360 degrees.

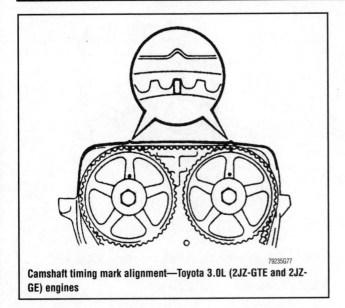

Camshaft timing mark alignment—Toyota 3.0L (2JZ-GTE and 2JZ-GE) engines

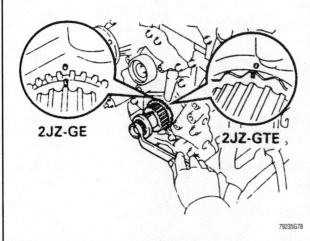

Crankshaft timing marks; notice the timing mark difference between the engines—Toyota 3.0L (2JZ-GTE and 2JZ-GE) engines

6. Alternately loosen the 2 bolts holding the timing belt tensioner. Remove the bolts and remove the tensioner.

7. Remove the timing belt from the camshaft pulleys. If the belt is to be reused, place matchmarks on the belt and gears before removing the belt. Mark the belt with an arrow to show direction of rotation.

8. Using Toyota tool SST 09960-10010 or equivalent, remove the bolts for the camshaft timing gears.

9. Remove the camshaft gears from the engine.

10. If necessary, disconnect the oil cooler tubes from the front of the engine by removing the 2 bolts and hose clamps.

11. Remove the crankshaft pulley by using Toyota tool 09330-0021 or equivalent, to hold the pulley and using tool 09213-70010 or equivalent, to remove the pulley bolt.

12. Remove the lower (No. 1) timing belt cover and the timing belt guide.

13. Remove the timing belt. If the belt is to be reused, protect it from contact with oil, grease or fluids.

To install:

14. Use the crankshaft pulley bolt to turn the crankshaft (clock-

wise) until the mark on the gear aligns with the oil pump body. Check all the pulleys and gears for cleanliness; remove any grease, oil or coolant. Install the timing belt onto the crankshaft gear and idler pulleys.

15. Install the timing belt guide with the cupped side facing outward.

16. Install the No. 1 timing belt cover.

17. Install the crankshaft pulley. Align the set key with the groove. Hold the pulley with the proper tool and tighten the pulley bolt to 239 ft. lbs. (324 Nm).

18. If equipped with automatic transmission, connect the oil cooler tubes with the clamps and 2 bolts.

19. Install the camshaft gears as follows:

a. Align the camshaft knock pin with the groove on the gear and slide on the timing gear.

b. Temporarily install the timing gear bolt.

c. Using the same tools as removal, tighten the camshaft gear bolts to 59 ft. lbs. (79 Nm).

d. Turn the crankshaft pulley and align its groove with the timing mark, **0** on the No. 1 timing belt cover.

e. Align the timing marks on the camshaft timing gears and the No. 4 timing belt cover.

20. Finish installing the timing belt.

21. Double check that all the timing marks for the crankshaft pulley and the camshaft gears are aligned as they were during disassembly.

22. Set the timing belt tensioner:

a. Use a press to slowly push in the pushrod on the tensioner. This will require between 220–2200 lbs. (100–1000 kg) of pressure.

b. Align the holes of the pushrod and housing. Place a 0.06 in. (1.5mm) hex wrench through the holes to keep the pushrod retracted.

c. Release the press and install the dust boot onto the tensioner.

23. Install the tensioner; alternately tighten the bolts to 20 ft. lbs. (26 Nm).

24. Remove the hex wrench from the tensioner with a pair of pliers.

25. Turn the crankshaft pulley 2 full turns clockwise. Check that each pulley's timing marks align correctly after the 2 turns. If any mark does not align, remove the timing belt and reinstall it.

26. Install the drive belt tensioner and tighten the bolts to 15 ft. lbs. (21 Nm).

27. Install the Nos. 2 and 3 timing belt covers.

28. Install all remaining components. During assembly be sure to tighten the drive belt tensioner damper nuts to 14 ft. lbs. (20 Nm).

29. Disconnect the negative battery cable.

Volkswagen

1.4L & 1.7L ENGINES

1980–84

➡**The timing belt is designed to last for more than 60,000 miles and does not normally require tension adjustments. If the belt is removed or replaced, the basic valve timing must be checked and the belt retensioned.**

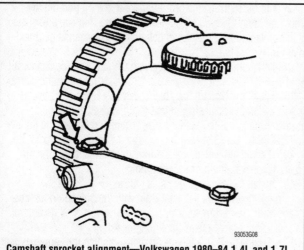

Camshaft sprocket alignment—Volkswagen 1980–84 1.4L and 1.7L and 1985–89 2.2L engines—Dasher, Golf, GTI, Jetta, Quantum, Rabbit, Scirocco

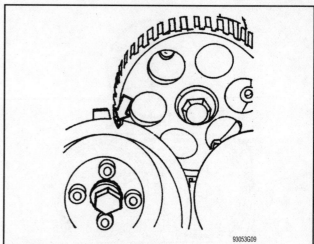

Crankshaft and intermediate sprocket alignment—Volkswagen 1980–84 1.4L and 1.7L and 1985–89 2.2L engines—Dasher, Golf, GTI, Jetta, Quantum, Rabbit, Scirocco

1. Alternator belt
2. Belt pulleys
3. Timing gear cover
4. Crankshaft sprocket
5. Intermediate sprocket
6. Drive belt
7. Tensioner
8. Camshaft sprocket

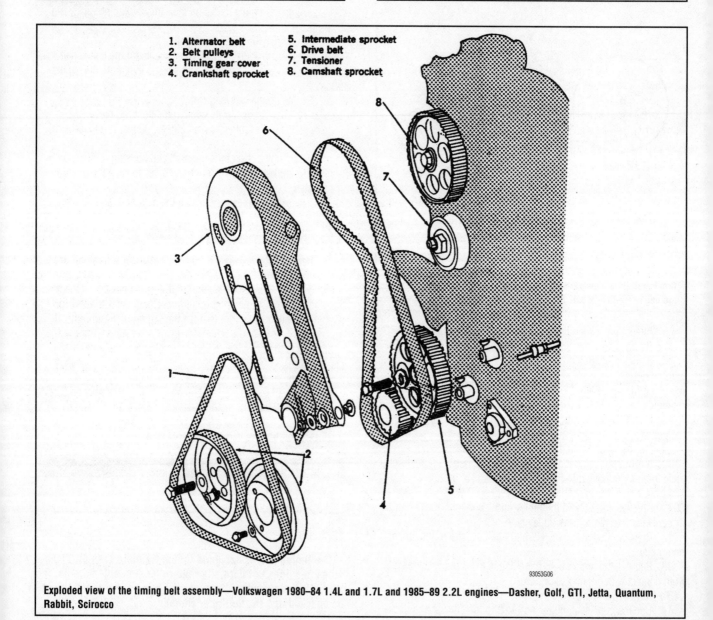

Exploded view of the timing belt assembly—Volkswagen 1980–84 1.4L and 1.7L and 1985–89 2.2L engines—Dasher, Golf, GTI, Jetta, Quantum, Rabbit, Scirocco

1. Disconnect the negative battery cable.
2. Loosen the alternator mounting bolts and if equipped, the power steering pump and air conditioner compressor bolts, if their drive belts interfere with the cover removal.
3. Pivot the alternator or driven component and slip the drive belt from the pulleys.
4. Reposition the spacers and nuts on the mounting studs so they will not get lost.
5. Rotate the crankshaft until the **0** degree mark on the flywheel is aligned with the stationary pointer on the bell housing. Turn the camshaft or make sure the camshaft sprocket is turned until the mark on the rear of the sprocket is aligned with the upper edge of the rear drive belt cover (or valve cover edge, depending on the year of the vehicle) on the left side (spark plug side) of the engine. The notch on the crankshaft pulley should align with the dot on the intermediate shaft sprocket and the distributor rotor (remove the distributor cap) should be pointing toward the mark on the rim of the distributor housing.
6. Remove the crankshaft drive pulley(s).
7. Hold the large nut on the tensioner pulley and loosen the smaller pulley locknut. Turn the tensioner counterclockwise to relieve the tension on the timing belt.
8. Slide the timing belt from the pulleys.

To install:

9. Install the timing belt and retension with the pulley or water pump. Reinstall the crankshaft pulley(s). Recheck the alignment of the timing marks.

✷✷ WARNING

If the timing marks are not correctly aligned with the No. 1 piston at TDC of the compression stroke and the belt is installed, valve timing will be incorrect. Poor performance and possible engine damage can result from improper valve timing.

10. Check the timing belt tension. The tension is correct when the belt can be twisted 90 degrees with the thumb and index finger along the straight run between the camshaft sprocket and the water pump.
11. Turn the engine 2 complete revolutions (clockwise) and align the flywheel mark at TDC. Recheck the belt tension and timing marks. Readjust as required.
12. Install the timing belt cover and drive belts.
13. Connect the negative battery cable.

2.2L ENGINES

1985–89

➡**The timing belt is designed to last for more than 60,000 miles and does not normally require tension adjustments. If the belt is removed or replaced, the basic valve timing must be checked and the belt retensioned.**

1. Disconnect the negative battery cable.
2. Loosen the alternator mounting bolts and if equipped, the power steering pump and air conditioner compressor bolts, if their drive belts interfere with the cover removal.
3. Pivot the alternator or driven component and slip the drive belt from the pulleys.
4. Reposition the spacers and nuts on the mounting studs so they will not get lost.

5. Rotate the crankshaft until the **0** degree mark on the flywheel is aligned with the stationary pointer on the bell housing. Turn the camshaft or make sure the camshaft sprocket is turned until the mark on the rear of the sprocket is aligned with the left side edge of the camshaft housing. The notch on the crankshaft pulley should align with the dot on the intermediate shaft sprocket and the distributor rotor (remove the distributor cap) should be pointing toward the mark on the rim of the distributor housing.
6. Remove the crankshaft drive pulley(s).
7. Loosen the water pump bolts and turn the pump clockwise to relieve the timing belt tension.
8. Slide the timing belt from the pulleys.

To install:

9. Install the timing belt and retension with the pulley or water pump. Reinstall the crankshaft pulley(s). Recheck the alignment of the timing marks.

✷✷ WARNING

If the timing marks are not correctly aligned with the No. 1 piston at TDC of the compression stroke and the belt is installed, valve timing will be incorrect. Poor performance and possible engine damage can result from improper valve timing.

10. Check the timing belt tension. The tension is correct when the belt can be twisted 90 degrees with the thumb and index finger along the straight run between the camshaft sprocket and the water pump.
11. Turn the engine 2 complete revolutions (clockwise) and align the flywheel mark at TDC. Recheck the belt tension and timing marks. Readjust as required.
12. Install the timing belt cover and drive belts.
13. Connect the negative battery cable.

1.5L & 1.6L DIESEL ENGINES

1980–86 and 1989–92

Some special tools are required. A flat bar, VW tool 2065A, or equivalent, is used to secure the camshaft in position. A pin, VW tool 2064, or equivalent, is used to fix the pump position while the timing belt is removed. The camshaft and pump work against spring pressure and will move out of position when the timing belt is removed. It is not difficult to find substitutes but do not remove the timing belt without these tools.

➡**The timing belt is designed to last for more than 60,000 miles and does not normally require tension adjustments. If the belt is removed or replaced, the basic valve timing must be checked and the belt retensioned.**

1. Disconnect the negative battery cable.
2. Loosen the alternator mounting bolts.
3. Pivot the alternator and slip the drive belt off the pulleys.
4. Unscrew the timing cover retaining nuts and remove the cover.
5. Remove the rocker arm cover.
6. Rotate the crankshaft so that the No. 1 cylinder is at Top Dead Center (TDC) of its compression stroke.
7. Using Volkswagen tool 2065A, or equivalent, retain the camshaft in this position by performing the following procedure:
    a. Turn the camshaft until one end of the tool touches the cylinder head.
    b. Measure the gap at the other end of the tool with a feeler gauge.

c. Take ½ of the measurement and insert a feeler gauge of this thickness between the tool and the cylinder head. Turn the camshaft so that the tool rests on the feeler gauge.

d. Insert a second feeler gauge of the same thickness between the other end of the tool and the cylinder head.

8. Using Volkswagen tool 2064 or equivalent, lock the injection pump sprocket in position.

9. Check that the marks on the sprocket, bracket and pump body are in alignment (engine at TDC).

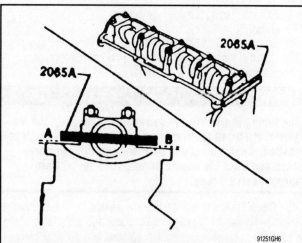

Using the special tool to retain the camshaft in position—Volkswagen 1980 1.5L and 1981–86 and 1989–92 1.6L diesel engines—Dasher, Golf, GTI, Jetta, Quantum, Rabbit

10. Loosen the timing belt tensioner.
11. Remove the timing belt.

To install:
12. Check that the TDC mark on the flywheel is aligned with the reference marks.

13. Install the timing belt and remove pin 2064 from the injection pump sprocket.

14. Tension the belt by turning the tensioner to the right.

15. Adjust the timing belt tension by performing the following procedure:

a. Using Volkswagen Tensioner Gauge tool VW210 or equivalent, install it midway on the timing belt between the longest span between 2 pulleys.

b. Tension the timing belt by turning the tensioner to the right.

c. Proper tension is when the scale reads 12–13.

16. Tighten the timing belt tensioner bolt to 33 ft. lbs. (45 Nm).

17. Torque the crankshaft damper pulley bolt to 145 ft. lbs. (196 Nm)

18. Remove the tool from the camshaft.

19. Rotate the crankshaft 2 turns in the direction of engine rotation (clockwise) and then strike the belt once with a rubber mallet between the camshaft sprocket and the injection pump sprocket.

20. Check the belt tension again. Check the injection pump timing.

21. Install the timing belt cover.

22. Reposition the spacers on the studs and install the washers and nuts.

23. Install the alternator belt and adjust its tension.

24. Connect the negative battery cable.

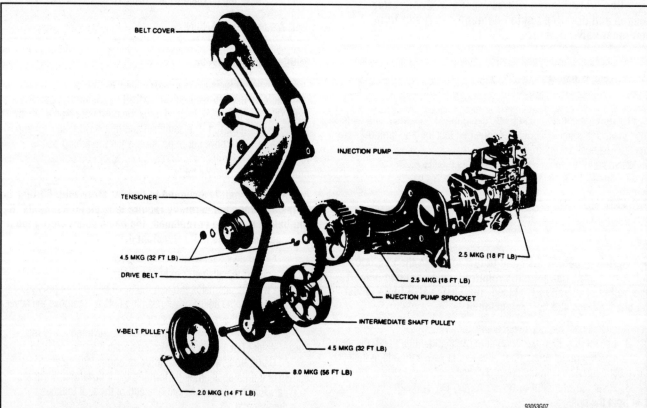

Exploded view of the timing belt and related components—Volkswagen 1980 1.5L and 1981–86 and 1989–92 1.6L diesel engines—Dasher, Golf, GTI, Jetta, Quantum, Rabbit

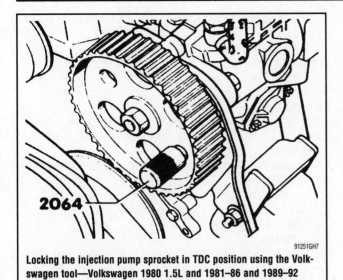

Locking the injection pump sprocket in TDC position using the Volkswagen tool—Volkswagen 1980 1.5L and 1981–86 and 1989–92 1.6L diesel engines—Dasher, Golf, GTI, Jetta, Quantum, Rabbit

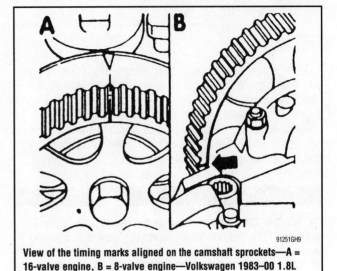

View of the timing marks aligned on the camshaft sprockets—A = 16-valve engine, B = 8-valve engine—Volkswagen 1983–00 1.8L engine

1.8L ENGINE

1983–00

➡Do not turn the engine or camshaft with the camshaft drive belt removed. The pistons will contact the valves and cause internal engine damage.

1. Disconnect the negative battery cable.
2. Remove the accessory drive belts.
3. To remove the crankshaft accessory drive pulley, hold the center crankshaft sprocket bolt with a socket and loosen the pulley bolts.
4. The cover is now accessible. It comes off in 2 pieces; remove the upper ½ first. Take note of any special spacers or other hardware.
5. Temporarily reinstall the crankshaft pulley bolt, if removed and turn the crankshaft to Top Dead Center (TDC) of the compression stroke of No. 1 piston. The mark on the camshaft sprocket should be aligned with the mark on the inner drive belt cover, if equipped, or the edge of the cylinder head.

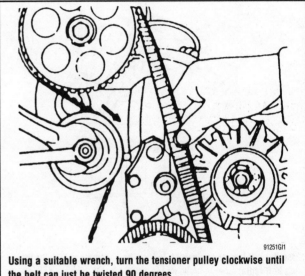

Using a suitable wrench, turn the tensioner pulley clockwise until the belt can just be twisted 90 degrees

6. On 8-valve engines, the notch on the crankshaft pulley should align with the dot on the intermediate shaft sprocket. With the distributor cap removed, the rotor should be pointing toward the No. 1 mark on the rim of the distributor housing.
7. Loosen the locknut on the tensioner pulley and turn the tensioner counterclockwise to relieve the tension on the timing belt.
8. Slide the timing belt from the sprockets.
To install:
9. Install the new timing belt and tension the belt so it can be twisted 90 degrees at the middle of its longest section, between the camshaft and intermediate sprockets.
10. Recheck the alignment of the timing marks and, if correct, turn the engine 2 full revolutions to return to TDC of No. 1 piston. Recheck belt tension and timing marks. Readjust as required. Tighten the tensioner nut to 33 ft. lbs. (45 Nm).
11. Reinstall the belt cover and accessory drive belts.
12. When running the engine, there will be a growling noise that rises and falls with engine speed if the belt is too tight.

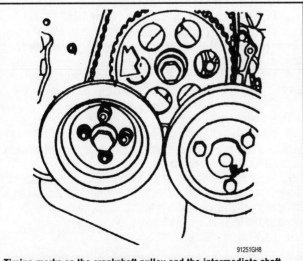

Timing marks on the crankshaft pulley and the intermediate shaft sprocket (arrow)—Volkswagen 1983–00 1.8L engine

1.9L DIESEL ENGINE

1997–00

Some special tools are required to perform this procedure properly. A flat bar, VW tool 2065A or equivalent, is used to secure the camshaft in position. A pin, VW tool 2064 or equivalent, is used to fix the pump position while the timing belt is removed. The camshaft and pump work against spring pressure and will move out of position when the timing belt is removed. It is not difficult to find substitutes but do not remove the timing belt without these tools.

✳✳ WARNING

Do not turn the engine or camshaft with the timing belt removed. The pistons will contact the valves and cause internal engine damage.

1. Disconnect the negative battery cable and remove the accessory drive belts, crankshaft pulley and the timing belt cover(s).

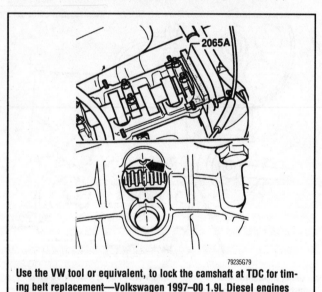

79235G79

Use the VW tool or equivalent, to lock the camshaft at TDC for timing belt replacement—Volkswagen 1997–00 1.9L Diesel engines

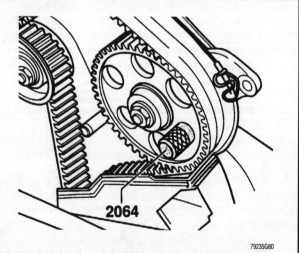

79235G80

Locking the injection pump with the VW tool or equivalent—Volkswagen 1997–00 1.9L Diesel engines

Remove the camshaft cover and rubber plug at the back end of the camshaft.

2. Temporarily reinstall the crankshaft pulley bolt and turn the crankshaft to Top Dead Center (TDC) of the compression stroke of No. 1 piston. The mark on the camshaft sprocket should be aligned with the mark on the inner timing belt cover or the edge of the cylinder head.

3. With the engine at TDC, insert the bar into the slot at the back of the camshaft. The bar rests on the cylinder head to will hold the camshaft in position.

4. Insert the pin into the injection pump drive sprocket to hold the pump in position.

5. Loosen the locknut on the tensioner pulley and turn the tensioner counterclockwise to relieve the tension on the timing belt. Slide the timing belt from the sprockets.

To install:

6. Install the new timing belt and adjust the tension so the belt can be twisted 45 degrees at the half-way point between the camshaft and pump sprockets. Tighten the tensioner nut to 33 ft. lbs. (45 Nm).

7. Remove the holding tools.

8. Turn the engine 2 full revolutions to return to TDC for the No. 1 cylinder. Recheck belt tension and timing mark alignment, readjust as required.

9. Install the belt cover and accessory drive belts.

➡**If the belt is too tight, there will be a growling noise that rises and falls with engine speed.**

2.0L ENGINE

1990–00

➡**Do not turn the engine or camshaft with the camshaft drive belt removed. The pistons will contact the valves and cause internal engine damage.**

1. Disconnect the negative battery cable and remove the accessory drive belts, crankshaft pulley and the timing belt cover(s).

2. Temporarily reinstall the crankshaft pulley bolt, if removed, and turn the crankshaft to Top Dead Center (TDC) of the compression stroke of No. 1 piston. The mark on the camshaft sprocket should be aligned with the mark on the inner drive belt cover, if equipped, or the edge of the cylinder head.

3. On 8-valve engines, the notch on the crankshaft pulley should align with the dot on the intermediate shaft sprocket. With the distributor cap removed, the rotor should be pointing toward the No. 1 mark on the rim of the distributor housing.

4. Loosen the locknut on the tensioner pulley and turn the tensioner counterclockwise to relieve the tension on the timing belt.

5. Slide the timing belt off the sprockets.

To install:

6. Install the new timing belt and tension the belt so that it can be twisted 90 degrees at the middle of its longest section, between the camshaft and intermediate sprockets.

7. Recheck the alignment of the timing marks, if correct, turn the engine 2 full revolutions to return to TDC of No. 1 piston. Recheck belt tension and timing marks. Readjust as required. Tighten the tensioner nut to 33 ft. lbs. (45 Nm).

8. Reinstall the belt cover and accessory drive belts.

➡**When running the engine, there will be a growling noise that rises and falls with engine speed if the belt is too tight.**

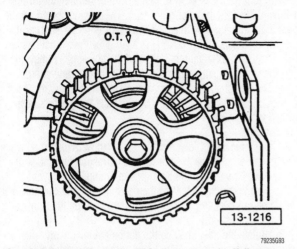

Camshaft timing belt sprocket TDC alignment mark—Volkswagen 1990–00 2.0L engine

79235G93

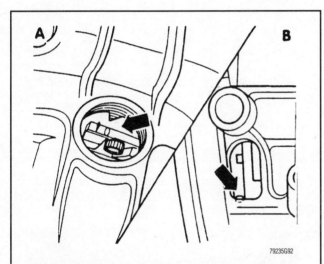

Position the flywheel (A) or driveplate (B) as shown for TDC alignment for cylinder No. 1—Volkswagen 1990–00 2.0L engine

79235G92

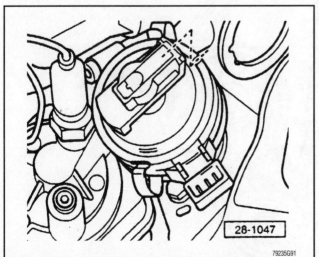

When the No. 1 cylinder is at TDC, the ignition rotor should face the notch in the distributor housing—Volkswagen 1990–00 2.0L engine

79235G91

Volvo

2.4L (D24 & D24T) DIESEL ENGINES

1981–86

1. Disconnect the negative battery cable.
2. Drain the cooling system.
3. Loosen the alternator, air pump, power steering pump (if equipped), air conditioning compressor (if equipped) and remove their drive belts.
4. Remove the water pump pulley.
5. Remove the timing belt cover bolts and the timing cover.
6. If necessary, remove the valve cover.
7. Using a 1¹⁄₁₆ in. (27mm) socket on the crankshaft pulley bolt, rotate the crankshaft to position the No. 1 cylinder at Top Dead Center (TDC) of its compression stroke. Both cam lobes should point up at equally large angles. The flywheel timing mark should be set at **0**.
8. Remove the vibration damper center bolt. It may be necessary to use Volvo special wrenches 5187 (to hold) and 5188 (to remove). The engine may have to be turned slightly to allow the holding wrench to rest temporarily on the cooling fan.
9. Check to make sure the No. 1 cylinder is at TDC. If necessary, adjust the flywheel to the **0** mark.
10. Remove the vibration damper by removing the 4 6mm Allen bolts.

➡ **The vibration damper and the crankshaft gear may be stuck together. You may have tap them apart.**

11. Remove the camshaft gear belt by removing the lower belt shield and releasing the retainer bolts for the water pump.
12. Pull the timing belt straight out and off of the gears.
13. Remove the idler pulley center bolt. Using a puller, remove the idler pulley.
   To install:

➡ **The idler pulley MUST be replaced when replacing the timing belt.**

14. Tap the new idler pulley into position. Install the idler pulley center bolt.

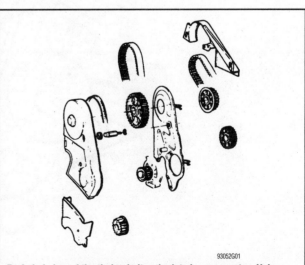

Exploded view of the timing belt and related components—Volvo 1981–86 2.4L (D24 and D24T) diesel engines

93052G01

15. With the timing marks aligned, install the timing belt.

16. Using the special tools, install the vibration damper pulley.

17. Apply tension to the timing belt.

18. Rotate the crankshaft 2 complete revolutions and realign the timing marks. If the timing marks do not align, reinstall the timing belt.

19. To complete the installation, reverse the removal procedures.

20. Refill the cooling system.

21. Connect the negative battery cable.

2.1L & 2.3L ENGINES

1980–00

1. Disconnect the negative battery cable. Loosen the fan shroud and remove the fan. Remove the shroud.

2. Loosen the alternator, power steering pump and air conditioning compressor and remove the drive belts.

3. Remove the water pump pulley.

4. Remove the 4 retaining bolts and lift off the timing belt cover.

5. To remove the tension from the belt, loosen the nut for the tensioner and press the idler roller back. The tension spring can be locked in this position by inserting the shank end of a 3mm drill through the pusher rod.

6. Remove the 6 retaining bolts and the crankshaft pulley.

7. Remove the belt, taking care not to bend it at any sharp angles. The belt should be replaced at 45,000 mile intervals, if it becomes oil soaked or frayed or if on a vehicle that has not been operated for any length of time.

To install:

8. If the crankshaft, idler shaft or camshaft were disturbed while the belt was out, align each shaft with its corresponding index mark to assure proper valve timing and ignition timing, as follows:

   a. Rotate the crankshaft so the notch in the convex crankshaft gear belt guide aligns with the embossed mark on the front cover (12 o'clock position).

   b. Rotate the idler shaft so the dot on the idler shaft drive sprocket aligns with the notch on the timing belt rear cover (4 o'clock position).

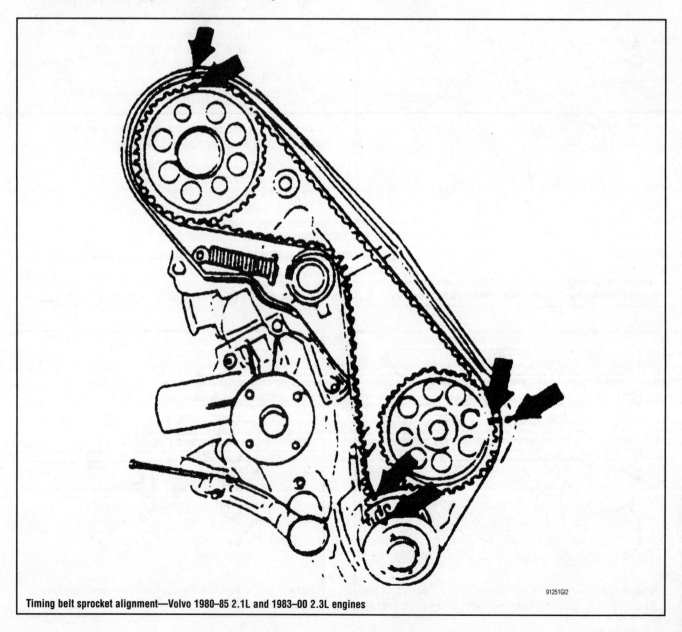

Timing belt sprocket alignment—Volvo 1980–85 2.1L and 1983–00 2.3L engines

91251GI2

c. Rotate the camshaft so the notch in the camshaft sprocket inner belt guide aligns with the notch in the forward edge of the valve cover (12 o'clock position).

9. Install the timing belt (don't use any sharp tools) over the sprockets and then over the tensioner roller. New belts have yellow marks. The 2 lines on the drive belt should fit toward the crankshaft marks. The next mark should then fit toward the intermediate shaft marks, etc. Loosen the tensioner nut and let the spring tension automatically take up the slack. Tighten the tensioner nut to 37 ft. lbs. (51 Nm).

10. Rotate the crankshaft 1 full revolution clockwise and make sure the timing marks still align.

11. Clean all gasket mating surfaces thoroughly. Install the timing belt cover using a new gasket.

12. Install the water pump pulley, drive belts, air conditioning compressor, power steering pump and alternator.

13. Install the fan and shroud. Install the accessory drive belts. Connect the negative battery cable. Start the engine and check for leaks.

2.3L (B234F) ENGINE

1989–91

➡The B234F engine has 2 belts, one driving the camshafts and one driving the balance shafts. The camshaft belt may be removed separately. The balance shaft belt requires removal of the camshaft belt. During reassembly, the exact placement of the belts and pulleys must be observed.

1. Remove the negative battery cable and the alternator belt.

2. Remove the radiator fan, its pulley and the fan shroud.

3. Remove the drive belts for the power steering belts and the air conditioning compressor.

4. Beginning with the top cover, remove the retaining bolts and remove the timing belt covers.

5. Turn the engine to Top Dead Center (TDC) of the compression stroke on cylinder No. 1. Make sure the marks on the camshaft pulleys align with the marks on the backing plate and that the marking on the belt guide plate (on the crankshaft) is opposite the TDC mark on the engine block.

6. Remove the protective cap over the timing belt tensioner locknut. Loosen the locknut, compress the tensioner to release tension on the belts and re-tighten the locknut, holding the tensioner in place.

7. Remove the timing belt from the camshafts. Do not crease or fold the belt.

➡The camshafts and the crankshaft must not be moved when the belt is removed.

8. Check the tensioner by spinning it counterclockwise and listening for any bearing noise within. Check also that the belt contact surface is clean and smooth. In the same fashion, check the timing belt idler pulleys. Make sure the are tightened to 19 ft. lbs. (25 Nm).

9. If the balance shaft belt is to be removed:
   a. Remove the balance shaft belt idler pulley from the engine.
   b. Loosen the locknut on the tensioner and remove the belt. Slide the belt under the crankshaft pulley assembly. Check the tensioner and idler wheels carefully for any sign of contamination. Check the ends of the shafts for any sign of oil leakage.
   c. Check the position of the balance shafts and the crankshaft after belt removal. The balance shaft markings on the pulleys

should align with the markings on the backing plate and the crankshaft marking should still be aligned with the TDC mark on the engine block.

   d. When refitting the balance shaft belt, observe that the belt has colored dots on it. These marks assist in the critical placement of the belt. The yellow dot will align the right lower shaft, the blue dot will align on the crank and the other yellow dot will match to the upper left balance shaft.

   e. Carefully work the belt in under the crankshaft pulley. Make sure the blue dot is opposite the bottom (TDC) marking on the belt guide plate at the bottom of the crankshaft. Fit the belt around the left upper balance shaft pulley, making sure the yellow mark is opposite the mark on the pulley. Install the belt around the right lower balance shaft pulley and again check that the mark on the belt aligns with the mark on the pulley.

   f. Work the belt around the tensioner. Double check that all the markings are still aligned.

   g. Set the belt tension by inserting an Allen key into the adjusting hole in the tensioner. Turn the crankshaft carefully

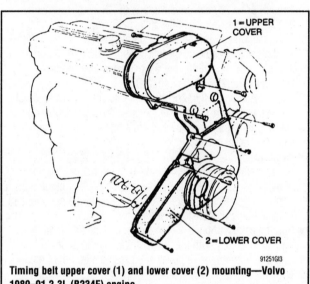

Timing belt upper cover (1) and lower cover (2) mounting—Volvo 1989–91 2.3L (B234F) engine

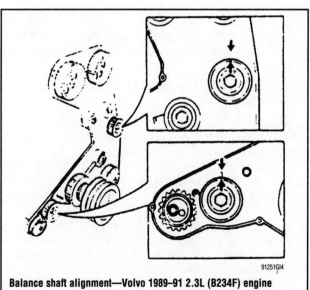

Balance shaft alignment—Volvo 1989–91 2.3L (B234F) engine

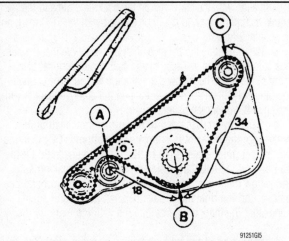

Balance shaft belt markings. There should be 18 teeth between A and B and 34 teeth between B and C—Volvo 1989–91 2.3L (B234F) engine

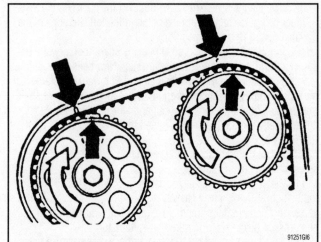

Rotate the crankshaft clockwise to position the camshaft 1½ teeth beyond the alignment marks to properly tension the belt—Volvo 1989–91 2.3L (B234F) engine

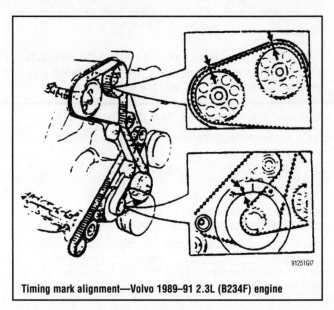

Timing mark alignment—Volvo 1989–91 2.3L (B234F) engine

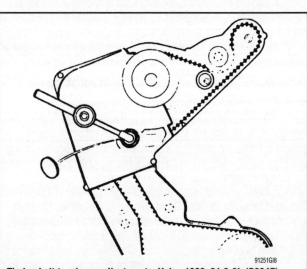

Timing belt tensioner adjustment—Volvo 1989–91 2.3L (B234F) engine

through a few degrees on either side of TDC to check that the belt has properly engaged the pulleys. Return the crank to the TDC position and set the adjusting hole just below the 3 o'clock position when tightening the adjusting bolt. Use the Allen wrench, in the adjusting hole, as a counter hold and tighten the locking bolt to 29.5 ft. lbs. (40 Nm).

  h. Use tool 998 8500 or equivalent, to check the tension of the belt. Install the gauge over the position of the removed idler pulley. The tension must be 1–4 units on the scale or the belt must be readjusted.

To install:

10. Reinstall the camshaft belt by aligning the double line marking on the belt with the top marking on the belt guide plate at the top of the crankshaft. Stretch the belt around the crank pulley and place it over the tensioner and the right-side idler. Place the belt on the camshaft pulleys. The single line marks on the belt should align exactly with the pulley markings. Route the belt around the oil pump drive pulley and press the belt onto the left-side idler.

11. Check that all the markings align and that the engine is still positioned at TDC, of the compression stroke, for cylinder No. 1.

12. Loosen the tensioner locknut.

13. Turn the crankshaft clockwise. The camshaft pulleys should rotate 1 full turn until the marks again align with the marks on the backing plate.

→The engine must not be rotated counterclockwise during this procedure.

14. Smoothly rotate the crankshaft further clockwise until the camshaft pulley markings are 1½ teeth beyond the marks on the backing plate. Tighten the tensioner locknut.

15. Check the tension on the balance shaft belt; it should now be 3.8 units on a suitable belt tension gauge. If the tension is too low, adjust the tensioner clockwise. If the tension is too high, repeat Step 9g.

16. Check the belt guide for the balance shaft belt and make sure it is properly seated. Install the center timing belt cover, the one that covers the tensioner, the fan shroud, fan pulley and fan. Install all the drive belts and connect the battery cable.

17. Double check all installation items, paying particular atten-

tion to loose hoses or hanging wires, loose nuts, poor routing of hoses and wires (too tight or rubbing) and tools left in the engine area.

18. Start the engine and allow it to run until the thermostat opens.

✴✴ CAUTION

The upper and lower timing belt covers are still removed. The belt and pulleys are exposed and moving at high speed.

19. Shut the engine OFF and bring the motor to TDC, of the compression stroke, on cylinder No. 1.

20. Check the tension of the camshaft belt. Position the gauge between the right (exhaust) camshaft pulley and the idler. Belt tension must be 5.5 ± 0.2 units on a suitable belt tension gauge. If the belt needs adjustment, remove the rubber cap over the tensioner locknut, cap is located on the timing belt cover, and loosen the locknut.

21. Insert a suitable tool between the tensioner wheel and the spring carrier pin to hold the tensioner. If the belt needs to be tightened, move the roller to adjust the tension to 6.0 units. If the belt is too tight, adjust to obtain a reading of 5.0 units on the gauge. Tighten the tensioner locknut.

22. Rotate the crankshaft so the camshaft pulleys move through 1 full revolution and recheck the tension on the camshaft belt. It should now be 5.5 ± 0.2 units. Install the plastic plug over the tensioner bolt.

23. Final check the tension on the balance shaft belt by fitting the gauge and turning the tensioner clockwise. Only small movements are needed. After any needed readjustments, rotate the crankshaft clockwise through 1 full revolution and recheck the balance shaft belt. The tension should now be on the final specification of 4.9 ± 0.2 units.

24. Install the idler pulley for the balance shaft belt. Reinstall the upper and lower timing belt covers.

25. Start the engine and final check performance.

2.3L (B5234) & 2.4L ENGINES

1989–00

1. Disconnect the negative battery cable.
2. Remove the coolant expansion tank and place it on top of the engine.
3. Remove the spark plug cover and drive belts.
4. Remove the timing belt cover.
5. Wait 5 minutes after aligning the marks, then install Volvo Gauge 998-8500 or equivalent, between the exhaust camshaft and water pump. Read the gauge using a mirror, while still installed. For 23mm belts, the tension should be 2.7–4.0 units.

➡ If the belt tension is incorrect, the tensioner must be replaced.

6. Remove the upper tensioner bolt and loosen the lower bolt, turning the tensioner to free up the pulley.
7. Remove the lower bolt and the tensioner. Remove the timing belt.

To install:

8. Turn all the pulleys listening for bearing noise. Check to see that the contact surfaces are clean and smooth. Remove the ten-

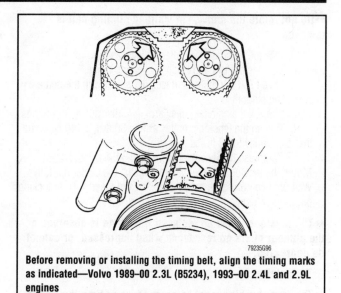

Before removing or installing the timing belt, align the timing marks as indicated—Volvo 1989–00 2.3L (B5234), 1993–00 2.4L and 2.9L engines

sioner pulley lever and idler pulley, lubricate the contact surfaces and bearing with grease. If the tensioner pulley lever or idler is seized replace it.

9. Install the tensioner pulley lever and idler pulley and tighten to 18 ft. lbs. (25 Nm).

10. Compress the tensioner with Volvo tool 999-5456 or equivalent, and insert a 0.079 in. (2.0mm) lockpin in the piston. If the tensioner leaks, has no resistance, or will not compress, replace it. Install the tensioner and tighten to 18 ft. lbs. (25 Nm).

11. Install the timing belt in the following order:
   a. Around the crankshaft sprocket.
   b. Around the right idler pulley
   c. Around the camshaft sprockets
   d. Around the water pump
   e. Onto the tensioner pulley

12. Pull the lockpin out from the tensioner and install the upper timing cover. Turn the crankshaft 2 complete revolutions and check to see that the timing marks on the crankshaft and camshaft pulleys are aligned.

13. Install the timing belt covers and the fuel line clips.
14. Install the accessory belts.
15. Install the vibration damper guard and the inner fender well.
16. Install the spark plug cover and any remaining components.

2.9L (6304F) ENGINES

1992–00

1. Disconnect the negative battery cable.
2. Remove the drive belts.
3. Remove the timing belt cover.
4. Remove the splash guard, vibration damper guard and ignition coil cover.
5. Rotate the crankshaft clockwise, until the timing marks on the camshaft pulleys and timing belt mounting plate and crankshaft pulley/oil pump housing are aligned.
6. Remove the tensioner upper mounting bolts. Loosen the tensioner lower mounting bolt and twist the tensioner to free the plunger. Remove the lower mounting bolt and remove the tensioner.
7. Remove the timing belt.

→Do not rotate the crankshaft while the timing belt is removed.

8. Check the tensioner and idler pulleys, as follows:
   a. Spin the pulleys and listen for bearing noise.
   b. Check that the pulley surfaces in contact with the belt are clean and smooth.
   c. Check the tensioner pulley arm and idler pulley mountings.
   d. Tighten the tensioner pulley arm to 30 ft. lbs. (40 Nm) and the idler pulley to 18 ft. lbs. (25 Nm).
   e. Compress the tensioner using tool 5456 or equivalent. Mount the tensioner in the tool and tighten the center nut fully. Wait until compression has taken place and insert a 2mm locking pin in the plunger.

→The tensioner must be replaced if leakage is observed or the plunger offers no resistance when depressed, or cannot be depressed.

To install:

→The lever bushing must be greased every time the belt is replace or the tensioner pulley removed. This is necessary to help prevent seizure of the bushing, with the possible risk of incorrect belt tension. Service the bushing, using the following procedure:

   • Remove the lever mounting bolt, tensioner pulley and sleeve behind the bolt.
   • Grease the surfaces of the bushing, bolt and sleeve, using part 1161246-2 or equivalent.
   • Install the sleeve, tensioner pulley and lever mounting bolt.
   • Tighten the bolt 30 ft. lbs. (40 Nm).

9. Place the belt around the crankshaft pulley and right-side idler. Place the belt over the camshaft pulleys. Position the belt around the water pump and press over the tensioner pulley.

10. Insert the tensioner mounting bolts. Tighten to 18 ft. lbs. (25 Nm).

11. Remove the locking pin from the tensioner. Install the front timing belt cover.

→The lever bushing must be greased every time the belt is replace or the pulley is removed. Service the bushing, using the following procedure:

   a. Remove the lever mounting bolt, tensioner pulley and sleeve.
   b. Grease the surfaces of the bushing, bolt and sleeve, using Volvo Part No. 1161246-2 or equivalent.
   c. Install the sleeve, tensioner pulley and lever mounting bolt.
   d. Tighten the bolt to 30 ft. lbs. (40 Nm).

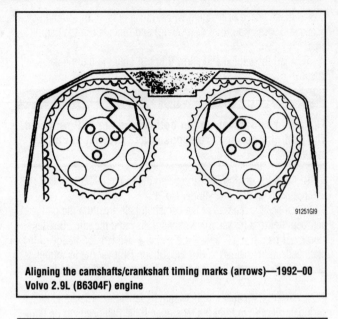

Aligning the camshafts/crankshaft timing marks (arrows)—1992–00 Volvo 2.9L (B6304F) engine

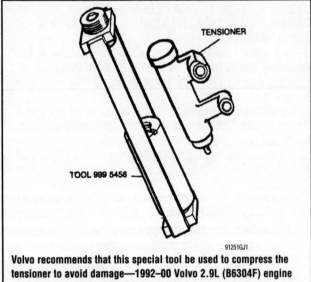

Volvo recommends that this special tool be used to compress the tensioner to avoid damage—1992–00 Volvo 2.9L (B6304F) engine

12. Turn the crankshaft 2 revolutions and check that the timing marks on the crankshaft and camshaft pulleys are correctly aligned.

13. Install the remaining components.

14. Connect the negative battery lead, start and check the engine operation.

3

TRUCKS, VANS AND SUVS

GENERAL INFORMATION

Whenever a vehicle with an unknown service history comes into your repair facility or is recently purchased, here are some points that should be asked to help prevent costly engine damage:
- Does the owner know if, or when the belt was replaced?
- If the vehicle purchased is used, or the condition and mileage

of the last timing belt replacement are unknown, it is recommended to inspect, replace or at least inform the owner that the vehicle is equipped with a timing belt.
- Note the mileage of the vehicle. The average replacement interval for a timing belt is approximately 60,000 miles (96,000 km).

INSPECTION

The average replacement interval for a timing belt is approximately 60,000 miles (96,000km). If, however, the timing belt is inspected earlier or more frequently than suggested, and shows signs of wear or defects, the belt should be replaced at that time.

✳✳ WARNING

Never allow antifreeze, oil or solvents to come into with a timing belt. If this occurs immediately wash the solution from the timing belt. Also, never excessive bend or twist the timing belt; this can damage the belt so that its lifetime is severely shortened.

Inspect both sides of the timing belt. Replace the belt with a new one if any of the following conditions exist:
- Hardening of the rubber—back side is glossy without resilience and leaves no indentation when pressed with a fingernail
- Cracks on the rubber backing
- Cracks or peeling of the canvas backing
- Cracks on rib root
- Cracks on belt sides
- Missing teeth or chunks of teeth
- Abnormal wear of belt sides—the sides are normal if they are sharp, as if cut by a knife.

If none of these conditions exist, the belt does not need replacement unless it is at the recommended interval. The belt MUST be replaced at the recommended interval.

✳✳ WARNING

On interference engines, it is very important to replace the timing belt at the recommended intervals, otherwise expensive engine damage will likely result if the belt fails.

Never bend or twist a timing belt excessively, and do not allow solvents, antifreeze, gasoline, acid or oil to come into contact with the belt

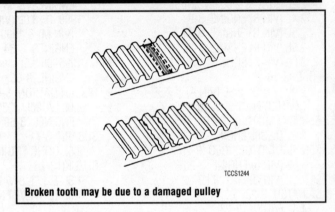

Broken tooth may be due to a damaged pulley

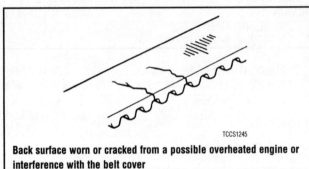

Back surface worn or cracked from a possible overheated engine or interference with the belt cover

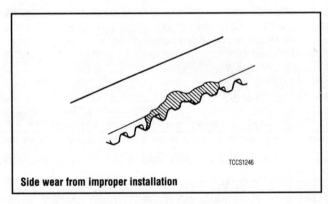

Side wear from improper installation

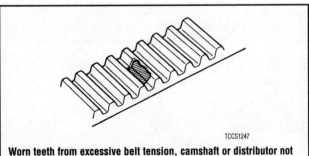

Worn teeth from excessive belt tension, camshaft or distributor not turning properly, or fluid leaking on the belt

REMOVAL & INSTALLATION

Acura

3.2L & 3.5L ENGINES

SLX

1996–00

1. Disconnect the negative battery cable.
2. Drain the engine coolant into a sealable container.
3. Remove the air cleaner assembly and intake air duct.
4. Disconnect the upper radiator hose from the coolant inlet.

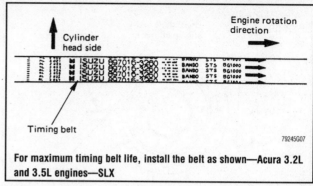

For maximum timing belt life, install the belt as shown—Acura 3.2L and 3.5L engines—SLX

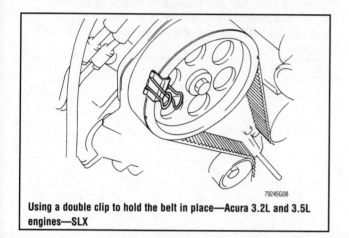

Using a double clip to hold the belt in place—Acura 3.2L and 3.5L engines—SLX

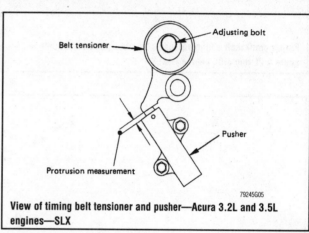

View of timing belt tensioner and pusher—Acura 3.2L and 3.5L engines—SLX

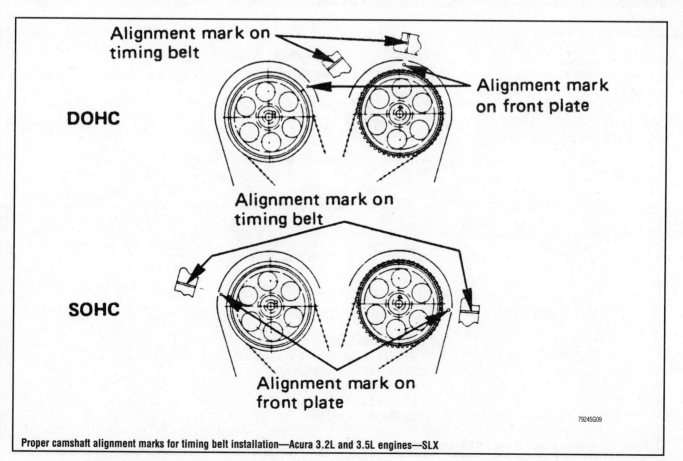

Proper camshaft alignment marks for timing belt installation—Acura 3.2L and 3.5L engines—SLX

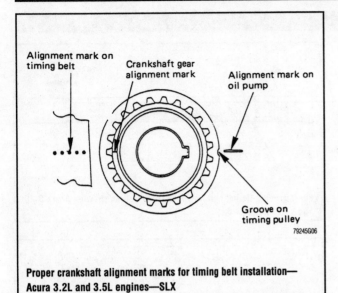

Proper crankshaft alignment marks for timing belt installation—Acura 3.2L and 3.5L engines—SLX

5. Remove the upper fan shroud from the radiator.
6. Remove the 4 nuts retaining the cooling fan assembly. Remove the cooling fan from the fan pulley.
7. Loosen and remove the drive belts.
8. Remove the upper timing belt covers.
9. Remove the fan pulley assembly.
10. Rotate the crankshaft to align the camshaft timing marks with the pointer dots on the back covers. Verify that the pointer on the crankshaft aligns with the mark on the lower timing cover.

➡ **When the timing marks are aligned, the No. 2 piston is at Top Dead Center (TDC) of the compression.**

❊❊ WARNING

Align the camshaft and crankshaft sprockets with their alignment marks before removing the timing belt. Failure to align the belt and sprocket marks may result In valve damage.

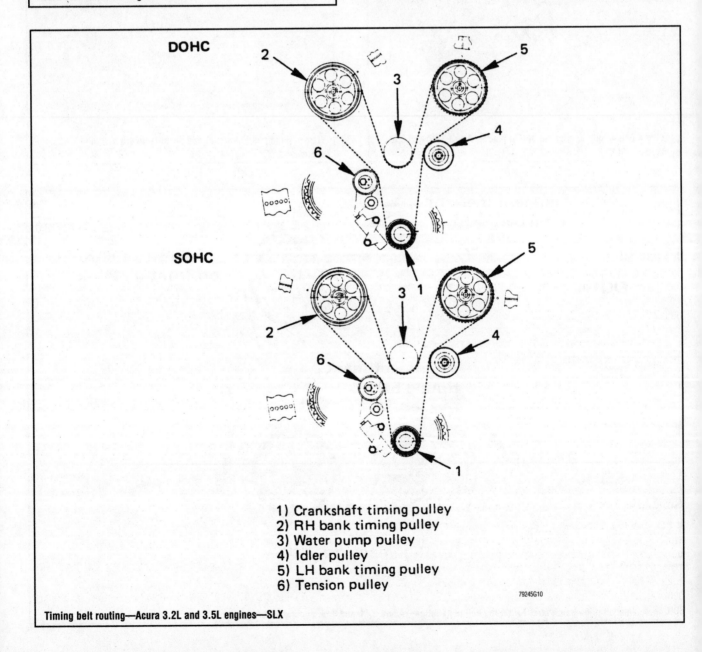

1) Crankshaft timing pulley
2) RH bank timing pulley
3) Water pump pulley
4) Idler pulley
5) LH bank timing pulley
6) Tension pulley

Timing belt routing—Acura 3.2L and 3.5L engines—SLX

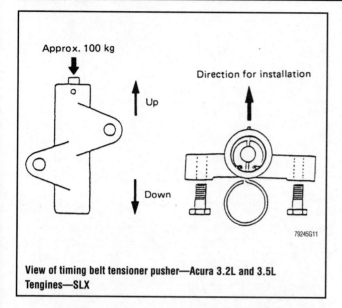

View of timing belt tensioner pusher—Acura 3.2L and 3.5L Tengines—SLX

11. Use tool No. J-8614-01 or a suitable pulley holding tool to remove the crankshaft pulley center bolt. Remove the crankshaft pulley.

12. If present, disconnect the 2 oil cooler hose bracket bolts on the timing cover. Move the oil cooler hoses and bracket off of the lower timing cover.

13. Remove the lower timing belt cover.

14. Remove the pusher assembly (tensioner) from below the belt tensioner pulley. The pusher rod must always face upward to prevent oil leakage. Depress the pusher rod, and insert a wire pin into the hole to keep the pusher rod retracted.

15. Remove the timing belt.

16. Inspect the water pump and replace it if there is any doubt about its condition.

17. Repair any oil or coolant leaks before installing a new timing belt. If the timing belt has been contaminated with oil or coolant, or is damaged, it must be replaced.

To install:

18. Verify that the sprocket timing marks are still aligned and that the groove and the keyway on the crankshaft timing sprocket align with the mark on the oil pump. The white pointers on the camshaft timing sprockets should align with the dots on the front plate.

19. Install the timing belt. Use clips to secure the belt onto each sprocket until the installation is complete. Align the dotted marks on the timing belt with the timing mark opposite the groove on the crankshaft sprocket.

➡**The arrows on the timing belt must follow the belt's direction of rotation. The manufacturer's trademark on the belt's spine should be readable left-to-right when the belt is installed.**

20. Align the white line on the timing belt with the alignment mark on the right bank camshaft timing pulley. Secure the belt with a clip.

❊❊ WARNING

If any binding is felt when adjusting the timing belt tension by turning the crankshaft, STOP turning the engine, because the pistons may be hitting the valves.

21. Rotate the crankshaft counterclockwise to remove the slack between the crankshaft sprocket and the right camshaft timing belt sprocket.

22. Install the belt around the water pump pulley.

23. Install the belt on the idler pulley.

24. Align the white alignment mark on the timing belt with the alignment mark on the left bank camshaft timing belt sprocket.

25. Install the crankshaft pulley and tighten the center bolt by hand. Rotate the crankshaft pulley clockwise to give slack between the crankshaft timing belt pulley and the right bank camshaft timing belt pulley.

26. Insert a 1.4mm piece of wire through the hole in the pusher to hold the rod in. Install the pusher assembly while pushing the tension pulley toward the belt.

27. Pull the pin out from the pusher to release the rod.

28. Remove the clamps from the sprockets. Rotate the crankshaft pulley clockwise 2 turns. Measure the rod protrusion to ensure it is between 0.16–0.24 in. (4–6mm).

29. If the tensioner pulley bracket pivot bolt was removed, tighten it to 31 ft. lbs. (42 Nm).

30. Tighten the pusher bolts to 14 ft. lbs. (19 Nm).

31. Remove the crankshaft pulley. Install the lower and upper timing belt covers and tighten their bolts to 12 ft. lbs. (17 Nm).

32. Fit the oil cooler hose onto the timing cover and tighten its mounting bracket bolts to 16 ft. lbs. (22 Nm).

33. Install the crankshaft pulley and tighten the pulley bolt to 123 ft. lbs. (167 Nm).

34. Install fan pulley assembly and tighten the bolts to 16 ft. lbs. (22 Nm).

35. Install and adjust the accessory drive belts.

36. Install the cooling fan assembly and tighten the bolts to 72 inch lbs. (8 Nm).

37. Install the upper fan shroud.

38. Install the air cleaner assembly and intake air duct.

39. Connect the upper radiator hose to the coolant inlet.

40. Refill and bleed the cooling system.

41. Connect the negative battery cable.

Chrysler

2.3L DIESEL ENGINE

1983–84 D-50 and Ram 50

1. Disconnect the negative battery cable.

2. Rotate the crankshaft to position the No. 1 cylinder to Top Dead Center (TDC) of its compression stroke. Make sure that the timing marks are aligned.

3. Remove the drive belt(s) and/or accessories that may interfere with the timing belt removal.

4. Remove the timing belt covers and the crankshaft pulley.

5. If reusing the timing belt, mark its direction of rotation of the back of the belt.

6. Slightly loosen both belt tensioner bolts, move the tensioners toward the water pump and secure them by tightening the tensioner bolts.

7. Remove the timing belt(s) and inspect for damage, wear or deterioration. Do not use solvents or detergent to clean the timing belts, sprockets or tensioners. Replace any component that is excessively contaminated with dirt, oil or grease.

To install:

8. Align the timing marks of the crankshaft and silent shaft drive sprockets.

9. Noting the direction of rotation, install the silent shaft timing belt.

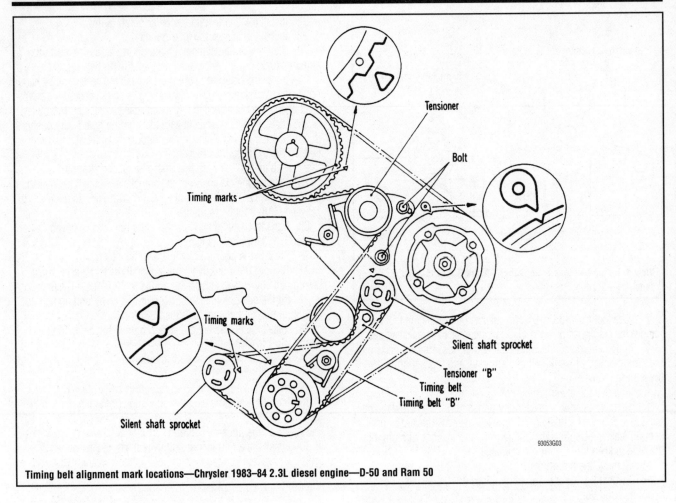

Timing belt alignment mark locations—Chrysler 1983–84 2.3L diesel engine—D-50 and Ram 50

10. Adjust the silent shaft timing belt by performing the following procedure:

a. Loosen the silent shaft belt tensioner and allow the spring tension to take up any belt slack. Do not attempt to move the tensioner to obtain more tension. Tighten the tensioner bolt to 25–28 ft. lbs. (34–38 Nm).

b. Standard belt deflection "play" when properly tensioned is

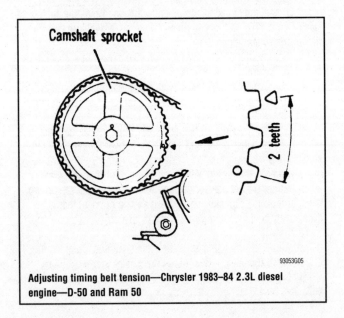

Adjusting timing belt tension—Chrysler 1983–84 2.3l diesel engine—D-50 and Ram 50

0.16–0.20 in. 4–5mm as measured at the midpoint between the upper silent shaft and crankshaft sprockets.

11. Align the timing marks for the camshaft, injection pump and crankshaft sprockets. Install the timing belt on the crankshaft sprocket, the injection pump sprocket and camshaft sprocket.

➡ With the tension side kept taut, engage the timing belt teeth with each sprocket. The injection pump sprocket tends to rotate itself, so it must be held while installing the belt.

12. Loosen the tensioner belt and allow the spring tension to take up the slack in the timing belt. Do not attempt to apply more tension by manually moving the tensioner. Once the belt is tensioned, tighten the tensioner bolt to 16–21 ft. lbs. (22–29 Nm).

➡ Tighten the slot side bolt before tightening the fulcrum bolt. If the fulcrum side is tightened earlier, the tension bracket will be turned together and the belt will be too tight.

13. Make sure the timing marks on all sprockets are still in correct alignment.

14. Turn the crankshaft in the normal direction of rotation through 2 camshaft sprocket teeth and hold it there. Reverse the direction to align the timing marks, and push down on the belt at a point halfway between the camshaft and injection pump sprocket to check the belt tension. Normal deflection is 0.16–0.20 in. (4–5mm).

15. Reset the timing belt switch by depressing its knob unit it is flush with the base and mount the timing belt covers.

16. To complete the installation, reverse the removal procedures.

17. Connect the negative battery cable.

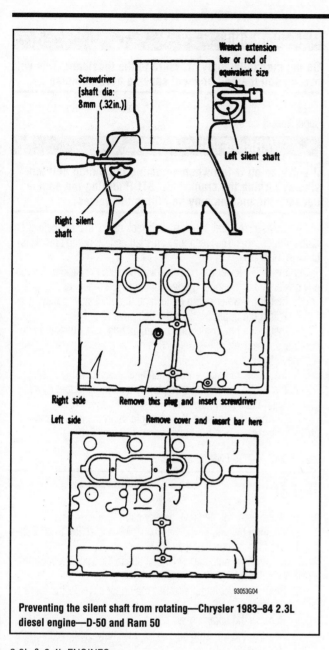

Preventing the silent shaft from rotating—Chrysler 1983–84 2.3L
diesel engine—D-50 and Ram 50

2.0L & 2.4L ENGINES

1985–90 Ram Van and Ram 50

1. Be sure that the engine's No. 1 piston is at Top Dead Center
(TDC) of the compression stroke.

✲✲✲ CAUTION

**Wait at least 90 seconds after the negative battery cable is
disconnected to prevent possible deployment of the air bag.**

2. Disconnect the negative battery cable.
3. Remove the spark plug wires from the tree on the upper cover.
4. Drain the cooling system.
5. Remove the shroud, fan and accessory drive belts.
6. Remove the radiator as required.
7. Remove the power steering pump, alternator, air conditioning
compressor, tension pulley and accompanying brackets, as
required.

8. Remove the upper front timing belt cover.
9. Remove the water pump pulley and the crankshaft pulley(s).
10. Remove the lower timing belt cover mounting screws and
remove the cover.
11. If the belt(s) are to be reused, mark the direction of rotation
on the belt.
12. Remove the timing (outer) belt tensioner and remove the
belt. Unbolt the tensioner from the block and remove.
13. Remove the outer crankshaft sprocket and flange.
14. Remove the silent shaft (inner) belt tensioner and remove the
inner belt. Unbolt the tensioner from the block and remove it.
15. To remove the camshaft sprockets, use SST MB990767-01
and MIT308239 or their equivalents.

To install:

16. Install the camshaft sprockets and tighten the center bolt to
65 ft. lbs. (90 Nm).
17. Align the timing mark of the silent shaft belt sprockets on the
crankshaft and silent shaft with the marks on the front case. Wrap
the silent shaft belt around the sprockets so there is no slack in the
upper span of the belt and the timing marks are still aligned.
18. Install the tensioner initially so the actual center of the pulley
is above and to the left of the installation bolt.
19. Move the pulley up by hand so the center span of the long
side of the belt deflects about ¼ in. (6mm).
20. Hold the pulley tightly so it does not rotate when the bolt is
tightened. Tighten the bolt to 15 ft. lbs. (20 Nm). If the pulley has
moved, the belt will be too tight.
21. Install the timing belt tensioner fully toward the water pump
and temporarily tighten the bolts. Place the upper end of the spring
against the water pump body. Align the timing marks of the cam,
crankshaft and oil pump sprockets with the corresponding marks on
the front case or head.

➡**If the following steps are not followed exactly, there is a
chance that the silent shaft alignment will be 180 degrees
off. This will cause a noticeable vibration in the engine and
the entire procedure will have to be repeated.**

22. Before installing the timing belt, ensure that the left-side
silent shaft is in the correct position.

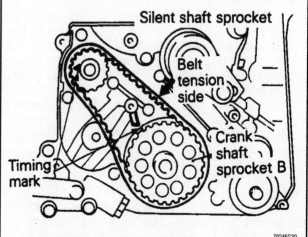

Silent shaft alignment marks. Notice the tension side of the inner
(silent shaft) belt—Chrysler 2.0L and 2.4L (VIN G) engines—Ram
Van and Ram 50

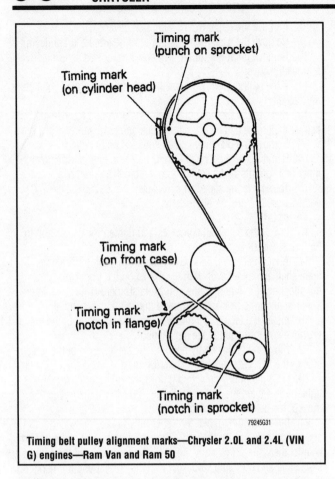

Timing mark (punch on sprocket)

Timing mark (on cylinder head)

Timing mark (on front case)

Timing mark (notch in flange)

Timing mark (notch in sprocket)

79245G31

Timing belt pulley alignment marks—Chrysler 2.0L and 2.4L (VIN G) engines—Ram Van and Ram 50

➥It is possible to align the timing marks on the camshaft sprocket, crankshaft sprocket and the oil pump sprocket with the left balance shaft out of alignment.

23. With the timing mark on the oil pump pulley aligned with the mark on the front case, check the alignment of the left balance shaft to assure correct shaft timing.

    a. Remove the plug located on the left-side of the block in the area of the starter.

    b. Insert a tool having a shaft diameter of 0.3 in. (8mm) into the hole.

    c. With the timing marks still aligned, the tool must be able to go in at least 2⅓ in. If it can only go in about 1 in. turn the oil pump sprocket 1 complete revolution.

    d. Recheck the position of the balance shaft with the timing marks realigned. Leave the tool in place to hold the silent shaft while continuing.

24. Install the belt to the crankshaft sprocket, oil pump sprocket and the camshaft sprocket, in that order. While doing so, be sure there is no slack between the sprockets except where the tensioner will take it up when released.

25. Recheck the timing marks' alignment.

26. If all are aligned, loosen the tensioner mounting bolt and allow the tensioner to apply tension to the belt.

27. Remove the tool that is holding the silent shaft in place and turn the crankshaft clockwise a distance equal to 2 teeth of the camshaft sprocket. This will allow the tensioner to automatically tension the belt the proper amount.

✳✳ WARNING

Do not manually apply pressure to the tensioner. This will overtighten the belt and will cause a howling noise.

28. First tighten the lower mounting bolt and then tighten the upper spacer bolt.

✳✳ WARNING

If any binding is felt when adjusting the timing belt tension by turning the crankshaft, STOP turning the engine, because the pistons may be hitting the valves.

29. To verify that belt tension is correct, check that the deflection of the longest span (between the camshaft and oil pump sprockets) is ½ in. (13mm).

30. Install the lower timing belt cover. Be sure the packing is properly positioned in the inner grooves of the covers when installing.

31. Install the water pump pulley and the crankshaft pulley(s).

32. Install the upper front timing belt cover.

33. Install the power steering pump, alternator, air conditioning compressor, tension pulley and accompanying brackets, as required.

34. Install the radiator, shroud, fan and accessory drive belts.

35. Install the spark plug wires to the tree on the upper cover.

36. Refill the cooling system.

37. Connect the negative battery cable. Start the engine and check for leaks.

2.2L SOHC & 2.5L ENGINES

1981–94

1. Disconnect the negative battery cable.

2. Position the engine so the No. 1 piston is at Top Dead Center (TDC) of the compression stroke.

3. Remove the nuts and bolts that attach the upper cover to the valve cover, block or cylinder head.

4. Remove the bolt that attaches the upper cover to the lower cover.

5. Remove the upper cover.

6. Remove the right tire and wheel assembly. Remove the right-side inner splash shield.

7. Remove the crankshaft pulley, water pump pulley and the accessory drive belt(s).

8. Remove the lower cover attaching bolts and the cover from the engine.

9. Remove the timing belt tensioner and allow the belt to hang free.

10. Place a floor jack under the engine and separate the right motor mount.

11. Remove the air conditioning compressor belt idler pulley, if equipped, and remove the mounting stud. Remove the compressor/alternator bracket as follows:

    a. Remove the alternator pivot bolt and remove the alternator from the bracket. Turn the alternator so the wire connections are facing up and disconnect the harness connectors from the rear of the alternator.

    b. Remove the air conditioning compressor belt idler.

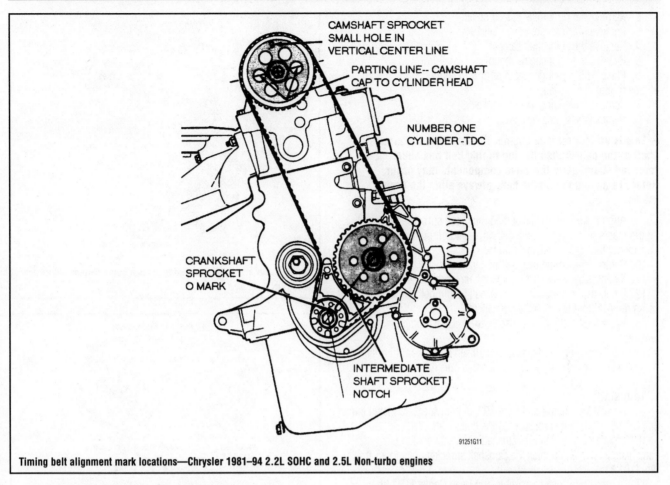

CAMSHAFT SPROCKET
SMALL HOLE IN
VERTICAL CENTER LINE

PARTING LINE-- CAMSHAFT
CAP TO CYLINDER HEAD

NUMBER ONE
CYLINDER -TDC

CRANKSHAFT
SPROCKET
O MARK

INTERMEDIATE
SHAFT SPROCKET
NOTCH

91251G11

Timing belt alignment mark locations—Chrysler 1981–94 2.2L SOHC and 2.5L Non-turbo engines

c. Remove the right engine mount yoke screw securing engine mount support strut to the engine.

d. Remove the 5 side mounting bolts retaining the bracket to the front of the engine.

e. Remove the front mounting nut. Remove the front bolt and strut and rotate the solid mount bracket away from the engine. Slide the bracket on the stud until free of the mounting studs and remove from the engine.

f. Remove the timing belt from the vehicle.

To install:

12. Turn the crankshaft sprocket and intermediate shaft sprocket until the marks are aligned. Use a straight-edge from bolt to bolt to confirm alignment.

13. Turn the camshaft until the small hole in the sprocket is at the top and the arrows on the hub are aligned with the camshaft cap to cylinder head mounting lines. When looking through the hole on top of the camshaft sprocket, the uppermost center nipple of the valve cover end seal should be at the center of the hole. Use a mirror to check the alignment of the arrows so it is viewed straight on and not at an angle from above. Install the belt but let it hang free at this point.

14. Install the air conditioning compressor/alternator bracket, idler pulley and motor mount. Remove the floor jack. Raise the vehicle and support safely. Have the tensioner at an arm's reach because the timing belt will have to be held in position with one hand.

15. To properly install the timing belt, reach up and engage it with the camshaft sprocket. Turn the intermediate shaft counterclockwise slightly, then engage the belt with the intermediate shaft sprocket. Hold the belt against the intermediate shaft sprocket and

turn clockwise to take up all tension; if the timing marks are out of alignment, repeat until alignment is correct.

16. Using a wrench, turn the crankshaft sprocket counterclockwise slightly and wrap the belt around it. Turn the sprocket clockwise so there is no slack in the belt between sprockets; if the timing marks are out of alignment, repeat until alignment is correct.

➡**If the timing marks are aligned but slack exists in the belt between either the camshaft and intermediate shaft sprockets or the intermediate and crankshaft sprockets, the timing will be incorrect when the belt is tensioned. All slack must be only between the crankshaft and camshaft sprockets.**

17. Install the tensioner and install the mounting bolt loosely. Place the special tensioning tool C-4703 on the hex of the tensioner so the weight is at about the 9 o'clock position (parallel to the ground, hanging toward the rear of the vehicle) ± 15 degrees.

18. Hold the tool in position and tighten the bolt to 45 ft. lbs. (61 Nm). Do not pull the tool past the 9 o'clock position; this will make the belt too tight and will cause it to howl or possibly break.

19. Lower the vehicle and recheck the camshaft sprocket positioning. If it is correct install the timing belt covers and all related parts.

20. Connect the negative battery cable and road test the vehicle.

2.4L (VIN B) ENGINE

➡**You may need DRB scan tool to perform the crankshaft and camshaft relearn alignment procedure.**

1. Disconnect the negative battery cable remote connection, located on the left strut tower.

2. Remove the right inner splash shield.

3. Remove the accessory drive belts.

4. Remove the crankshaft damper.

5. Remove the right engine mount.

6. Place a floor jack under the engine to support it while the engine mount is removed.

7. Remove the engine mount bracket.

8. Remove the timing belt cover.

➡This is an interference engine. Do not rotate the crankshaft or the camshafts after the timing belt has been removed. Damage to the valve components may occur. Before removing the timing belt, always align the timing marks.

9. Align the timing marks of the timing belt sprockets to the timing marks on the rear timing belt cover and oil pump cover. Loosen the timing belt tensioner bolts.

10. Remove the timing belt and the tensioner.

11. If necessary, remove the camshaft timing belt sprockets.

12. If necessary, remove the crankshaft timing belt sprocket using removal tool No. 6793 or equivalent.

13. Place the tensioner into a soft-jawed vise to compress the tensioner.

14. After compressing the tensioner, insert a pin (5⁄64 in. Allen wrench will also work) into the plunger side hole to retain the plunger until installation.

To install:

15. If necessary, using tool No. 6792, or equivalent, to install the crankshaft timing belt sprocket onto the crankshaft.

16. If necessary, install the camshaft sprockets onto the camshafts. Install and tighten the camshaft sprocket bolts to 75 ft. lbs. (101 Nm).

17. Set the crankshaft sprocket to Top Dead Center (TDC) by aligning the notch on the sprocket with the arrow on the oil pump housing.

18. Set the camshafts to align the timing marks on the sprockets.

19. Move the crankshaft to ½ notch before TDC.

20. Install the timing belt starting at the crankshaft, then around the water pump sprocket, idler pulley, camshaft sprockets and around the tensioner pulley.

21. Move the crankshaft sprocket to TDC to take up the belt slack.

22. Install the tensioner on the engine block but do not tighten.

23. Using a torque wrench on the tensioner pulley, apply 250 inch lbs. (28 Nm) of torque to the tensioner pulley.

24. With torque being applied to the tensioner pulley, move the tensioner up against the tensioner pulley bracket and tighten the fasteners to 23 ft. lbs. (31 Nm).

25. Remove the tensioner plunger pin, the tension is correct when the plunger pin can be removed and reinserted easily.

✳✳ WARNING

If any binding is felt when adjusting the timing belt tension by turning the crankshaft, STOP turning the engine, because the pistons may be hitting the valves.

26. Rotate the crankshaft 2 revolutions and recheck the timing marks. Wait several minutes and then recheck that the plunger pin can easily be removed and installed.

27. Install the front timing belt cover.

28. Install the engine mount bracket.

29. Install the right engine mount.

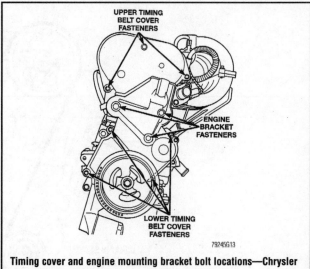

Timing cover and engine mounting bracket bolt locations—Chrysler 2.4L (VIN B) engine

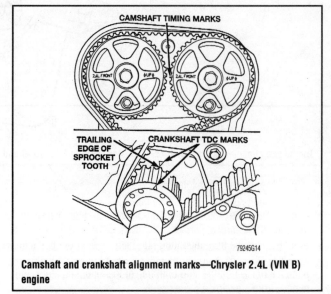

Camshaft and crankshaft alignment marks—Chrysler 2.4L (VIN B) engine

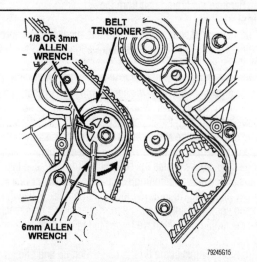

To lock the timing belt tensioner, be sure to fully insert the smaller Allen wrench into the tensioner as shown—Chrysler 2.4L (VIN B) engine

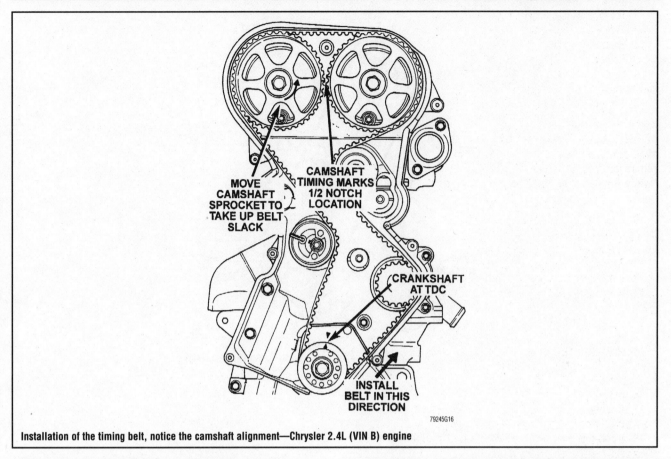

Installation of the timing belt, notice the camshaft alignment—Chrysler 2.4L (VIN B) engine

30. Remove the floor jack from under the vehicle.

31. Install the crankshaft damper and tighten it to 105 ft. lbs. (142 Nm).

32. Install the accessory drive belts and adjust to the proper tension.

33. Install the right inner splash shield.

34. Reconnect the negative battery cable.

35. Perform the crankshaft and camshaft relearn alignment procedure, using the DRB scan tool or equivalent.

2.5L (VIN G) ENGINE

1. Position the engine so that the No. 1 piston is at Top Dead Center (TDC) of the compression stroke.

2. Disconnect the negative battery cable.

3. Remove the timing belt covers.

4. Remove the timing belt tensioner and allow the belt to hang free.

5. Remove the air conditioning compressor belt idler pulley, if equipped, and remove the mounting stud. Unbolt the compressor/alternator bracket and position it to the side.

6. Remove the timing belt from the vehicle.

To install:

7. Turn the crankshaft sprocket and intermediate shaft sprocket until the marks are aligned. Use a straightedge from bolt-to-bolt to confirm alignment.

8. Turn the camshaft until the small hole in the sprocket is at the top and rows on the hub are aligned with the camshaft cap-to-cylinder head mounting lines. Use a mirror to see the alignment so it is viewed straight on and not at an angle from above. Install the belt, but let it hang free at this point.

9. Install the air conditioning compressor/alternator bracket, idler pulley and motor mount. Raise the vehicle and support safely. Have the tensioner within an arm's reach because the timing belt will have to be held in position with one hand.

10. To properly install the timing belt, reach up and engage it with the camshaft sprocket. Turn the intermediate shaft counterclockwise slightly, then engage the belt with the intermediate shaft sprocket. Hold the belt against the intermediate shaft sprocket and turn the sprocket clockwise to take up all tension; if the timing marks are out of alignment, repeat the installation until alignment is correct.

11. Using a 13mm wrench, turn the crankshaft sprocket counterclockwise slightly and wrap the belt around it. Turn the sprocket clockwise so there is no slack in the belt between the sprockets, if the timing marks are out of alignment, repeat the installation until alignment is correct.

✳✳ WARNING

If any binding is felt when adjusting the timing belt tension by turning the crankshaft, STOP turning the engine, because the pistons may be hitting the valves.

➡**If the timing marks are aligned, but slack exists in the belt between either the camshaft and intermediate shaft sprockets or the intermediate and crankshaft sprockets, the timing will be incorrect when the belt is tensioned. All slack must be between the crankshaft and camshaft sprockets only.**

12. Install the tensioner and install the mounting bolt loosely. Place special tensioning tool C-4703 or equivalent, on the hex of the tensioner so the weight is approximately at the 9 o'clock position (parallel to the ground, hanging to the left) ± 15 degrees.

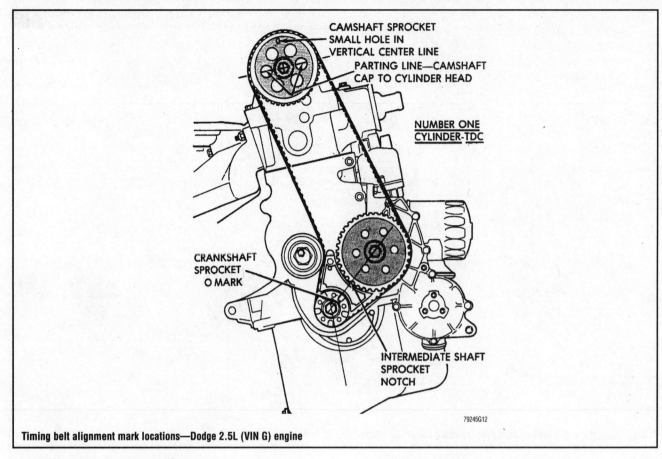

CAMSHAFT SPROCKET
SMALL HOLE IN
VERTICAL CENTER LINE
PARTING LINE—CAMSHAFT
CAP TO CYLINDER HEAD

NUMBER ONE
CYLINDER-TDC

CRANKSHAFT
SPROCKET
O MARK

INTERMEDIATE SHAFT
SPROCKET
NOTCH

79245G12

Timing belt alignment mark locations—Dodge 2.5L (VIN G) engine

13. Hold the tool in position and tighten the bolt to 45 ft. lbs. (61 Nm). Do not pull the tool past the 9 o'clock position; this will make the belt too tight and will cause it to howl or possibly break during engine use.

14. Lower the vehicle and recheck the camshaft sprocket positioning. If it is correct, install the timing belt covers and all related parts.

15. Connect the negative battery cable and road test the vehicle.

2.5L (VIN K) ENGINE

1986–95

Working on any engine (especially overhead camshaft engines) requires much care be given to valve timing. It is good practice to set the engine up to Top Dead Center (TDC) of the compression stroke of the No. 1 cylinder firing position. Verify that all timing marks on the crankshaft and camshaft sprockets are properly aligned before removing the timing belt. This serves as a point of reference for all work that follows. Valve timing is most important and engine damage will result if the work is incorrect.

➡This is an interference engine. Do not rotate the crankshaft or the camshafts after the timing belt has been removed. Damage to the valve components may occur. Before removing the timing belt, always align the timing marks.

1. Disconnect the negative battery cable.
2. Remove the accessory drive belts.
3. Remove the right engine mount yoke screw.
4. Remove the air conditioning compressor and set it aside. Remove the solid mount compressor bracket mounting bolts.

5. Turn the solid mount bracket away from the engine and slide it on the No. 2 stud until it is free. The front bolt and spacer will be removed with the bracket.

6. Remove the alternator and the drive belt idler.

7. Raise and safely support the vehicle. Remove the right inner fender splash shield.

8. Loosen and remove the 3 water pump pulley mounting bolts and remove the pulley.

9. Remove the 4 crankshaft pulley retaining bolts and the crankshaft pulley.

10. Remove the nuts at the upper portion of the timing cover and the bolts from the lower portion, then remove both halves of the cover.

11. Remove the timing belt covers.

12. Position a jack under the engine.

13. Separate the right engine mount and raise the engine slightly.

14. Loosen the timing belt tensioner bolt, rotate the hex nut, and remove timing belt.

15. Remove the timing belt tensioner, if necessary.

16. Remove the crankshaft sprocket with a suitable puller tool and a bolt approximately 6 in. (15 cm) long.

17. Remove the camshaft sprocket and intermediate shaft sprocket, if necessary.

To install:

18. Clean all parts well. A small amount of white paint on the sprocket timing marks may make alignment easier.

19. Install the crankshaft sprocket. Tighten the crankshaft sprocket bolt to 85 ft. lbs. (115 Nm).

20. If necessary, turn the crankshaft and intermediate shaft until markings on both sprockets are aligned.

21. Rotate the camshaft so the arrows on the hub are aligned with the No. 1 camshaft cap-to-cylinder head line. The small hole in the cam sprocket should be centered in the vertical center line.

22. If removed, install the timing belt tensioner.

23. Install the timing belt over the drive sprockets and adjust.

24. Tighten the tensioner by turning the tensioner hex to the right. Tension should be correct when the belt can be twisted 90 degrees with the thumb and forefinger, midway between the camshaft and intermediate sprocket.

✳✳ WARNING

If any binding is felt when adjusting the timing belt tension by turning the crankshaft, STOP turning the engine, because the pistons may be hitting the valves.

25. Turn the engine clockwise from TDC, 2 complete revolutions with the crankshaft bolt. Check the timing marks for correct alignment.

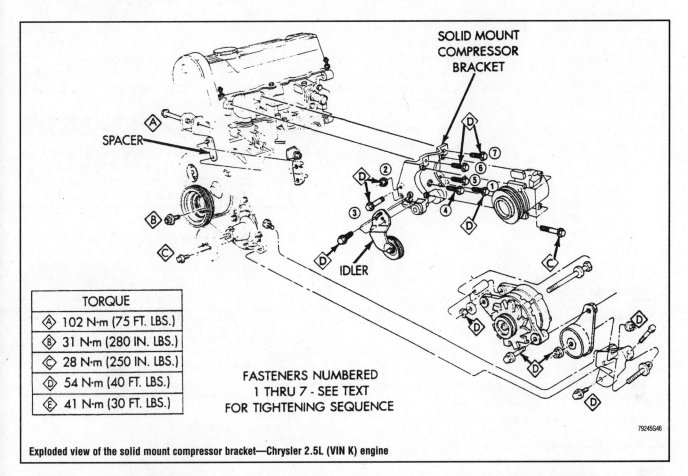

TORQUE	
Ⓐ	102 N·m (75 FT. LBS.)
Ⓑ	31 N·m (280 IN. LBS.)
Ⓒ	28 N·m (250 IN. LBS.)
Ⓓ	54 N·m (40 FT. LBS.)
Ⓔ	41 N·m (30 FT. LBS.)

FASTENERS NUMBERED 1 THRU 7 - SEE TEXT FOR TIGHTENING SEQUENCE

Exploded view of the solid mount compressor bracket—Chrysler 2.5L (VIN K) engine

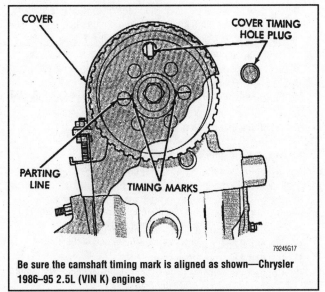

Be sure the camshaft timing mark is aligned as shown—Chrysler 1986–95 2.5L (VIN K) engines

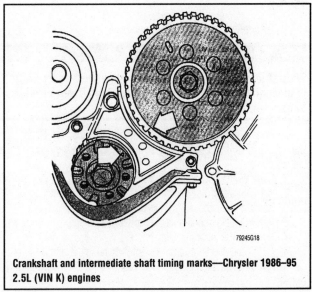

Crankshaft and intermediate shaft timing marks—Chrysler 1986–95 2.5L (VIN K) engines

❋❋ WARNING

Do not use the camshaft or intermediate shaft to rotate the engine. Do not allow oil or solvent to contact the timing belt as they will deteriorate the belt and cause slipping.

26. Tighten the locknut on the tensioner, while holding the weighted wrench tool C-4503 or equivalent, in position, to 45 ft. lbs. (61 Nm).

27. Lower the engine onto the right engine mount and install the fasteners. Remove the support from the engine.

28. Some engines use a foam stuffer block inside the timing belt housing. Inspect the foam block's condition and position. The stuffer block should be intact and secure within the engine bracket tunnel.

29. Install the timing belt cover. Secure the upper section to the cylinder head with nuts and the lower section to the cylinder block with screws. Tighten all of the timing belt cover fasteners to 40 inch lbs. (4 Nm).

30. Check valve timing again. With the timing belt cover installed, and with No. 1 cylinder at TDC, the small hole in the sprocket must be centered in the timing belt cover hole. If the hole is not aligned correctly, perform the timing belt installation procedure again.

31. Install the water pump pulley and the crankshaft pulley. Tighten the water pump pulley bolts to 250 inch lbs. (28 Nm). Tighten the crankshaft pulley bolts to 280 inch lbs. (31 Nm).

32. Install the inner fender splash shield. Lower the vehicle.

33. Install the solid mount compressor bracket. The bracket mounting fasteners must be tightened to 40 ft. lbs. (54 Nm).

34. Install the alternator and drive belt idler. Tighten mounting bolts to 40 ft. lbs. (54 Nm).

35. Install the right engine mount yoke bolt and tighten to 100 ft. lbs. (133 Nm).

36. Install the accessory drive belts and adjust them to the proper tension.

➡**With the timing belt cover installed and the piston in the No. 1 cylinder at TDC, the small hole in the cam sprocket should be centered in timing belt cover hole.**

37. Reconnect the negative battery cable. Road test the vehicle.

3.0L (VIN 3) SOHC ENGINE

1986–94

1. Disconnect the negative battery cable.

❋❋ CAUTION

Wait at least 90 seconds after the negative battery cable is disconnected to prevent possible deployment of the air bag.

2. Position the engine so the No. 1 cylinder is at Top Dead Center (TDC) of the compression stroke. Disconnect the negative battery cable.

3. Remove the engine undercover.

4. Remove the cruise control actuator.

5. Remove the accessory drive belts.

6. Remove the air conditioner compressor tension pulley assembly.

7. Remove the tension pulley bracket.

8. Using the proper equipment, slightly raise the engine to take the weight off the side engine mount. Remove the engine mounting bracket.

9. Disconnect the power steering pump pressure switch connector. Remove the power steering pump with hoses attached and wire aside.

10. Remove the engine support bracket.

11. Remove the crankshaft pulley.

12. Remove the timing belt cover cap.

13. Remove the timing belt upper and lower covers.

14. If the same timing belt will be reused, mark the direction of the timing belt's rotation for installation in the same direction. Make sure the engine is positioned so the No. 1 cylinder is at the Top Dead Center (TDC) of its compression stroke and the timing marks are aligned with the engine's timing mark indicators.

15. Loosen the timing belt tensioner bolt and remove the belt. If the tensioner is not being removed, position it as far away from the center of the engine as possible and tighten the bolt.

16. If the tensioner is being removed, paint the outside of the spring to ensure that it is not installed backwards. Unbolt the tensioner and remove it along with the spring.

❋❋ WARNING

Do not rotate the camshafts when the timing belt is removed from the engine. Turning the camshaft when the timing belt is removed could cause the valves to interfere with the pistons thus causing severe internal engine damage.

To install:

17. Install the tensioner, if removed, and hook the upper end of the spring to the water pump pin and the lower end to the tensioner in exactly the same position as originally installed. If not already done, position both camshafts so the marks align with those on the rear. Rotate the crankshaft so the timing mark aligns with the mark on the front cover.

18. Install the timing belt on the crankshaft sprocket and while keeping the belt tight on the tension side, install the belt on the front camshaft sprocket.

19. Install the belt on the water pump pulley, then the rear camshaft sprocket and the tensioner.

20. Rotate the front camshaft counterclockwise to tension the belt between the front camshaft and the crankshaft. If the timing marks became misaligned, repeat the procedure.

21. Install the crankshaft sprocket flange.

22. Loosen the tensioner bolt and allow the spring to apply tension to the belt.

23. Turn the crankshaft 2 full turns in the clockwise direction until the timing marks align again. Now that the belt is properly tensioned, torque the tensioner lock bolt to 21 ft. lbs. (29 Nm). Measure the belt tension between the rear camshaft sprocket and the crankshaft with belt tension gauge. The specification is 46–68 lbs. (210–310 N).

24. Install the timing covers. Make sure all pieces of packing are positioned in the inner grooves of the covers when installing.

25. Install the crankshaft pulley. Tighten the bolt to 108–116 ft. lbs. (150–160 Nm).

26. Install the engine support bracket.

27. Install the power steering pump and reconnect wire harness at the power steering pump pressure switch.

28. Install the engine mounting bracket and remove the engine support fixture.

29. Install the tension pulleys and drive belts.

30. Install the cruise control actuator.

31. Install the engine undercover.

32. Connect the negative battery cable and road test the vehicle.

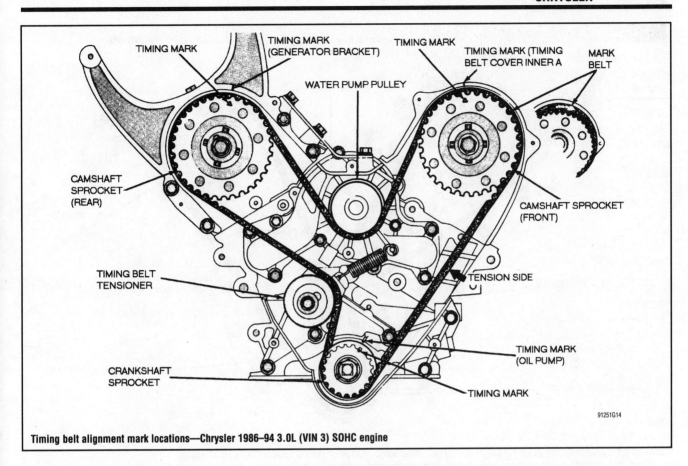

Timing belt alignment mark locations—Chrysler 1986–94 3.0L (VIN 3) SOHC engine

1995–00

The timing belt can be inspected by removing the upper front outer timing belt cover.

Working on any engine (especially overhead camshaft engines) requires much care be given to valve timing. It is good practice to set the engine up at Top Dead Center (TDC) of the compression stroke of the No. 1 cylinder firing position before beginning work. Verify that all timing marks on the crankshaft and camshaft sprockets are properly aligned before removing the timing belt and starting camshaft service. This serves as a point of reference for all work that follows. Valve timing is very important and engine damage will result if the work is incorrect.

1. Disconnect the negative battery cable.

❋❋ CAUTION

Wait at least 90 seconds after the negative battery cable is disconnected to prevent possible deployment of the air bag.

2. Remove the accessory drive belts. Remove the engine mount insulator from the engine support bracket.

3. Remove the engine support bracket. Remove the crankshaft pulleys and torsional damper. Remove the timing belt covers.

4. Rotate the crankshaft until the sprocket timing marks are aligned. The crankshaft sprocket timing mark should align with the oil pump timing mark. The rear camshaft sprocket timing mark should align with the generator bracket timing mark and the front camshaft sprocket timing mark should align with the inner timing belt cover timing mark.

5. If the belt is to be reused, mark the direction of rotation on the belt for installation reference.

6. Loosen the timing belt tensioner bolt and remove the timing belt.

❋❋ WARNING

Do not rotate the camshafts when the timing belt is removed from the engine. Turning the camshaft when the timing belt is removed could cause the valves to interfere with the pistons thus causing severe internal engine damage.

7. If necessary, remove the timing belt tensioner.

8. Remove the crankshaft sprocket flange shield and crankshaft sprocket.

9. Hold the camshaft sprocket using spanner tool MB990775 or equivalent, and remove the camshaft sprocket bolt and washer. Remove the camshaft sprocket.

To install:

10. Install the camshaft sprocket on the camshaft with the retaining bolt and washer. Hold the camshaft sprocket using spanner tool MB990775 or equivalent, and tighten the bolt to 70 ft. lbs. (95 Nm).

11. Install the crankshaft sprocket.

12. If removed, install the timing belt tensioner and tensioner spring. Hook the spring upper end to the water pump pin and the lower end to the tensioner bracket with the hook out.

13. Turn the timing belt tensioner counterclockwise full travel in the adjustment slot and tighten the bolt to temporarily hold it in this position.

14. Rotate the crankshaft sprocket until its timing mark is aligned with the oil pump timing mark.

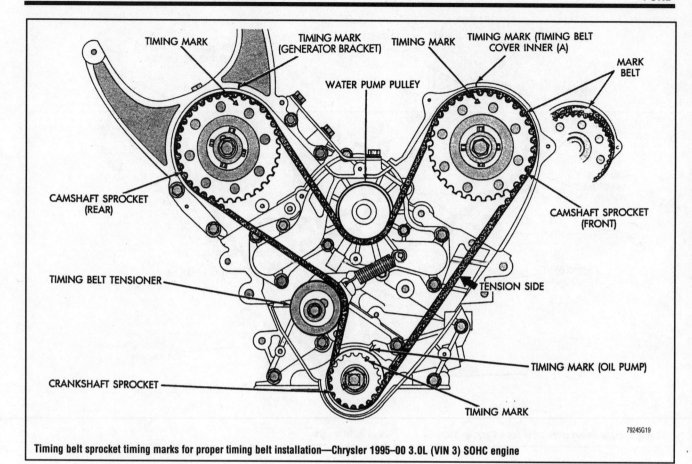

Timing belt sprocket timing marks for proper timing belt installation—Chrysler 1995–00 3.0L (VIN 3) SOHC engine

79245G19

15. Rotate the rear camshaft sprocket until its timing mark is aligned with the timing mark on the generator bracket.

16. Rotate the front (radiator side) camshaft sprocket until its mark is aligned with the timing mark on the inner timing belt cover.

17. Install the timing belt on the crankshaft sprocket while keeping the belt tight on the tension side.

➡If the original belt is being reused, be sure to install it in the same rotational direction.

18. Position the timing belt over the front camshaft sprocket (radiator side). Next, position the belt under the water pump pulley, then over the rear camshaft sprocket and finally over the tensioner.

☼☼ WARNING

If any binding is felt when adjusting the timing belt tension by turning the crankshaft, STOP turning the engine, because the pistons may be hitting the valves.

19. Apply rotating force in the opposite direction to the front camshaft sprocket (radiator side) to create tension on the timing belt tension side. Check that all timing marks are aligned.

20. Install the crankshaft sprocket flange.

21. Loosen the tensioner bolt and allow the tensioner spring to tension the belt.

22. Rotate the crankshaft 2 full turns in a clockwise direction. Turn the crankshaft smoothly and in a clockwise direction only.

23. Again align the timing marks. If all marks are aligned, tighten the tensioner bolt to 250 inch lbs. (28 Nm). Otherwise repeat the installation procedure.

24. Install the timing belt covers. Install the engine support bracket. Tighten the support bracket mounting bolts to 35 ft. lbs. (47 Nm).

25. Install the engine mount insulator, torsional damper and crankshaft pulleys. Tighten the crankshaft pulley bolt to 112 ft. lbs. (151 Nm).

26. Install the accessory drive belts and adjust them to the proper tension.

27. Reconnect the negative battery cable.

28. Run the engine and check for proper operation. Road test the vehicle.

Ford

2.0L (VIN C), 2.3L (VIN A)
& 2.5L (VIN C) ENGINES

1983–94

1. Disconnect the negative battery cable and drain the cooling system. Remove the 4 water pump pulley bolts.

2. Remove the automatic belt tensioner and accessory drive belt. Remove the upper radiator hose.

3. Remove the crankshaft pulley bolt and pulley. Remove the thermostat housing and gasket.

4. Remove the timing belt outer cover retaining bolt(s). Release the cover interlocking tabs, if equipped, and remove the cover.

5. Loosen the belt tensioner adjustment screw, position belt tensioner tool T74P-6254-A or equivalent, on the tension spring roll

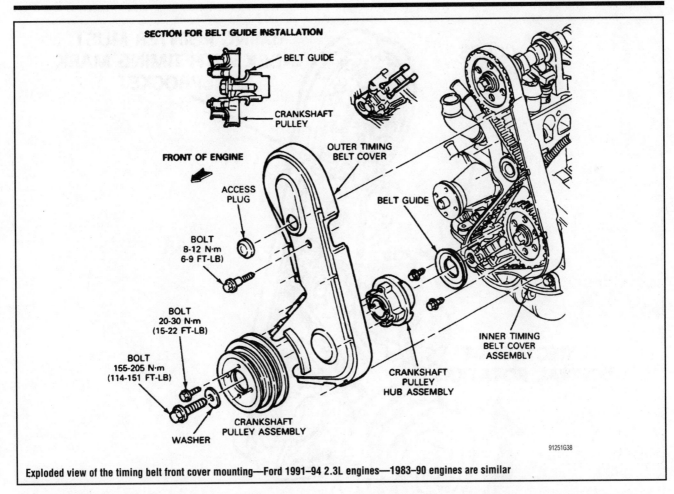

Exploded view of the timing belt front cover mounting—Ford 1991–94 2.3L engines—1983–90 engines are similar

pin and release the belt tensioner. Tighten the adjustment screw to hold the tensioner in the released position.

6. On 1991–94 vehicles, remove the bolts holding the timing sensor in place and pull the sensor assembly free of the dowel pin.

7. Remove the crankshaft pulley, hub and belt guide. Remove the timing belt. If the belt is to be reused, mark the direction of rotation so it may be reinstalled in the same direction.

To install:

8. Position the crankshaft sprocket to align with the TDC mark and the camshaft sprocket to align with the camshaft timing pointer. On 1983–90 vehicles, remove the distributor cap and set the rotor to the No. 1 firing position by turning the auxiliary shaft.

9. Install the timing belt over the crankshaft sprocket and then counterclockwise over the auxiliary and camshaft sprockets. Align the belt fore-and-aft on the sprockets.

10. Loosen the tensioner adjustment bolt to allow the tensioner to move against the belt. If the spring does not have enough tension to move the roller against the belt, it may be necessary to manually push the roller against the belt and tighten the bolt.

11. To make sure the belt does not jump time during rotation in Step 10, remove a spark plug from each cylinder.

12. Rotate the crankshaft 2 complete turns in the direction of normal rotation to remove the slack from the belt. Tighten the tensioner adjustment to 29–40 ft. lbs. (40–55 Nm) and pivot bolts to 14–22 ft. lbs. (20–30 Nm). Check the alignment of the timing marks.

13. Install the crankshaft belt guide.

14. On 1983–90 vehicles, install the crankshaft pulley and

tighten the retaining bolt to 103–133 ft. lbs. (140–180 Nm). On 1991–94 vehicles, proceed as follows:

a. Install the timing sensor onto the dowel pin and tighten the 2 longer bolts to 14–22 ft. lbs. (20–30 Nm).

b. Rotate the crankshaft 45 degrees counterclockwise and install the crankshaft pulley and hub assembly. Tighten the bolt to 114–151 ft. lbs. (155–205 Nm).

c. Rotate the crankshaft 90 degrees clockwise so the vane of the crankshaft pulley engages with timing sensor positioner tool T89P-6316-A or equivalent. Tighten the 2 shorter sensor bolts to 14–22 ft. lbs. (20–30 Nm).

d. Rotate the crankshaft 90 degrees counterclockwise and remove the sensor positioner tool.

e. Rotate the crankshaft 90 degrees clockwise and measure the outer vane to sensor air gap. The air gap must be 0.018–0.039 in. (0.458–0.996mm).

15. Position the timing belt front cover. Snap the interlocking tabs into place, if necessary. Install the timing belt outer cover retaining bolt(s) and tighten to 71–106 inch lbs. (8–12 Nm).

16. Install the thermostat housing and a new gasket. Install the upper radiator hose.

17. Install the crankshaft pulley and retaining bolt. Tighten to 103–133 ft. lbs. (140–180 Nm) on 1983–90 vehicles or 114–151 ft. lbs. (155–205 Nm) on 1991–94 vehicles.

18. Install the water pump pulley and the automatic belt tensioner. Install the accessory drive belt.

19. Install the spark plugs and remaining components.

20. Connect the negative battery cable, start the engine and check the ignition timing.

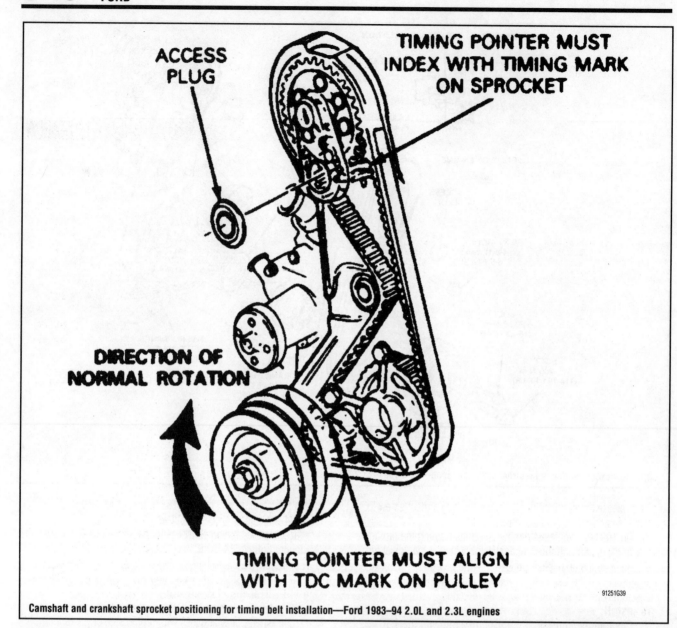

ACCESS PLUG

TIMING POINTER MUST INDEX WITH TIMING MARK ON SPROCKET

DIRECTION OF NORMAL ROTATION

TIMING POINTER MUST ALIGN WITH TDC MARK ON PULLEY

91251G39

Camshaft and crankshaft sprocket positioning for timing belt installation—Ford 1983–94 2.0L and 2.3L engines

1995–00

1. Rotate the engine so that No. 1 cylinder is at Top Dead Center (TDC) of the compression stroke. Check that the timing marks are aligned on the camshaft and crankshaft pulleys. An access plug is provided in the cam belt cover so that the camshaft timing can be checked without removal of the cover or any other parts. Set the crankshaft to TDC by aligning the timing mark on the crank pulley with the TDC mark on the belt cover. Look through the access hole in the belt cover to be sure that the timing mark on the cam drive sprocket is aligned with the pointer on the inner belt cover.

➡Always turn the engine in the normal direction of rotation. Backward rotation may cause the timing belt to jump time, due to the arrangement of the belt tensioner.

2. Drain cooling system. Remove the upper radiator hose as necessary. Remove the fan blade and water pump pulley bolts.

✳✳ CAUTION

Always drain the coolant into a sealable container. Coolant should be reused unless it is contaminated or several years old.

3. Loosen the alternator retaining bolts and remove the drive belt from the pulleys. Remove the water pump pulley.

4. Remove the power steering pump and set it aside.

5. Remove the 4 timing belt outer cover retaining bolts and remove the cover. Remove the crankshaft pulley and belt guide.

6. Loosen the belt tensioner pulley assembly, then position a Camshaft Belt Adjuster tool T74P-6254-A or equivalent, on the tension spring rollpin and retract the belt tensioner away from the timing belt. Tighten the adjustment bolt to lock the tensioner in the retracted position.

7. If the belt is to be reused, mark the direction of rotation on the belt for installation reference.

8. Remove the timing belt.
To install:
9. Install the new belt over the crankshaft sprocket and then counterclockwise over the auxiliary and camshaft sprockets, making sure the lugs on the belt properly engage the sprocket teeth on the pulleys. Be careful not to rotate the pulleys when installing the belt.
10. Release the timing belt tensioner pulley, allowing the tensioner to take up the belt slack. If the spring does not have enough tension to move the roller against the belt (belt hangs loose), it might be necessary to manually push the roller against the belt and tighten the bolt.

➡ The spring cannot be used to set belt tension; a wrench must be used on the tensioner assembly.

⁜ WARNING

If any binding is felt when adjusting the timing belt tension by turning the crankshaft, STOP turning the engine, because the pistons may be hitting the valves.

11. Rotate the crankshaft 2 complete turns by hand (in the normal direction of rotation) to remove slack from the belt. Tighten the tensioner adjustment to 26–33 ft. lbs. (35–45 Nm) and pivot bolts to 30–40 ft. lbs. (40–55 Nm). Be sure the belt is seated properly on the pulleys and that the timing marks are still in alignment when No. 1 cylinder is again at TDC/compression.
12. Install the crankshaft pulley and belt guide.

13. Install the timing belt cover.
14. Install the water pump pulley and fan blades. Install the upper radiator hose if necessary. Refill the cooling system.
15. Install the accessory drive belts.
16. Start the engine and check the ignition timing. Adjust the timing, if necessary.

General Motors

2.2L DIESEL ENGINE

1982–85

1. Disconnect the negative battery cable.
2. Drain the cooling system.
3. Remove the radiator.
4. If equipped, remove the air conditioning compressor drive belt by moving the power steering pump or idler pulley.
5. Loosen the alternator adjusting plate bolt and fixing bolt, then remove the fan belt.
6. Remove the 4 crankshaft pulley-to-crankshaft bolts and the pulley.
7. Remove the timing cover bolts and the covers.

Aligning the timing marks—General Motors 2.2L diesel engine

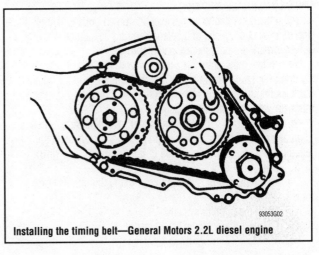
Installing the timing belt—General Motors 2.2L diesel engine

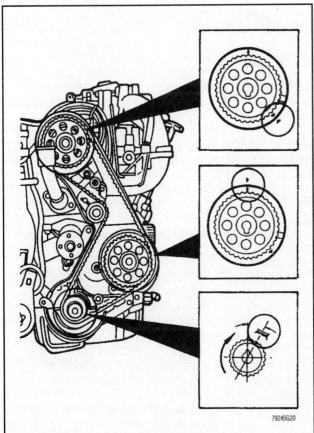
Camshaft, auxiliary shaft and crankshaft timing belt sprocket alignment mark locations—Ford 1995–00 2.3L and 2.5L (VIN C) engines

8. Remove the injection pump timing pulley flange bolts and the flange.

9. Remove the tension spring.

➡ **When removing the tension spring, avoid using excess force or distortion of the spring will result.**

10. Remove the tension pulley nut, the tension pulley and the tension center.

11. If reusing the timing belt, mark its rotational direction on the back of the belt.

12. Remove the timing belt.

To install:

✳✳ WARNING

No attempt should be made to readjust the belt tension. If the belt has been loosened through service of the timing system, it should be replaced with a new one.

13. Rotate the crankshaft sprocket, the injection pump timing sprocket and camshaft sprocket so that they are aligned with their timing marks.

14. Noting the direction of rotation, install the timing belt in the following sequence: crankshaft sprocket, camshaft sprocket and injection pump sprocket.

➡ **Make an adjustment so that the belt slack is taken up by the tension pulley.**

15. Install the tension center and tension pulley by making sure that the end of the tension center is in proper contact with 2 pins on the timing pulley housing.

16. Hand-tighten the nut, so that tension pulley can slide freely.

17. Install the tension spring correctly and semi-tighten the tension pulley fixing nut.

18. Turn the crankshaft 2 turns in the normal direction of rotation to permit seating of the belt. Further rotate the crankshaft 90 degrees beyond Top Dead Center (TDC) to settle the injection pump.

✳✳ WARNING

Never attempt to turn the crankshaft in the reverse direction.

19. Loosen the tension pulley fixing nut completely, allowing the pulley to take up looseness of the belt. Then, tighten the nut to 78–95 ft. lbs. (106–129 Nm).

20. Install the flange on the injection pump pulley; the hole in the outer circumference of the flange should be aligned with the triangular mark on the injection pump sprocket.

21. Rotate the crankshaft 2 turns in normal direction of rotation to bring the No. 1 piston to TDC on the compression stroke and recheck that the triangular mark on the injection pump sprocket is aligned with the hole in the flange.

22. Using tool J-29771 or equivalent, check the timing belt tension at a point between the injection pump and crankshaft sprockets; belt tension should be 33–55 ft. lbs. (45<\#208>75 Nm).

23. Adjust the valve clearances.

24. Complete the installation by reversing the removal procedures.

25. Refill the cooling system.

26. Connect the negative battery cable.

General Motors/GEO

1.6L (VIN 6) ENGINE

The 1.6L 16-valve engine is known as an interference motor, because it is fabricated with such close tolerances between the pistons and valves that, if the timing belt is incorrectly positioned, jumps teeth on one of the sprockets or breaks, the valve and pistons will come into contact. This can cause severe internal engine damage.

➡ **Do not rotate the crankshaft counterclockwise or attempt to rotate the crankshaft by turning the camshaft sprocket.**

1. Remove the timing belt cover.

2. If the timing belt is not already marked with a directional arrow, use white paint, a grease pencil or correction fluid to do so.

3. Rotate the crankshaft clockwise until the timing mark on the camshaft sprocket and the **V** mark on the timing belt inside cover are aligned, and the punch mark on the crankshaft sprocket is aligned with the mark on the engine.

✳✳ WARNING

Do not rotate the crankshaft or camshaft once the timing belt is removed, because the valves and pistons can come into contact, which may cause internal engine damage.

4. Disconnect one end of the tensioner spring. Loosen the timing belt tensioner bolt and stud, then, using your finger, press the tensioner plate up and remove the timing belt from the crankshaft and camshaft sprockets.

5. Remove the timing belt tensioner, tensioner plate and spring from the engine.

6. Install Suzuki tool 09917-68220 or equivalent, onto the camshaft sprocket to hold the camshaft from rotating. Loosen the camshaft sprocket retaining bolt, then pull the camshaft sprocket off of the end of the camshaft.

7. Remove the crankshaft timing belt sprocket by loosening the center bolt, while preventing the crankshaft from rotating. To hold the crankshaft from turning, use Suzuki tool 09927-56010 or equivalent, or a large prybar inserted in the transmission housing slot and the flywheel teeth. Pull the sprocket off of the end of the crankshaft. Be sure to retain the crankshaft sprocket key and belt guide for assembly.

8. If necessary, remove the timing belt inside cover from the cylinder head.

To install:

9. If necessary, install the timing belt inside cover.

10. Slide the timing belt guide on the crankshaft so that the concave side faces the oil pump, then install the sprocket key in the groove in the crankshaft.

11. Slide the pulley onto the crankshaft, and install the center retaining bolt. Tighten the center bolt to 80 ft. lbs. (110 Nm). To hold the crankshaft from turning, use Suzuki tool 09927-56010 or equivalent, or a large prybar inserted in the transmission housing slot and the flywheel teeth.

12. Install the timing belt camshaft sprocket, ensuring that the slot in the sprocket engages the camshaft (pulley) pin; this ensures that the sprocket is properly positioned on the end of the camshaft. Secure the camshaft with the holding tool used during removal, then tighten the sprocket bolt to 44 ft. lbs. (60 Nm).

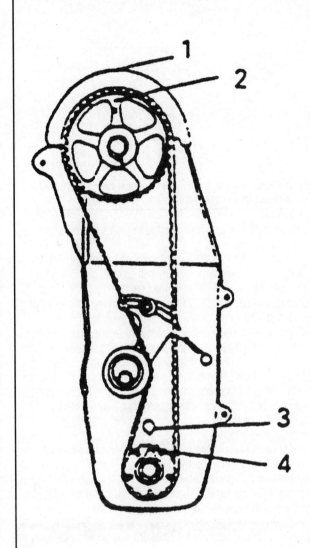

1. "V" mark on cylinder head cover
2. Timing mark by "E" on camshaft timing belt pulley
3. Arrow mark on oil pump case
4. Punch mark on crankshaft timing belt pulley

79245G47

Rotate the crankshaft clockwise until the camshaft and crankshaft timing marks are aligned—General Motors/GEO 1.6L 16-valve engine

13. Assemble the timing belt tensioner plate and the tensioner, making sure that the lug of the tensioner plate engages the tensioner.

⁕⁕ WARNING

If any binding is felt when adjusting the timing belt tension by turning the crankshaft, STOP turning the engine, because the pistons may be hitting the valves.

14. Install the timing belt tensioner, tensioner plate and spring on the engine. Tighten the mounting bolt and stud only finger-tight at this time. Ensure that when the tensioner is moved in a counter-clockwise direction, the tensioner moves in the same direction. If the tensioner does not move, remove it and the tensioner plate to reassemble them properly.

15. Loosen all rocker arm valve lash locknuts and adjusting screws. This will permit movement of the camshaft without any rocker arm associated drag, which is essential for proper timing belt tensioning. If the camshaft does not rotate freely (free of rocker arm

drag), the belt will not be properly tensioned.

16. Rotate the camshaft sprocket clockwise until the timing mark on the sprocket and the V mark on the timing belt inside cover are aligned.

17. Using a wrench, or socket and breaker bar, on the crankshaft sprocket center bolt, turn the crankshaft clockwise until the punch mark on the sprocket is aligned with the arrow mark on the oil pump.

18. With the camshaft and crankshaft marks properly aligned, push the tensioner up with your finger and install the timing belt on the 2 sprockets, ensuring that the drive side of the belt is free of all slack. Release the tensioner. Be sure to install the timing belt so that the directional arrow is pointing in the appropriate direction.

➥**In this position, the No. 4 cylinder is at Top Dead Center (TDC) on the compression stroke.**

19. Rotate the crankshaft clockwise 2 full revolutions, then tighten the tensioner stud to 97 inch lbs. (11 Nm). Then, tighten the tensioner bolt to 18 ft. lbs. (24 Nm).

20. Ensure that all 4 timing marks are still aligned as before; if they are not, remove the timing belt, and install and tension it again.

21. Install the timing belt cover and all related components.

1.6L (VIN U) ENGINE

1989-95

➡**Do not rotate the crankshaft counterclockwise or attempt to rotate the crankshaft by turning the camshaft sprocket.**

1. Remove the timing belt cover.
2. Remove rocker arm cover.
3. If the timing belt is not already marked with a directional arrow, use white paint, a grease pencil or correction fluid to do so.
4. Disconnect one end of the tensioner spring. Loosen the timing belt tensioner bolt and stud, then, using your finger, press the tensioner plate up and remove the timing belt from the crankshaft and camshaft sprockets.
5. Remove the timing belt tensioner, tensioner plate and spring from the engine.
6. Insert a metal rod through the hole in the camshaft to lock the camshaft from rotating. Loosen the camshaft sprocket retaining bolt, then pull the camshaft sprocket off of the end of the camshaft.
7. Remove the crankshaft timing belt sprocket by loosening the center bolt, while preventing the crankshaft from rotating. To hold the crankshaft from turning, use Suzuki tool 09927-56010 or equivalent, or a large prybar inserted in the transmission housing slot and the flywheel teeth. Pull the sprocket off of the end of the crankshaft. Be sure to retain the crankshaft sprocket key and belt guide for assembly.
8. If necessary, remove the timing belt inside cover from the cylinder head.

To install:

9. If necessary, install the timing belt inside cover.
10. Slide the timing belt guide on the crankshaft so that the concave side faces the oil pump, then install the sprocket key in the groove in the crankshaft.
11. Slide the pulley onto the crankshaft, and install the center retaining bolt. Tighten the center bolt to 58–65 ft. lbs. (80–90 Nm). To hold the crankshaft from turning, use Suzuki tool 09927-56010

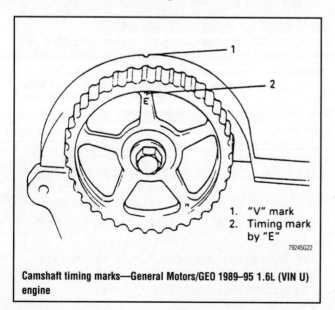

1. "V" mark
2. Timing mark by "E"

79245G22

Camshaft timing marks—General Motors/GEO 1989-95 1.6L (VIN U) engine

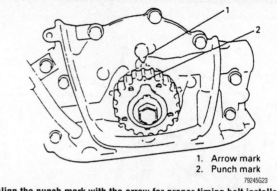

1. Arrow mark
2. Punch mark

79245G23

Align the punch mark with the arrow for proper timing belt installation—General Motors/GEO 1989-95 1.6L (VIN U) engine

or equivalent, or a large prybar inserted in the transmission housing slot and the flywheel teeth.

12. Install the timing belt camshaft sprocket, ensuring that the slot in the sprocket engages the camshaft (pulley) pin; this ensures that the sprocket is properly positioned on the end of the camshaft. Secure the camshaft with the metal rod used during removal, then tighten the sprocket bolt to 41–46 ft. lbs. (56–64 Nm).

13. Assemble the timing belt tensioner plate and the tensioner, making sure that the lug of the tensioner plate engages the tensioner.

14. Install the timing belt tensioner, tensioner plate and spring on the engine. Tighten the mounting bolt and stud only finger-tight at this time. Ensure that when the tensioner is moved in a counterclockwise direction, the tensioner moves in the same direction. If the tensioner does not move, remove it and the tensioner plate to reassemble them properly.

15. Loosen all rocker arm valve lash locknuts and adjusting screws. This will permit movement of the camshaft without any rocker arm associated drag, which is essential for proper timing belt tensioning. If the camshaft does not rotate freely (free of rocker arm drag), the belt will not be properly tensioned.

✳✳ WARNING

If any binding is felt when adjusting the timing belt tension by turning the crankshaft, STOP turning the engine, because the pistons may be hitting the valves.

16. Rotate the camshaft sprocket clockwise until the timing mark on the sprocket and the V mark on the timing belt inside cover are aligned.

17. Using a 17mm wrench, or socket and breaker bar, on the crankshaft sprocket center bolt, turn the crankshaft clockwise until the punch mark on the sprocket is aligned with the arrow mark on the oil pump.

18. With the camshaft and crankshaft marks properly aligned, push the tensioner up with your finger and install the timing belt on the 2 sprockets, ensuring that the drive side of the belt is free of all slack. Release the tensioner. Be sure to install the timing belt so that the directional arrow is pointing in the appropriate direction.

➡**In this position, the No. 4 cylinder is at Top Dead Center (TDC) on the compression stroke.**

19. Rotate the crankshaft clockwise 2 full revolutions, then tighten the tensioner stud to 80–106 inch lbs. (9–12 Nm). Then, tighten the tensioner bolt to 18–21 ft. lbs. (24–30 Nm).

20. Ensure that all 4 timing marks are still aligned as before; if they are not, remove the timing belt, and install and tension it again.

21. Install the timing belt cover and all related components.

Daihatsu

1.6L ENGINE

1. Disconnect the negative battery cable.
2. Remove the air cleaner.
3. Remove the accessory drive belts.
4. Remove the cooling fan and fan shroud.
5. Drain the cooling system and remove the radiator.
6. Remove the water pump pulley.
7. If equipped, remove the power steering pump.
8. Remove the crankshaft pulley.
9. Remove the upper and lower timing covers.
10. Rotate the crankshaft so that the **F** on the camshaft pulley is facing up and the timing marks align.

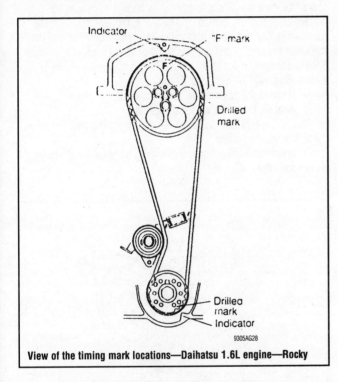

View of the timing mark locations—Daihatsu 1.6L engine—Rocky

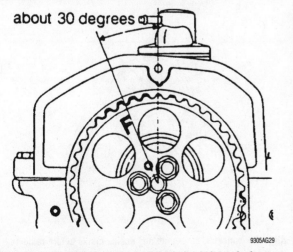

View of the "F" mark alignment for final timing belt tensioning—Daihatsu 1.6L engine—Rocky

11. Loosen the tensioner pulley lock bolt.
12. Pry the tensioner pulley away from the timing belt and tighten the lock bolt.
13. Remove the timing belt.
To install:
14. Ensure that the camshaft and crankshaft timing marks are aligned with the timing marks on the engine and install the timing belt.
15. Loosen the tensioner lock bolt and allow it to apply tension to the timing belt. Tighten the tensioner lock bolt.
16. Rotate the crankshaft clockwise until the **F** mark on the camshaft pulley is 3 teeth away from aligning with the indicator mark on the engine.
17. Loosen the tensioner lock bolt.
18. Continue to rotate the crankshaft clockwise until the **F** mark aligns with the indicator mark. Tighten the tensioner lock bolt to 22–33 ft. lbs. (29–44 Nm).
19. Rotate the crankshaft 2 complete turns clockwise and ensure that the timing marks align.
20. Install the upper and lower timing covers.
21. Install the crankshaft pulley. Tighten the bolts to 15–22 ft. lbs. (20–29 Nm).
22. Install the water pump pulley.
23. If removed, install the power steering pump.
24. Install the radiator.
25. Install the cooling fan and shroud.
26. Install the accessory drive belts.
27. Install the air cleaner.
28. Connect the negative battery cable and fill the cooling system.

Honda

2.0L (B20B4) ENGINE

CR-V

1. Disconnect the negative battery cable.
2. Position crankshaft so that No. 1 piston is at Top Dead Center (TDC) of the compression stroke.
3. Remove the splash guard.
4. Remove the accessory drive belts.
5. If equipped, remove the cruise control actuator.
6. Place a piece of wood between the oil pan and the jack, support the engine with a jack.
7. Remove upper engine bracket.
8. Remove the valve cover.
9. Remove the timing belt covers.
10. Loosen the adjusting bolt 180 degrees. Release the tension from the belt by pushing on the tensioner, then retighten the adjusting bolt.
11. Remove the timing belt.
To install:
12. Be sure the timing marks are properly aligned.
13. Install the timing belt on the pulleys following this sequence:
    a. Crankshaft pulley.
    b. Adjusting pulley.
    c. Water pump pulley.
    d. Exhaust camshaft pulley.
    e. Intake camshaft pulley.
14. Loosen and retighten the adjusting bolt to allow tension to be applied to the belt.
15. Install the lower and middle timing covers.

16. Install the crankshaft pulley and tighten the bolt to 130 ft. lbs. (177 Nm).

⁕⁕ WARNING

If any binding is felt when adjusting the timing belt tension by turning the crankshaft, STOP turning the engine, because the pistons may be hitting the valves.

17. Rotate the crankshaft about 5–6 times counterclockwise to seat the timing belt.
18. Position the No. 1 piston to TDC.
19. Loosen the adjusting bolt ½ turn.
20. Rotate the crankshaft counterclockwise 3 teeth on the camshaft pulley.
21. Tighten the adjusting bolt to 40 ft. lbs. (54Nm).
22. Retighten the crankshaft pulley bolt to 130 ft. lbs. (177 Nm).
23. Install the valve cover.
24. Install the engine mounting bracket, then remove the jack.
25. If removed, install the cruise control actuator.
26. Install the accessory drive belts.

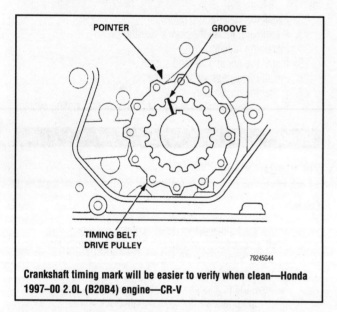

Crankshaft timing mark will be easier to verify when clean—Honda 1997–00 2.0L (B20B4) engine—CR-V

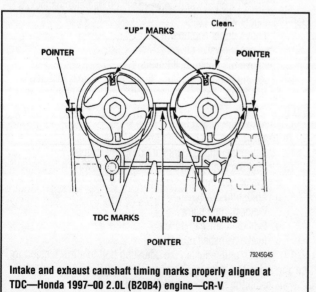

Intake and exhaust camshaft timing marks properly aligned at TDC—Honda 1997–00 2.0L (B20B4) engine—CR-V

27. Install the splash guard.
28. Connect the negative battery cable.
29. Check the engine operation and road test.

2.2L (F22B6) & 2.3L (F23A7) ENGINES

Odyssey

➡ **The radio may contain a coded theft protection circuit. Always make note the code number before disconnecting the battery.**

1. Disconnect the negative and positive battery cables.
2. Remove the cylinder head cover.
3. Remove the upper timing belt cover.
4. Turn the crankshaft to align the timing marks and set cylinder No. 1 to Top Dead Center (TDC) for the compression stroke. The white mark on the crankshaft pulley should align with the pointer on the timing belt cover. The words **UP** embossed on the camshaft pulley should be aligned in the upward position and the marks on the edge of the pulley should be aligned with the cylinder head or the back cover upper edge. Once in this position, the engine must NOT be turned or disturbed.
5. Remove the splash shield from below the engine.
6. Remove the wheel well splash shield.
7. Loosen and remove the power steering pump belt. Remove the power steering pump.
8. Loosen the adjusting and mounting bolts for the alternator and remove the drive belt.

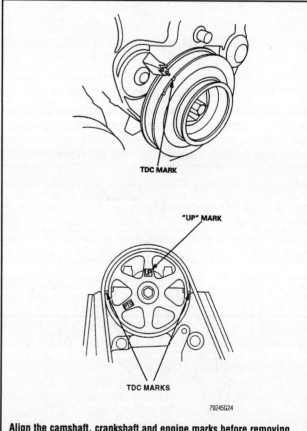

Align the camshaft, crankshaft and engine marks before removing the timing belt and pulleys—Honda 2.2L and 2.3L engines—Odyssey

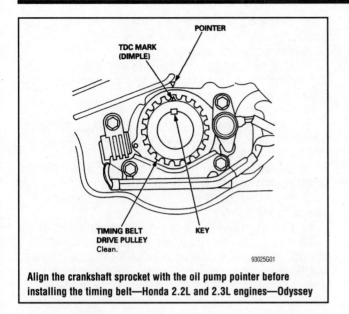

Align the crankshaft sprocket with the oil pump pointer before installing the timing belt—Honda 2.2L and 2.3L engines—Odyssey

9. Support the engine with a floor jack cushioned with a piece of wood under the oil pan.

10. Remove the dipstick and the dipstick tube.

11. Remove the through-bolt for the side engine mount and remove the mount.

12. Remove the crankshaft pulley bolt and remove the crankshaft pulley. Use a Crank Pulley Holder tool No. 07MAB-PY3010A and Holder Handle tool No. 07JAB-001020A or their equivalents, to hold the crankshaft pulley while removing the bolt.

13. Remove the lower timing belt cover.

14. Loosen the timing belt/timing balancer belt adjuster nut ⅔–1 turn. Move the tension adjuster to release the belt tension and retighten the adjuster nut.

15. Remove the balancer shaft belt and its drive pulley.

16. Insert a suitable tool into the maintenance hole in the front balancer shaft. Unbolt and remove the balancer driven pulley.

→For servicing the balance shafts, front refers to the side of the engine facing the radiator. Rear refers to the side of the engine facing the firewall.

17. Remove the timing belt.

18. If equipped with a TDC sensor assembly at the crankshaft sprocket, unbolt the assembly and move it to the side before removing the sprocket.

19. Remove the key and the spacers to remove the crankshaft timing sprocket.

20. Unbolt and remove the camshaft timing sprocket.

To install:

21. Install the camshaft timing sprocket so that the **UP** mark is up and the TDC marks are parallel to the cylinder head gasket surface. Install the key and tighten the bolt to 27 ft. lbs. (37 Nm).

22. Install the crankshaft sprocket so that the TDC mark aligns with the pointer on the oil pump. Install the spacers with their concave surfaces facing in. Install the key. Install the TDC sensor assembly back into position before installing the timing belt.

23. Install and tension the timing belt.

24. Rotate the crankshaft counterclockwise 5–6 turns to be sure the belt is properly seated.

25. Set the No. 1 piston at TDC for its compression stroke.

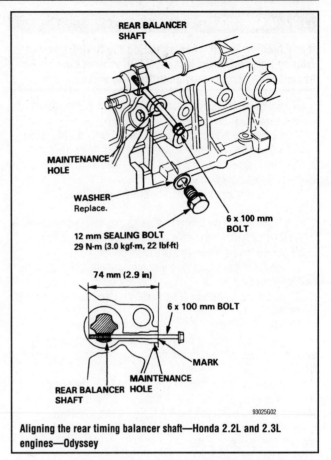

Aligning the rear timing balancer shaft—Honda 2.2L and 2.3L engines—Odyssey

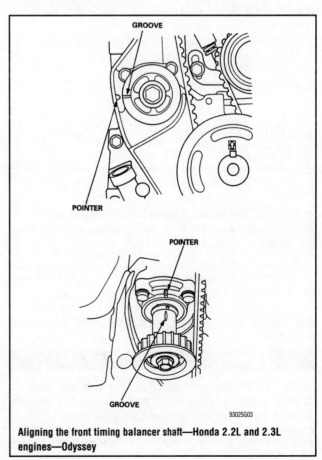

Aligning the front timing balancer shaft—Honda 2.2L and 2.3L engines—Odyssey

❋❋ WARNING

If any binding is felt when adjusting the timing belt tension by turning the crankshaft, STOP turning the engine, because the pistons may be hitting the valves.

26. Rotate the crankshaft counterclockwise so that the camshaft pulley moves only 3 teeth beyond its TDC mark.

27. Tighten the tensioner adjusting nut to 33 ft. lbs. (45 Nm).

28. Tighten the crankshaft pulley bolt to 181 ft. lbs. (245 Nm).

29. Align the rear balancer shaft into position by performing the following procedures:

    a. Scribe a 3 in. (74mm) line from the end of a 6 x 1.0mm bolt.

    b. Remove the maintenance hole sealing bolt and insert the 6 x 1.0mm bolt into the maintenance hole to the scribed line.

30. Install the balancer shaft belt drive pulley.

31. Align the groove on the pulley edge with the pointer on the balancer gear case.

32. Check the alignment of the pointer on the balancer pulley to the pointer on the oil pump.

33. Install and tension the timing balancer shaft belt.

34. Be sure the timing belts have been tensioned correctly and that all TDC and alignment marks are in their proper positions.

35. Install the lower timing cover and the crankshaft pulley. Apply engine oil to the pulley bolt threads and washer surface. Install the pulley bolt and tighten it to 181 ft. lbs. (245 Nm).

36. Install the upper timing cover and the valve cover. Be sure the seals are properly seated.

37. Install the side engine mount. Tighten the through-bolt to 47 ft. lbs. (64 Nm). Tighten the mount nut and bolt to 40 ft. lbs. (55 Nm) each.

38. Remove the floor jack.

39. Install and tension the alternator belt.

40. Install the power steering pump and tension its belt.

41. Install the splash shields.

42. Reconnect the positive and negative battery cables. Enter the radio security code.

43. Check engine operation.

3.5L (J35A1) V6 ENGINE

Odyssey

➡**The radio may contain a coded theft protection circuit. Always make note of the code number before disconnecting the battery.**

1. Disconnect the negative battery terminal.

2. Turn the crankshaft so the white mark on the crankshaft pulley aligns with the pointer on the oil pump housing cover.

3. Open the inspection plugs on the upper timing belt covers and check that the camshaft sprocket marks align with the upper cover marks.

❋❋ WARNING

Align the camshaft and crankshaft sprockets with their alignment marks before removing the timing belt. Failure to align the timing marks correctly may result in valve damage.

4. Raise and safely support the vehicle and remove both front tires/wheels.

5. Remove the front lower splash shield.

6. Move the alternator tensioner with a Belt Tensioner Release Arm tool YA9317 or equivalent, to release tension from the belt and remove the alternator drive belt.

7. Remove the alternator belt tensioner release arm.

8. Loosen the power steering pump adjustment nut, adjustment locknut and mounting bolt, then remove the power steering pump with the hoses attached.

9. Support the weight of the engine by placing a wood block on a floor jack and carefully lift on the oil pan.

10. Remove the bolts from the side engine mount bracket and remove the bracket.

11. Remove the dipstick, the dipstick tube and discard the O-ring.

12. Hold the crankshaft pulley with the Handle tool 07JAB-001020A and Crankshaft Holding tool 07MAB-PY3010A or equivalent. While holding the crankshaft pulley, remove the crankshaft pulley bolt using a heavy duty ¾ in. (19mm) socket and breaker bar.

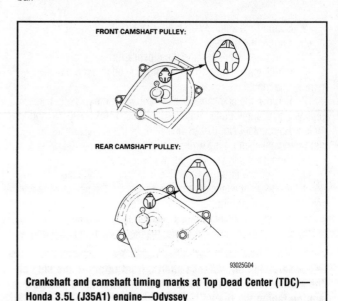

93025G04

Crankshaft and camshaft timing marks at Top Dead Center (TDC)—Honda 3.5L (J35A1) engine—Odyssey

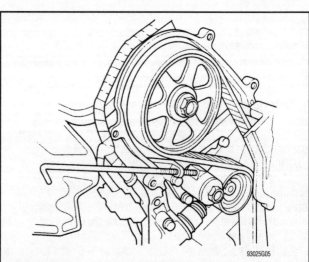

93025G05

Battery hold-down bolt installed to hold auto-tensioner—Honda 3.5L (J35A1) engine—Odyssey

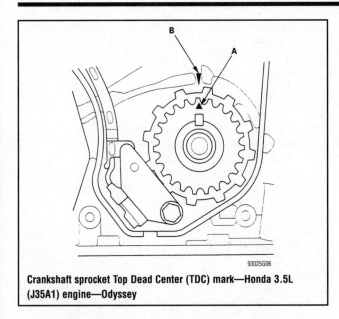

Crankshaft sprocket Top Dead Center (TDC) mark—Honda 3.5L (J35A1) engine—Odyssey

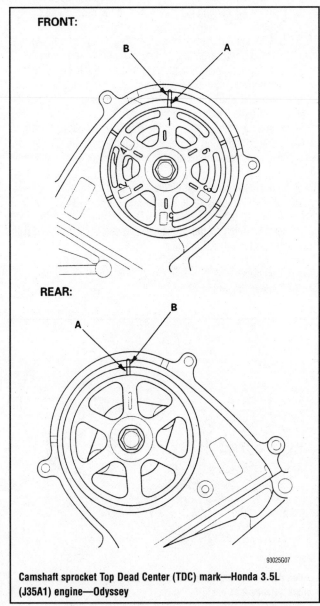

Camshaft sprocket Top Dead Center (TDC) mark—Honda 3.5L (J35A1) engine—Odyssey

13. Remove the crankshaft pulley, the upper timing belt covers and the lower timing belt cover.

14. Remove one of the battery clamp fasteners from the battery tray and grind a 45 degree bevel on the threaded end of the battery clamp bolt.

15. Screw in the battery hold-down bolt into the threaded bracket just above the auto-tensioner (automatic timing belt adjuster) and tighten the bolt hand-tight to hold the auto-tensioner adjuster in its current position.

16. Remove the engine mount bracket bolts and the bracket.

17. Loosen the timing belt idler pulley bolt (located on the right side across from the auto-tensioner pulley) about 5–6 revolutions and remove the timing belt.

To install:

18. Clean the timing belt sprockets and the timing belt covers.

❋❋ **WARNING**

Align the camshaft and crankshaft sprockets with their alignment marks before installing the timing belt. Failure to align the timing marks correctly may result in valve damage.

19. Align the timing mark on the crankshaft sprocket with the oil pump pointer.

20. Align the camshaft sprocket TDC timing marks with the pointers on the rear cover.

21. If installing a new belt or if the auto-tensioner has extended or if the timing belt cannot be reinstalled easily, the auto-tensioner must be collapsed before installation of the timing belt, perform the following procedures:

   a. Remove the battery hold-down bolt from the auto-tensioner bracket.

   b. Remove the timing belt auto-tensioner bolts and the auto-tensioner.

   c. Secure the auto-tensioner in a soft jawed vise, clamping onto the flat surface of one of the mounting bolt holes with the maintenance bolt facing upward.

   d. Remove the maintenance bolt and use caution not to spill oil from the tensioner assembly.

   e. Should oil spill from the tensioner, be sure the tensioner is filled with 0.22 ounces (6.5 ml) of fresh engine oil.

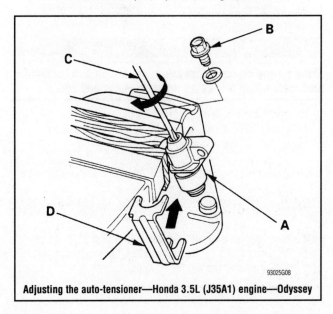

Adjusting the auto-tensioner—Honda 3.5L (J35A1) engine—Odyssey

f. Using care not to damage the threads or the gasket sealing surface, insert a flat-blade screwdriver through the tensioner maintenance hole and turn the screwdriver clockwise to compress the auto-tensioner bottom while the Tensioner Holder tool 14540-P8A-A01 or equivalent, is installed on the auto-tensioner assembly.

g. Install the auto-tensioner maintenance bolt with a new gasket and tighten to a torque 72 inch lbs. (8 Nm).

h. Install the auto-tensioner on the engine with the tensioner holder tool installed and torque the mounting bolts to 104 inch lbs. (12 Nm).

22. Install the timing belt in a counterclockwise pattern starting with the crankshaft drive sprocket. Install the timing belt counterclockwise in the following sequence:
- Crankshaft drive sprocket.
- Idler pulley.
- Left side camshaft sprocket.
- Water pump.
- Right side camshaft sprocket.
- Auto-tensioner adjustment pulley.

23. Torque the timing belt idler pulley bolt to 33 ft. lbs. (44 Nm).

24. Remove the auto-tensioner holding tool to allow the tensioner to extend.

25. Install the engine mount bracket to the engine and torque the bolts to 33 ft. lbs. (44 Nm).

26. Install the lower timing belt cover and both upper timing belt covers.

27. Hold the crankshaft pulley with special tools 07JAB-001020A handle and 07MAB-PY3010A crankshaft holding tool or equivalent tools. While holding the crankshaft pulley, install the crankshaft pulley bolt using a heavy duty ¾ in. (19mm) socket and a commercially available torque wrench and torque the bolt to 181 ft. lbs. (245 Nm).

❊❊ WARNING

If any binding is felt while moving the crankshaft pulley, STOP turning the crankshaft pulley immediately because the pistons may be hitting the valves.

28. Rotate the crankshaft pulley clockwise 5–6 revolutions to allow the timing belt to be seated in the pulleys.

29. Move the crankshaft pulley to the white TDC mark and inspect the camshaft TDC marks to ensure proper timing of the camshafts.

❊❊ WARNING

If the timing marks do not align, the timing belt removal and installation procedure must be performed again.

30. Install the engine dipstick tube using a new O-ring.

31. Install the power steering pump, and loosely install the mounting bolt, adjustment locknut and adjustment nut.

32. Adjust the power steering belt to a tension such that a 22 lb. (98 N) pull halfway between the 2 drive pulleys will allow the belt to move 0.51–0.65 in. (13.0–16.5mm).

33. Tighten the power steering pump mounting bolt and adjustment locknut.

➡ **If a new belt is used, set the deflection to 0.33–0.43 in. (8.5–11.0mm) and after engine has run for 5 minutes, readjust the new belt to the used belt specification.**

34. Install the alternator belt tensioner arm.

35. Move the alternator tensioner with a Belt Tensioner Release Arm tool YA9317 or equivalent, to release tension from the belt and install the alternator drive belt.

36. Install both engine mount bracket bolts and torque to 33 ft. lbs. (44 Nm).

37. Install the bushing through bolt and tighten to 40 ft. lbs. (54 Nm).

38. Release and carefully remove the floor jack.

39. Install the front lower splash shield.

40. Install both front tires/wheels.

41. Carefully lower the vehicle.

42. Install the battery hold-down bolt in the battery tray.

43. Install the negative battery cable.

44. Enter the radio security code.

2.6L (4ZE1/E) ENGINE

Passport

1. Disconnect the negative battery cable.

2. Loosen and remove the engine accessory drive belts.

3. Remove the cooling fan assembly and the water pump pulley.

4. Drain the fluid from the power steering reservoir.

5. Unbolt and remove the power steering pump. Unbolt the hydraulic line brackets from the upper timing cover and move the pump from the work area without disconnecting the hydraulic lines.

6. Disconnect and remove the starter motor if a Flywheel Holder tool No. J-38674 or equivalent, is to be used.

7. Remove the upper timing belt cover.

8. Rotate the crankshaft to set the engine at Top Dead Center (TDC) of the compression for the No. 1 cylinder. The arrow mark on the camshaft sprocket will be aligned with the mark on the rear timing cover.

9. Remove the crankshaft pulley.

10. Remove the lower timing belt cover.

11. Verify that the engine is set at TDC/compression for the No. 1 cylinder. The notch on the crankshaft sprocket will be aligned with the pointer on the oil seal retainer.

12. Release and remove the tensioner spring to release the timing belt's tension.

13. Remove the timing belt.

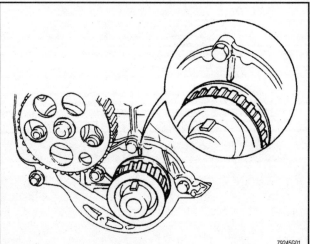

79245G01

Align the crankshaft pulley timing mark the with oil retainer setting mark—Honda 2.6L engine—Passport

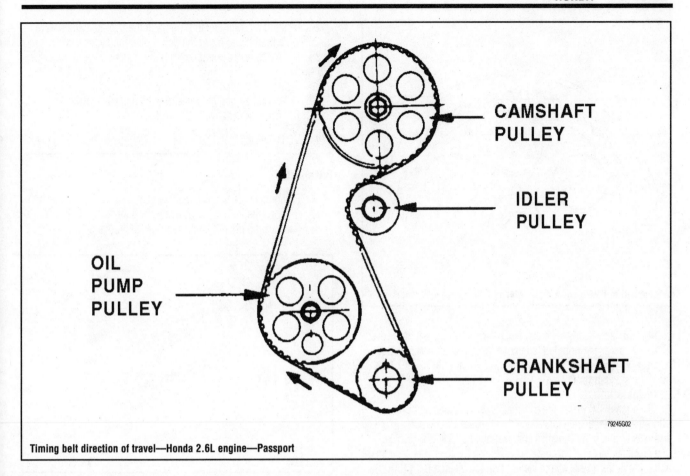

CAMSHAFT PULLEY

IDLER PULLEY

OIL PUMP PULLEY

CRANKSHAFT PULLEY

79245G02

Timing belt direction of travel—Honda 2.6L engine—Passport

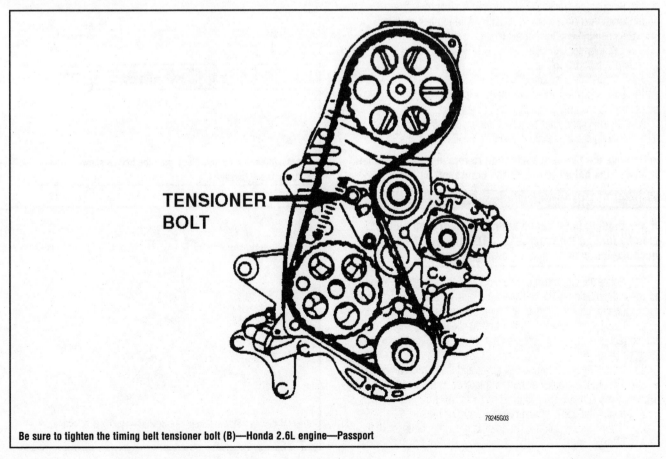

TENSIONER BOLT

79245G03

Be sure to tighten the timing belt tensioner bolt (B)—Honda 2.6L engine—Passport

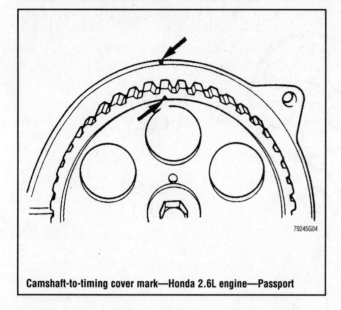

Camshaft-to-timing cover mark—Honda 2.6L engine—Passport

14. Unbolt the tensioner pulley bracket from the engine's front cover.

15. If necessary, unbolt and remove the camshaft sprockets. If necessary, use a puller to remove the crankshaft pulley. Don't lose the crankshaft sprocket key.

To install:

16. If removed, install the camshaft and crankshaft sprockets. Align the camshaft and crankshaft timing marks and be sure to install any keys. Tighten the camshaft sprocket bolt to 43 ft. lbs. (59 Nm).

17. Install the tensioner assembly. Tighten the tensioner mounting bolt to 14 ft. lbs. (19 Nm) and the cap bolt to 108 inch lbs. (13 Nm).

18. Be sure the crankshaft and the camshaft sprockets are aligned with their timing marks. Install the timing belt onto the sprockets using the following sequence:
   - Crankshaft sprocket
   - Oil pump sprocket
   - Camshaft sprocket

19. Loosen the tensioner mounting bolt. This will allow the tensioner spring to apply pressure to the timing belt.

20. After the spring has pulled the timing belt as far as possible, temporarily tighten the tensioner mounting bolt to 14 ft. lbs. (19 Nm).

➡**Remove the flywheel holder before rotating the crankshaft. Reinstall the holder to tighten the crankshaft pulley bolt.**

※※ WARNING

If any binding is felt when adjusting the timing belt tension by turning the crankshaft, STOP turning the engine, because the pistons may be hitting the valves.

21. Rotate the crankshaft counterclockwise 2 complete revolutions to check the rotation of the belt and the alignment of the timing marks. Listen for any rubbing noises that may mean the belt is binding.

22. Loosen the tensioner pulley bolt to allow the spring to adjust the correct tension. Then, retighten the tensioner pulley bolt to 14 ft. lbs. (19 Nm).

23. Install the lower timing belt cover and the crankshaft pulley.

24. Tighten the crankshaft pulley bolt to 87 ft. lbs. (118 Nm). Tighten the small pulley bolts to 72 inch lbs. (8 Nm).

25. Install the upper timing cover.

26. If removed, install the starter and tighten the bolts to 30 ft. lbs. (40 Nm).

27. Install the power steering pump. If the hydraulic lines were disconnected, refill and bleed the power steering system.

28. Install the water pump pulley and tighten its nut to 20 ft. lbs. (26 Nm).

29. Install the cooling fan assembly.

30. Install and adjust the accessory drive belts.

31. Connect the negative battery cable.

3.2L (6VD1) ENGINES

Passport

1. Disconnect the negative battery cable.
2. Remove the air cleaner assembly and intake air duct.
3. Remove the upper fan shroud from the radiator.

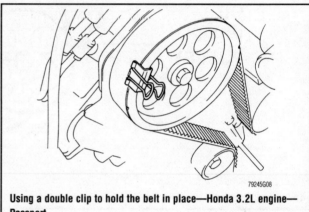

Using a double clip to hold the belt in place—Honda 3.2L engine—Passport

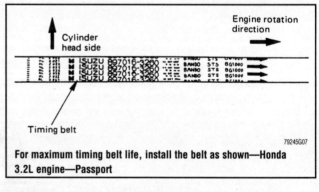

For maximum timing belt life, install the belt as shown—Honda 3.2L engine—Passport

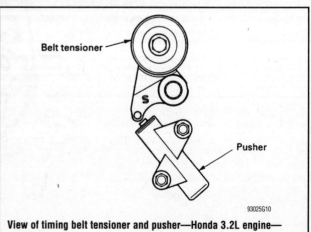

View of timing belt tensioner and pusher—Honda 3.2L engine—Passport

4. Remove the 4 nuts retaining the cooling fan assembly. Remove the cooling fan from the fan pulley.

5. Loosen and remove the drive belts.

6. Remove the upper timing belt covers.

7. Remove the fan pulley assembly.

8. Rotate the crankshaft to align the camshaft timing marks with the pointer dots on the back covers. Verify that the pointer on the crankshaft aligns with the mark on the lower timing cover.

➡When the timing marks are aligned, the No. 2 piston is at Top Dead Center (TDC) of the compression.

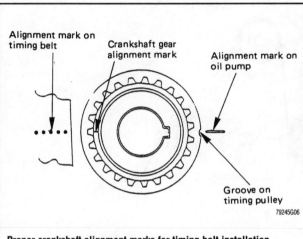

Proper crankshaft alignment marks for timing belt installation—Honda 3.2L SOHC (VIN V) engine—Passport

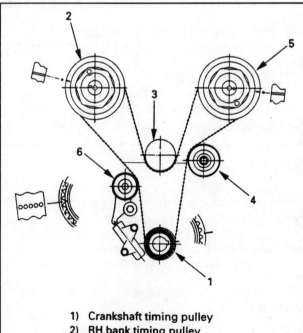

1) Crankshaft timing pulley
2) RH bank timing pulley
3) Water pump pulley
4) Idler pulley
5) LH bank timing pulley
6) Tension pulley

Timing mark alignment and timing belt routing —Honda 3.2L SOHC (VIN V) engine—Passport

✳✳ WARNING

Align the camshaft and crankshaft sprockets with their alignment marks before removing the timing belt. Failure to align the belt and sprocket marks may result in valve damage.

9. Use tool No. J-8614-01 or a suitable pulley holding tool to remove the crankshaft pulley center bolt. Remove the crankshaft pulley.

10. If present, disconnect the 2 oil cooler hose bracket bolts on the timing cover. Move the oil cooler hoses and bracket off of the lower timing cover.

11. Remove the lower timing belt cover.

12. Remove the pusher assembly (tensioner) from below the belt tensioner pulley. The pusher rod must always face upward to prevent oil leakage. Depress the pusher rod and insert a wire pin into the hole to keep the pusher rod retracted.

13. Remove the timing belt.

14. Inspect the water pump and replace it if there is any doubt about its condition.

15. Repair any oil or coolant leaks before installing a new timing belt. If the timing belt has been contaminated with oil or coolant, or is damaged, it must be replaced.

To install:

16. Verify that the sprocket timing marks are still aligned and that the groove and the keyway on the crankshaft timing sprocket align with the mark on the oil pump. The white pointers on the camshaft timing sprockets should align with the dots on the front plate.

17. Install the timing belt. Use clips to secure the belt onto each sprocket until the installation is complete. Align the dotted marks on the timing belt with the timing mark opposite the groove on the crankshaft sprocket.

➡The arrows on the timing belt must follow the belt's direction of rotation. The manufacturer's trademark on the belt's spine should be readable left-to-right when the belt is installed.

18. Align the white line on the timing belt with the alignment mark on the right bank camshaft timing pulley. Secure the belt with a clip.

✳✳ WARNING

If any binding is felt when adjusting the timing belt tension by turning the crankshaft, STOP turning the engine, because the pistons may be hitting the valves.

19. Rotate the crankshaft counterclockwise to remove the slack between the crankshaft sprocket and the right camshaft timing belt sprocket.

20. Install the belt around the water pump pulley.

21. Install the belt on the idler pulley.

22. Align the white alignment mark on the timing belt with the alignment mark on the left bank camshaft timing belt sprocket.

23. Install the crankshaft pulley and tighten the center bolt by hand. Rotate the crankshaft pulley clockwise to give slack between the crankshaft timing belt pulley and the right bank camshaft timing belt pulley.

24. Insert a 1.4mm piece of wire through the hole in the pusher to hold the rod in. Install the pusher assembly while pushing the tension pulley toward the belt.

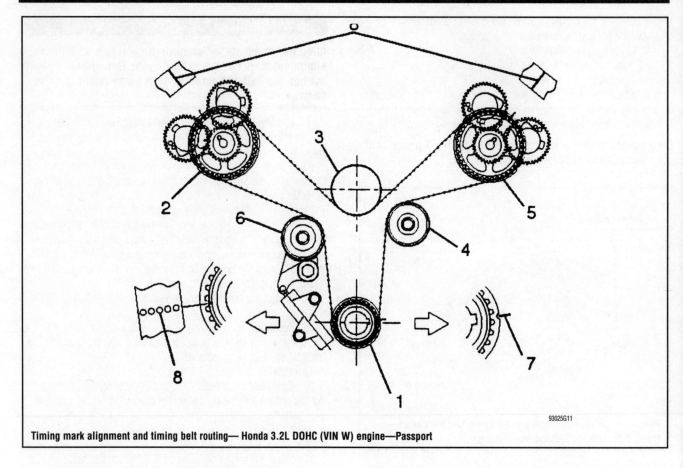

Timing mark alignment and timing belt routing— Honda 3.2L DOHC (VIN W) engine—Passport

25. Pull the pin out from the pusher to release the rod.

26. Remove the clamps from the sprockets. Rotate the crankshaft pulley clockwise 2 turns. Measure the rod protrusion to ensure it is between 0.16–0.24 in. (4–6mm).

27. If the tensioner pulley bracket pivot bolt was removed, tighten it to 31 ft. lbs. (42 Nm).

28. Tighten the pusher bolts to 14 ft. lbs. (19 Nm).

29. Remove the crankshaft pulley. Install the lower and upper timing belt covers and tighten their bolts to 12 ft. lbs. (17 Nm).

30. Fit the oil cooler hose onto the timing cover and tighten its mounting bracket bolts to 16 ft. lbs. (22 Nm).

31. Install the crankshaft pulley and tighten the pulley bolt to 123 ft. lbs. (167 Nm).

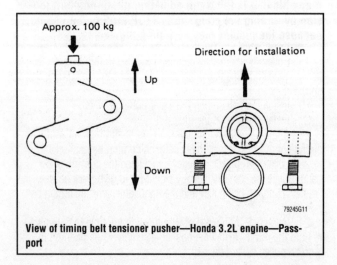

View of timing belt tensioner pusher—Honda 3.2L engine—Passport

32. Install fan pulley assembly and tighten the bolts to 16 ft. lbs. (22 Nm).

33. Install and adjust the accessory drive belts.

34. Install the cooling fan assembly and tighten the bolts to 72 inch lbs. (8 Nm).

35. Install the upper fan shroud.

36. Install the air cleaner assembly and intake air duct.

37. Connect the negative battery cable.

2.2L (X22SE/D) ENGINE

Passport

1. Disconnect the negative battery cable.

2. Using a box-end wrench on the drive belt adjuster, turn the adjuster clockwise and remove the drive belt.

3. From the left rear of the engine compartment, disconnect the 3 electrical connectors from the chassis harness.

4. Remove the crankshaft pulley-to-crankshaft bolts and remove the pulley.

5. From the front of the engine, remove the nut and the engine harness cover.

6. Remove the timing belt cover.

7. Rotate the crankshaft to position the timing marks at Top Dead Center (TDC) of the No. 1 cylinder's compression stroke.

➡ **Mark the rotational direction of the timing belt for reinstallation purposes.**

8. Remove the timing belt tensioner adjusting bolt and the tensioner from the engine.

9. Remove the timing belt.

J-43037

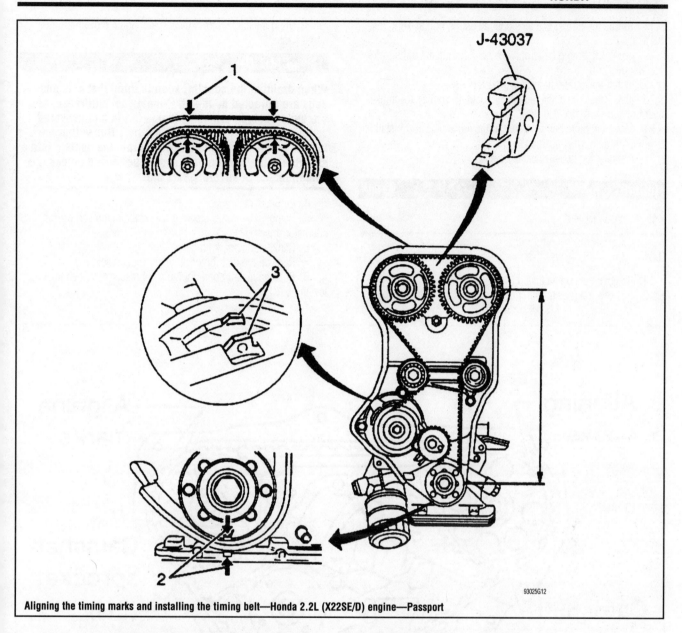

Aligning the timing marks and installing the timing belt—Honda 2.2L (X22SE/D) engine—Passport

93025G12

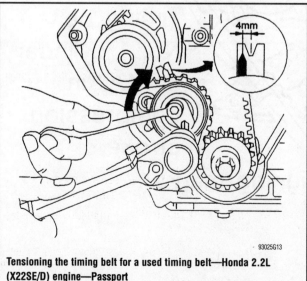

Tensioning the timing belt for a used timing belt—Honda 2.2L (X22SE/D) engine—Passport

93025G13

To install:

10. Install the timing belt tensioner and finger-tighten the tensioner bolt.

11. Inspect the timing marks to be sure that the engine is positioned at TDC of the No. 1 cylinder's compression stroke.

12. Using tool J-43037 or equivalent, place it between the intake and exhaust sprockets to prevent the camshaft gear from moving during the timing belt installation.

13. Install the timing belt.

14. Position the timing belt to ensure that the tension side of the belt is taut and move the timing belt tension adjusting lever clockwise until the tensioner pointer is flowing.

15. If installing a used timing belt (used over 60 min. from new), the pointer should be positioned approximately 0.16 in. (4mm) to the left of the **V** notch when viewed from the front of the engine.

16. If installing a new timing belt, the pointer should be positioned at the center of the **V** notch when viewed from the front of the engine.

17. Torque the timing belt tensioner adjusting bolt to 18 ft. lbs. (25 Nm).

18. Install the timing belt front cover and torque the bolts to 53 inch lbs. (6 Nm).

19. Install the engine harness connectors.

20. Install the crankshaft pulley and toque the pulley-to-crankshaft bolts to 14 ft. lbs. (20 Nm).

21. Move the drive belt tensioner to the loose side and install the drive belt to its normal position.

22. Connect the negative battery cable.

Infiniti

3.3L (VG33E) ENGINE

QX4

1. Remove the engine undercover.
2. Remove the radiator shroud, the fan and the pulleys.

3. Drain the coolant from the radiator and remove the water pump hose.

☀☀ CAUTION

When draining the coolant, keep in mind that cats and dogs are attracted by the ethylene glycol antifreeze, and are quite likely to drink any that is left in an uncovered container, or in puddles on the ground. This will prove fatal in sufficient quantity. Always drain the coolant into a sealable container. Coolant should be reused unless it is contaminated or several years old.

4. Remove the radiator.
5. Remove the power steering, air conditioning compressor and alternator drive belts.
6. Remove the spark plugs.
7. Remove the distributor protector (dust shield).
8. Remove the air conditioning compressor drive belt idler pulley and bracket.

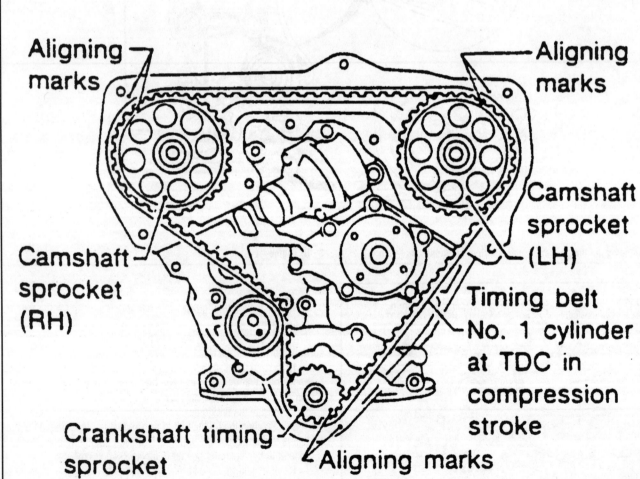

Timing belt alignment mark locations—Infiniti 3.3L engine—QX4

9. Remove the fresh air intake tube at the cylinder head cover.

10. Disconnect the radiator hose at the thermostat housing.

11. Remove the crankshaft pulley bolt, then pull off the pulley with a suitable puller.

12. Remove the bolts, then remove the front upper and lower timing belt covers.

13. Set the No. 1 piston at Top Dead Center (TDC) of its compression stroke. Align the punchmark on the left camshaft sprocket with the punchmark on the timing belt upper rear cover. Align the punchmark on the crankshaft sprocket with the notch on the oil pump housing. Temporarily install the crank pulley bolt so the crankshaft can be rotated if necessary.

14. Loosen the timing belt tensioner and return spring, then remove the timing belt.

To install:

❄❄ CAUTION

Before installing the timing belt, confirm that the No. 1 cylinder is set at the TDC of the compression stroke.

15. Remove both cylinder head covers and loosen all rocker arm shaft retaining bolts.

➡**The rocker arm shaft bolts MUST be loosened so that the correct belt tension can be obtained.**

16. Install the tensioner and the return spring. Using a hexagon wrench, turn the tensioner clockwise and temporarily tighten the locknut.

17. Be sure that the timing belt is clean and free from oil or water.

18. When installing the timing belt, align the white lines on the belt with the punchmarks on the camshaft and crankshaft sprockets. Have the arrow on the timing belt pointing toward the front belt covers.

➡**A good way to check for proper timing belt installation is to count the number of belt teeth between the timing marks. There are 133 teeth on the belt; there should be 40 teeth between the timing marks on the left and right side camshaft sprockets, and 43 teeth between the timing marks on the left side camshaft sprocket and the crankshaft sprocket.**

19. While keeping the tensioner steady, loosen the locknut with a hex wrench.

20. Turn the tensioner approximately 70–80 degrees clockwise with the wrench, then tighten the locknut.

❄❄ WARNING

If any binding is felt when adjusting the timing belt tension by turning the crankshaft, STOP turning the engine, because the pistons may be hitting the valves.

21. Turn the crankshaft in a clockwise direction several times, then **slowly** set the No. 1 piston to TDC of the compression stroke.

22. Apply 22 lbs. (10 kg) of pressure to the center span of the timing belt between the right side camshaft sprocket and the tensioner pulley, then loosen the tensioner locknut.

23. Using a 0.0138 in. (0.35mm) thick feeler gauge (the actual width of the blade **must** be ½ in. or 13mm!), turn the crankshaft clockwise (**slowly!**). The timing belt should move approximately 2½ teeth. Tighten the tensioner locknut, turn the crankshaft slightly and remove the feeler gauge.

24. Slowly rotate the crankshaft clockwise several more times, then set the No. 1 piston to TDC of the compression stroke.

25. Position the 2 timing covers on the block, then tighten the mounting bolts to 24 ft. lbs. (35 Nm).

26. Press the crankshaft pulley onto the shaft, then tighten the bolt to 90–98 ft. lbs. (123–132 Nm).

27. Connect the radiator hose to the thermostat housing.

28. Reconnect the fresh air intake tube at the cylinder head cover.

29. Install the air conditioning compressor drive belt idler pulley and bracket.

30. Install the distributor protector (dust shield).

31. Install the spark plugs.

32. Install the power steering, air conditioning compressor and alternator drive belts.

33. Install the radiator.

34. Reconnect the water pump hose and fill the engine with coolant. Install the fan shroud and pulleys.

35. Install the engine undercover.

36. Start the engine and check for any leaks.

Isuzu

2.2L (C223 & C223-T) DIESEL ENGINES

1. Disconnect the negative battery cable.

2. Drain the cooling system.

3. Remove the radiator.

4. If equipped, remove the air conditioning compressor drive belt by moving the power steering pump or idler pulley.

5. Loosen the alternator adjusting plate bolt and fixing bolt, then remove the fan belt.

6. Remove the 4 crankshaft pulley-to-crankshaft bolts and the pulley.

7. Remove the timing cover bolts and the covers.

8. Remove the injection pump timing pulley flange bolts and the flange.

9. Remove the tension spring.

➡**When removing the tension spring, avoid using excess force or distortion of the spring will result.**

10. Remove the tension pulley nut, the tension pulley and the tension center.

11. If reusing the timing belt, mark its rotational direction on the back of the belt.

12. Remove the timing belt.

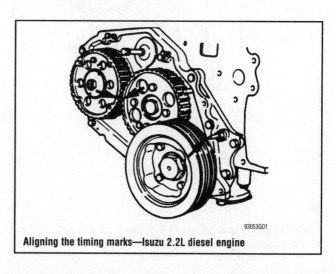

93053G01

Aligning the timing marks—Isuzu 2.2L diesel engine

To install:

⁂ WARNING

No attempt should be made to readjust the belt tension. If the belt has been loosened through service of the timing system, it should be replaced with a new one.

13. Rotate the crankshaft sprocket, the injection pump timing sprocket and camshaft sprocket so that they are aligned with their timing marks.

14. Noting the direction of rotation, install the timing belt in the following sequence: crankshaft sprocket, camshaft sprocket and injection pump sprocket.

➡**Make an adjustment so that the belt slack is taken up by the tension pulley.**

15. Install the tension center and tension pulley by making sure that the end of the tension center is in proper contact with 2 pins on the timing pulley housing.

16. Hand-tighten the nut, so that tension pulley can slide freely.

17. Install the tension spring correctly and semi-tighten the tension pulley fixing nut.

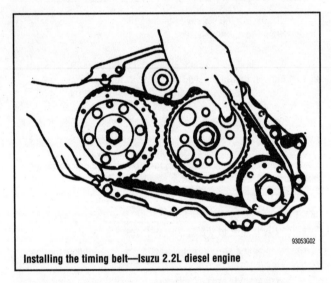

Installing the timing belt—Isuzu 2.2L diesel engine

93053G02

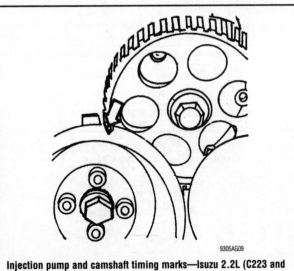

Injection pump and camshaft timing marks—Isuzu 2.2L (C223 and C223-T) diesel engines

9305AG09

18. Turn the crankshaft 2 turns in the normal direction of rotation to permit seating of the belt. Further rotate the crankshaft 90 degrees beyond Top Dead Center (TDC) to settle the injection pump.

⁂ WARNING

Never attempt to turn the crankshaft in the reverse direction.

19. Loosen the tension pulley fixing nut completely, allowing the pulley to take up looseness of the belt. Then, tighten the nut to 78–95 ft. lbs. (106–129 Nm).

20. Install the flange on the injection pump pulley; the hole in the outer circumference of the flange should be aligned with the triangular mark on the injection pump sprocket.

21. Rotate the crankshaft 2 turns in normal direction of rotation to bring the No. 1 piston to TDC on the compression stroke and recheck that the triangular mark on the injection pump sprocket is aligned with the hole in the flange.

22. Using tool J-29771 or equivalent, check the timing belt tension at a point between the injection pump and crankshaft sprockets; belt tension should be 33–55 ft. lbs. (45–75 Nm).

23. Adjust the valve clearances.

24. Complete the installation by reversing the removal procedures.

25. Refill the cooling system.

26. Connect the negative battery cable.

2.2L (X22SE/D) ENGINE

1998–00

1. Disconnect the negative battery cable.

2. Using a box-end wrench on the drive belt adjuster, turn the adjuster clockwise and remove the drive belt.

3. From the left rear of the engine compartment, disconnect the 3 electrical connectors from the chassis harness.

4. Remove the crankshaft pulley-to-crankshaft bolts and remove the pulley.

5. From the front of the engine, remove the nut and the engine harness cover.

6. Remove the timing belt cover.

7. Rotate the crankshaft to position the timing marks at Top Dead Center (TDC) of the No. 1 cylinder's compression stroke.

➡**Mark the rotational direction of the timing belt for reinstallation purposes.**

8. Remove the timing belt tensioner adjusting bolt and the tensioner from the engine.

9. Remove the timing belt.

To install:

10. Install the timing belt tensioner and finger-tighten the tensioner bolt.

11. Inspect the timing marks to be sure that the engine is positioned at TDC of the No. 1 cylinder's compression stroke.

12. Using tool J-43037 or equivalent, place it between the intake and exhaust sprockets to prevent the camshaft gear from moving during the timing belt installation.

13. Install the timing belt.

14. Position the timing belt to ensure that the tension side of the belt is taut and move the timing belt tension adjusting lever clockwise until the tensioner pointer is flowing.

15. If installing a used timing belt (used over 60 min. from new),

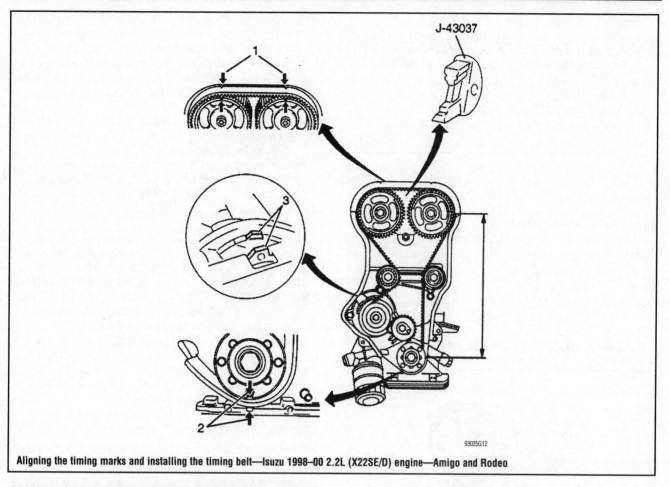

Aligning the timing marks and installing the timing belt—Isuzu 1998–00 2.2L (X22SE/D) engine—Amigo and Rodeo

the pointer should be positioned approximately 0.16 in. (4mm) to the left of the **V** notch when viewed from the front of the engine.

16. If installing a new timing belt, the pointer should be positioned at the center of the **V** notch when viewed from the front of the engine.

17. Torque the timing belt tensioner adjusting bolt to 18 ft. lbs. (25 Nm).

18. Install the timing belt front cover and torque the bolts to 53 inch lbs. (6 Nm).

19. Install the engine harness connectors.

20. Install the crankshaft pulley and toque the pulley-to-crankshaft bolts to 14 ft. lbs. (20 Nm).

21. Move the drive belt tensioner to the loose side and install the drive belt to its normal position.

22. Connect the negative battery cable.

2.3L (4ZD1) AND 2.6L (4ZE1) ENGINES

1. Disconnect the negative battery cable.
2. Loosen and remove the engine accessory drive belts.
3. Remove the cooling fan assembly and the water pump pulley.
4. Drain the fluid from the power steering reservoir.

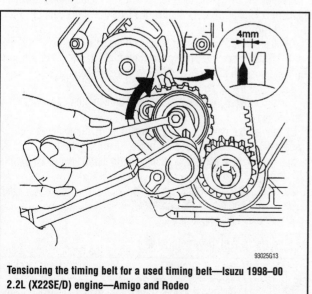

Tensioning the timing belt for a used timing belt—Isuzu 1998–00 2.2L (X22SE/D) engine—Amigo and Rodeo

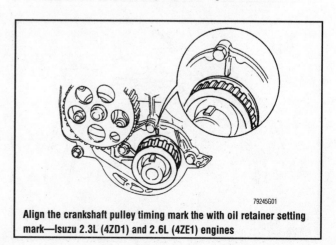

Align the crankshaft pulley timing mark the with oil retainer setting mark—Isuzu 2.3L (4ZD1) and 2.6L (4ZE1) engines

5. Unbolt and remove the power steering pump. Unbolt the hydraulic line brackets from the upper timing cover and move the pump out of the work area without disconnecting the hydraulic lines.

6. Disconnect and remove the starter motor if a flywheel holder tool No. J-38674 or equivalent, is to be used.

7. Remove the upper timing belt cover.

8. Rotate the crankshaft to set the engine at Top Dead Center (TDC) of its compression stroke for the No. 1 cylinder. The arrow

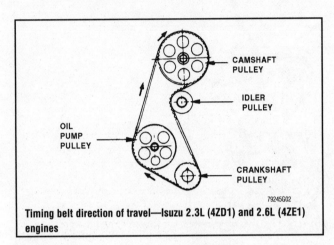

Timing belt direction of travel—Isuzu 2.3L (4ZD1) and 2.6L (4ZE1) engines

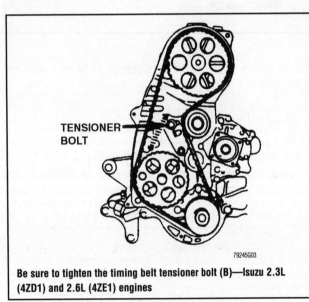

Be sure to tighten the timing belt tensioner bolt (B)—Isuzu 2.3L (4ZD1) and 2.6L (4ZE1) engines

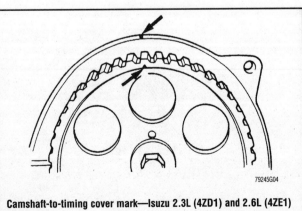

Camshaft-to-timing cover mark—Isuzu 2.3L (4ZD1) and 2.6L (4ZE1) engines

mark on the camshaft sprocket will be aligned with the mark on the rear timing cover.

9. Remove the crankshaft pulley.

10. Remove the lower timing belt cover.

11. Verify that the engine is set at TDC/compression for the No. 1 cylinder. The notch on the crankshaft sprocket will be aligned with the pointer on the oil seal retainer.

12. Release and remove the tensioner spring to release the timing belt's tension.

13. Remove the timing belt.

14. Unbolt the tensioner pulley bracket from the engine's front cover.

15. If necessary, unbolt and remove the camshaft sprockets. Use a puller to remove the crankshaft pulley if necessary. Don't lose the crankshaft sprocket key.

To install:

16. If removed, install the camshaft and crankshaft sprockets. Align the camshaft and crankshaft timing marks and be sure to install any keys. Tighten the camshaft sprocket bolt to 43 ft. lbs. (59 Nm).

17. Install the tensioner assembly. Tighten the tensioner mounting bolt to 14 ft. lbs. (19 Nm) and the cap bolt to 108 inch lbs. (13 Nm).

18. Be sure the crankshaft and the camshaft sprockets are aligned with their timing marks. Install the timing belt onto the sprockets using the following sequence: first around the crankshaft sprocket; second around the oil pump sprocket; third around the camshaft sprocket.

19. Loosen the tensioner mounting bolt. This will allow the tensioner spring to apply pressure to the timing belt.

20. After the spring has pulled the timing belt as far as possible, temporarily tighten the tensioner mounting bolt to 14 ft. lbs. (19 Nm).

➡Remove the flywheel holder before rotating the crankshaft. Reinstall the holder to tighten the crankshaft pulley bolt.

✳✳ WARNING

If any binding is felt when adjusting the timing belt tension by turning the crankshaft, STOP turning the engine, because the pistons may be hitting the valves.

21. Rotate the crankshaft counterclockwise 2 complete revolutions to check the rotation of the belt and the alignment of the timing marks.

22. Loosen the tensioner pulley bolt to allow the spring to adjust the correct tension. Then, retighten the tensioner pulley bolt to 14 ft. lbs. (19 Nm).

23. Install the lower timing cover and the crankshaft pulley.

24. Tighten the crankshaft pulley bolt to 87 ft. lbs. (118 Nm). Tighten the small pulley bolts to 72 inch lbs. (8 Nm).

25. Install the upper timing cover.

26. If removed, install the starter and tighten the bolts to 30 ft. lbs. (40 Nm).

27. Install the power steering pump. If the hydraulic lines were disconnected, refill and bleed the power steering system.

28. Install the water pump pulley and tighten its nut to 20 ft. lbs. (26 Nm).

29. Install the cooling fan assembly.

30. Install and adjust the accessory drive belts.

31. Connect the negative battery cable.

3.2L (6VD1) & 3.5L (6VE1) ENGINES

1. Disconnect the negative battery cable.

2. Remove the air cleaner assembly and intake air duct.

3. Remove the upper fan shroud from the radiator.

4. Remove the 4 nuts retaining the cooling fan assembly. Remove the cooling fan from the fan pulley.

5. Loosen and remove the drive belts.

6. Remove the upper timing belt covers.

7. Remove the fan pulley assembly.

8. Rotate the crankshaft to align the camshaft timing marks with the pointer dots on the back covers. Verify that the pointer on the crankshaft aligns with the mark on the lower timing cover.

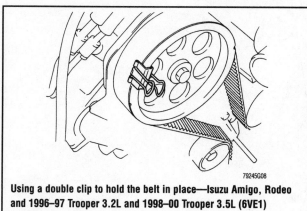

Using a double clip to hold the belt in place—Isuzu Amigo, Rodeo and 1996–97 Trooper 3.2L and 1998–00 Trooper 3.5L (6VE1) engines

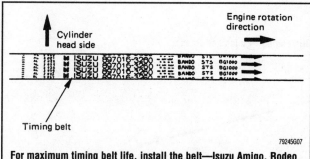

For maximum timing belt life, install the belt—Isuzu Amigo, Rodeo and 1996–97 Trooper 3.2L (6VD1) and 1998–00 Trooper 3.5L (6VE1) engines

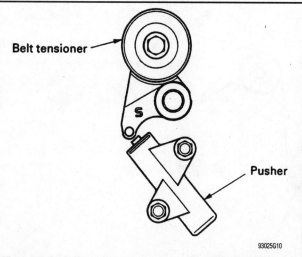

View of timing belt tensioner and pusher—Isuzu Amigo, Rodeo and 1996–97 Trooper 3.2L (6VD1) and 1998–00 Trooper 3.5L (6VE1) engines

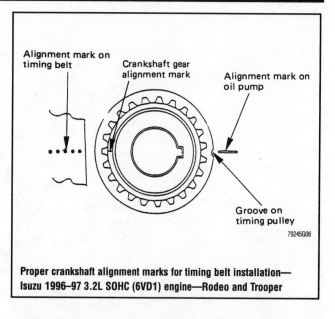

Proper crankshaft alignment marks for timing belt installation—Isuzu 1996–97 3.2L SOHC (6VD1) engine—Rodeo and Trooper

➡ When the timing marks are aligned, the No. 2 piston is at Top Dead Center (TDC) of the compression stroke.

❋❋ WARNING

Align the camshaft and crankshaft sprockets with their alignment marks before removing the timing belt. Failure to align the belt and sprocket marks may result in valve damage.

9. Use tool No. J-8614-01, or a suitable pulley holding tool to remove the crankshaft pulley center bolt. Remove the crankshaft pulley.

10. If present, disconnect the 2 oil cooler hose bracket bolts on the timing cover. Move the oil cooler hoses and bracket off of the lower timing cover.

11. Remove the lower timing belt cover.

12. Remove the pusher assembly (tensioner) from below the belt tensioner pulley. The pusher rod must always face upward to prevent oil leakage. Depress the pusher rod, and insert a wire pin into the hole to keep the pusher rod retracted.

13. Remove the timing belt.

14. Inspect the water pump and replace it if there is any doubt about its condition.

15. Repair any oil or coolant leaks before installing a new timing belt. If the timing belt has been contaminated with oil or coolant, or is damaged, it must be replaced.

To install:

16. Verify that the sprocket timing marks are still aligned and that the groove and the keyway on the crankshaft timing sprocket align with the mark on the oil pump. The white pointers on the camshaft timing sprockets should align with the dots on the front plate.

17. Install the timing belt. Use clips to secure the belt onto each sprocket until the installation is complete. Align the dotted marks on the timing belt with the timing mark opposite the groove on the crankshaft sprocket.

➡ The arrows on the timing belt must follow the belt's direction of rotation. The manufacturer's trademark on the belt's spine should be readable left-to-right when the belt is installed.

18. Align the white line on the timing belt with the alignment mark on the right bank camshaft timing pulley. Secure the belt with a clip.

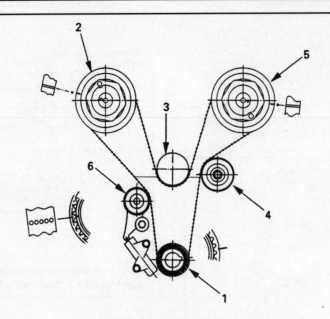

1) Crankshaft timing pulley
2) RH bank timing pulley
3) Water pump pulley
4) Idler pulley
5) LH bank timing pulley
6) Tension pulley

93025G09

Timing mark alignment and timing belt routing—Isuzu 1996–97 3.2L (6VD1) SOHC engine—Rodeo and Trooper

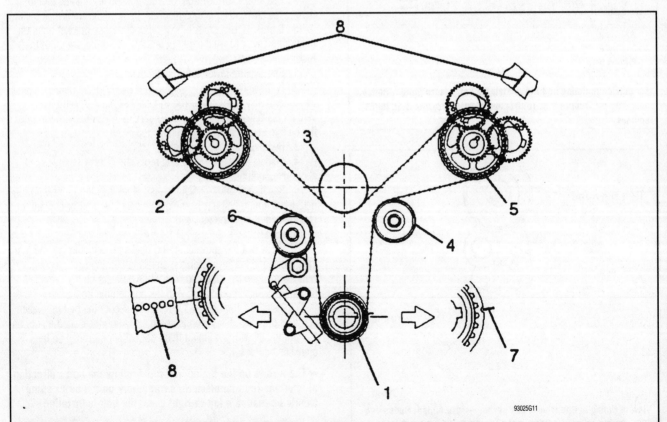

93025G11

Timing mark alignment and timing belt routing—Isuzu 1998–00 Amigo and Rodeo 3.2L (6VD1) DOHC (VIN W) and 1998–00 Trooper 3.5L DOHC (6VE1) engines

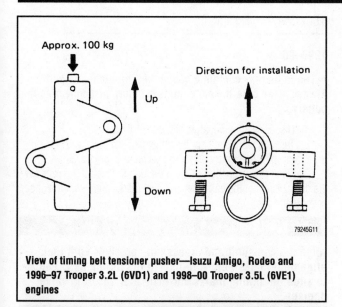

View of timing belt tensioner pusher—Isuzu Amigo, Rodeo and 1996–97 Trooper 3.2L (6VD1) and 1998–00 Trooper 3.5L (6VE1) engines

✳✳ WARNING

If any binding is felt when adjusting the timing belt tension by turning the crankshaft, STOP turning the engine, because the pistons may be hitting the valves.

19. Rotate the crankshaft counterclockwise to remove the slack between the crankshaft sprocket and the right camshaft timing belt sprocket.
20. Install the belt around the water pump pulley.
21. Install the belt on the idler pulley.
22. Align the white alignment mark on the timing belt with the alignment mark on the left bank camshaft timing belt sprocket.
23. Install the crankshaft pulley and tighten the center bolt by hand. Rotate the crankshaft pulley clockwise to give slack between the crankshaft timing belt pulley and the right bank camshaft timing belt pulley.
24. Insert a 1.4mm piece of wire through the hole in the pusher to hold the rod in. Install the pusher assembly while pushing the tension pulley toward the belt.
25. Pull the pin out from the pusher to release the rod.
26. Remove the clamps from the sprockets. Rotate the crankshaft pulley clockwise 2 turns. Measure the rod protrusion to ensure it is between 0.16–0.24 in. (4–6mm).
27. If the tensioner pulley bracket pivot bolt was removed, tighten it to 31 ft. lbs. (42 Nm).
28. Tighten the pusher bolts to 14 ft. lbs. (19 Nm).
29. Remove the crankshaft pulley. Install the lower and upper timing belt covers and tighten their bolts to 12 ft. lbs. (17 Nm).
30. Fit the oil cooler hose onto the timing cover and tighten its mounting bracket bolts to 16 ft. lbs. (22 Nm).
31. Install the crankshaft pulley and tighten the pulley bolt to 123 ft. lbs. (167 Nm).
32. Install fan pulley assembly and tighten the bolts to 16 ft. lbs. (22 Nm).
33. Install and adjust the accessory drive belts.
34. Install the cooling fan assembly and tighten the bolts to 72 inch lbs. (8 Nm).
35. Install the upper fan shroud.
36. Install the air cleaner assembly and intake air duct.
37. Connect the negative battery cable.

2.2L (F22B6) & 2.3L (F23A7) ENGINES

Oasis

➡The radio may contain a coded theft protection circuit. Always make note of the code number before disconnecting the battery.

1. Disconnect the negative and positive battery cables.
2. Remove the valve cover.
3. Remove the upper timing belt cover.
4. Turn the engine to align the timing marks and set cylinder No. 1 to Top Dead Center (TDC) for the compression stroke. The white mark on the crankshaft pulley should align with the pointer on the timing belt cover. The words **UP** embossed on the camshaft pulley should be aligned in the upward position. The marks on the edge of the pulley should be aligned with the cylinder head or the back cover upper edge. Once in this position, the engine must NOT be turned or disturbed.
5. Remove the splash shield from below the engine.
6. Remove the wheel well splash shield.
7. Loosen and remove the power steering pump belt. Remove the power steering pump.
8. Loosen the adjusting and mounting bolts for the alternator and remove the drive belt.
9. Support the engine with a floor jack cushioned with a piece of wood.
10. Remove the through-bolt for the side engine mount and remove the mount.
11. Remove the crankshaft pulley bolt and remove the crankshaft pulley. Use a Crank Pulley Holder tool No. 07MAB-PY3010A and

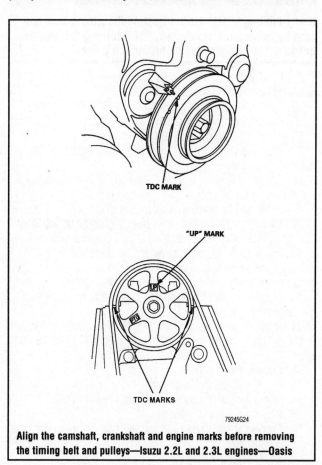

Align the camshaft, crankshaft and engine marks before removing the timing belt and pulleys—Isuzu 2.2L and 2.3L engines—Oasis

Holder Handle tool No. 07JAB-001020A or their equivalents, to hold the crankshaft pulley in place while removing the bolt.

12. Remove the lower timing belt cover.

13. Remove the balancer shaft belt and its drive pulley.

14. Insert a suitable tool into the maintenance hole in the front balancer shaft. Unbolt and remove the balancer driven pulley.

➡ **For servicing the balance shafts, front refers to the side of the engine facing the radiator. Rear refers to the side of the engine facing the firewall.**

15. Remove the timing belt.

16. If equipped with a TDC sensor assembly at the crankshaft sprocket, unbolt the assembly and move it aside before removing the sprocket.

17. Remove the key and the spacers to remove the crankshaft timing sprocket.

18. Unbolt and remove the camshaft timing sprocket.

To install:

19. Install the camshaft timing sprocket so that the **UP** mark is up and the TDC marks are parallel to the cylinder head gasket surface. Install the key and tighten the bolt to 27 ft. lbs. (37 Nm).

20. Install the crankshaft timing sprocket so that the TDC mark aligns with the pointer on the oil pump. Install the spacers with their concave surfaces facing in. Install the key. Install the TDC sensor assembly back into position before installing the timing belt.

21. Install and tension the timing belt.

22. Rotate the crankshaft counterclockwise 5–6 turns to be sure the belt is properly seated.

23. Set the No. 1 piston at TDC for its compression stroke.

❊❊ WARNING

If any binding is felt when adjusting the timing belt tension by turning the crankshaft, STOP turning the engine, because the pistons may be hitting the valves.

24. Rotate the crankshaft counterclockwise so that the camshaft pulley moves only 3 teeth beyond its TDC mark.

25. Tighten the tensioner adjusting nut to 33 ft. lbs. (45 Nm).

26. Tighten the crankshaft pulley bolt to 181 ft. lbs. (245 Nm).

27. Install the balancer shaft belt drive pulley.

28. Align the groove on the pulley edge to the pointer on the balancer gear case.

29. Check the alignment of the pointer on the balancer pulley to the pointer on the oil pump.

30. Install and tension the balancer shaft belt.

31. Be sure the timing belts have been tensioned correctly and that all TDC and alignment marks are in their proper positions.

32. Install the lower timing cover and the crankshaft pulley. Apply engine oil to the pulley bolt threads and washer surface. Install the pulley bolt and tighten it to 181 ft. lbs. (245 Nm).

33. Install the upper timing cover and the valve cover. Be sure the seals are properly seated.

34. Install the side engine mount. Tighten the through-bolt to 47 ft. lbs. (64 Nm). Tighten the mount nut and bolt to 40 ft. lbs. (55 Nm) each.

35. Remove the floor jack.

36. Install and tension the alternator belt.

37. Install the power steering pump and tension its belt.

38. Install the splash shields.

39. Reconnect the positive and negative battery cables. Enter the radio security code.

40. Check engine operation.

3.5L (J35A1) V6 ENGINE

1999–00

➡ **The radio may contain a coded theft protection circuit. Always make note the code number before disconnecting the battery.**

1. Disconnect the negative battery terminal.

2. Turn the crankshaft so the white mark on the crankshaft pulley aligns with the pointer on the oil pump housing cover.

3. Open the inspection plug on the upper cover of the cam covers and check that the camshaft sprocket marks align with the upper cover marks.

❊❊ WARNING

Align the camshaft and crankshaft sprockets with their alignment marks before removing the timing belt. Failure to align the timing marks correctly may result in valve damage.

4. Raise and safely support the vehicle and remove both front tires/wheels.

5. Remove the front lower splash shield.

6. Move the alternator tensioner with a Belt Tensioner Release Arm tool YA9317 or equivalent, to release tension from the belt and remove the alternator belt.

7. Remove the alternator belt tensioner release arm.

8. Loosen the power steering pump adjustment nut, adjustment locknut, and mounting bolt, then remove the power steering pump with the hoses attached.

9. Support the weight of the engine by placing a wood block on a floor jack and carefully lift on the oil pan.

10. Remove the bolts from the side engine mount bracket and remove the bracket.

11. Remove the dipstick, the dipstick tube and discard the O-ring.

12. Hold the crankshaft pulley with Handle tool 07JAB-001020A and Crankshaft Holding tool 07MAB-PY3010A or equivalent. While holding the crankshaft pulley, remove the crankshaft pulley bolt using a heavy duty ¾ in. (19mm) socket and breaker bar.

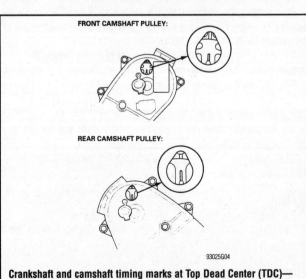

93025G04

Crankshaft and camshaft timing marks at Top Dead Center (TDC)— Isuzu 1999–00 3.5L (J35A1) engine—Oasis

13. Remove the crankshaft pulley, the upper timing belt covers and the lower timing belt cover.

14. Remove one of the battery hold-down fasteners from the battery tray and grind a 45 degree bevel on the threaded end of the battery hold-down bolt.

15. Screw in the battery hold-down bolt into the threaded bracket just above the auto-tensioner, (automatic timing belt adjuster), and tighten the bolt hand-tight to hold the auto-tensioner adjuster in its current position.

16. Remove the engine mount bracket bolts and the bracket.

17. Loosen the timing belt idler pulley bolt (located on the right side across from the auto-tensioner pulley) about 5–6 revolutions and remove the timing belt.

To install:

18. Clean the timing belt sprockets and the timing belt covers.

✳✳ WARNING

Align the camshaft and crankshaft sprockets with their alignment marks before installing the timing belt. Failure to align the timing marks correctly may result in valve damage.

19. Align the timing mark on the crankshaft sprocket with the oil pump pointer.

20. Align the camshaft sprocket TDC timing marks with the pointers on the rear cover.

21. If installing a new belt or if the auto-tensioner has extended or if the timing belt cannot be reinstalled easily, the auto-tensioner must be collapsed before installation of the timing belt, perform the following procedures:

  a. Remove the battery hold-down bolt from the auto-tensioner bracket.

  b. Remove the timing belt auto-tensioner bolts and the auto-tensioner.

  c. Secure the auto-tensioner in a soft jawed vise, clamping onto the flat surface of one of the mounting bolt holes with the maintenance bolt facing upward.

  d. Remove the maintenance bolt and use caution not to spill oil from the tensioner assembly.

  e. Should oil spill from the tensioner, be sure the tensioner is filled with 0.22 ounces (6.5 ml) of fresh engine oil.

  f. Using care not to damage the threads or the gasket sealing surface, insert a flat-blade screwdriver through the tensioner maintenance hole and turn the screwdriver clockwise to compress the auto-tensioner bottom while the tensioner holder tool 14540-P8A-A01 or equivalent, is installed on the auto-tensioner assembly.

  g. Install the auto-tensioner maintenance bolt with a new gasket and tighten to a torque 72 inch lbs. (8 Nm).

  h. Install the auto-tensioner on the engine with the tensioner

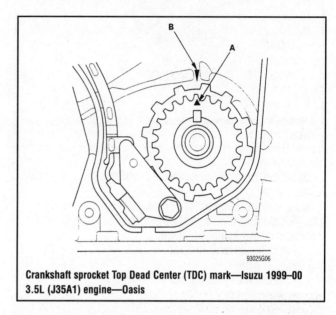

Crankshaft sprocket Top Dead Center (TDC) mark—Isuzu 1999–00 3.5L (J35A1) engine—Oasis

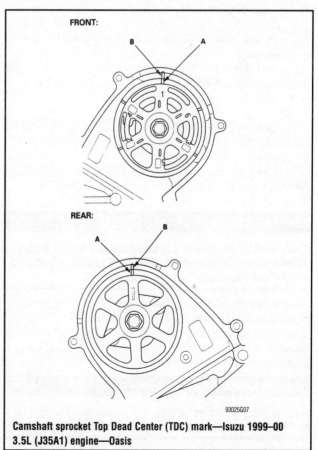

Camshaft sprocket Top Dead Center (TDC) mark—Isuzu 1999–00 3.5L (J35A1) engine—Oasis

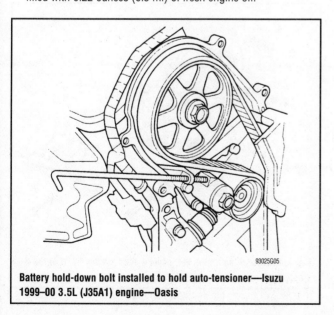

Battery hold-down bolt installed to hold auto-tensioner—Isuzu 1999–00 3.5L (J35A1) engine—Oasis

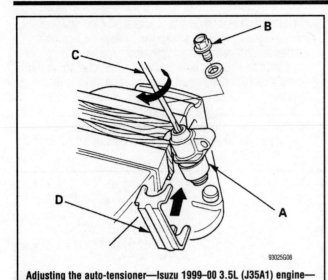

Adjusting the auto-tensioner—Isuzu 1999–00 3.5L (J35A1) engine—Oasis

holder tool installed and torque the mounting bolts to 104 inch lbs. (12 Nm).

22. Install the timing belt in a counterclockwise pattern starting with the crankshaft drive sprocket. Install the timing belt counterclockwise in the following sequence:
- Crankshaft drive sprocket.
- Idler pulley.
- Left side camshaft sprocket.
- Water pump.
- Right side camshaft sprocket.
- Auto-tensioner adjustment pulley.

23. Torque the timing belt idler pulley bolt to 33 ft. lbs. (44 Nm).

24. Remove the auto-tensioner holding tool to allow the tensioner to extend.

25. Install the engine mount bracket to the engine and torque the bolts to 33 ft. lbs. (44 Nm).

26. Install the lower timing belt cover and both upper timing belt covers.

27. Hold the crankshaft pulley with Handle tool 07JAB-001020A and Crankshaft Holding tool 07MAB-PY3010A or equivalent tools. While holding the crankshaft pulley, install the crankshaft pulley bolt using a heavy duty ¾ in. (19mm) socket and a commercially available torque wrench and torque the bolt to 181 ft. lbs. (245 Nm).

❊❊❊ WARNING

If any binding is felt while moving the crankshaft pulley, STOP turning the crankshaft pulley immediately because the pistons may be hitting the valves.

28. Rotate the crankshaft pulley clockwise 5–6 revolutions to allow the timing belt to be seated in the pulleys.

29. Move the crankshaft pulley to the white TDC mark and inspect the camshaft TDC marks to ensure proper timing of the camshafts.

❊❊❊ WARNING

If the timing marks do not align, the timing belt removal and installation procedure must be performed again.

30. Install the engine dipstick tube using a new O-ring.

31. Install the power steering pump, and loosely install the mounting bolt, adjustment locknut and adjustment nut.

32. Adjust the power steering belt to a tension such that a 22 lb. (98 N) pull halfway between the 2 drive pulleys will allow the belt to move 0.51–0.65 in. (13.0–16.5mm).

33. Tighten the power steering pump mounting bolt and adjustment locknut.

➡**If a new belt is used, set the deflection to 0.33–0.43 in. (8.5–11.0mm) and after engine has run for 5 minutes, readjust the new belt to the used belt specification.**

34. Install the alternator belt tensioner arm.

35. Move the alternator tensioner with a Belt Tensioner Release Arm tool YA9317 or equivalent, to release tension from the belt and install the alternator drive belt.

36. Install the 2 bolts for the engine mount bracket and torque to 33 ft. lbs. (44 Nm).

37. Install the bushing through bolt and tighten to 40 ft. lbs. (54 Nm).

38. Release and carefully remove the floor jack.

39. Install the front lower splash shield.

40. Install both front tires/wheels.

41. Carefully lower the vehicle.

42. Install the battery clamp bolt in the battery tray.

43. Install the negative battery cable.

44. Enter the radio security code.

Jeep

2.1L TURBO DIESEL ENGINE

1985–87

1. Disconnect the negative battery cable.

2. Rotate the crankshaft to position the No. 1 cylinder on the Top Dead Center (TDC) of the compression stroke.

3. Remove all of the necessary components in order to gain access to the timing belt cover bolts.

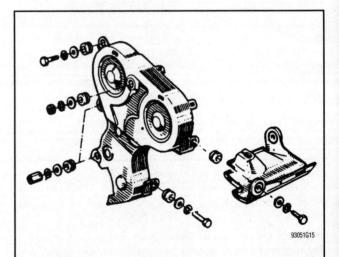

Exploded view of the timing belt cover—Jeep 1985–87 2.1L turbo diesel engine

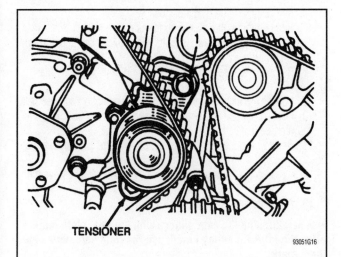

TENSIONER

93051G16

View of the timing belt tensioner bolts—Jeep 1985–87 2.1L turbo diesel engine

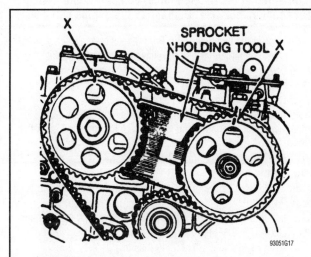

X

SPROCKET HOLDING TOOL X

93051G17

Camshaft and injection pump timing marks aligned—Jeep 1985–87 2.1L turbo diesel engine

4. Remove the timing belt cover retaining bolts and the cover.
5. Install the Camshaft Sprocket Holding tool MOT-854 or equivalent, to retain the camshaft and injection pump sprockets.
6. Loosen the timing belt tensioner bolts, move the tensioner away from the timing belt and secure the bolts.
7. If reusing the timing belt, mark its direction of rotation.
8. Remove the timing belt.
To install:
9. Remove the access plug on the left side of the block and install the Crankshaft Holding tool MOT-861 in the hole.
10. Slowly, rotate the crankshaft clockwise until the tool drops into the TDC locating slot into the crankshaft's counterweight.

➡Do not use this tool as a crankshaft holding tool when tightening or loosening gear train fasteners, use a Flywheel Holding tool MOT-582 or equivalent.

11. If removed, install the Camshaft Sprocket Holding tool MOT-854 or equivalent, to retain the camshaft and injection pump

sprockets. Make sure that the timing marks are positioned correctly.
12. Noting the direction of rotation, install the timing belt. There should be a total of 19 belt teeth between the camshaft and injection pump timing marks.
13. Temporarily position the timing cover over the sprockets. The camshaft and injection pump timing marks must index with the pointer in the cover's timing slots.
14. Remove the cover.
15. Remove the Camshaft Holding tool MOT-854 or equivalent.
16. Loosen the timing belt tensioner bolts are ½ turn, maximum.
17. The tensioner should, automatically, bear against the belt, giving the proper belt tension. Tighten the tensioner bolts.
18. Remove the TDC locating tool and install the plug.
19. Slowly, rotate the crankshaft clockwise 2 complete revolutions.

⚹⚹ WARNING

Never rotate the crankshaft counterclockwise while adjusting the belt tension.

20. Check the belt deflection at a point midway between the camshaft and injection pump sprockets. The belt deflection should be 3–5mm.
21. Install the timing belt cover.
22. To complete the installation, reverse the remove procedures.
23. Connect the negative battery cable.

KIA

2.0L SOHC ENGINE

1995

1. Disconnect the negative battery cable.
2. Remove the intake fresh air duct.
3. Drain the cooling system and remove the upper radiator hose.
4. Remove the accessory drive belts.
5. Remove the cooling fan and fan shroud.
6. Remove the fan pulley.
7. Remove the upper timing cover.
8. Rotate the crankshaft to align the camshaft pulley No. **2** mark with the timing mark on the front housing.
9. Remove the crankshaft pulley.
10. Remove the lower timing cover.
11. Loosen the tensioner lock bolt.
12. Pry the tensioner away from the timing belt and tighten the lock bolt.
13. Remove the timing belt.
To install:
14. Insure that the crankshaft timing mark is aligned with the timing mark on the engine, and that the camshaft pulley No. **2** mark is aligned with the timing mark on the front housing.
15. Install the timing belt.
16. Loosen the tensioner lock bolt to apply tension to the timing belt.
17. Turn the crankshaft 2 complete turns clockwise and check that the timing marks align correctly.
18. Tighten the tensioner lock bolt to 33 ft. lbs. (45 Nm).

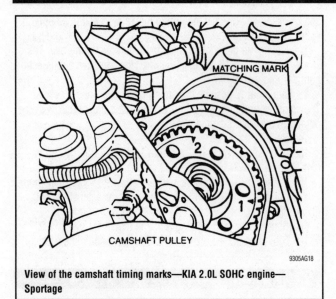

View of the camshaft timing marks—KIA 2.0L SOHC engine—Sportage

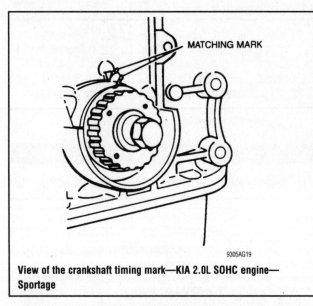

View of the crankshaft timing mark—KIA 2.0L SOHC engine—Sportage

19. Check the timing belt deflection by applying 22 lbs. (10 kg) pressure to the belt between the crankshaft and camshaft pulleys. Belt deflection should be 0.43–0.51 inches. (11–13mm).

20. Install the upper and lower timing covers.

21. Install the crankshaft pulley.

22. Install the cooling fan pulley. Install the cooling fan and shroud.

23. Install the accessory drive belts.

24. Install the upper radiator hose.

25. Install the intake fresh air duct.

26. Connect the negative battery cable and fill the cooling system.

2.0L DOHC ENGINE

1995–00

1. Disconnect the negative battery cable.

2. Properly relieve the fuel system pressure.

3. Remove the alternator drive belt.

4. Remove the fresh air duct from the top of the radiator.

5. Remove the upper radiator hose.

6. Remove the 4 attaching nuts to the clutch fan.

7. Remove the 5 fan shroud bolts. Remove the fan and shroud as an assembly.

8. Remove the 4 splash guard mounting bolts and the splash guard.

9. Loosen the lockbolts and loosen the air conditioning drive belt.

10. Loosen the power steering lock and mounting bolt. Remove the power steering belt.

11. Remove the 5 upper timing belt cover bolts and remove the cover.

12. Remove the 2 lower timing belt cover bolts and remove the cover.

13. Align the timing marks.

➡When aligning the cam pulleys with the seal plate marks, align the left cam pulley I mark and the right cam pulley on the E mark.

❊❊ WARNING

When aligning the timing marks, do not turn the timing gear counterclockwise. Damage to the engine will occur.

14. Loosen the tensioner bolt. Pry the tensioner away from the belt. Tighten the tensioner bolt to relieve the pressure against the timing belt.

15. Remove the timing belt.

16. Remove the camshaft pulley attaching bolts. Use a driver placed through one of the holes in the pulley to prevent it from moving when the attaching bolt is removed. Remove and mark the pulleys.

17. Remove the lower timing belt pulley and locking bolt.

To install:

18. Install the camshaft pulleys. Tighten the bolts to 35–48 ft. lbs. (47–65 Nm).

19. Install the lower timing belt pulley and locking bolt. Tighten the bolt to 120 ft. lbs. (162 Nm).

20. If necessary, align the timing marks.

➡When aligning the cam pulleys with the seal plate marks, align the left cam pulley I mark and the right cam pulley on the E mark.

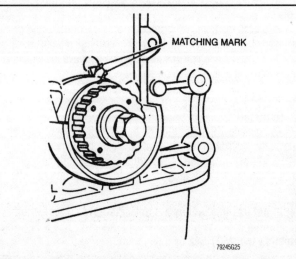

Align the crankshaft marks before removing the timing belt —KIA 2.0L (DOHC) engine—Sportage

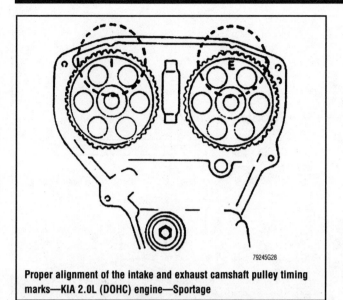

Proper alignment of the intake and exhaust camshaft pulley timing marks—KIA 2.0L (DOHC) engine—Sportage

✳✳ WARNING

When aligning the timing marks, do not turn the timing gear counterclockwise. Damage to the engine will occur.

21. Loosen the tensioner bolt. Pry the tensioner away from the belt. Tighten tensioner bolt to relieve the pressure against the timing belt.
22. Install the timing belt.

✳✳ WARNING

If any binding is felt when adjusting the timing belt tension by turning the crankshaft, STOP turning the engine, because the pistons may be hitting the valves.

23. Loosen the tensioner bolt and allow the tensioner to tighten the timing belt. Tighten the tensioner bolt 27–38 ft. lbs. (37–52 Nm).
24. Check the timing belt deflection. If there is more than 0.30–0.33 in. (7.5–8.5mm) replace the tensioner spring.
25. Install the 2 lower timing belt cover bolts to the cover.
26. Install the 5 upper timing belt cover bolts to the cover.
27. Install and adjust the air conditioning and power steering drive belts.
28. Install the splash guard.
29. Install and tighten the alternator belt.
30. Install the upper radiator hose.
31. Install the fan and shroud as an assembly.
32. Install the 4 attaching nuts to the clutch fan.
33. Install the 5 fan shroud bolts.
34. Install the fresh air duct to the top of the radiator.
35. Properly fill the cooling system.
36. Connect the negative battery cable.
37. Start the engine and check for leaks.
38. Road test the vehicle.

Lexus

4.7L (2UZ-FE) ENGINE

1. Disconnect the negative battery cable.
2. Raise and safely support the vehicle.

3. Remove the oil pan protector and the engine under cover.
4. Drain the cooling system and store the coolant for refilling purposes.
5. Lower the vehicle and remove the battery clamp cover.
6. From the top of the engine, remove the fuel return hose, the engine cover nuts/bolts and the cover.
7. Remove the air cleaner and the intake air connector assembly.
8. Remove the cooling fan pulley by performing the following procedures:
   a. Loosen the 4 fan clutch-to-fan pulley nuts.
   b. Using a box-end wrench on the serpentine drive belt tensioner bolt, rotate the tensioner counterclockwise and remove the drive belt.

➡ **The serpentine drive belt tensioner bolt is a left-hand thread.**

   c. Remove the fan clutch-to-fan pulley nuts, the fan, the clutch assembly and the fan pulley.
9. Remove the radiator by performing the following procedures:
   a. Disconnect the upper, lower and reservoir hoses from the radiator.
   b. Disconnect and plug the automatic transmission oil cooler at the radiator. Disconnect the automatic transmission oil cooler hoses from the fan shroud clamp.
   c. Remove the radiator reservoir tank.
   d. Remove the fan shroud-to-radiator bolts and the shroud.
   e. Remove the 2 upper radiator-to-chassis nuts.
   f. Remove the middle radiator-to-chassis nut/bolts and brackets.
   g. Carefully, lift the radiator from the vehicle.
10. Remove the serpentine drive belt idler pulley bolt, cover plate and pulley.
11. Remove the right side (No. 3) timing belt cover.
12. Remove the left side (No. 3) timing belt cover by performing the following procedures:
   a. Disconnect the engine wire from both wire clamps.
   b. Disconnect the camshaft position sensor wire from the wire clamp on the left-side (No. 3) timing belt cover.
   c. Disconnect the sensor connector from the connector bracket.
   d. Disconnect the sensor connector.
   e. Remove the wire grommet from the left-side (No. 3) timing belt cover.
   f. Remove the oil cooler tube bolts and tube.
13. Remove the middle (No. 2) timing belt cover bolts and cover.
14. Remove the cooling fan bracket nuts/bolts and bracket.

➡ **If reusing the timing belt, make sure that there are 3 installation marks on the belt; if there are none, install them.**

15. Using the Crankshaft Pulley Holding tool 09213-70010, Bolt tool 90105-08076 and Companion Flange Holding tool 09330-00021or equivalent, loosen the crankshaft pulley bolt.
16. Position the No. 1 cylinder to approximately 50 degrees After Top Dead Center (ATDC) of the compression stroke by performing the following procedures:
   a. Rotate the crankshaft pulley (CLOCKWISE) to align its groove with the timing mark **0** on the lower (No. 1) timing belt cover.
   b. Check that the camshaft sprocket timing marks are aligned with the rear timing belt plate marks; if not, rotate the crankshaft 1 revolution (360 degrees).
   c. Rotate the crankshaft pulley approximately 50 degrees (CLOCKWISE) and align the crankshaft pulley timing mark between the centers of the crankshaft pulley bolt and the idler pulley bolt.

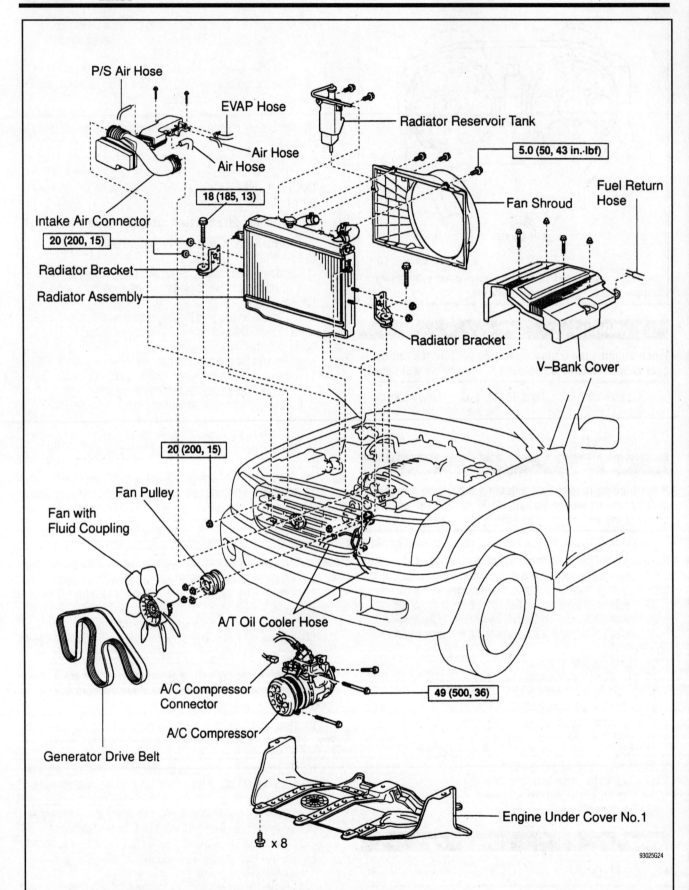

P/S Air Hose

EVAP Hose

Radiator Reservoir Tank

Air Hose

Air Hose

5.0 (50, 43 in.-lbf)

Fuel Return Hose

Fan Shroud

18 (185, 13)

Intake Air Connector

20 (200, 15)

Radiator Bracket

Radiator Assembly

Radiator Bracket

V–Bank Cover

20 (200, 15)

Fan Pulley

Fan with Fluid Coupling

A/T Oil Cooler Hose

A/C Compressor Connector

49 (500, 36)

A/C Compressor

Generator Drive Belt

Engine Under Cover No.1

x 8

93025G24

Exploded view of vehicle components for timing belt replacement—Lexus 4.7L (2UZ-FE) engine—LX 470

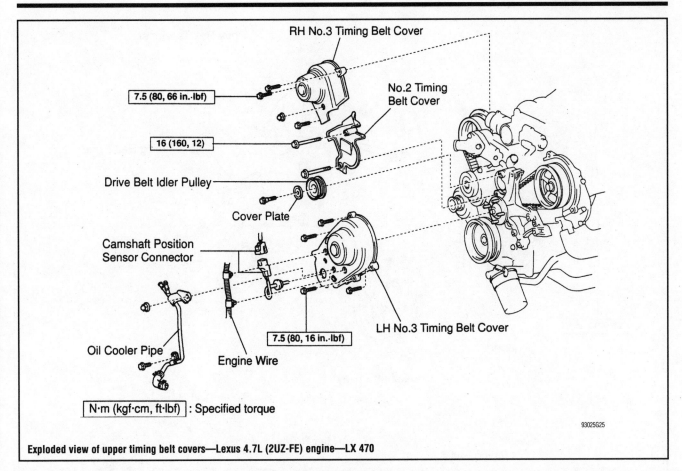

RH No.3 Timing Belt Cover

7.5 (80, 66 in.·lbf)

No.2 Timing
Belt Cover

16 (160, 12)

Drive Belt Idler Pulley

Cover Plate

Camshaft Position
Sensor Connector

LH No.3 Timing Belt Cover

7.5 (80, 16 in.·lbf)

Oil Cooler Pipe

Engine Wire

N·m (kgf·cm, ft·lbf) : Specified torque

93025G25

Exploded view of upper timing belt covers—Lexus 4.7L (2UZ-FE) engine—LX 470

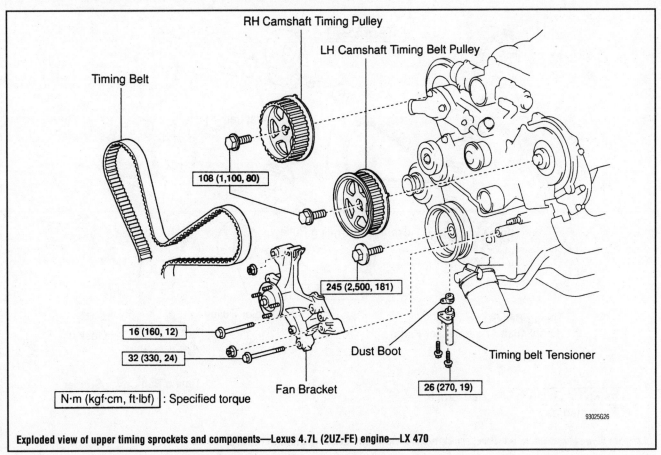

RH Camshaft Timing Pulley

LH Camshaft Timing Belt Pulley

Timing Belt

108 (1,100, 80)

245 (2,500, 181)

16 (160, 12)

32 (330, 24)

Dust Boot

Timing belt Tensioner

Fan Bracket

26 (270, 19)

N·m (kgf·cm, ft·lbf) : Specified torque

93025G26

Exploded view of upper timing sprockets and components—Lexus 4.7L (2UZ-FE) engine—LX 470

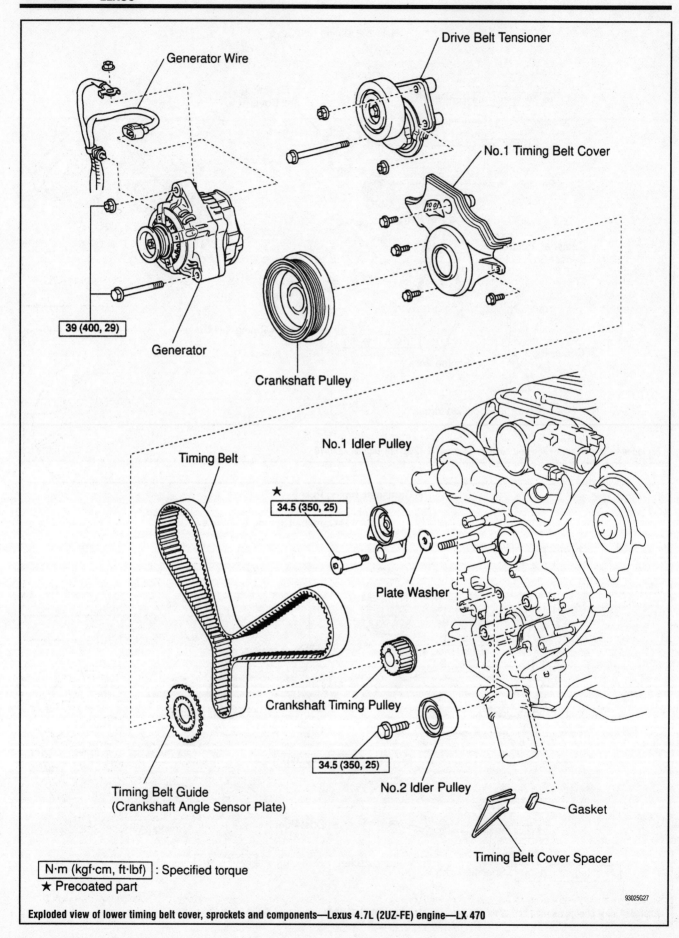

Generator Wire

Drive Belt Tensioner

No.1 Timing Belt Cover

39 (400, 29)

Generator

Crankshaft Pulley

Timing Belt

No.1 Idler Pulley

★ 34.5 (350, 25)

Plate Washer

Crankshaft Timing Pulley

Timing Belt Guide
(Crankshaft Angle Sensor Plate)

34.5 (350, 25)

No.2 Idler Pulley

Gasket

Timing Belt Cover Spacer

N·m (kgf·cm, ft·lbf) : Specified torque
★ Precoated part

93025G27

Exploded view of lower timing belt cover, sprockets and components—Lexus 4.7L (2UZ-FE) engine—LX 470

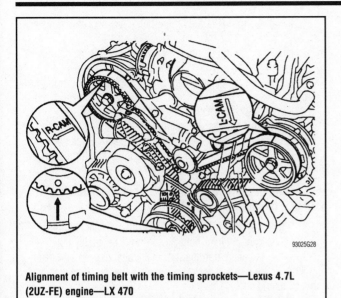

Alignment of timing belt with the timing sprockets—Lexus 4.7L (2UZ-FE) engine—LX 470

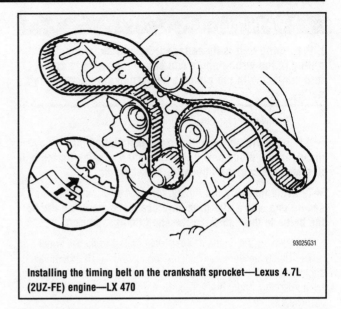

Installing the timing belt on the crankshaft sprocket—Lexus 4.7L (2UZ-FE) engine—LX 470

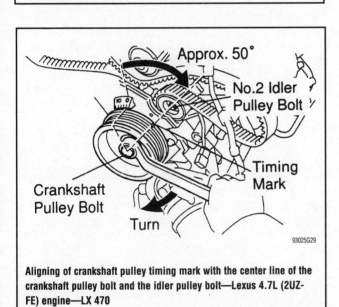

Aligning of crankshaft pulley timing mark with the center line of the crankshaft pulley bolt and the idler pulley bolt—Lexus 4.7L (2UZ-FE) engine—LX 470

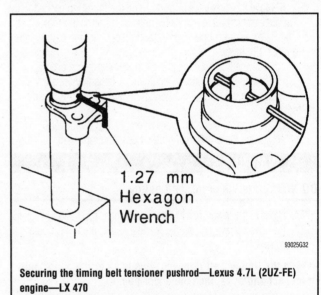

Securing the timing belt tensioner pushrod—Lexus 4.7L (2UZ-FE) engine—LX 470

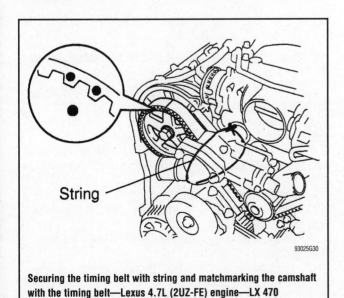

Securing the timing belt with string and matchmarking the camshaft with the timing belt—Lexus 4.7L (2UZ-FE) engine—LX 470

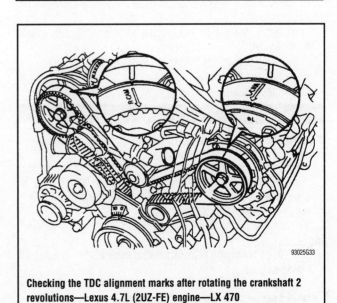

Checking the TDC alignment marks after rotating the crankshaft 2 revolutions—Lexus 4.7L (2UZ-FE) engine—LX 470

✳ WARNING

If the timing belt is disengaged, having the crankshaft pulley in the wrong angle can cause the valve to come into contact with the piston when removing the camshaft pulley.

17. Remove the crankshaft pulley bolt.

➡ **If reusing the timing belt and the installation marks have disappeared, place new installation marks on the timing belt to match the camshaft timing sprocket marks.**

➡ **To avoid meshing the timing sprocket and the timing belt, secure one with a string; then, place matchmarks on the timing belt and the right-side camshaft timing sprocket.**

18. Remove the timing belt tensioner bolts and the tensioner.

19. Using the Camshaft Holding tool 09960-10010 or equivalent, slightly turn the left-side camshaft sprocket clockwise to loosen the tension spring. Then, disconnect the timing belt from the camshaft sprockets.

20. Remove the alternator by performing the following procedures:
   a. Disconnect the electrical connector from the alternator.
   b. Remove the rubber cap/nut and disconnect the battery wire from the alternator.
   c. Disconnect the wire clamp from the alternator cord clip.
   d. Remove the alternator-to-engine nuts/bolts and the alternator.

21. Remove the serpentine drive belt tensioner nuts/bolts and the tensioner.

22. Using the Crankshaft Puller Assembly tool 09950-50012 or equivalent, press the crankshaft pulley from the crankshaft.

✳ WARNING

DO NOT rotate the crankshaft pulley.

23. Remove the lower (No. 1) timing belt cover bolts and the cover.
24. Remove the timing belt guide, spacer and the timing belt.
To install:

➡ **With the timing belt removed, this is a perfect opportunity to inspect and/or replace the water pump.**

25. Inspect the timing belt tensioner by performing the following procedures:
   a. Inspect the seal for leakage; if leakage is suspected, replace the tensioner.
   b. Using both hands to hold the tensioner facing upward, strongly press the pushrod against a solid surface. If the pushrod moves, replace the tensioner.

✳ WARNING

Never hold the tensioner with the pushrod facing downward.

   c. Measure the pushrod's protrusion from the housing end, it should be 0.413–0.453 in. (10.5–11.5mm). If the protrusion is not as specified, replace the tensioner.
26. Temporarily install the timing belt by performing the following procedures:
   a. Align the timing belt's installation mark with the crankshaft timing sprocket.
   b. Install the timing belt on the crankshaft timing sprocket, the No. 1 idler pulley and the No. 2 idler pulley.

27. Install the gasket to the timing belt cover spacer and install the cover spacer.
28. Install the timing belt guide with the cup side facing outward.
29. Install the lower (No. 1) timing belt cover.
30. Install the crankshaft pulley by performing the following procedures:
   a. Align the crankshaft pulley with the crankshaft key.
   b. Using the Crankshaft Installer tool 09223-46011 or equivalent, and a hammer, tap the crankshaft pulley into position.
31. Install the serpentine drive belt tensioner and torque the tensioner-to-engine bolts to 12 ft. lbs. (16 Nm).

➡ **To install the serpentine drive belt tensioner, use a bolt 4.18 in. (106mm) in length.**

32. Check that the crankshaft pulley's timing mark is aligned with the centers of the idler pulley and crankshaft pulley bolts.
33. Install the alternator and torque the alternator-to-engine nuts/bolts to 29 ft. lbs. (39 Nm). Connect the alternator's electrical connectors and clip.
34. Install the timing belt to the left-side camshaft by performing the following procedures:
   a. Rotate the left-side camshaft pulley to align the timing belt installation mark with the camshaft sprocket's timing mark and slide the belt onto the camshaft timing sprocket.
   b. Using the Camshaft Holding tool 09960-10010 or equivalent, slightly turn the left-side camshaft sprocket counterclockwise to place tension on the timing belt between the crankshaft sprocket and the camshaft sprocket.
35. Rotate the right-side camshaft pulley to align the timing belt installation mark with the camshaft sprocket's timing mark and slide the belt onto the camshaft timing sprocket.
36. Using a vertical press, slowly press the pushrod into the housing using 200–2205 lbs. (981–9807 N) until the holes align, then, install a 1.27mm Allen® wrench to secure the pushrod and release the press. Install the dust boot on the tensioner housing.
37. Install the timing belt tensioner and torque the bolts to 19 ft. lbs. (26 Nm).
38. Using a pair of pliers, remove the Allen® wrench from the tensioner housing.
39. Check the valve timing by performing the following procedure:
   a. Temporarily install the crankshaft pulley bolt.
   b. Slowly, rotate the crankshaft pulley 2 revolutions (CLOCKWISE) and realign the TDC marks.

➡ **If the pulley/sprocket timing marks do not realign, remove the timing belt and reinstall it.**

40. Using the Crankshaft Pulley Holding tool 09213-70010, Bolt tool 90105-08076 and Companion Flange Holding tool 09330-00021 or equivalent, torque the crankshaft pulley bolt to 181 ft. lbs. (245 Nm).
41. Install the cooling fan bracket and torque the 12mm (head size) bolt to 12 ft. lbs. (16 Nm) and the 14mm (head size) bolt to 24 ft. lbs. (32 Nm).
42. Install the air conditioning compressor.
43. Install the middle (No. 2) timing belt cover and torque the bolts to 12 ft. lbs. (16 Nm).
44. Install the upper right-side (No. 3) timing belt cover and torque the bolts to 66 inch lbs. (7.5 Nm).
45. Install the upper left-side (No. 3) timing belt cover by performing the following procedures:
   a. Install the oil cooler tube and bolt.

b. Feed the Camshaft Position Sensor (CPS) through the left-side (No. 3) timing belt cover hole.

c. Install the left-side (No. 3) timing belt cover and torque the bolts to 66 inch lbs. (7.5 Nm).

d. Install the wire grommet to the left-side (No. 3) timing belt cover.

e. Install the sensor connector to the connector bracket and connect the sensor connector.

f. Install the sensor wire and the engine wire to the clamps on the left-side (No. 3) timing belt cover.

46. Install the drive belt idler pulley and cover plate; then, torque the pulley bolt to 27 ft. lbs. (37 Nm).

47. To complete the installation, reverse the removal procedures.

48. Refill the cooling system and connect the negative battery cable.

Mazda

1987–1993 2.2L (F2) ENGINES

1. Disconnect the negative battery cable.
2. Drain the cooling system.
3. Remove the cooling fan and the fan shroud.
4. Remove the accessory drive belts.
5. Remove the cooling fan pulley.
6. Remove the cooling fan bracket.
7. Remove the secondary air pipe assembly.
8. Remove the crankshaft pulley retaining bolts and remove the crankshaft pulley.
9. Remove the upper and lower timing belt covers and remove the baffle plate from in front of the crankshaft sprocket.
10. Turn the crankshaft clockwise until the "arrow" **1** mark on the camshaft is aligned with the mark on top of the seal plate.
11. Loosen the tensioner lockbolt. Position the tensioner with the spring in a fully extended position. Torque the lockbolt.
12. Remove the timing belt.
13. If the timing belt is to be reused, mark the direction of rotation on the timing belt.

➡**Do not rotate the engine after the timing belt has been removed.**

14. Inspect the belt for wear, peeling, cracking, hardening or signs of oil contamination. Inspect the tensioner pulley for free and smooth rotation. Measure the tensioner spring free length; it should not exceed 2.480 in. (63mm) from end to end. Check the camshaft and crankshaft sprockets for worn teeth or other damage.

To install:

15. Make sure the crankshaft and camshaft timing marks are aligned.

16. Install the timing belt tensioner and spring. Move the tensioner until the spring is fully extended and temporarily torque the tensioner bolt to hold it in place.

17. Install the timing belt. Make sure there is no slack at the side of the water pump and idler pulleys. If reusing the old belt, be sure it is installed in the original direction of rotation as marked.

18. Turn the crankshaft 2 turns in the direction of rotation and make sure the timing marks are aligned.

19. Loosen the tensioner lockbolt to apply tension to the belt. Torque the tensioner bolt to 38 ft. lbs. (52 Nm).

20. Turn the crankshaft again 2 turns in the direction of rotation and make sure the timing marks are aligned.

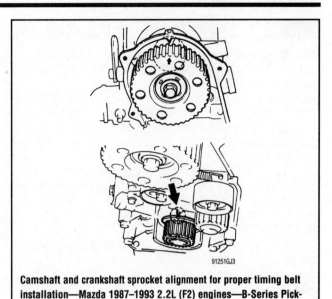

Camshaft and crankshaft sprocket alignment for proper timing belt installation—Mazda 1987–1993 2.2L (F2) engines—B-Series Pick-Ups

21. Apply approximately 22 lbs. (98 N) pressure to the timing belt at a point midway between the idler pulley and camshaft sprocket. A new belt should deflect 0.31–0.35 in. (8–9mm). A used belt should deflect 0.35–0.39 in. (9–10mm). If the deflection is not as specified, replace the tensioner spring.

22. Install the baffle plate with the dished side facing away from the engine.

23. Install the timing belt covers and torque the bolts to 87 inch lbs. (10 Nm).

24. Install the crankshaft pulley and torque the bolts to 13 ft. lbs. (17 Nm).

25. Install the secondary air pipe assembly.
26. Install the cooling fan bracket.
27. Install the cooling fan pulley.
28. Install the cooling fan and the fan shroud.
29. Install the accessory drive belts and adjust the belt tension.
30. Refill and bleed the cooling system.
31. Connect the negative battery cable.
32. Start the engine and check for proper operation.
33. Check the ignition timing.

1994–97 2.3L (VIN A) & 1998–00 2.5L (VIN C) ENGINES

1. Disconnect the negative battery cable.
2. Rotate the engine so that No. 1 cylinder is at Top Dead Center (TDC) on the compression stroke. Check that the timing marks are aligned on the camshaft and crankshaft pulleys. An access plug is provided in the cam belt cover so that the camshaft timing can be checked without removal of the cover. Set the crankshaft to TDC by aligning the timing mark on the crank pulley with the TDC mark on the belt cover. Look through the access hole in the belt cover to be sure that the timing mark on the cam drive sprocket is aligned with the pointer on the inner belt cover.

❄❄ WARNING

Always turn the engine in the normal direction of rotation. Backward rotation may cause the timing belt to jump time, due to the arrangement of the belt tensioner.

3. Drain the cooling system.

4. Remove the upper radiator hose.

5. Remove the cooling fan and water pump pulley bolts.

6. Loosen the alternator retaining bolts and remove the drive belt from the pulleys.

7. Remove the water pump pulley.

8. Remove the power steering pump and set it aside.

9. Remove the 4 timing belt outer cover retaining bolts and remove the cover.

10. Remove the crankshaft pulley and belt guide.

11. Loosen the belt tensioner pulley assembly, then position a Camshaft Belt Adjuster tool T74P-6254-A or equivalent, on the tension spring roll pin and retract the belt tensioner away from the timing belt. Tighten the adjustment bolt to lock the tensioner in the retracted position.

12. If the belt is to be reused, mark the direction of rotation on the timing belt for installation reference.

13. Remove the timing belt.

To install:

14. Install the new belt over the crankshaft sprocket and then counterclockwise over the auxiliary and camshaft sprockets, making sure the lugs on the belt properly engage the sprocket teeth on the pulleys. Be careful not to rotate the pulleys when installing the belt.

15. Release the timing belt tensioner pulley, allowing the tensioner to take up the belt slack. If the spring does not have enough tension to move the roller against the belt, it may be necessary to manually push the roller against the belt and tighten the bolt.

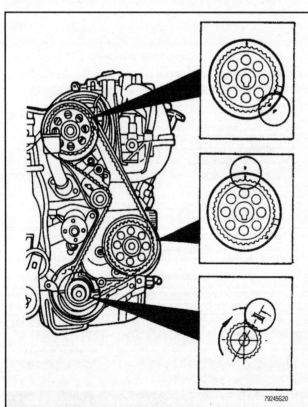

79245G20

Camshaft, auxiliary shaft and crankshaft timing belt sprocket alignment mark locations—Mazda 1996–97 2.3L (VIN A) and 1998–00 2.5L (VIN C) engines—B-Series Pick-Ups

➡The spring cannot be used to set belt tension; a wrench must be used on the tensioner assembly.

⁕⁕ WARNING

If any binding is felt when adjusting the timing belt tension by turning the crankshaft, STOP turning the engine, because the pistons may be hitting the valves.

16. Rotate the crankshaft 2 complete turns by hand (in the normal direction of rotation) to remove slack from the belt. Torque the tensioner adjustment to 26–33 ft. lbs. (35–45 Nm) and pivot bolts to 30–40 ft. lbs. (40–55 Nm). Be sure the belt is seated properly on the pulleys and that the timing marks are in alignment.

17. Install the crankshaft pulley and belt guide. Torque the bolt to 93–121 ft. lbs. (125–165 Nm).

18. Install the timing belt cover.

19. Install the water pump pulley and the cooling fan. Torque the bolts to 13–17 ft. lbs. (17–23 Nm).

20. Install the accessory drive belts and adjust tension.

21. Install the upper radiator hose.

22. Refill and bleed the cooling system.

23. Disconnect the negative battery cable.

24. Start the engine and check for proper operation.

25. Check ignition timing and adjust if necessary.

3.0L (JE) ENGINE

1. Disconnect the negative battery cable.

2. Drain the cooling system.

3. Tag and disconnect the spark plug wires. Remove the spark plugs.

4. Remove the fresh air duct.

5. Remove the cooling fan and the fan shroud.

6. Remove the accessory drive belts and the air conditioning compressor idler pulley.

7. Remove the crankshaft pulley and baffle plate.

8. Remove the coolant bypass hose and the upper radiator hose.

9. Remove the timing belt covers and gaskets.

10. Turn the crankshaft, in the normal direction of rotation, and align the crankshaft and camshaft sprocket timing marks. Mark the direction of rotation on the timing belt.

11. Remove the upper idler pulley and remove the timing belt. Remove the automatic tensioner.

➡Do not rotate the engine after the timing belt has been removed.

12. Inspect the belt for wear, peeling, cracking, hardening or signs of oil contamination. Inspect the tensioner pulley for free and smooth rotation. Check the automatic tensioner for oil leakage. Check the tensioner rod projection (free length); it should be 0.47–0.55 in. (12–14mm). Inspect the sprocket teeth for wear or damage. Replace parts, as necessary.

To install:

13. Position the automatic tensioner in a suitable press. Place a flat washer under the tensioner body to prevent damage to the body plug.

14. Slowly press in the tensioner rod but do not exceed 2200 lbs. (998 kg) force. Insert a pin into the tensioner body to hold the rod in place.

15. Install the tensioner and torque the bolts to 19 ft. lbs. (25 Nm).

16. Make sure the crankshaft and camshaft sprocket timing marks are aligned.

17. Install the timing belt in the following order:
  a. Crankshaft sprocket
  b. Lower idler pulley
  c. Left camshaft sprocket
  d. Right camshaft sprocket
  e. Tensioner pulley

18. If reusing the old timing belt, make sure it is installed in the same direction of rotation as marked.

19. Install the upper idler pulley and torque the bolt to 38 ft. lbs. (52 Nm).

20. Turn the crankshaft 2 turns, in the normal direction of rotation, and align the timing marks.

21. Remove the pin from the automatic tensioner. Turn the crankshaft 2 turns, in the normal direction of rotation, and make sure the timing marks are aligned.

22. Apply approximately 22 lbs. (98 N) pressure to the timing belt at a point midway between the right camshaft sprocket and tensioner pulley. The belt should deflect 0.20–0.28 in. (5–7mm). If the deflection is not as specified, replace the automatic tensioner.

23. Install the timing belt covers with new gaskets. Torque the bolts to 95 inch lbs. (11 Nm).

24. Install the upper radiator hose and coolant bypass hose.

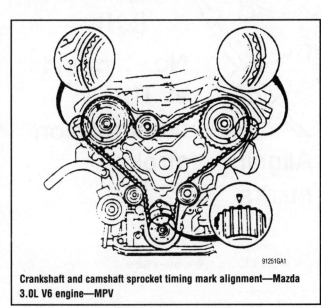

Crankshaft and camshaft sprocket timing mark alignment—Mazda 3.0L V6 engine—MPV

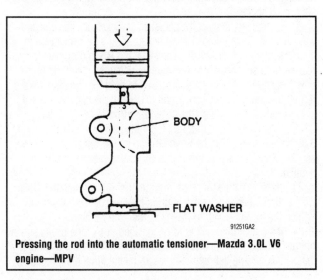

Pressing the rod into the automatic tensioner—Mazda 3.0L V6 engine—MPV

25. Install the baffle plate and crankshaft pulley and torque the bolts to 130 inch lbs. (15 Nm).

26. Install the air conditioning compressor idler pulley and torque the bolts to 19 ft. lbs. (25 Nm).

27. Install the fan shroud and cooling fan.

28. Install the accessory drive belts and adjust the tension.

29. Install the fresh air duct.

30. Install the spark plugs and connect the spark plug wires.

31. Connect the negative battery cable.

32. Refill and bleed the cooling system.

33. Start the engine and bring to normal operating temperature.

34. Check for leaks and for proper operation.

35. Check the ignition timing.

Mercury

3.0L (VIN W) & 3.3L (VIN T) ENGINES

On this vehicle, right side refers to the "rear" components (near the firewall) and left side refers to the "front" components (near the radiator).

1. If the timing belt is to be removed, it is good practice to turn the crankshaft until the engine is at Top Dead Center (TDC) of the No. 1 cylinder, compression stroke (firing position), before beginning work. This should align all timing marks and serve as a reference for all work that follows. After verifying that the engine is at TDC for the No. 1 cylinder, do not crank the engine or allow the crankshaft or camshaft sprockets to be turned otherwise engine timing will be lost.

2. Drain the cooling system.

3. Disconnect the negative battery cable.

4. Remove the alternator drive belt, water pump and power steering pump belt and the air conditioning compressor belt, if equipped, using the recommended drive belt removal procedure.

5. If equipped with air conditioning, remove the 3 air conditioning compressor drive belt idler pulley bolts and remove the idler pulley.

6. Remove the upper radiator hose bracket bolt. Remove the upper hose with the bracket from the vehicle.

7. Remove the water bypass hose from between the thermostat housing and the lower water hose connection.

8. Remove the main wiring harness from the upper engine front cover.

9. Remove the 8 upper engine front cover bolts and remove the upper cover.

10. Raise and safely support the vehicle.

11. Remove the right side front wheel and tire assembly.

12. Remove the 4 right side engine and transmission splash shield bolts and 2 screws, and remove the right side outer engine and transaxle splash shield.

13. Use a strap wrench to hold the water pump pulley. Remove the 4 pulley bolts, and the water pump pulley.

14. Use a strap wrench to hold the crankshaft pulley. Remove the center pulley bolt, and the crankshaft pulley using a harmonic balancer (damper) puller to draw the pulley from the front of the crankshaft.

15. Remove the 5 lower engine front cover bolts, then remove the lower engine front cover.

16. Be sure that the timing marks between the crankshaft sprocket and the oil pump housing align.

17. If the timing belt is to be reused, mark an arrow on the belt indicating the direction of rotation. The directional arrow is necessary to ensure that the timing belt, if it to be reused, can reinstalled in the same direction.

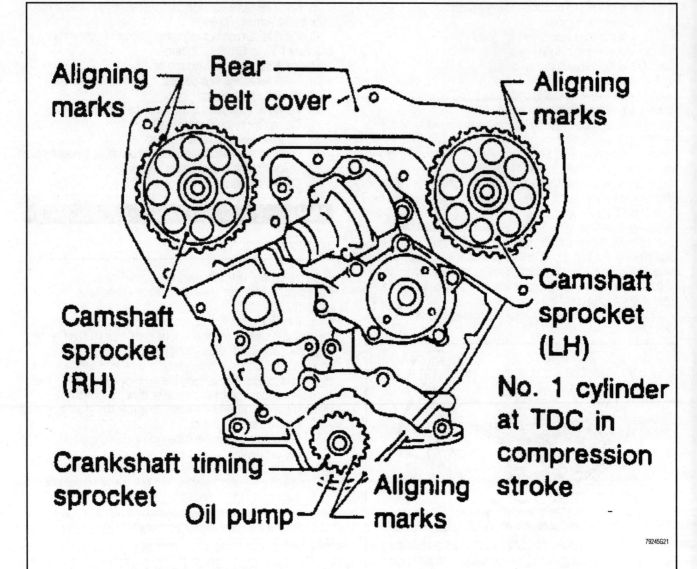

Aligning marks

Rear belt cover

Aligning marks

Camshaft sprocket (RH)

Camshaft sprocket (LH)

No. 1 cylinder at TDC in compression stroke

Crankshaft timing sprocket

Oil pump

Aligning marks

79245G21

Use a shop rag to clean the alignment marks for the timing belt—Mercury 3.0L (VIN W) and 3.3L (VIN T) engines—Villager

18. Loosen the timing belt tensioner nut and slip the timing belt off of the sprockets.

19. If necessary, the camshaft sprockets can be removed. A special spanner tool is designed to hold the sprocket to keep it from turning while the center bolt is being loosened. Use care if using substitutes.

➡The sprockets are not interchangeable.

20. If necessary, the crankshaft sprocket can be removed. The outer timing belt guide (looks like a large washer) and the crankshaft sprocket simply pull off the front of the crankshaft.

➡Be careful, there are 2 crankshaft keys. Use care not to loose them.

To install:

21. Clean all parts well. If removed, inspect the crankshaft sprocket for warping or abnormal wear. Check the sprocket teeth for wear, deformation, chipping or other damage. Replace as necessary. Clean the sprocket mounting surface to ease installation. Install the

key. Slip the sprocket onto the crankshaft. Tap it in place with a suitably-sized socket.

22. If removed, inspect the camshaft sprockets for damage and wear. Replace as required. The sprockets should be marked **L3** to designate the front, or left side camshaft and **R3** to designate the rear, or right side camshaft. Use care to install the sprockets properly. A special spanner tool is designed to hold the sprocket to keep it from turning while the center bolt is being tightened. Use care if using a substitute. Tighten the camshaft sprocket center bolts to 58–65 ft. lbs. (78–88 Nm) for 3.0L engine or 61 ft. lbs. (83 Nm) for 3.3L engine. Verify that the timing marks on the camshaft sprockets and the timing marks on the rear cover (called the seal plate) are aligned.

23. Use an Allen wrench to turn the timing belt tensioner clockwise until the belt tensioner spring is fully extended. Temporarily tighten the tensioner nut to 32–43 ft. lbs. (43–58 Nm).

24. If a new timing belt is to be installed, look for a printed arrow on the belt. Be sure the arrow is pointing away from the engine. If the original timing belt is to be reused, be sure that the directional

arrow that was marked at disassembly is facing the correct direction.

25. A new Original Equipment Manufacture (OEM) timing belt should have 3 white timing marks on it that indicate the correct timing positions of the camshafts and the crankshaft. These marks are to help ensure that the engine is properly timed. When the engine is properly timed, each white timing mark on the timing belt will be aligned with the corresponding camshaft and crankshaft timing mark on the sprocket. Because the white timing marks are not evenly spaced, the technician needs to use care in installing the belt. There should be 40 timing belt teeth between the timing marks on the front and rear camshaft sprockets and 43 teeth between the timing mark on the front camshaft sprocket and the timing mark on the crankshaft sprocket.

26. Verify that the camshaft timing marks are aligned with the timing marks on the rear cover (seal plate) and that the crankshaft sprocket timing mark is aligned with the timing mark on the oil pump housing.

27. Install the timing belt starting at the crankshaft sprocket and moving around the camshaft sprockets following a counterclockwise path. Do not allow any slack in the timing belt between the sprockets. After all of the timing marks are aligned with the timing belt installed, slip the timing belt onto the belt tensioner.

28. While holding the timing belt tensioner with an Allen wrench, loosen the tensioner nut. Allow the tensioner to put pressure on the timing belt. Use an Allen wrench to turn the timing belt tensioner 70–80 degrees clockwise and tighten the timing belt tensioner nut to 32–43 ft. lbs. (43–58 Nm).

✳✳ WARNING

If any binding is felt when adjusting the timing belt tension by turning the crankshaft, STOP turning the engine, because the pistons may be hitting the valves.

29. Rotate the crankshaft clockwise twice and align the No. 1 piston to TDC on the compression stroke (firing position).

30. Apply 22 lbs. (10kg) of force on the timing belt between the rear camshaft sprocket and the timing belt tensioner. An assistant may be needed. While holding the timing belt tensioner steady with an Allen wrench, loosen the timing belt tensioner nut. Remove the Allen wrench and adjust the timing belt tensioner using the following procedure:

   a. Install a 0.0138 in. (0.35mm) thick and 0.500 in. (12.7mm) wide feeler gauge where the timing belt just starts to go around the tensioner (approximately the 4 o'clock position, looking at the tensioner).

   b. Turn the crankshaft sprocket clockwise, which should force the feeler gauge between the timing belt and the tensioner, up to a position on the tensioner of about 1 o'clock.

   c. Tighten the timing belt tensioner nut to 32–43 ft. lbs. (43–58 Nm) for the 3.0L engine or 61 ft. lbs. (83 Nm) for the 3.3L engine.

   d. Turn the crankshaft clockwise to rotate the feeler gauge out from between the timing belt tensioner and the timing belt.

31. Rotate the crankshaft clockwise twice, and once again align the No. 1 piston to TDC on the compression stroke (firing position).

32. Apply 22 lbs. (10kg) of force on the timing belt between the front and rear camshaft sprockets. Measure the amount of belt deflection. Belt deflection should be between 0.51–0.59 in.

(13–15mm). If belt deflection is out of specification, repeat Steps 29 through 33. If the timing belt deflection cannot be adjusted into specification, the timing belt will have to be replaced.

33. Position the lower engine front cover and install the 5 lower cover bolts. Do not over tighten. Tighten to 27–44 inch lbs. (3–5 Nm).

34. Install the outer timing belt guide next to the crankshaft sprocket with the dished side facing away from the cylinder block. Install the crankshaft pulley. Use a strap wrench to keep the crankshaft pulley from turning and tighten the center bolt to 90–98 ft. lbs. (123–132 Nm) for the 3.0L engine or 148 ft. lbs. (201 Nm) for the 3.3L engine.

35. Position the water pump pulley on the pump. Install the 4 bolts. Use a strap wrench to keep the water pump pulley from turning and tighten the 4 water pump pulley bolts to 12–15 ft. lbs. (16–21 Nm) for the 3.0L engine or 89 inch lbs. (10 Nm) for the 3.3L engine.

36. Position the right side outer engine and transaxle splash shield, and secure with the 4 bolts and 2 screws.

37. Install the right side front wheel. Tighten the lug nuts to 72–87 ft. lbs. (98–118 Nm).

38. Lower the vehicle.

39. Position the upper engine timing belt front cover, and tighten the 8 bolts to 27–44 inch lbs. (3–5 Nm).

40. Install the main wiring harness on the upper engine front cover.

41. Position the water bypass hose between the thermostat housing and water connection. Install the upper radiator hose between the radiator and the water hose connection. Secure the hoses with clamps. Install the upper radiator hose bracket. Tighten the bracket bolt to 34–58 ft. lbs. (46–65 Nm).

42. If equipped, position the air conditioning compressor drive belt idler pulley and install the 3 bolts. Tighten to 15 ft. lbs. (21 Nm).

43. Install and adjust the alternator drive belt, the water pump and power steering pump drive belt and the air conditioning compressor drive belt, if equipped.

44. Connect the battery cable.

45. Fill the cooling system.

46. Start the engine and allow it to warm to operating temperature. Check and adjust the ignition timing. Road test to verify correct engine operation.

Mitsubishi

2.3L DIESEL ENGINE

1983–84 Pick-Up

1. Disconnect the negative battery cable.

2. Rotate the crankshaft to position the No. 1 cylinder to Top Dead Center (TDC) of its compression stroke. Make sure that the timing marks are aligned.

3. Remove the drive belt(s) and/or accessories that may interfere with the timing belt removal.

4. Remove the timing belt covers and the crankshaft pulley.

5. If reusing the timing belt, mark its direction of rotation of the back of the belt.

6. Slightly loosen both belt tensioner bolts, move the tensioners toward the water pump and secure them by tightening the tensioner bolts.

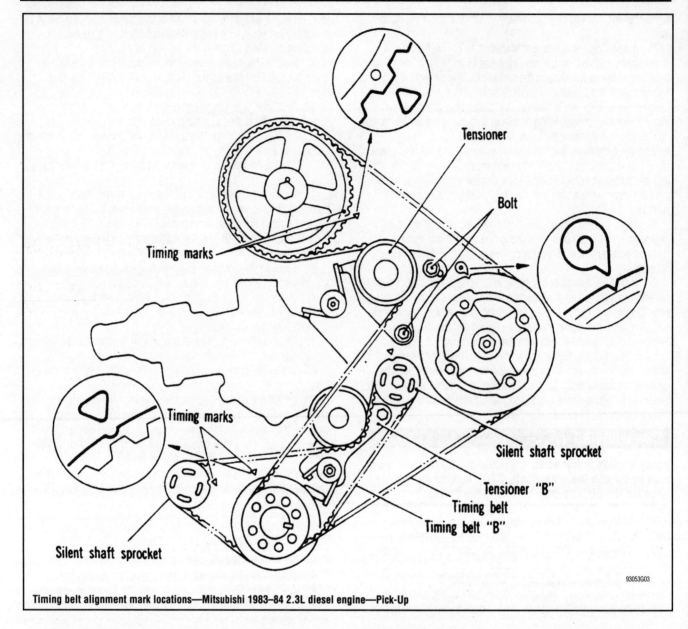

Timing belt alignment mark locations—Mitsubishi 1983–84 2.3L diesel engine—Pick-Up

93053G03

7. Remove the timing belt(s) and inspect for damage, wear or deterioration. Do not use solvents or detergent to clean the timing belts, sprockets or tensioners. Replace any component that is excessively contaminated with dirt, oil or grease.

To install:

8. Align the timing marks of the crankshaft and silent shaft drive sprockets.

9. Noting the direction of rotation, install the silent shaft timing belt.

10. Adjust the silent shaft timing belt by performing the following procedure:

   a. Loosen the silent shaft belt tensioner and allow the spring tension to take up any belt slack. Do not attempt to move the tensioner to obtain more tension. Tighten the tensioner bolt to 25–28 ft. lbs. (34–38 Nm).

   b. Standard belt deflection "play" when properly tensioned is 0.16–0.20 in. 4–5mm as measured at the midpoint between the upper silent shaft and crankshaft sprockets.

11. Align the timing marks for the camshaft, injection pump and crankshaft sprockets. Install the timing belt on the crankshaft sprocket, the injection pump sprocket and camshaft sprocket.

➡ **With the tension side kept taut, engage the timing belt teeth with each sprocket. The injection pump sprocket tends to rotate itself, so it must be held while installing the belt.**

12. Loosen the tensioner belt and allow the spring tension to take up the slack in the timing belt. Do not attempt to apply more tension by manually moving the tensioner. Once the belt is tensioned, tighten the tensioner bolt to 16–21 ft. lbs. (22–29 Nm).

➡ **Tighten the slot side bolt before tightening the fulcrum bolt. If the fulcrum side is tightened earlier, the tension bracket will be turned together and the belt will be too tight.**

13. Make sure the timing marks on all sprockets are still in correct alignment.

14. Turn the crankshaft in the normal direction of rotation through 2 camshaft sprocket teeth and hold it there. Reverse the direction to align the timing marks and push down on the belt at a

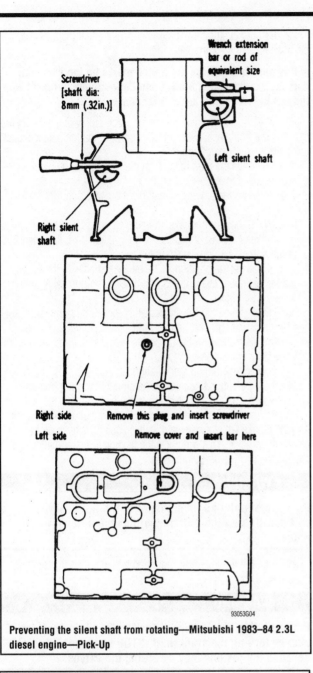

Preventing the silent shaft from rotating—Mitsubishi 1983–84 2.3L diesel engine—Pick-Up

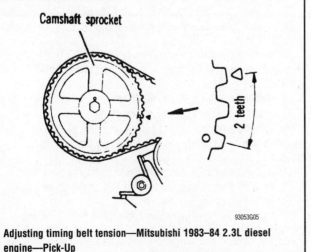

Adjusting timing belt tension—Mitsubishi 1983–84 2.3L diesel engine—Pick-Up

point halfway between the camshaft and injection pump sprocket to check the belt tension. Normal deflection is 0.16–0.20 in. (4–5mm).

15. Reset the timing belt switch by depressing its knob unit it is flush with the base and mount the timing belt covers.

16. To complete the installation, reverse the removal procedures.

17. Connect the negative battery cable.

2.0L & 2.4L ENGINES

1985–97

1. Be sure that the engine's No. 1 piston is at Top Dead Center (TDC) of the compression stroke.

⁂ **CAUTION**

Wait at least 90 seconds after the negative battery cable is disconnected to prevent possible deployment of the air bag.

2. Disconnect the negative battery cable.

3. Remove the spark plug wires from the tree on the upper cover.

4. Drain the cooling system.

5. Remove the shroud, fan and accessory drive belts.

6. Remove the radiator as required.

7. Remove the power steering pump, alternator, air conditioning compressor, tension pulley and accompanying brackets, as required.

8. Remove the upper front timing belt cover.

9. Remove the water pump pulley and the crankshaft pulley(s).

10. Remove the lower timing belt cover mounting screws and remove the cover.

11. If the belt(s) are to be reused, mark the direction of rotation on the belt.

12. Remove the timing (outer) belt tensioner and remove the belt. Unbolt the tensioner from the block and remove.

13. Remove the outer crankshaft sprocket and flange.

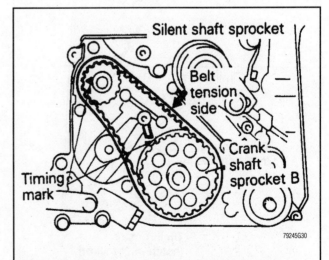

Silent shaft alignment marks. Notice the tension side of the inner (silent shaft) belt—Mitsubishi 2.0L and 2.4L engines—Pick-Up, Mighty Max and Montero Sport

14. Remove the silent shaft (inner) belt tensioner and remove the inner belt. Unbolt the tensioner from the block and remove it.

15. To remove the camshaft sprockets, use SST MB990767-01 and MIT308239 or their equivalents.

To install:

16. Install the camshaft sprockets and tighten the center bolt to 65 ft. lbs. (90 Nm).

17. Align the timing mark of the silent shaft belt sprockets on the crankshaft and silent shaft with the marks on the front case. Wrap the silent shaft belt around the sprockets so there is no slack in the upper span of the belt and the timing marks are still aligned.

18. Install the tensioner initially so the actual center of the pulley is above and to the left of the installation bolt.

19. Move the pulley up by hand so the center span of the long side of the belt deflects about ¼ in. (6mm).

20. Hold the pulley tightly so it does not rotate when the bolt is tightened. Tighten the bolt to 15 ft. lbs. (20 Nm). If the pulley has moved, the belt will be too tight.

21. Install the timing belt tensioner fully toward the water pump and temporarily tighten the bolts. Place the upper end of the spring against the water pump body. Align the timing marks of the cam, crankshaft and oil pump sprockets with the corresponding marks on the front case or head.

➡**If the following steps are not followed exactly, there is a chance that the silent shaft alignment will be 180 degrees off. This will cause a noticeable vibration in the engine and the entire procedure will have to be repeated.**

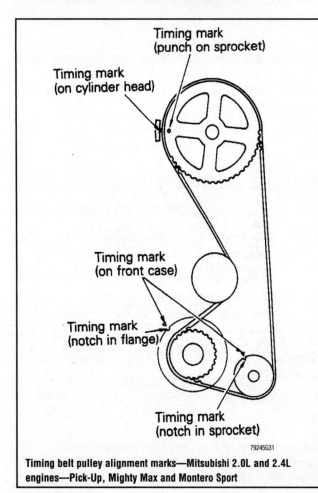

Timing belt pulley alignment marks—Mitsubishi 2.0L and 2.4L engines—Pick-Up, Mighty Max and Montero Sport

22. Before installing the timing belt, ensure that the left-side silent shaft is in the correct position.

➡**It is possible to align the timing marks on the camshaft sprocket, crankshaft sprocket and the oil pump sprocket with the left balance shaft out of alignment.**

23. With the timing mark on the oil pump pulley aligned with the mark on the front case, check the alignment of the left balance shaft to assure correct shaft timing.

   a. Remove the plug located on the left-side of the block in the area of the starter.

   b. Insert a tool having a shaft diameter of 0.3 in. (8mm) into the hole.

   c. With the timing marks still aligned, the tool must be able to go in at least 2⅓ in. If it can only go in about 1 in. turn the oil pump sprocket 1 complete revolution.

   d. Recheck the position of the balance shaft with the timing marks realigned. Leave the tool in place to hold the silent shaft while continuing.

24. Install the belt to the crankshaft sprocket, oil pump sprocket and the camshaft sprocket, in that order. While doing so, be sure there is no slack between the sprockets except where the tensioner will take it up when released.

25. Recheck the timing marks' alignment.

26. If all are aligned, loosen the tensioner mounting bolt and allow the tensioner to apply tension to the belt.

27. Remove the tool that is holding the silent shaft in place and turn the crankshaft clockwise a distance equal to 2 teeth of the camshaft sprocket. This will allow the tensioner to automatically tension the belt the proper amount.

❋❋ **WARNING**

Do not manually apply pressure to the tensioner. This will overtighten the belt and will cause a howling noise.

28. First tighten the lower mounting bolt and then tighten the upper spacer bolt.

❋❋ **WARNING**

If any binding is felt when adjusting the timing belt tension by turning the crankshaft, STOP turning the engine, because the pistons may be hitting the valves.

29. To verify that belt tension is correct, check that the deflection of the longest span (between the camshaft and oil pump sprockets) is ½ in. (13mm).

30. Install the lower timing belt cover. Be sure the packing is properly positioned in the inner grooves of the covers when installing.

31. Install the water pump pulley and the crankshaft pulley(s).

32. Install the upper front timing belt cover.

33. Install the power steering pump, alternator, air conditioning compressor, tension pulley and accompanying brackets, as required.

34. Install the radiator, shroud, fan and accessory drive belts.

35. Install the spark plug wires to the tree on the upper cover.

36. Refill the cooling system.

37. Connect the negative battery cable. Start the engine and check for leaks.

3.0L (VIN S) SOHC ENGINE

1989–92

1. Disconnect the negative battery cable.

❊❊ CAUTION

Wait at least 90 seconds after the negative battery cable is disconnected to prevent possible deployment of the air bag.

2. Position the engine so the No. 1 cylinder is at Top Dead Center (TDC) of the compression stroke. Disconnect the negative battery cable.
3. Remove the engine undercover.
4. Remove the cruise control actuator.
5. Remove the accessory drive belts.
6. Remove the air conditioner compressor tension pulley assembly.
7. Remove the tension pulley bracket.
8. Using the proper equipment, slightly raise the engine to take the weight off the side engine mount. Remove the engine mounting bracket.
9. Disconnect the power steering pump pressure switch connector. Remove the power steering pump with hoses attached and wire aside.
10. Remove the engine support bracket.
11. Remove the crankshaft pulley.
12. Remove the timing belt cover cap.
13. Remove the timing belt upper and lower covers.

14. If the same timing belt will be reused, mark the direction of the timing belt's rotation for installation in the same direction. Make sure the engine is positioned so the No. 1 cylinder is at the Top Dead Center (TDC) of its compression stroke and the timing marks are aligned with the engine's timing mark indicators.
15. Loosen the timing belt tensioner bolt and remove the belt. If the tensioner is not being removed, position it as far away from the center of the engine as possible and tighten the bolt.
16. If the tensioner is being removed, paint the outside of the spring to ensure that it is not installed backwards. Unbolt the tensioner and remove it along with the spring.

❊❊ WARNING

Do not rotate the camshafts when the timing belt is removed from the engine. Turning the camshaft when the timing belt is removed could cause the valves to interfere with the pistons thus causing severe internal engine damage.

To install:

17. Install the tensioner, if removed, and hook the upper end of the spring to the water pump pin and the lower end to the tensioner in exactly the same position as originally installed. If not already done, position both camshafts so the marks align with those on the rear. Rotate the crankshaft so the timing mark aligns with the mark on the front cover.
18. Install the timing belt on the crankshaft sprocket and while keeping the belt tight on the tension side, install the belt on the front camshaft sprocket.
19. Install the belt on the water pump pulley, then the rear camshaft sprocket and the tensioner.

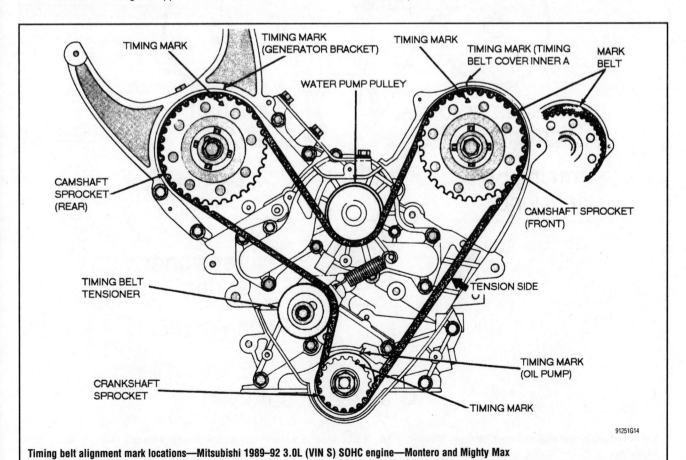

Timing belt alignment mark locations—Mitsubishi 1989–92 3.0L (VIN S) SOHC engine—Montero and Mighty Max

20. Rotate the front camshaft counterclockwise to tension the belt between the front camshaft and the crankshaft. If the timing marks became misaligned, repeat the procedure.

21. Install the crankshaft sprocket flange.

22. Loosen the tensioner bolt and allow the spring to apply tension to the belt.

23. Turn the crankshaft 2 full turns in the clockwise direction until the timing marks align again. Now that the belt is properly tensioned, torque the tensioner lock bolt to 21 ft. lbs. (29 Nm). Measure the belt tension between the rear camshaft sprocket and the crankshaft with belt tension gauge. The specification is 46–68 lbs. (210–310 N).

24. Install the timing covers. Make sure all pieces of packing are positioned in the inner grooves of the covers when installing.

25. Install the crankshaft pulley. Tighten the bolt to 108–116 ft. lbs. (150–160 Nm).

26. Install the engine support bracket.

27. Install the power steering pump and reconnect wire harness at the power steering pump pressure switch.

28. Install the engine mounting bracket and remove the engine support fixture.

29. Install the tension pulleys and drive belts.

30. Install the cruise control actuator.

31. Install the engine undercover.

32. Connect the negative battery cable and road test the vehicle.

1993–96 3.0L (VIN H) 12-VALVE ENGINE

1. Position the engine with No. 1 cylinder at Top Dead Center (TDC) of the compression stroke.

2. Disconnect the negative battery cable.

3. Drain the cooling system. Remove the drive belts.

✳✳ CAUTION

Never open, service or drain the radiator or cooling system when hot; serious burns can occur from the steam and hot coolant. Also, when draining engine coolant, keep in mind that cats and dogs are attracted to ethylene glycol antifreeze and could drink any that is left in an uncovered container or in puddles on the ground. This

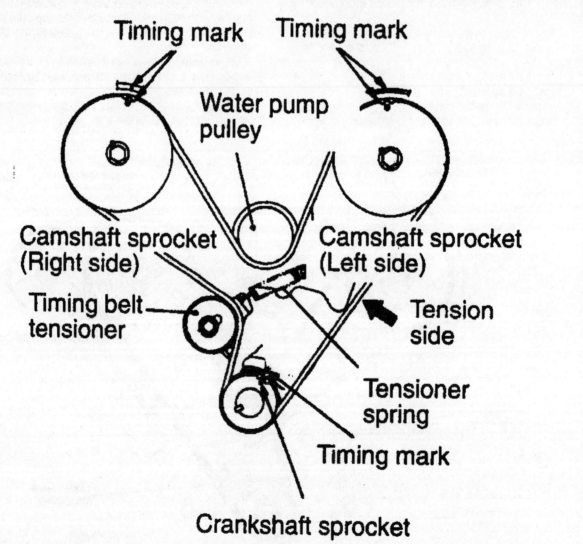

Timing belt routing and alignment mark locations–Mitsubishi 1993–96 3.0L 12-valve engine—note the tension side of the belt—Montero and Mighty Max

will prove fatal in sufficient quantities. Always drain coolant into a sealable container. Coolant should be reused unless it is contaminated or is several years old.

4. Remove the upper radiator shroud.
5. Remove the fan and fan pulley.
6. Without disconnecting the lines, remove the power steering pump from its bracket and position it to the side. Remove the pump brackets.
7. Remove the belt tensioner pulley bracket.
8. Without releasing the refrigerant, remove the air conditioning compressor from its bracket and position it to the side. Remove the bracket.
9. Remove the cooling fan bracket.
10. On some vehicles, it may be necessary to remove the pulley from the crankshaft to access the lower cover bolts.
11. Remove the timing belt cover bolts and the upper and lower covers from the engine.
12. If the same timing belt will be reused, mark the direction of timing belt's rotation, for installation in the same direction. Be sure the engine is positioned so that the No. 1 cylinder is at TDC and the sprockets timing marks are aligned with the engine's timing mark indicators.
13. Loosen the timing belt tensioner bolt and remove the belt. If not removing the tensioner, position it as far away from the center of the engine as possible and tighten the bolt.
14. If the tensioner is being removed, mark the outside of the spring to ensure that it is not installed backwards. Unbolt the tensioner and remove it along with the spring.
15. Slide the timing belt off of the sprockets.
To install:
16. Install the tensioner, if removed, and hook the upper end of the spring to the water pump pin. Install the lower end of the spring to the tensioner in exactly the same position as originally installed.
17. If not already done, position both camshafts so the timing marks align with those on the alternator bracket (rear bank) and inner timing cover (front bank). Rotate the crankshaft so the timing mark aligns with the mark on the oil pump.
18. Install the timing belt on the crankshaft sprocket and, while keeping the belt tight on the tension side (right side), install the belt on the front camshaft sprocket.
19. Install the belt on the water pump pulley, then the rear camshaft sprocket and the tensioner.
20. Rotate the front camshaft counterclockwise to tension the belt between the front camshaft and the crankshaft. If the timing marks came out of line, repeat the procedure.
21. Install the crankshaft sprocket flange.
22. Loosen the tensioner bolt and allow the spring to tension the belt.

☀ WARNING

If any binding is felt when adjusting the timing belt tension by turning the crankshaft, STOP turning the engine, because the pistons may be hitting the valves.

23. Slowly turn the crankshaft 2 full turns in the clockwise direction until the timing marks align. Now that the belt is properly tensioned, tighten the tensioner lockbolt to 35 ft. lbs. (48 Nm).
24. Install the upper and lower covers to the engine and secure with the retaining screws. Be sure the packing is positioned in the inner grooves of the covers properly when installing.

25. Install the crankshaft pulley if it was removed. Tighten the bolt to 110 ft. lbs. (150 Nm).
26. Install the air conditioning bracket and compressor to the engine. Install the belt tensioner.
27. Install the power steering pump in position. Install the fan pulley and fan.
28. Install the fan shroud to the radiator.
29. Refill the cooling system.
30. Connect the negative battery cable. Start the engine and check for fluid leaks.

1996 3.0L (VIN H) 24-VALVE
1997–98 3.0L (VIN P) ENGINES

1. Position the engine with No. 1 cylinder at Top Dead Center (TDC) of the compression stroke.
2. Disconnect the negative battery cable.
3. Drain the cooling system. Remove the drive belts.

☀ CAUTION

Never open, service or drain the radiator or cooling system when hot; serious burns can occur from the steam and hot coolant. Also, when draining engine coolant, keep in mind that cats and dogs are attracted to ethylene glycol antifreeze and could drink any that is left in an uncovered container or in puddles on the ground. This will prove fatal in sufficient quantities. Always drain coolant into a sealable container. Coolant should be reused unless it is contaminated or is several years old.

4. Remove the upper radiator shroud.
5. Remove the fan and fan pulley.
6. Without disconnecting the lines, remove the power steering pump from its bracket and position it aside. Remove the pump brackets.
7. Remove the belt tensioner pulley bracket.
8. Without releasing the refrigerant, remove the air conditioning compressor from its bracket and position it aside. Remove the bracket.
9. Remove the cooling fan bracket.
10. On some vehicles, it may be necessary to remove the pulley from the crankshaft to access the lower cover bolts.
11. Remove the timing belt cover bolts, and the upper and lower covers from the engine.
12. If the same timing belt will be reused, mark the direction of the timing belt's rotation, for installation in the same direction. Be sure engine is positioned so that the No. 1 cylinder is at TDC of its compression stroke and the sprockets timing marks are aligned with the engine's timing mark indicators.
13. Loosen the timing belt tensioner bolt and remove the belt. If not removing the tensioner, position it as far away from the center of the engine as possible and tighten the bolt.
14. If tensioner is being removed, mark outside of the spring to ensure that it is not installed backwards. Unbolt the tensioner and remove it along with the spring.
15. Using SST MB990767-01 and MIT308239 or their equivalents, remove the camshaft sprockets.
To install:
16. Hold the hexagonal portion of the camshaft with a wrench when tightening the camshaft sprocket bolt and tighten to 64 ft. lbs. (88 Nm).
17. If removed, install the tensioner and hook the upper end of

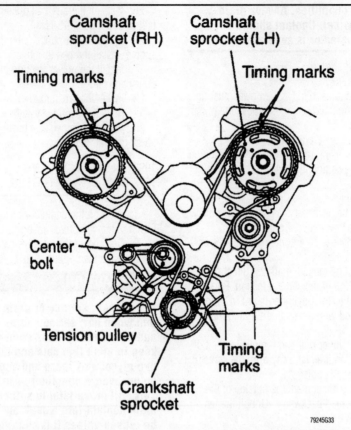

Camshaft sprocket (RH)

Timing marks

Camshaft sprocket (LH)

Timing marks

Center bolt

Tension pulley

Timing marks

Crankshaft sprocket

79245G33

Be sure to align the timing marks before removing or installing the timing belt—Mitsubishi 1996 3.0L (VIN H) 24-valve and 1997–98 3.0L (VIN P) engines–Montero and Montero Sport

the spring to the water pump pin. Install the lower end of the spring to the tensioner in exactly the same position as originally installed.

18. Position both camshafts so the timing marks align with those on the alternator bracket (rear bank) and inner timing cover (front bank). Rotate the crankshaft so the timing mark aligns with the mark on the oil pump.

19. Install the timing belt on the crankshaft sprocket, and while keeping the belt tight on the tension side (right side), install the belt on the front camshaft sprocket.

20. Install the belt on the water pump pulley, then the rear camshaft sprocket and the tensioner.

21. Rotate the front camshaft counterclockwise to tension the belt between the front camshaft and the crankshaft. If the timing marks came out of line, repeat the procedure.

22. Install the crankshaft sprocket flange.

23. Loosen the tensioner bolt and allow the spring to tension the belt.

⁂ WARNING

If any binding is felt when adjusting the timing belt tension by turning the crankshaft, STOP turning the engine, because the pistons may be hitting the valves.

24. Slowly turn the crankshaft 2 full turns in the clockwise direction until the timing marks align. Now that the belt is properly tensioner, tighten the tensioner lockbolt to 35 ft. lbs. (48 Nm).

25. Install the upper and lower covers to the engine and secure with the retaining screws. Be sure the packing is positioned in the inner grooves of the covers properly when installing.

26. Install the crankshaft pulley if it was removed. Tighten the bolt to 110 ft. lbs. (150 Nm).

27. Install the air conditioning bracket and compressor on the engine. Install the belt tensioner.

28. Install the power steering pump into position. Install the fan pulley and fan.

29. Install the fan shroud to the radiator.

30. Refill the cooling system.

31. Connect the negative battery cable. Start the engine and check for fluid leaks.

1999–00 3.0L (VIN H)
1997–98 3.5L (VIN M) &
1999–00 3.5L (VIN R) ENGINES

1. Disconnect the negative battery cable.

2. Drain the engine coolant and store it for reinstallation. Remove the upper radiator hose.

3. Remove the cooling fan shroud assembly.

4. Remove the cooling fan-to-clutch bolts and the fan.

5. Remove the cooling fan clutch-to-water pump nuts and the clutch assembly.

6. Remove the drive belts for the alternator, power steering pump and air conditioning compressor.

7. Disconnect the electrical connectors from the alternator.

8. Remove the alternator-to-engine bolts and the alternator bracket-to-engine bolts; then, remove the alternator and bracket from the engine.

9. Remove the power steering pump cover. Remove the power

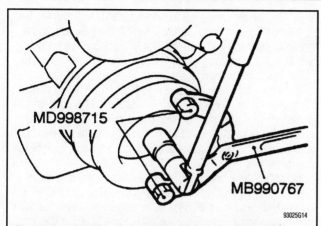

Removing or installing the crankshaft pulley bolt—Mitsubishi 1999–00 3.0L (VIN H), 1997–98 3.5L (VIN M) and 1999–00 3.5L (VIN R) engines—Montero and Montero Sport

steering pump-to-engine bolts and move the pump aside with the hoses and electrical connector attached.

10. Remove the air conditioning compressor-to-bracket bolts and move the compressor aside with the lines and electrical connector attached.

11. Remove the air conditioning compressor bracket-to-engine bolts and the bracket.

12. Remove the timing indicator bracket (near crankshaft pulley) bolts and the bracket.

13. Remove the accessory mount assembly-to-engine bolts and the mount assembly.

14. Remove the upper timing belt cover assembly.

15. Using the End Yoke Holder tool MD990767 and 2 Crankshaft

Pulley Holder Pin tools MD998715, or equivalent to hold the crankshaft pulley, and a socket wrench, remove the crankshaft pulley bolt and the pulley.

16. Remove the lower timing belt cover.

17. Rotate the crankshaft clockwise to align the timing marks to position the No. 1 cylinder at the Top Dead Center (TDC) of its compression stroke.

18. Use chalk to mark the rotating (clockwise) direction of the timing belt for reinstallation purposes.

19. Loosen the auto-tensioner pulley center bolt and remove the timing bolt.

20. Remove the auto-tensioner pulley and the auto-tensioner arm assembly.

To install:

21. Press the end of the auto-tensioner inward with 72–145 ft. lbs. (98–196 Nm) of force and measure the distance that the pushrod is pushed in. If the standard distance is not 0.04 in. (1mm), replace the auto-tensioner.

22. Position the auto-tensioner in a soft-jawed vise and SLOWLY compress the pushrod until the pushrod and housing holes align; then, install a setting pin to secure the auto-tensioner in the retracted position.

23. Align the camshaft and crankshaft TDC timing marks.

24. Install the timing belt (noting its rotational direction) so that there is no deflection between the sprockets and pulleys in the following manner:
- Crankshaft sprocket
- Idler pulley
- Left camshaft sprocket
- Water pump pulley
- Right camshaft sprocket
- Tension pulley

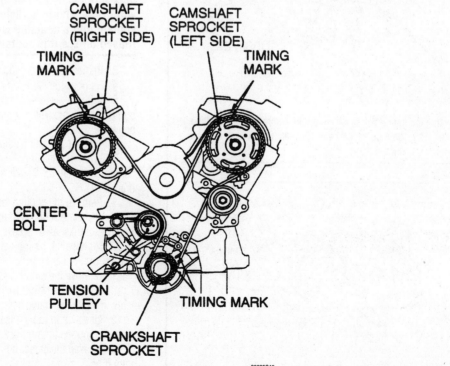

View of the timing belt alignment marks—Mitsubishi 1999–00 3.0L (VIN H), 1997–98 3.5L (VIN M) and 1999–00 3.5L (VIN R) engines—Montero and Montero Sport

25. Turn the camshaft sprocket counterclockwise until the tension side of the timing belt is firmly stretched, then, recheck the timing marks.

26. Using the Tension Pulley Socket Wrench tool MD998767, or equivalent, push the tensioner pulley into the timing belt and secure the center bolt.

27. Using the Crankshaft Pulley Spacer tool MD998769, or equivalent, rotate the crankshaft ¼ turn counterclockwise, then, turn it again clockwise to align the timing marks.

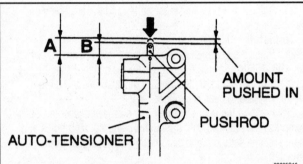

Inspecting the auto-tensioner movement—Mitsubishi 1999–00 3.0L (VIN H), 1997–98 3.5L (VIN M) and 1999–00 3.5L (VIN R) engines—Montero and Montero Sport

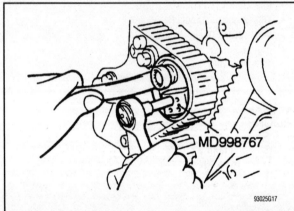

Adjusting the timing belt tensioner pulley—Mitsubishi 1999–00 3.0L (VIN H), 1997–98 3.5L (VIN M) and 1999–00 3.5L (VIN R) engines—Montero and Montero Sport

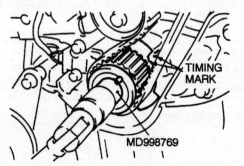

Using crankshaft spacer tool to rotate the crankshaft—Mitsubishi 1999–00 3.0L (VIN H), 1997–98 3.5L (VIN M) and 1999–00 3.5L (VIN R) engines—Montero and Montero Sport

28. Loosen the timing belt tensioner center bolt. Using the Tension Pulley Socket Wrench tool MD998767, or equivalent, and a torque wrench, apply 39 inch lbs. (4.4 Nm) pressure on the timing belt. Torque the tensioner pulley center bolt to 35 ft. lbs. (48 Nm).

29. Remove the setting pin from the auto-tensioner.

30. Rotate the crankshaft 2 complete revolutions and realign the timing marks. Then, wait for 5 minutes until the auto-tensioner's pushrod extends to its standard value. If the standard value is not 0.15–0.20 in. (3.8–5.0mm), repeat the adjustment procedure. If the standard value is still not achieved, replace the auto-tensioner.

31. Install the lower timing belt cover and crankshaft pulley.

32. Using the End Yoke Holder tool MD990767 and 2 Crankshaft Pulley Holder Pin tools MD998715, or equivalent to hold the crankshaft pulley, and a socket torque wrench, torque the crankshaft pulley bolt to 134 ft. lbs. (181 Nm).

33. Install the upper timing belt cover assembly.

34. Install the remaining items by reversing the removal procedures.

35. Refill the cooling system.

36. Connect the negative battery cable.

1995–96 3.5L (VIN M) DOHC ENGINE

1. Disconnect the negative battery cable.
2. Drain the cooling system. Remove the drive belts.

✳✳ CAUTION

Never open, service or drain the radiator or cooling system when hot; serious burns can occur from the steam and hot coolant. Also, when draining engine coolant, keep in mind that cats and dogs are attracted to ethylene glycol antifreeze and could drink any that is left in an uncovered container or in puddles on the ground. This will prove fatal in sufficient quantities. Always drain coolant into a sealable container. Coolant should be reused unless it is contaminated or is several years old.

3. Remove the upper radiator shroud.
4. Remove the fan and fan pulley.
5. Without disconnecting the lines, remove the power steering pump from its bracket and position it aside. Remove the pump brackets.
6. Remove the belt tensioner pulley bracket.
7. Without releasing the refrigerant, remove the air conditioning compressor from its bracket and position it aside. Remove the bracket.
8. Remove the cooling fan bracket.
9. On some vehicles, it may be necessary to remove the pulley from the crankshaft to access the lower cover bolts.
10. Remove the timing belt cover bolts and the upper and lower covers from the engine.
11. Remove the crankshaft position sensor connector.
12. Using SST MB990767-01 and MD998754 or their equivalents, remove the crankshaft pulley from the crankshaft.
13. Use a shop rag to clean the timing marks to assist in properly aligning the timing marks.
14. Loosen the center bolt on the tension pulley and remove the timing belt.

➡If the same timing belt will be reused, mark the direction of timing belt's rotation, for installation in the same direc-

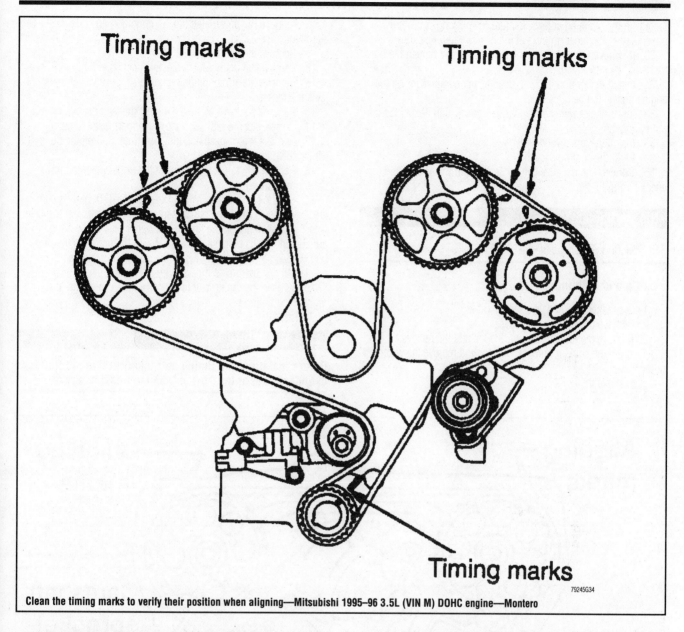

Timing marks

Timing marks

Timing marks

79245G34

Clean the timing marks to verify their position when aligning—Mitsubishi 1995–96 3.5L (VIN M) DOHC engine—Montero

tion. Be sure engine is positioned so No. 1 cylinder is at the **Top Dead Center (TDC) of its compression stroke and the sprockets timing marks are aligned with the engine's timing mark indicators.**

15. Remove the auto-tensioner, the tension pulley and the tension arm assembly.

16. Remove the sprockets by holding the hexagonal portion of the camshaft with a wrench while removing the sprocket bolt.

To install:

17. Install the crankshaft pulley and turn the crankshaft sprocket timing mark forward (clockwise) 3 teeth to move the piston slightly past No. 1 cylinder TDC.

18. If removed, install the camshaft sprockets and tighten the bolts to 64 ft. lbs. (88 Nm).

19. Align the timing mark of the left bank side camshaft sprocket.

20. Align the timing mark of the right bank side camshaft sprocket, and hold the sprocket with a wrench so that it doesn't turn.

21. Set the timing belt onto the water pump pulley.

22. Check that the camshaft sprocket timing mark of the left bank side is aligned and clamp the timing belt with double clips.

23. Set the timing belt onto the idler pulley.

❄❄ WARNING

If any binding is felt when adjusting the timing belt tension by turning the crankshaft, STOP turning the engine, because the pistons may be hitting the valves.

24. Turn the crankshaft 1 turn counterclockwise and set the timing belt onto the crankshaft sprocket.

25. Set the timing belt on the tension pulley.

26. Place the tension pulley pin hole so that it is towards the top. Press the tension pulley onto the timing belt, and then provisionally tighten the fixing bolt. Tighten the bolt to 35 ft. lbs. (48 Nm).

27. Slowly turn the crankshaft 2 full turns in the clockwise direction until the timing marks align. Remove the 4 double clips.

28. Install the crankshaft position sensor connector.

29. Install the upper and lower covers on the engine and secure

them with the retaining screws. Be sure the packing is properly positioned in the inner grooves of the covers when installing.

30. If removed, install the crankshaft pulley and tighten the bolt to 110 ft. lbs. (150 Nm).

31. Install the air conditioning bracket and compressor on the engine. Install the belt tensioner.

32. Install the power steering pump into position. Install the fan pulley and fan.

33. Install the fan shroud on the radiator.

34. Refill the cooling system.

35. Connect the negative battery cable. Start the engine and check for fluid leaks.

Nissan/Datsun

3.0L (VG30I, VG30E) & 3.3L (VG33E) ENGINES

Pick-Up and Pathfinder

1. Disconnect the negative battery cable.
2. Remove the engine undercover.
3. Remove the radiator shroud, the fan and the pulleys.
4. Drain the coolant from the radiator and remove the water pump hose.
5. Remove the radiator.

6. Remove the power steering, air conditioning compressor and alternator drive belts.

7. Remove the spark plugs.

8. Remove the distributor protector (dust shield).

9. Remove the air conditioning compressor drive belt idler pulley and bracket.

10. Remove the fresh air intake tube at the cylinder head cover.

11. Disconnect the radiator hose at the thermostat housing.

12. Remove the crankshaft pulley bolt, then pull off the pulley with a suitable puller.

13. Remove the bolts, then remove the front upper and lower timing belt covers.

14. Set the No. 1 piston at Top Dead Center (TDC) of its compression stroke. Align the punchmark on the left camshaft sprocket with the punchmark on the timing belt upper rear cover. Align the punchmark on the crankshaft sprocket with the notch on the oil pump housing. Temporarily install the crank pulley bolt so the crankshaft can be rotated if necessary.

15. Loosen the timing belt tensioner and return spring, then remove the timing belt.

To install:

✷✷ CAUTION

Before installing the timing belt, confirm that the No. 1 cylinder is set at the TDC of the compression stroke.

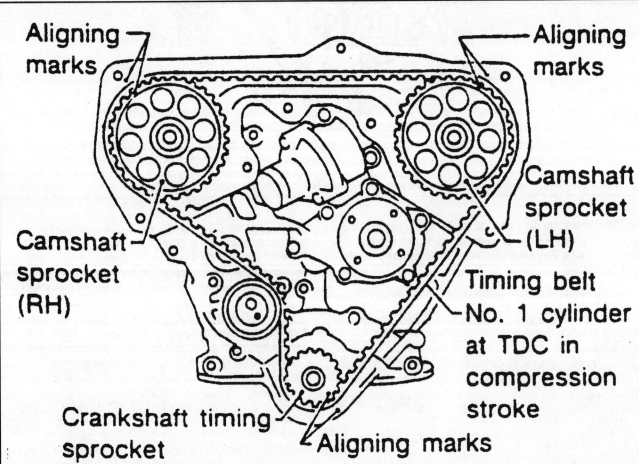

Timing belt alignment mark locations—Nissan 3.0L (VG30I, VG30E) and 3.3L (VG33E) engines—Pick-Up and Pathfinder

79245G35

16. Remove both cylinder head covers and loosen all rocker arm shaft retaining bolts.

➡**The rocker arm shaft bolts MUST be loosened so that the correct belt tension can be obtained.**

17. Install the tensioner and the return spring. Using a hexagon wrench, turn the tensioner clockwise and temporarily tighten the locknut.

18. Be sure that the timing belt is clean and free from oil or water.

19. When installing the timing belt, align the white lines on the belt with the punchmarks on the camshaft and crankshaft sprockets. Have the arrow on the timing belt pointing toward the front belt covers.

➡**A good way (although rather tedious!) to check for proper timing belt installation is to count the number of belt teeth between the timing marks. There are 133 teeth on the belt; there should be 40 teeth between the timing marks on the left and right side camshaft sprockets, and 43 teeth between the timing marks on the left side camshaft sprocket and the crankshaft sprocket.**

20. While keeping the tensioner steady, loosen the locknut with a hex wrench.

21. Turn the tensioner approximately 70–80 degrees clockwise with the wrench, then tighten the locknut.

✳✳ WARNING

If any binding is felt when adjusting the timing belt tension by turning the crankshaft, STOP turning the engine, because the pistons may be hitting the valves.

22. Turn the crankshaft in a clockwise direction several times, then **slowly** set the No. 1 piston to TDC of the compression stroke.

23. Apply 22 lbs. (10 kg) of pressure (push it in!) to the center span of the timing belt between the right side camshaft sprocket and the tensioner pulley, then loosen the tensioner locknut.

24. Using a 0.0138 in. (0.35mm) thick feeler gauge (the actual width of the blade **must** be ½ in. or 13mm!), turn the crankshaft clockwise (**slowly!**). The timing belt should move approximately 2½ teeth. Tighten the tensioner locknut, turn the crankshaft slightly and remove the feeler gauge.

25. Slowly rotate the crankshaft clockwise several more times, then set the No. 1 piston to TDC of the compression stroke.

26. Position the 2 timing covers on the block, then tighten the mounting bolts to 24 ft. lbs. (35 Nm).

27. Press the crankshaft pulley onto the shaft, then tighten the bolt to 90–98 ft. lbs. (123–132 Nm).

28. Connect the radiator hose to the thermostat housing.

29. Reconnect the fresh air intake tube at the cylinder head cover.

30. Install the air conditioning compressor drive belt idler pulley and bracket.

31. Install the distributor protector (dust shield).

32. Install the spark plugs.

33. Install the power steering, air conditioning compressor and alternator drive belts.

34. Install the radiator.

35. Reconnect the water pump hose and refill the engine with coolant. Install the fan shroud and pulleys.

36. Install the engine undercover.

37. Start the engine and check for any leaks.

38. Connect the negative battery cable.

Quest

On this vehicle, right side refers to the "rear" components (near the firewall) and left side refers to the "front" components (near the radiator).

1. Disconnect the negative battery cable.

2. If the timing belt is to be removed, it is good practice to turn the crankshaft until the engine is at Top Dead Center (TDC) of the No. 1 cylinder, compression stroke (firing position), before beginning work. This should align all timing marks and serve as a reference for all work that follows. After verifying that the engine is at TDC for the No. 1 cylinder, do not crank the engine or allow the crankshaft or camshaft sprockets to be turned otherwise engine timing will be lost.

3. Drain the cooling system.

4. Disconnect the negative battery cable.

5. Remove the alternator drive belt, water pump and power steering pump belt and the air conditioning compressor belt, if equipped, using the recommended drive belt removal procedure.

6. If equipped with air conditioning, remove the 3 air conditioning compressor drive belt idler pulley bolts and remove the idler pulley.

7. Remove the upper radiator hose bracket bolt. Remove the upper hose with the bracket from the vehicle.

8. Remove the water bypass hose from between the thermostat housing and the lower water hose connection.

9. Remove the main wiring harness from the upper engine front cover.

10. Remove the 8 upper engine front cover bolts and remove the upper cover.

11. Raise and safely support the vehicle.

12. Remove the right side front wheel and tire assembly.

13. Remove the 4 right side engine and transmission splash shield bolts and 2 screws, and remove the right side outer engine and transaxle splash shield.

14. Use a strap wrench to hold the water pump pulley. Remove the 4 pulley bolts, and the water pump pulley.

15. Use a strap wrench to hold the crankshaft pulley. Remove the center pulley bolt, and the crankshaft pulley using a harmonic balancer (damper) puller to draw the pulley from the front of the crankshaft.

16. Remove the 5 lower engine front cover bolts, then remove the lower engine front cover.

17. Be sure that the timing marks between the crankshaft sprocket and the oil pump housing align.

18. If the timing belt is to be reused, mark an arrow on the belt indicating the direction of rotation. The directional arrow is necessary to ensure that the timing belt, if it to be reused, can reinstalled in the same direction.

19. Loosen the timing belt tensioner nut and slip the timing belt off of the sprockets.

20. If necessary, the camshaft sprockets can be removed. A special spanner tool is designed to hold the sprocket to keep it from turning while the center bolt is being loosened. Use care if using substitutes.

➡**The sprockets are not interchangeable.**

21. If necessary, the crankshaft sprocket can be removed. The outer timing belt guide (looks like a large washer) and the crankshaft sprocket simply pull off the front of the crankshaft.

➡**Be careful, there are 2 crankshaft keys. Use care not to loose them.**

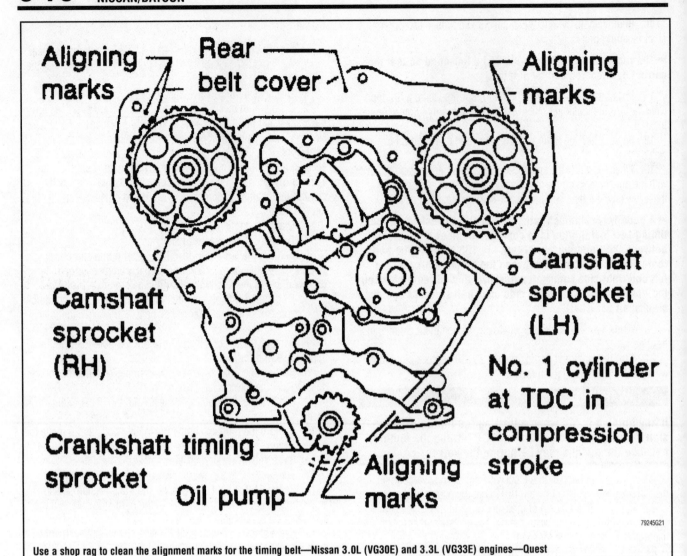

Aligning marks — **Rear belt cover**

Aligning marks

Camshaft sprocket (RH)

Camshaft sprocket (LH)

No. 1 cylinder at TDC in compression stroke

Crankshaft timing sprocket

Oil pump — **Aligning marks**

79245G21

Use a shop rag to clean the alignment marks for the timing belt—Nissan 3.0L (VG30E) and 3.3L (VG33E) engines—Quest

To install:

22. Clean all parts well. If removed, inspect the crankshaft sprocket for warping or abnormal wear. Check the sprocket teeth for wear, deformation, chipping or other damage. Replace as necessary. Clean the sprocket mounting surface to ease installation. Install the key. Slip the sprocket onto the crankshaft. Tap it in place with a suitably-sized socket.

23. If removed, inspect the camshaft sprockets for damage and wear. Replace as required. The sprockets should be marked **L3** to designate the front, or left side camshaft and **R3** to designate the rear, or right side camshaft. Use care to install the sprockets properly. A special spanner tool is designed to hold the sprocket to keep it from turning while the center bolt is being tightened. Use care if using a substitute. Tighten the camshaft sprocket center bolts to 58–65 ft. lbs. (78–88 Nm) for the 3.0L engine or 61 ft. lbs. (83 Nm) for the 3.3L engine. Verify that the timing marks on the camshaft sprockets and the timing marks on the rear cover (called the seal plate) are aligned.

24. Use an Allen wrench to turn the timing belt tensioner clockwise until the belt tensioner spring is fully extended. Temporarily tighten the tensioner nut to 32–43 ft. lbs. (43–58 Nm).

25. If a new timing belt is to be installed, look for a printed arrow on the belt. Be sure the arrow is pointing away from the engine. If the original timing belt is to be reused, be sure that the directional arrow that was marked at disassembly is facing the correct direction.

26. A new Original Equipment Manufacture (OEM) timing belt should have 3 white timing marks on it that indicate the correct timing positions of the camshafts and the crankshaft. These marks are to help ensure that the engine is properly timed. When the engine is properly timed, each white timing mark on the timing belt will be aligned with the corresponding camshaft and crankshaft timing mark on the sprocket. Because the white timing marks are not evenly spaced, the technician needs to use care in installing the belt. There should be 40 timing belt teeth between the timing marks on the front and rear camshaft sprockets and 43 teeth between the timing mark on the front camshaft sprocket and the timing mark on the crankshaft sprocket.

27. Verify that the camshaft timing marks are aligned with the timing marks on the rear cover (seal plate) and that the crankshaft sprocket timing mark is aligned with the timing mark on the oil pump housing.

28. Install the timing belt starting at the crankshaft sprocket and moving around the camshaft sprockets following a counterclockwise path. Do not allow any slack in the timing belt between the sprockets. After all of the timing marks are aligned with the timing belt installed, slip the timing belt onto the belt tensioner.

29. While holding the timing belt tensioner with an Allen wrench, loosen the tensioner nut. Allow the tensioner to put pressure on the timing belt. Use an Allen wrench to turn the timing belt tensioner 70–80 degrees clockwise and tighten the timing belt tensioner nut to 32–43 ft. lbs. (43–58 Nm).

※※ **WARNING**

If any binding is felt when adjusting the timing belt tension by turning the crankshaft, STOP turning the engine, because the pistons may be hitting the valves.

30. Rotate the crankshaft clockwise twice and align the No. 1 piston to TDC on the compression stroke (firing position).
31. Apply 22 lbs. (10kg) of force on the timing belt between the rear camshaft sprocket and the timing belt tensioner. An assistant may be needed. While holding the timing belt tensioner steady with an Allen wrench, loosen the timing belt tensioner nut. Remove the Allen wrench and adjust the timing belt tensioner using the following procedure:

  a. Install a 0.0138 in. (0.35mm) thick and 0.500 in. (12.7mm) wide feeler gauge where the timing belt just starts to go around the tensioner (approximately the 4 o'clock position, looking at the tensioner).

  b. Turn the crankshaft sprocket clockwise, which should force the feeler gauge between the timing belt and the tensioner, up to a position on the tensioner of about 1 o'clock.

  c. Tighten the timing belt tensioner nut to 32–43 ft. lbs. (43–58 Nm) for the 3.0L engine or 61 ft. lbs. (83 Nm) for the 3.3L engine.

  d. Turn the crankshaft clockwise to rotate the feeler gauge out from between the timing belt tensioner and the timing belt.

32. Rotate the crankshaft clockwise twice, and once again align the No. 1 piston to TDC on the compression stroke (firing position).
33. Apply 22 lbs. (10 kg) of force on the timing belt between the front and rear camshaft sprockets. Measure the amount of belt deflection. Belt deflection should be between 0.51–0.59 in. (13–15mm). If belt deflection is out of specification, repeat Steps 29 through 33. If the timing belt deflection cannot be adjusted into specification, the timing belt will have to be replaced.
34. Position the lower engine front cover and install the 5 lower cover bolts. Do not over tighten. Tighten to 27–44 inch lbs. (3–5 Nm).
35. Install the outer timing belt guide next to the crankshaft sprocket with the dished side facing away from the cylinder block. Install the crankshaft pulley. Use a strap wrench to keep the crankshaft pulley from turning and tighten the center bolt to 90–98 ft. lbs. (123–132 Nm) for the 3.0L engine or 148 ft. lbs. (201 Nm) for the 3.3L engine.
36. Position the water pump pulley on the pump. Install the 4 bolts. Use a strap wrench to keep the water pump pulley from turning and tighten the 4 water pump pulley bolts to 12–15 ft. lbs. (16–21 Nm) for the 3.0L engine or 89 inch lbs. (10 Nm) for the 3.3L engine.
37. Position the right side outer engine and transaxle splash shield, and secure with the 4 bolts and 2 screws.
38. Install the right side front wheel and tire assembly. Tighten the lug nuts to 72–87 ft. lbs. (98–118 Nm).
39. Lower the vehicle.
40. Position the upper engine timing belt front cover and tighten the 8 bolts to 27–44 inch lbs. (3–5 Nm).
41. Install the main wiring harness on the upper engine front cover.

42. Position the water bypass hose between the thermostat housing and water connection. Install the upper radiator hose between the radiator and the water hose connection. Secure the hoses with clamps. Install the upper radiator hose bracket. Tighten the bracket bolt to 34–58 ft. lbs. (46–65 Nm).
43. If equipped, position the air conditioning compressor drive belt idler pulley and install the 3 bolts. Tighten to 15 ft. lbs. (21 Nm).
44. Install and adjust the alternator drive belt, the water pump and power steering pump drive belt and the air conditioning compressor drive belt, if equipped.
45. Connect the battery cable.
46. Fill the cooling system.
47. Start the engine and allow it to warm to operating temperature. Check and adjust the ignition timing. Road test to verify correct engine operation.
48. Connect the negative battery cable.

Subaru

2.5L DOHC ENGINE

Forester

When servicing the timing belt, note the following:
• The intake and exhaust camshafts can be rotated independently when the timing belt is removed. If the intake and exhaust valves are lifted off of their seats simultaneously, their heads will contact each other, possibly causing damage.
• When the timing belt is removed, the camshafts are positioned so that none of the valves are lifted off of their seats, resulting in a "zero-lift" position.
• The left-hand cylinder head camshafts must be rotated from the "zero-lift" position as little as possible when orienting it for timing belt installation, otherwise possible valve head interference may occur.
• Never allow the camshafts to rotate in the direction shown in the accompanying illustration, which would cause both the intake and exhaust valves to lift simultaneously, causing interference.

1. Disconnect the negative battery cable.
2. Remove all necessary components to gain access to the timing belt.
3. If equipped with a manual transmission, loosen the 2 timing belt guide mounting bolts, then separate the guide from the engine block.
4. If the directional arrow and alignment marks on the timing belt are faded and the belt is to be reused, remark the belt with white paint or a grease pencil as follows:

  a. Using a Subaru tool No. ST-499987500 Crankshaft Socket or equivalent, installed on the crankshaft sprocket, rotate the crankshaft until the crankshaft sprocket, left-hand exhaust camshaft sprocket, left-hand intake camshaft sprocket, right-hand intake camshaft sprocket and right-hand exhaust camshaft sprocket timing mark notches are aligned with the respective marks on the belt cover and engine block.

  b. Make alignment and/or arrow marks on the timing belt in relation to the sprockets as indicated in the accompanying illustration.
    • Z1: 54.5 tooth length
    • Z2: 51 tooth length
    • Z3: 28 tooth length

5. Loosen the center bolt from the timing belt idler pulley, then remove the idler pulley from the engine block.

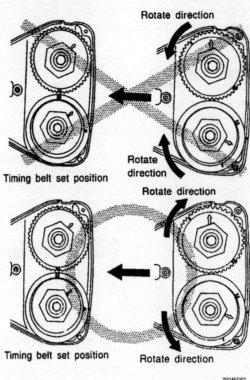

Rotate direction

Rotate direction

Timing belt set position

Rotate direction

Rotate direction

Timing belt set position

Rotate direction

79245G62

If the camshafts must be rotated, do not turn them in toward each other (upper diagram); only rotate them away from each other (lower diagram)—Subaru 2.5L DOHC engine—Forester

✳ WARNING

After removing the timing belt, DO NOT rotate the camshafts. Damage to the valves may occur.

6. Carefully remove the timing belt from all of the sprockets.
7. Remove the automatic belt tension adjuster assembly as follows:
    a. Remove the 2 timing belt idler pulleys, as indicated in the accompanying illustration.
    b. Loosen the automatic tension adjuster assembly mounting bolts, then separate the adjuster assembly from the engine block.
To install:

✳ WARNING

Do not allow oil, grease or coolant to come in contact with the timing belt. If this occurs, quickly and thoroughly remove all traces of the compound. Also, never bend the timing belt sharply; the minimum bending radius is 2.36 in. (60mm).

8. Inspect the camshaft and crankshaft sprocket teeth for abnormal or excessive wear or scratches. Ensure there is no free-play between the sprocket and the key. Inspect the crankshaft sprocket sensor notch for damage or contamination with debris or dirt.

➡When preparing the automatic tension adjuster assembly for installation, adhere to the following points:

• Always use a vertical press, rather than a horizontal press or vise, to depress the adjuster assembly rod

• Depress the adjuster rod in a vertical position ONLY
• Depress the adjuster rod slowly (taking more than 3 minutes) with a force of 66 lbs. (30 kg)
• Do not allow the press force to exceed 2205 lbs. (1000 kg)
• Press the adjuster rod in as far as the end surface of the cylinder—do not press the rod into the cylinder, which may cause damage to the assembly
• Do not release the press force from the rod until the stopper pin is completely inserted in the cylinder

9. Prepare the automatic timing belt tension adjuster assembly for installation as follows:
    a. Position the adjuster assembly in a vertical press.
    b. Slowly depress the adjuster rod with a force of 66 lbs. (30 kg) until the hole in the rod is aligned with the hole in the adjuster cylinder housing.
    c. Insert a 0.08 in. (2mm) diameter stopper pin or Allen wrench through the hole in the cylinder housing and rod, then slowly release the press force from the adjuster rod.

10. Install the adjuster assembly onto the engine block.
11. Install timing belt idler pulley No. 2 on the engine block.
12. Install the timing belt idler pulley No. 1 on the engine block.
13. If the camshaft and crankshaft timing marks are no longer aligned, perform the following:
    a. Position the crankshaft sprocket so that its mark is aligned with the mark on the oil pump cover on the engine block.
    b. Align the single line mark on the right-hand exhaust camshaft sprocket with the notch on the belt cover.
    c. Rotate the right-hand intake camshaft so that the single line mark is aligned with the notch on the belt cover.

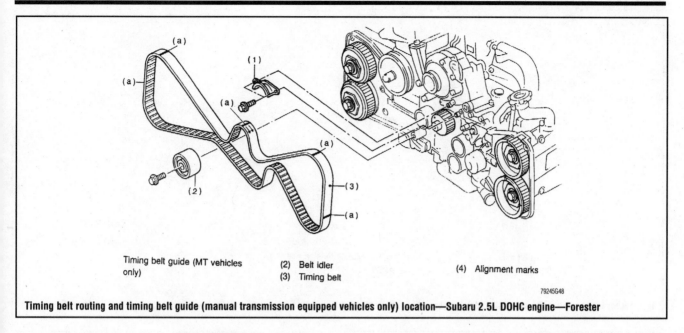

Timing belt guide (MT vehicles only)
(2) Belt idler
(3) Timing belt
(4) Alignment marks

79245G48

Timing belt routing and timing belt guide (manual transmission equipped vehicles only) location—Subaru 2.5L DOHC engine—Forester

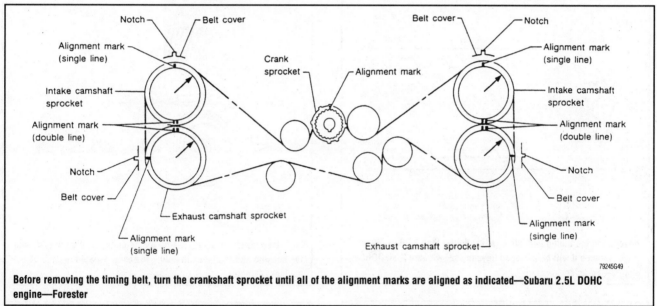

79245G49

Before removing the timing belt, turn the crankshaft sprocket until all of the alignment marks are aligned as indicated—Subaru 2.5L DOHC engine—Forester

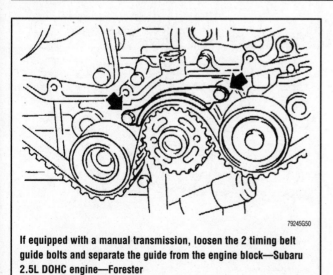

79245G50

If equipped with a manual transmission, loosen the 2 timing belt guide bolts and separate the guide from the engine block—Subaru 2.5L DOHC engine—Forester

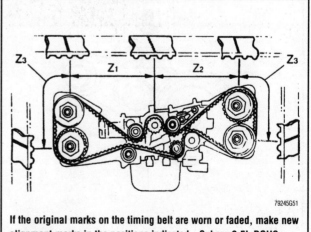

79245G51

If the original marks on the timing belt are worn or faded, make new alignment marks in the positions indicated—Subaru 2.5L DOHC engine—Forester

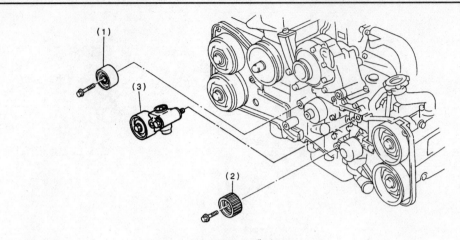

(1) Belt idler
(2) Belt idler No. 2
(3) Automatic belt tension adjuster
    ASSY

79245G52

It is necessary to remove the automatic adjuster assembly and reset the pushrod for timing belt installation—Subaru 2.5L DOHC engine—Forester

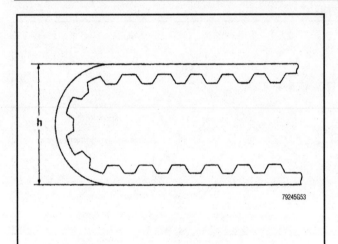

79245G53

Never bend the timing belt into a radius tighter than 2.36 in./60mm (h), otherwise it will be damaged beyond use—Subaru 2.5L DOHC engine—Forester

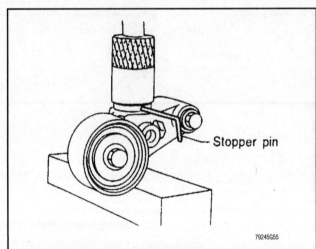

Stopper pin

79245G55

. . . then insert a 0.08 in. (2mm) diameter pin or Allen wrench into the housing and rod holes to hold it in position—Subaru 2.5L DOHC engine—Forester

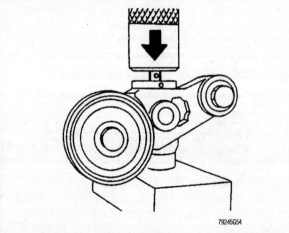

79245G54

Use a vertical press to push the adjuster rod into its housing until it is flush with the assembly's outer surface . . .—Subaru 2.5L DOHC engine—Forester

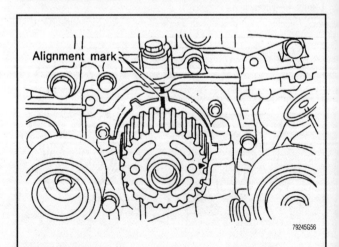

Alignment mark

79245G56

If the camshaft sprockets are no longer aligned, rotate the crankshaft sprocket until the marks are aligned . . .—Subaru 2.5L DOHC engine—Forester

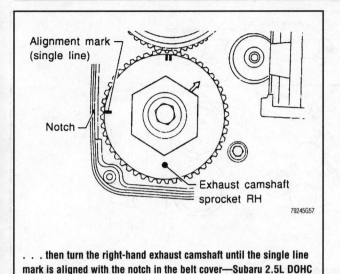

. . . then turn the right-hand exhaust camshaft until the single line mark is aligned with the notch in the belt cover—Subaru 2.5L DOHC engine—Forester

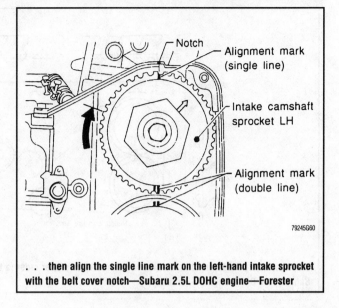

. . . then align the single line mark on the left-hand intake sprocket with the belt cover notch—Subaru 2.5L DOHC engine—Forester

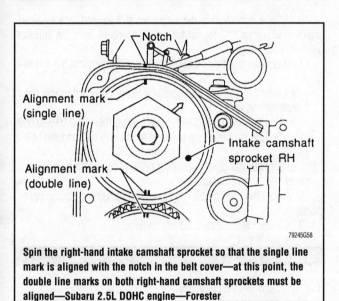

Spin the right-hand intake camshaft sprocket so that the single line mark is aligned with the notch in the belt cover—at this point, the double line marks on both right-hand camshaft sprockets must be aligned—Subaru 2.5L DOHC engine—Forester

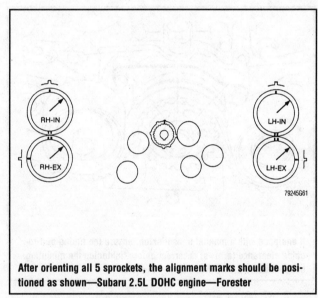

After orienting all 5 sprockets, the alignment marks should be positioned as shown—Subaru 2.5L DOHC engine—Forester

➤At this point, the double line marks on both right-hand camshaft sprockets should be aligned.

    d. Turn the left-hand exhaust (lower) camshaft counterclockwise (as viewed from the front of the engine) until the single line mark is aligned with the notch on the belt cover.

    e. Position the single line mark on the left-hand intake camshaft sprocket so that it is aligned with the notch on the belt cover. When rotating the camshaft, do so only in a clockwise direction (as viewed from the front of the engine).

➤At this point, the double line marks on both left-hand camshaft sprockets should be aligned.

    f. Ensure the timing marks are aligned as shown in the accompanying illustration. If they are not, repeat Substeps 12a through 12e until they are properly aligned.

    14. Install the timing belt around the camshaft, crankshaft and idler pulleys so that the positioning marks on the timing belt are aligned with the marks on the sprockets as follows:

    a. Position the timing belt on the crankshaft sprocket so that the marks are aligned.

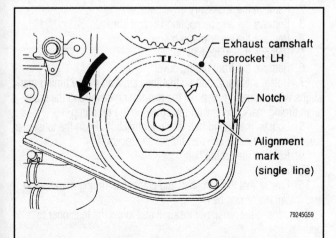

Rotate the left-hand exhaust camshaft until the single line mark is aligned with the notch in the belt cover . . .—Subaru 2.5L DOHC engine—Forester

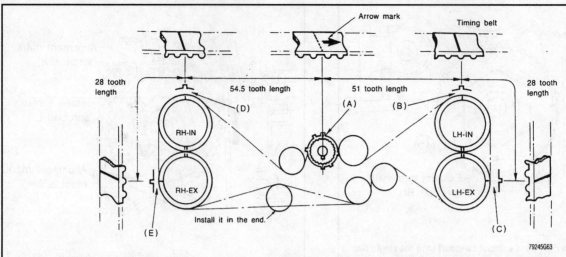

When installing the timing belt, be sure to route it in the proper order (a through e), and ensure that all of the matchmarks are properly aligned—Subaru 2.5L DOHC engine—Forester

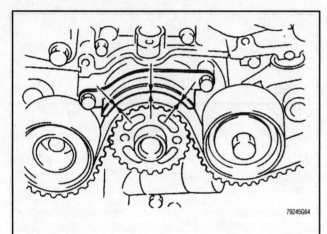

If equipped with a manual transmission, ensure the timing belt-to-guide clearance (arrows) is correct before tightening the mounting bolts—Subaru 2.5L DOHC engine—Forester

    b. Route the belt down and under the left-hand, upper idler pulley, then up and around the left-hand intake camshaft sprocket, ensuring the camshaft sprocket mark is aligned with the mark on the belt.

    c. Route the belt down and around the left-hand exhaust camshaft sprocket, making sure the marks are properly aligned, then up and over the first lower idler pulley and down and around the second lower idler pulley.

    d. While holding the timing belt on the inner, left-hand, lower idler pulley, route the other side of the timing belt (from the crankshaft sprocket) down and under the right-hand upper idler pulley.

    e. Route the timing belt up and around the right-hand intake camshaft sprocket so that the belt and sprocket marks are aligned.

    f. Position the belt down and around the right-hand exhaust camshaft sprocket, ensuring the positioning marks are aligned.

15. Install the right-hand lower idler pulley so that the timing belt is routed over the top side of it.

➡**Once the belt is completely installed on all of the pulleys and sprockets, ensure that the positioning marks are still all aligned.**

16. After ensuring all of the marks are still aligned, use a pair of pliers to withdraw the stopper pin or Allen wrench from the adjuster assembly housing.

17. If equipped with a manual transmission, perform the following:

    a. Install the timing belt guide by temporarily tightening the mounting bolts.

    b. Position the timing belt guide so that there is 0.019–0.059 in. (0.5–1.5mm) clearance between the timing belt and the belt guide.

    c. Tighten the guide mounting bolts securely, then double check the guide clearance.

18. Install the timing belt covers and all remaining engine components.

19. Connect the negative battery cable.

Suzuki

1.3L & 1.6L 8-VALVE ENGINES

1. Disconnect the negative battery cable.
2. Remove the accessory drive belts.
3. Remove the engine cooling fan and shroud.
4. Remove the water pump pulley.
5. Remove the crankshaft pulley.
6. Remove the front timing cover.
7. Rotate the crankshaft so that the camshaft pulley timing mark aligns with the timing mark on the rear timing cover and the crankshaft timing mark aligns with the arrow on the oil pump.
8. Loosen the timing belt tensioner lockbolt. Push the tensioner away from the timing belt and tighten the lockbolt.
9. Remove the timing belt.

To install:

10. Ensure that the camshaft and crankshaft timing marks align and install the timing belt.
11. Loosen the tensioner lockbolt and allow the tensioner to operate.
12. Rotate the crankshaft 2 complete turns and check that the timing marks align.
13. Tighten the tensioner lockbolt.
14. Install the front timing cover.

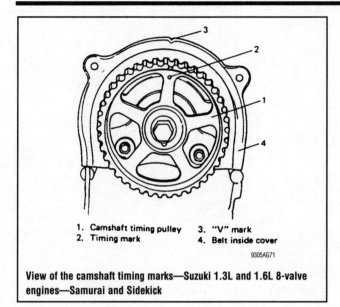

1. Camshaft timing pulley   3. "V" mark
2. Timing mark   4. Belt inside cover

9305AG71

View of the camshaft timing marks—Suzuki 1.3L and 1.6L 8-valve engines—Samurai and Sidekick

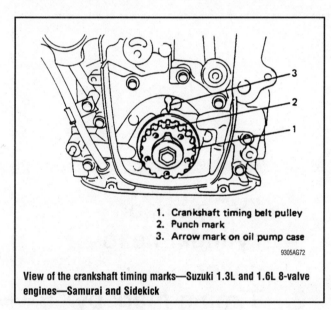

1. Crankshaft timing belt pulley
2. Punch mark
3. Arrow mark on oil pump case

9305AG72

View of the crankshaft timing marks—Suzuki 1.3L and 1.6L 8-valve engines—Samurai and Sidekick

15. Install the crankshaft pulley and tighten the bolts to 80–106 inch lbs. (9–12 Nm).
16. Install the water pump pulley.
17. Install the cooling fan and shroud.
18. Install the accessory drive belts.
19. Connect the negative battery cable.

1.6L 16-VALVE ENGINE

1. Disconnect the negative battery cable.

✳✳ WARNING

Do not rotate the crankshaft counterclockwise or attempt to rotate the crankshaft by turning the camshaft sprocket.

2. Remove the timing belt cover.
3. If the timing belt is not already marked with a directional arrow, use white paint, a grease pencil or correction fluid to do so.
4. Rotate the crankshaft clockwise until the timing mark on the camshaft sprocket and the "V" mark on the timing belt inside cover

are aligned, and the punch mark on the crankshaft sprocket is aligned with the mark on the engine.

✳✳ WARNING

Do not rotate the crankshaft or camshaft once the timing belt is removed.

5. Disconnect one end of the tensioner spring. Loosen the timing belt tensioner bolt and stud, then, using your finger, press the tensioner plate up and remove the timing belt from the crankshaft and camshaft sprockets.
6. Remove the timing belt tensioner, tensioner plate and spring from the engine.
7. Install Suzuki tool 09917-68220 or equivalent, onto the camshaft sprocket to hold the camshaft from rotating. Loosen the camshaft sprocket retaining bolt, then pull the camshaft sprocket off of the end of the camshaft.
8. Remove the crankshaft timing belt sprocket by loosening the center bolt, while preventing the crankshaft from rotating. To hold the crankshaft from turning, use Suzuki tool 09927-56010 or equivalent, or a large prybar inserted in the transmission housing slot and the flywheel teeth. Pull the sprocket off of the end of the crankshaft. Be sure to retain the crankshaft sprocket key and belt guide for assembly.
9. If necessary, remove the timing belt inside cover from the cylinder head.

To install:
10. If necessary, install the timing belt inside cover.
11. Slide the timing belt guide on the crankshaft so that the concave side faces the oil pump, then install the sprocket key in the groove in the crankshaft.
12. Slide the pulley onto the crankshaft and install the center retaining bolt. Tighten the center bolt to 80 ft. lbs. (110 Nm). To hold the crankshaft from turning, use Suzuki tool 09927-56010 or equivalent, or a large prybar inserted in the transmission housing slot and the flywheel teeth.
13. Install the timing belt camshaft sprocket, ensuring that the slot in the sprocket engages the camshaft (pulley) pin; this ensures that the sprocket is properly positioned on the end of the camshaft. Secure the camshaft with the holding tool used during removal, then tighten the sprocket bolt to 44 ft. lbs. (60 Nm).
14. Assemble the timing belt tensioner plate and the tensioner, making sure that the lug of the tensioner plate engages the tensioner.

✳✳ WARNING

If any binding is felt when adjusting the timing belt tension by turning the crankshaft, STOP turning the engine, because the pistons may be hitting the valves.

15. Install the timing belt tensioner, tensioner plate and spring on the engine. Tighten the mounting bolt and stud only finger-tight at this time. Ensure that when the tensioner is moved in a counterclockwise direction, the tensioner moves in the same direction. If the tensioner does not move, remove it and the tensioner plate to reassemble them properly.
16. Loosen all rocker arm valve lash locknuts and adjusting screws. This will permit movement of the camshaft without any rocker arm associated drag, which is essential for proper timing belt tensioning. If the camshaft does not rotate freely (free of rocker arm drag), the belt will not be properly tensioned.

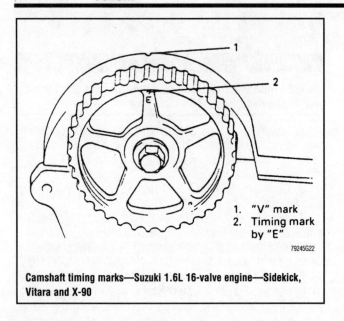

1. "V" mark
2. Timing mark by "E"

79245G22

Camshaft timing marks—Suzuki 1.6L 16-valve engine—Sidekick, Vitara and X-90

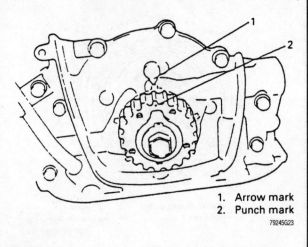

1. Arrow mark
2. Punch mark

79245G23

Align the punch mark with the arrow for proper timing belt installation—Suzuki 1.6L 16-valve engine—Sidekick, Vitara and X-90

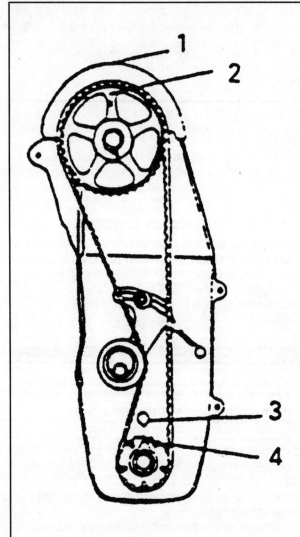

1. "V" mark on cylinder head cover
2. Timing mark by "E" on camshaft timing belt pulley
3. Arrow mark on oil pump case
4. Punch mark on crankshaft timing belt pulley

79245G47

Rotate the crankshaft clockwise until the camshaft and crankshaft timing marks are aligned—Suzuki 1.6L 16-valve engine—Samurai, Sidekick, Vitara and X-90

17. Rotate the camshaft sprocket clockwise until the timing mark on the sprocket and the "V" mark on the timing belt inside cover are aligned.

18. Using a wrench or socket and breaker bar, on the crankshaft sprocket center bolt, turn the crankshaft clockwise until the punch mark on the sprocket is aligned with the arrow mark on the oil pump.

19. With the camshaft and crankshaft marks properly aligned, push the tensioner up with your finger and install the timing belt on the 2 sprockets, ensuring that the drive side of the belt is free of all slack. Release your finger from the tensioner. Be sure to install the timing belt so that the directional arrow is pointing in the appropriate direction.

➡**In this position, the No. 4 cylinder is at Top Dead Center (TDC) on the compression stroke.**

20. Rotate the crankshaft clockwise 2 full revolutions, then tighten the tensioner stud to 97 inch lbs. (11 Nm). Then, tighten the tensioner bolt to 18 ft. lbs. (24 Nm).

21. Ensure that all 4 timing marks are still aligned as before; if they are not, remove the timing belt and install and tension it again.

22. Install the timing belt cover and all related components.

23. Connect the negative battery cable.

Toyota

2.0L (3S-FE) ENGINE

Toyota RAV4

The timing belt is not adjustable.

1. Disconnect the negative battery cable.

2. Disconnect the power steering reservoir tank and remove the reservoir bracket.

3. Detach the wiring harness bracket for the Data Link Connector 1 (DLC1).

4. Remove the alternator and alternator bracket.

5. If equipped with ABS brakes, remove the ABS actuator.

6. Remove the right front wheel and the fender apron seal.

7. Remove the power steering drive belt.

8. Slightly raise the engine using a block of wood and floor jack under the oil pan to prevent damage.

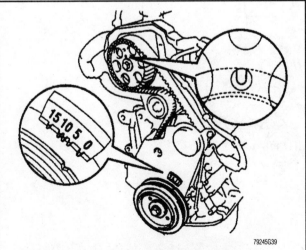

It is necessary to align the timing reference indicators prior to removing the timing belt—Toyota 2.0L (3S-FE) engine—RAV4

9. Remove the 4 bolts, 2 nuts and right-hand mounting bracket.

10. Remove the spark plugs.

11. Using SST 09213-54015, or equivalent, loosen the crankshaft pulley bolt and remove it by pulling it straight off the crankshaft.

12. Using SST 09249-63010, or equivalent, loosen the retaining bolts and remove the right engine mounting bracket.

13. Remove the upper (No. 2) timing belt cover.

14. Install the crankshaft pulley to the crankshaft and temporarily install the retaining bolt.

15. Turn the crankshaft pulley and align its groove with the timing mark **0** of the No. 1 timing belt cover. Check that the hole of the camshaft timing pulley is aligned with the timing mark of the bearing cap. If not, turn the crankshaft 360 degrees and align the marks.

➡**If the timing belt is to be reused, matchmark the timing belt to the timing pulleys and timing belt covers so the belt can be reinstalled in its original position. Also, be sure to mark an arrow on the belt to indicate which direction it was turning.**

16. Remove the timing belt from the camshaft timing pulley.

17. Hold the camshaft sprocket with a spanner wrench and remove the mounting bolt. Remove the camshaft pulley.

18. Remove the crankshaft pulley bolt and remove the crankshaft pulley.

19. Remove the No. 1 timing belt cover.

20. Remove the timing belt guide and the timing belt.

21. Remove the No. 1 idler pulley and tension spring.

22. Remove the No. 2 idler pulley.

23. Remove the crankshaft timing pulley.

24. Support the oil pump sprocket with a spanner wrench, then remove the mounting bolt and remove the sprocket.

To install:

25. Install the oil pump pulley. Tighten the nut to 18 ft. lbs. (24 Nm).

26. Install the crankshaft timing pulley. Align the pulley set key with the key groove of the pulley. Slide on the pulley facing the flange side inward.

27. Install the No. 2 idler pulley and tighten the mounting bolt to 31 ft. lbs. (42 Nm). Be sure that the pulley moves smoothly.

28. Install the No. 1 idler pulley with the bolt and the tension spring. Pry the pulley toward the left as far as it will go and tighten the bolt. Make sure that the pulley moves smoothly.

29. Temporarily install the timing belt. Using the crankshaft pulley bolt, turn the crankshaft and position the key groove of the crankshaft timing pulley upward. If reusing the timing belt, align the points marked during removal.

30. Install the timing belt on the crankshaft timing pulley, oil pump pulley, No. 1 idler pulley, water pump pulley and the No. 2 idler pulley.

31. Install the timing belt guide.

➡**If the old timing belt is being reinstalled, be sure the directional arrow is facing in the original direction and that the belt and sprocket/cover matchmarks are properly aligned.**

32. Install the lower (No. 1) timing belt cover and new gasket with the 4 bolts.

33. Align the crankshaft pulley set key with the pulley key groove. Temporarily install the crankshaft pulley and bolt.

34. Align the camshaft knock pin with the groove of the pulley, and slide the timing pulley onto the camshaft with the plate washer and set bolt.

35. Tighten the pulley set bolt to 40 ft. lbs. (54 Nm).

✳ WARNING

If any binding is felt when adjusting the timing belt tension by turning the crankshaft, STOP turning the engine, because the pistons may be hitting the valves.

36. Turn the crankshaft pulley and align the **0** mark on the lower (No. 1) timing belt cover.

37. Finish installing the timing belt and check the valve timing, as follows:

a. If reusing the old timing belt, align the matchmarks made previously and install the timing belt onto the camshaft pulley.

b. Align the marks on the timing belt with the marks on the camshaft pulley.

c. Loosen the No. 1 idler pulley set bolt ½ turn.

d. Turn the crankshaft pulley 2 complete revolutions TDC-to-TDC. ALWAYS turn the crankshaft CLOCKWISE. Check that the pulleys are still in alignment with the timing marks.

e. If the No. 1 idler pulley uses a green tension spring, slowly turn the crankshaft pulley 1⅞ revolutions, and align its groove with the mark at 45 degrees BTDC (for the No. 1 cylinder) of the No. 1 timing belt cover.

f. Tighten the No. 1 idler pulley set bolt to 31 ft. lbs. (42 Nm).

g. Be sure there is belt tension between the crankshaft and camshaft timing pulleys.

38. Place the right-hand engine mounting bracket in position but do not install the bolts.

39. Install the upper (No. 2) timing cover with a new gasket(s).

40. Remove the engine crankshaft pulley bolt and pulley.

41. Using SST 09249-63010 or equivalent, install the mounting

bolts for the right-hand mounting bracket. Tighten the mounting bolts to 38 ft. lbs. (52 Nm).

42. Align the crankshaft pulley set key with the pulley key groove. Install the pulley. Tighten the pulley bolt to 80 ft. lbs. (108 Nm).

43. Install the spark plugs.

44. Install the right-hand mounting insulator, as follows:

a. Attach the mounting insulator to the body and mounting bracket with the 4 bolts and 2 nuts.

b. Tighten the 3 bolts to hold the mounting insulator to the body. Tighten the bolts to 47 ft. lbs. (64 Nm).

c. Tighten the 2 nuts and bolt to hold the mounting insulator to the mounting bracket. Tighten the bolt to 27 ft. lbs. (37 Nm) and the nut to 38 ft. lbs. (52 Nm).

45. Install and adjust the power steering pump drive belt.

46. Install the right-hand engine undercover.

47. Install the right front wheel.

48. Lower the engine.

49. If equipped, install the ABS actuator.

50. Install the alternator and alternator bracket.

51. Install the wiring harness bracket for the DLC1.

52. Install the power steering reservoir bracket and reservoir.

53. Connect the negative battery cable.

2.2L (1L) & 2.4L (2L, 2L-T) ENGINES

Pick-Up

1. Disconnect the negative battery cables.

2. Drain the cooling system and remove the radiator hoses.

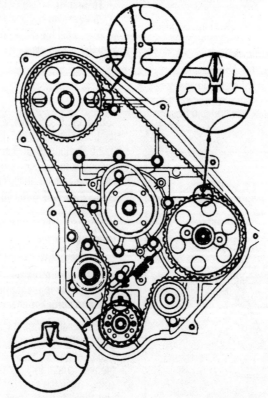

Alignment of the timing marks—Toyota 2.2L (1L) and 2.4L (2L, 2L-T) diesel engines—Pick-Up

9305AG74

3. Remove the accessory drive belts.
4. Remove the engine cooling fan and shroud.
5. Remove the water pump pulley.
6. If equipped with air conditioning, unbolt the compressor and position aside.
7. Remove the crankshaft pulley.
8. Remove the timing cover.
9. Rotate the crankshaft so that the camshaft, injection pump and crankshaft timing marks align with the timing marks on the engine and rear timing cover.
10. Loosen the timing belt tensioner locknut and pivot bolt.
11. Pry the tensioner pulley away from the timing belt and tighten the locknut.
12. Remove the timing belt.
To install:
13. Ensure that the camshaft and crankshaft timing marks are aligned.
14. Retard the injection pump timing mark by 1 tooth to allow for proper belt tensioning.
15. Install the timing belt.
16. Loosen the tensioner locknut.
17. Rotate the crankshaft 2 complete turns and check that the camshaft and crankshaft timing marks align.
18. Check that the injection pump timing marks are now aligned and tighten the tensioner locknut and pivot bolt.
19. Install the front timing cover.
20. Install the crankshaft pulley and tighten the bolt to 102 ft. lbs. (139 Nm).
21. Install the air conditioning compressor, if removed.
22. Install the water pump pulley.
23. Install the engine cooling fan and shroud.
24. Install the radiator hoses.
25. Connect the negative battery cables and refill the cooling system.

3.0L (3VZ-E) ENGINE

4Runner and Pick-Up

1989–92

1. Disconnect the negative battery cable.
2. Drain the cooling system.
3. Remove the radiator and fan shroud.
4. Remove the accessory drive belts.
5. Unbolt the power steering pump and position aside.
6. Remove the spark plugs.
7. Disconnect the No. 2 and No. 3 air hoses at the air pipe.
8. Disconnect the No. 1 water bypass hose at the air pipe and remove the water outlet.
9. Remove the upper timing cover.
10. Remove the cooling fan and the fan pulley bracket.
11. Rotate the crankshaft to align the timing marks on the camshaft pulleys with the timing marks on the rear timing cover.
12. Remove the crankshaft pulley and the lower timing cover.
13. Loosen the timing belt tensioner lockbolt.
14. Push the tensioner pulley away from the timing belt and tighten the lockbolt.
15. Remove the timing belt.
To install:
16. Hold the camshafts from turning and remove the camshaft pulley bolts.
17. Remove the camshaft pulley match pins and check that the pulleys rotate on the camshafts freely.
18. Reinstall the camshaft pulley bolts but do not tighten them. The pulleys must be able to rotate without binding to achieve proper timing belt tension. Make sure the bolts do not touch the pulleys.
19. Align the camshaft pulley matchmarks and check that the crankshaft timing pulley mark is aligned with the timing mark in the oil pump.

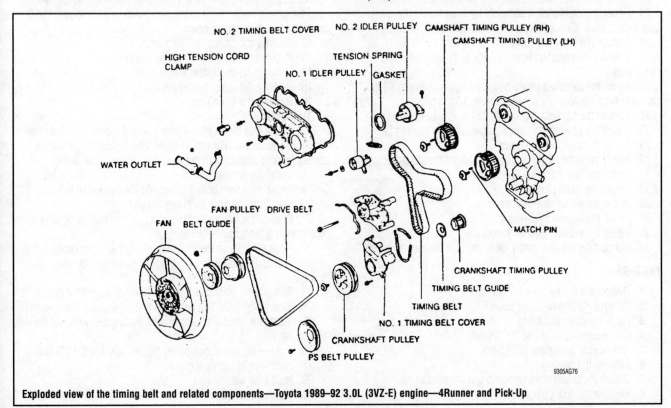

Exploded view of the timing belt and related components—Toyota 1989–92 3.0L (3VZ-E) engine—4Runner and Pick-Up

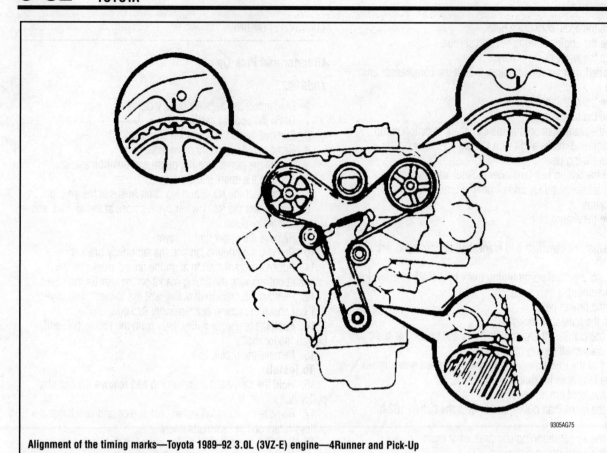

Alignment of the timing marks—Toyota 1989–92 3.0L (3VZ-E) engine—4Runner and Pick-Up

9305AG75

20. Install the timing belt.

21. Loosen the tensioner lockbolt and allow the spring pressure to tension the timing belt.

22. Install the lower timing cover and the crankshaft pulley.

23. Rotate the crankshaft 2 complete turns and align the crankshaft timing mark. Check that the camshaft timing marks align.

24. Tighten the tensioner lockbolt to 27 ft. lbs. (37 Nm).

25. Remove the camshaft timing pulley bolts and install the match pins.

26. Install the camshaft pulley bolts and tighten them to 80 ft. lbs. (109 Nm). Remove the camshaft pulley match pins.

27. Tighten the crankshaft pulley to 181 ft. lbs. (245 Nm).

28. Install the cooling fan pulley bracket and the cooling fan.

29. Install the upper timing cover.

30. Install the water outlet and the No. 1 water bypass hose.

31. Install the No. 2 and No. 3 air hoses.

32. Install the spark plugs.

33. Install the power steering pump.

34. Install the accessory drive belts.

35. Install the radiator and fan shroud.

36. Connect the negative battery cable and refill the cooling system.

1993–95

1. Disconnect the negative battery cable.

2. Remove the engine under cover.

3. Drain the cooling system.

4. Remove the radiator and fan shroud.

5. Remove the accessory drive belts.

6. Remove the cooling fan.

7. Unbolt the power steering pump and position aside.

8. Remove the spark plugs.

9. Disconnect the No. 2 and No. 3 air hoses at the air pipe.

10. Remove the upper timing cover.

11. Rotate the crankshaft to align the timing marks on the camshaft pulleys with the timing marks on the rear timing cover, and the timing mark on the crankshaft pulley is aligned with the **0** mark on the lower timing cover.

12. Remove the timing belt tensioner.

13. Remove the cooling fan pulley bracket.

14. Remove the crankshaft pulley.

15. Remove the lower timing cover.

16. Remove the timing belt.

To install:

17. Check that the camshaft pulley timing marks align with the timing marks on the rear timing cover, and the crankshaft timing pulley mark is aligned with the timing mark on the oil pump.

18. Install the timing belt.

19. Install the lower timing cover and the crankshaft pulley.

20. Install the cooling fan pulley bracket.

21. Prepare the timing belt tensioner for installation by slowly compressing the tensioner in a vise.

22. Align the hole in the tensioner plunger with the holes in the tensioner housing and insert a 1.5mm Allen wrench to retain the plunger.

23. Remove the tensioner from the vise and install the dust boot.

24. Install the timing belt tensioner and remove the Allen wrench.

25. Rotate the crankshaft 2 complete turns and check that the timing marks align.

26. Tighten the crankshaft pulley bolt to 181 ft. lbs. (245 Nm).

27. Install the upper timing cover.

28. Install the water outlet.

29. Install the No. 2 and No. 3 air hoses.

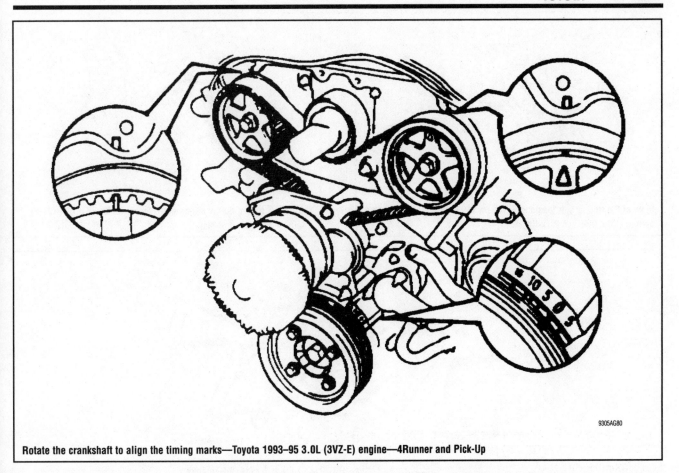

9305AG80

Rotate the crankshaft to align the timing marks—Toyota 1993–95 3.0L (3VZ-E) engine—4Runner and Pick-Up

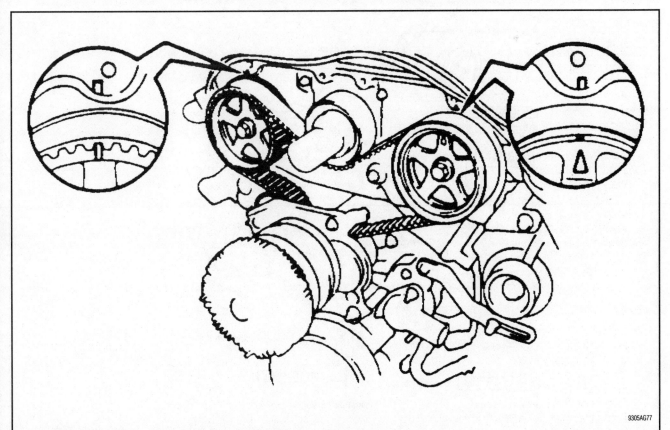

9305AG77

View of the camshaft timing marks—Toyota 1993–95 3.0L (3VZ-E) engine—4Runner and Pick-Up

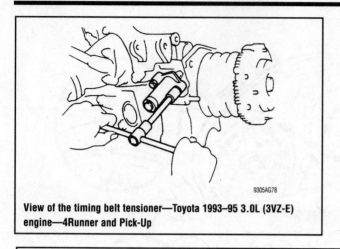

View of the timing belt tensioner—Toyota 1993–95 3.0L (3VZ-E) engine—4Runner and Pick-Up

9305AG78

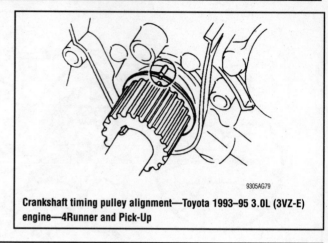

Crankshaft timing pulley alignment—Toyota 1993–95 3.0L (3VZ-E) engine—4Runner and Pick-Up

9305AG79

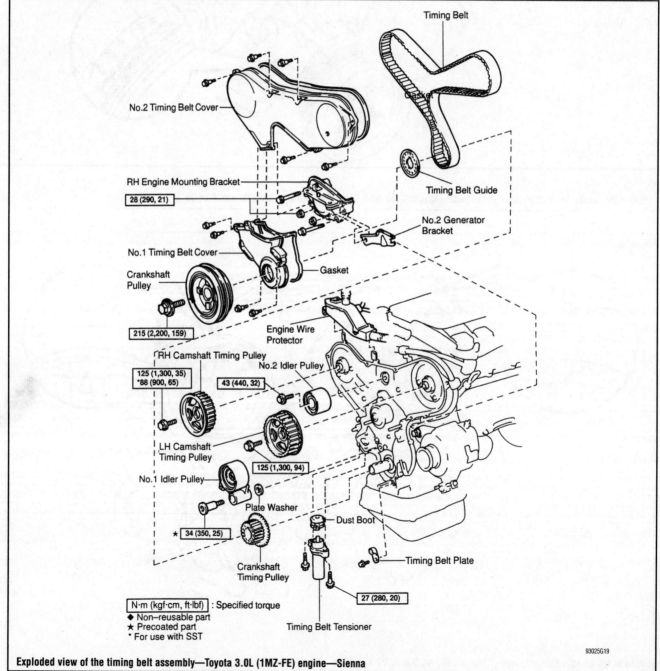

Timing Belt

No.2 Timing Belt Cover

RH Engine Mounting Bracket

28 (290, 21)

Timing Belt Guide

No.2 Generator Bracket

No.1 Timing Belt Cover

Gasket

Crankshaft Pulley

215 (2,200, 159)

Engine Wire Protector

RH Camshaft Timing Pulley

125 (1,300, 35)
*88 (900, 65)

No.2 Idler Pulley

43 (440, 32)

LH Camshaft Timing Pulley

125 (1,300, 94)

No.1 Idler Pulley

Plate Washer

★ 34 (350, 25)

Dust Boot

Crankshaft Timing Pulley

Timing Belt Plate

27 (280, 20)

Timing Belt Tensioner

N·m (kgf·cm, ft·lbf) : Specified torque
◆ Non–reusable part
★ Precoated part
* For use with SST

93025G19

Exploded view of the timing belt assembly—Toyota 3.0L (1MZ-FE) engine—Sienna

30. Install the spark plugs.
31. Install the power steering pump.
32. Install the accessory drive belts.
33. Install the cooling fan.
34. Install the radiator and fan shroud.
35. Replace the engine under cover.
36. Connect the negative battery cable and fill the cooling system.

3.0L (1MZ-FE) ENGINE

Sienna

1. Disconnect the negative battery cable.
2. Remove the outer front cowl top panel assembly by performing the following procedure:
   a. Remove the wiper arm/blade assemblies head caps, nuts and assemblies.
   b. Remove the head-to-cowl seal and the cowl panel hole cover.
   c. Disconnect the windshield washer clip and hose.
   d. Remove both (right and left) cowl top ventilator louvers.
   e. Disconnect the electrical connector from the windshield wiper motor.
   f. Remove the outer front cowl top panel assembly-to-cowl bolts and the panel.
3. Raise and safely support the vehicle.
4. Remove the right front wheel assembly and apron seal.
5. Remove the alternator by performing the following procedure:
   a. Loosen the pivot bolt, adjusting lockbolt and adjusting bolt, then, remove the drive belt.
   b. Disconnect the alternator's electrical connector.
   c. Remove the nut and the alternator wire.

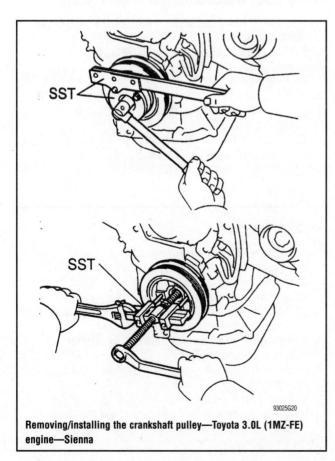

Removing/installing the crankshaft pulley—Toyota 3.0L (1MZ-FE) engine—Sienna

d. Disconnect the wiring harness from the clip.
e. Remove the alternator-to-bracket pivot bolt, the washer, adjusting lockbolt and alternator.
6. Loosen the power steering pump's mount and adjusting bolt, then, remove the drive belt.
7. Disconnect the hose from the engine coolant reservoir.
8. Disconnect the Diagnostic Link Connector 1 (DLC1) from the No. 2 right side engine mounting bracket.
9. Remove the right side engine mounting stay, the engine moving control rod and the No. 2 right side engine mounting bracket.

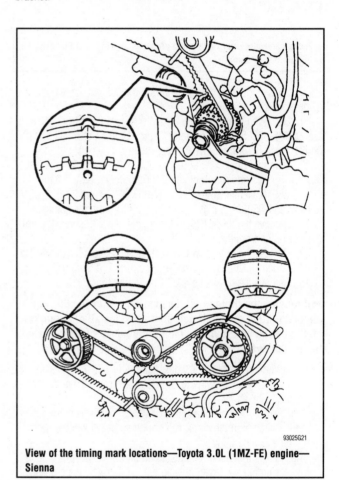

View of the timing mark locations—Toyota 3.0L (1MZ-FE) engine—Sienna

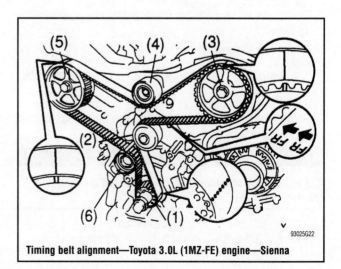

Timing belt alignment—Toyota 3.0L (1MZ-FE) engine—Sienna

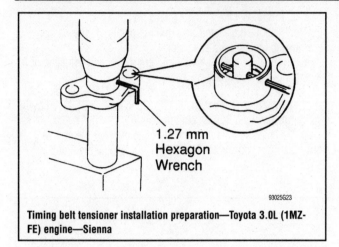

1.27 mm
Hexagon
Wrench

93025G23

Timing belt tensioner installation preparation—Toyota 3.0L (1MZ-FE) engine—Sienna

10. Loosen the alternator's pivot bolt, the nut and the No. 2 alternator bracket.

11. Using the Crankshaft Pulley Holding tool 09213-54015, Bolt tool 91651-60855 and Companion Flange Holding tool 09330-00021 or equivalent, remove the crankshaft pulley bolt.

12. Using a Puller "C" Set 09950-50011 (Hanger 150 tool 09951-05010, Slide Arm tool 09952-5010, Center Bolt 100 tool 09953-05010, Center Bolt 150 tool 09953-05020 and 2 No. 2 Claw tools 09954-05020), pull the crankshaft pulley from the crankshaft.

13. Remove the lower (No. 1) timing belt cover. Remove the timing belt guide from the crankshaft.

14. Remove the engine wire protector clamps from the upper (No. 2) timing belt cover and remove the upper (No. 2) timing belt cover.

15. Remove the right side engine mounting brace.

➡ **If reusing the timing belt, be sure that you can still read the installation marks. If not, place new installation marks on the timing belt to match the timing marks of the camshaft timing pulleys.**

16. Temporarily install the crankshaft pulley bolt.

17. Set the No. 1 cylinder to Top Dead Center (TDC) of the compression stroke, as follows:

 a. Rotate the crankshaft (CLOCKWISE) to align the timing marks: dimple on the crankshaft timing sprocket with the notch on the oil pump body.

 b. Check that the timing marks on the camshaft sprockets and the rear timing belt cover are aligned; if not, rotate the crankshaft 360 degrees (1 revolution) and align the marks.

18. Remove the timing belt tensioner and the timing belt.

To install:

19. Inspect the timing belt tensioner by performing the following procedures:

 a. Inspect the seal for leakage; if leakage is suspected, replace the tensioner.

 b. Using both hands to hold the tensioner facing upward, strongly press the pushrod against a solid surface. If the pushrod moves, replace the tensioner.

⁑ WARNING

Never hold the tensioner with the pushrod facing downward.

 c. Measure the pushrod protrusion from the housing end, it should be 0.394–0.425 in. (10.0–10.8mm). If the protrusion is not as specified, replace the tensioner.

20. Set the No. 1 cylinder to Top Dead Center (TDC) of the compression stroke, as follows:

 a. Rotate the crankshaft (CLOCKWISE) to align the timing marks: dimple on the crankshaft timing sprocket with the notch on the oil pump body.

 b. Check that the timing marks on the camshaft sprockets and the rear timing belt cover are aligned; if not, rotate the crankshaft 1 revolution (360 degrees) and align the marks.

21. Install the timing belt in the following order:

 a. Crankshaft timing sprocket

 b. Water pump pulley

 c. Left camshaft timing sprocket

 d. No. 2 idler pulley

 e. Right camshaft sprocket

 f. No. 1 idler pulley

22. Using a vertical press, slowly press the pushrod into the housing using 200–2205 lbs. (981–9807 N) until the holes align, then, install a 1.27mm Allen® wrench to secure the pushrod and release the press. Install the dust boot on the tensioner housing.

23. Install the timing belt tensioner and torque the bolts to 20 ft. lbs. (27 Nm).

24. Remove the Allen® wrench from the tensioner housing.

25. Slowly, rotate the crankshaft (CLOCKWISE) 2 complete revolutions and realign the timing marks. If the timing marks do not align, remove the timing belt and reinstall it.

26. Remove the crankshaft pulley bolt.

27. Install the right side engine mounting bracket and torque the bolts to 21 ft. lbs. (28 Nm).

28. Clean and install the upper (No. 2) timing belt cover.

➡ **If the gasket material on the timing belt covers is cracked, peeling or etc., replace it.**

29. Install the timing belt guide on the crankshaft with the cup side facing outward.

30. Clean and install the lower (No. 1) timing belt cover.

31. Install the crankshaft pulley.

32. Using the Crankshaft Pulley Holding tool 09213-54015, Bolt tool 91651-60855 and Companion Flange Holding tool 09330-00021 or equivalent, install the crankshaft pulley bolt and torque the bolt to 159 ft. lbs. (215 Nm).

33. To complete the installation, reverse the removal procedures.

34. Connect the negative battery cable.

35. Start the engine and check for leaks.

3.4L (5VZ-FE) ENGINE

T-100 and Tacoma

1. Disconnect the negative battery cable.

2. Raise and safely support the vehicle.

3. Remove the engine undercover.

4. Drain the engine coolant.

5. Disconnect the upper radiator hose from the engine.

6. Remove the power steering drive belt.

7. Remove the air conditioning drive belt by loosening the idler pulley nut and the adjusting bolt.

8. Loosen the lockbolt, pivot bolt, and the adjusting bolt and the alternator drive belt.

9. Remove the No. 2 fan shroud by removing the 2 clips.

10. Remove the fan with the fluid coupling and fan pulleys.

11. Disconnect the power steering pump from the engine and set aside. Do not disconnect the lines from the pump.

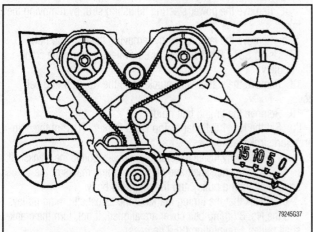

Turn the crankshaft clockwise to align the timing marks before removing the timing belt—Toyota 3.4L (5VZ-FE) engine—T-100 and Tacoma

79245G37

12. If equipped with air conditioning, disconnect the compressor from the engine and set aside. Do not disconnect the lines from the compressor.

13. If equipped with air conditioning, disconnect the air conditioning bracket.

14. Remove the No. 2 timing belt cover, as follows:

a. Detach the camshaft position sensor connector from the No. 2 timing belt cover.

b. Disconnect the 3 spark plug wire clamps from the No. 2 timing belt cover.

c. Remove the 6 bolts and remove the timing belt cover.

15. Remove the fan bracket, as follows:

a. Remove the power steering adjusting strut by removing the nut.

b. Remove the fan bracket by removing the bolt and nut.

16. Set the No. 1 cylinder at Top Dead Center (TDC) of the compression stroke, as follows:

a. Turn the crankshaft pulley and align its groove with the timing mark **0** of the No. 1 timing belt cover.

b. Check that the timing marks of the camshaft timing pulleys and the No. 3 timing belt cover are aligned. If not, turn the crankshaft pulley 1 revolution (360 degrees).

➡️**If reusing the timing belt, be sure that you can still read the installation marks. If not, place new installation marks on the timing belt to match the timing marks of the camshaft timing pulleys.**

17. Remove the timing belt tensioner by alternately loosening the 2 bolts.

18. Remove the camshaft timing pulleys, as follows:

a. Using SST 09960-10010, or equivalent, remove the pulley bolt, the timing pulley and the knock pin. Remove the 2 timing pulleys with the timing belt.

19. Remove the crankshaft pulley, as follows:

a. Using SST 09213-54015 and 09330-00021 or their equivalents, loosen the pulley bolt.

b. Remove the SST tool, the pulley bolt and the pulley.

20. Remove the starter wire bracket and the No. 1 timing belt cover.

21. Remove the timing belt guide and remove the timing belt.

22. Remove the bolt and the No. 2 idler pulley.

23. Remove the pivot bolt, the No. 1 idler pulley and the plate washer.

24. Remove the crankshaft gear.

To install:

25. Install the crankshaft timing gear.

a. Align the timing pulley set key with the key groove of the gear.

b. Using SST 09214-60010 or equivalent, and a hammer, tap in the timing gear with the flange side facing inward.

26. Install the plate washer and the No. 1 idler pulley with the pivot bolt and tighten it to 26 ft. lbs. (35 Nm). Check that the pulley bracket moves smoothly.

27. Install the No. 2 timing belt idler with the bolt. Tighten the bolt to 30 ft. lbs. (40 Nm). Check that the pulley bracket moves smoothly.

28. Temporarily install the timing belt, as follows:

a. Using the crankshaft pulley bolt, turn the crankshaft and align the timing marks of the crankshaft timing pulley and the oil pump body.

b. Align the installation mark on the timing belt with the dot mark of the crankshaft timing pulley.

c. Install the timing belt on the crankshaft timing pulley, No. 1 idler pulley and the water pump pulleys.

29. Install the timing belt guide with the cup side facing outward.

30. Install the No. 1 timing belt cover and starter wire bracket. Tighten the timing belt cover bolts to 80 inch lbs. (9 Nm).

❊❊ WARNING

If any binding is felt when adjusting the timing belt tension by turning the crankshaft, STOP turning the engine, because the pistons may be hitting the valves.

31. Install the crankshaft pulley, as follows:

a. Align the pulley set key with the key groove of the crankshaft pulley.

b. Install the pulley bolt and tighten it to 184 ft. lbs. (250 Nm).

32. Install the left camshaft timing pulley.

a. Install the knock pin to the camshaft.

b. Align the knock pin hose of the camshaft with the knock pin groove of the timing pulley.

c. Slide the timing belt pulley on the camshaft with the flange side facing outward. Tighten the pulley bolt to 81 ft. lbs. (110 Nm).

33. Set the No. 1 cylinder to TDC of the compression stroke, as follows:

a. Turn the crankshaft pulley, and align its groove with the timing mark **0** of the No. 1 timing belt cover.

b. Turn the camshaft to align the knock pin hole of the camshaft with the timing mark of the No. 3 timing belt cover.

c. Turn the camshaft timing pulley, and align the timing marks of the camshaft timing pulley and the No. 3 timing belt cover.

34. Connect the timing belt to the left camshaft timing pulley, as follows:

➡️**Check that the installation mark on the timing belt is aligned with the end of the No. 1 timing belt cover.**

a. Using SST 09960-01000, or equivalent, slightly turn the left camshaft timing pulley clockwise. Align the installation mark on the timing belt with the timing mark of the camshaft timing pulley and hang the timing belt on the left camshaft timing pulley.

b. Align the timing marks of the left camshaft pulley and the No. 3 timing belt cover.

c. Check that the timing belt has tension between the crankshaft timing pulley and the left camshaft timing pulley.

35. Install the right camshaft timing pulley and the timing belt, as follows:

a. Align the installation mark on the timing belt with the timing mark of the right camshaft timing pulley and hang the timing belt on the right camshaft timing pulley with the flange side facing inward.

b. Slide the right camshaft timing pulley on the camshaft. Align the timing marks on the right camshaft timing pulley and the No. 3 timing belt cover.

c. Align the knock pin hole of the camshaft with the knock pin groove of the pulley and install the knock pin. Install the bolt and tighten it to 81 ft. lbs. (110 Nm).

36. Set the timing belt tensioner, as follows:

a. Using a press, slowly press in the pushrod using 220–2205 lbs. (981–9807 N) of force.

b. Align the holes of the pushrod and housing, pass a 1.5mm hex wrench through the holes to keep the setting position of the pushrod.

c. Release the press and install the dust boot on the tensioner.

37. Install the timing belt tensioner and alternately tighten the bolts to 20 ft. lbs. (28 Nm). Using pliers, remove the 1.5mm hex wrench from the belt tensioner.

38. Check the valve timing, as follows:

a. Slowly turn the crankshaft pulley 2 revolutions from TDC-to-TDC. Always turn the crankshaft pulley clockwise.

b. Check that each pulley aligns with the timing marks. If the timing marks do not align, remove the timing belt and reinstall it.

39. Install the fan bracket with the bolt and nut.

40. Install the power steering adjusting strut with the nut.

41. Install the No. 2 timing belt cover. Tighten the bolts to 80 inch lbs. (9 Nm). Install the remaining components.

42. Fill the cooling system with coolant.

43. Connect the negative battery cable.

44. Start the engine and check for leaks.

4Runner

1. Disconnect the negative battery cable.
2. Raise and safely support the vehicle.
3. Remove the engine undercover.
4. Drain the engine coolant.
5. Disconnect the upper radiator hose from the engine.
6. Remove the power steering drive belt.
7. Remove the air conditioning drive belt by loosening the idler pulley nut and the adjusting bolt.
8. If equipped with air conditioning, disconnect the compressor from the engine and set aside. Do not disconnect the lines from the compressor.
9. If equipped with air conditioning, disconnect the air conditioning bracket.
10. Remove the fan with the fluid coupling and fan pulleys.
11. Loosen the lockbolt, pivot bolt, and the adjusting bolt and the alternator drive belt.
12. Remove the No. 2 fan shroud by removing the 2 clips.
13. Disconnect the power steering pump from the engine and set aside. Do not disconnect the lines from the pump.
14. Remove the oil dipstick and the guide.
15. Remove the No. 2 timing belt cover as follows:

a. Detach the camshaft position sensor connector from the No. 2 timing belt cover.

b. Disconnect the 4 spark plug wire clamps from the No. 2 timing belt cover.

c. Remove the 6 bolts and remove the timing belt cover.

16. Remove the fan bracket as follows:

a. Remove the power steering adjusting strut by removing the nut.

b. Remove the fan bracket by removing the bolt and nut.

17. Using SST 09213-54015, or equivalent, remove the crankshaft pulley.

18. Remove the starter wire bracket and the No. 1 timing belt cover.

19. Remove the timing belt guide.

20. Set the No. 1 cylinder at Top Dead Center (TDC) of the compression stroke, as follows:

a. Temporarily install the crankshaft pulley bolt to the crankshaft.

b. Turn the crankshaft and align the timing marks of the crankshaft timing pulley and the oil pump body.

c. Check that the timing marks of the camshaft timing pulleys and the No. 3 timing belt cover are aligned. If not, turn the crankshaft pulley 1 revolution (360 degrees).

➡**If reusing the timing belt, be sure that you can still read the installation marks. If not, place new installation marks on the timing belt to match the timing marks of the camshaft timing pulleys.**

21. Remove the timing belt tensioner by alternately loosening the 2 bolts.

22. Remove the right and left camshaft pulleys.

23. Remove the No. 2 idler pulley.

24. Using a 10mm hex wrench, remove the pivot bolt, No.1 idler pulley and the plate washer.

25. Remove the timing belt guide and remove the timing belt.

26. Remove the crankshaft timing pulley.

To install:

27. Install the crankshaft timing belt pulley, as follows:

a. Align the timing belt pulley set key with the key groove of the timing pulley and slide on the timing pulley.

b. Slide on the timing belt pulley with the flange side facing inward.

28. Install the plate washer and the No. 1 idler pulley with the pivot bolt and tighten it to 26 ft. lbs. (35 Nm). Check that the pulley bracket moves smoothly.

29. Install the No. 2 timing belt idler with the bolt. Tighten the bolt to 30 ft. lbs. (40 Nm). Check that the pulley bracket moves smoothly.

30. Install the left and right camshaft timing pulleys.

31. Set the No. 1 cylinder to TDC of the compression stroke, as follows:

a. Using the crankshaft pulley bolt, turn the crankshaft and align the timing marks of the crankshaft timing pulley and the oil pump body.

b. Using SST 09960-10010 or equivalent, to turn the camshaft pulley to align the marks of the camshaft timing belt pulley and the No. 3 timing belt cover.

32. Install the timing belt, as follows:

➡**The engine should be cold.**

a. Face the front mark on the timing belt forward.

b. Align the installation mark on the timing belt with the timing mark of the crankshaft timing pulley.

c. Align the installation marks on the timing belt with the timing marks of the camshaft pulleys.

33. Install the timing belt in the following order:

- Left camshaft pulley
- No. 2 idler pulley
- Right camshaft pulley

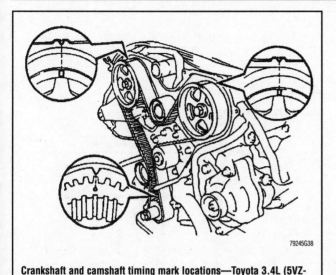

Crankshaft and camshaft timing mark locations—Toyota 3.4L (5VZ-FE) engine—4Runner

- Water pump pulley
- Crankshaft pulley
- No. 1 idler pulley

✳✳ WARNING

If any binding is felt when adjusting the timing belt tension by turning the crankshaft, STOP turning the engine, because the pistons may be hitting the valves.

34. Set the timing belt tensioner as follows:
    a. Using a press, slowly press in the pushrod using 220–2205 lbs. (981–9807 N) of force.
    b. Align the holes of the pushrod and housing, pass a 1.27mm wrench through the holes to keep the setting position of the pushrod.
    c. Release the press and install the dust boot to the tensioner.
35. Install the timing belt tensioner and alternately tighten the bolts to 20 ft. lbs. (27 Nm). Using pliers, remove the 1.27mm wrench from the belt tensioner.
36. Check the valve timing, as follows:
    a. Slowly turn the crankshaft and align the timing marks of the crankshaft timing pulley and the oil pump body. Always turn the crankshaft pulley clockwise.
    b. Check that the timing marks of the right and left timing pulleys align with the timing marks of the No. 3 timing belt cover. If the marks do not align, remove the timing belt and reinstall it.
37. Install the timing belt guide with the cup side facing outward.
38. Install the No. 1 timing belt cover and starter wire bracket. Tighten the timing belt cover fasteners to 80 inch lbs. (9 Nm).
39. Install the crankshaft pulley, as follows:
    a. Align the pulley set key with the key groove of the pulley and slide the pulley.
    b. Using SST 09213-54014, or equivalent, tighten the bolt to 184 ft. lbs. (250 Nm).
40. Install the fan bracket with the bolt and nut.
41. Install the No. 2 timing belt cover, and tighten the bolts to 80 inch lbs. (9 Nm). Install the remaining components.
42. Fill the cooling system with coolant.
43. Connect the negative battery cable.
44. Start the engine and check for leaks.
45. Check the ignition timing.

4.7L (2UZ-FE) ENGINE

Land Cruiser

1. Disconnect the negative battery cable.
2. Raise and safely support the vehicle.
3. Remove the oil pan protector and the engine under cover.
4. Drain the cooling system and store the coolant for refilling purposes.
5. Lower the vehicle and remove the battery clamp cover.
6. From the top of the engine, remove the fuel return hose, the engine cover nuts/bolts and the cover.
7. Remove the air cleaner and the intake air connector assembly.
8. Remove the cooling fan pulley by performing the following procedures:
   a. Loosen the 4 fan clutch-to-fan pulley nuts.
   b. Using a box-end wrench on the serpentine drive belt tensioner bolt, rotate the tensioner counterclockwise and remove the drive belt.

➡ **The serpentine drive belt tensioner bolt is a left-hand thread.**

   c. Remove the fan clutch-to-fan pulley nuts, the fan, the clutch assembly and the fan pulley.
9. Remove the radiator by performing the following procedures:
   a. Disconnect the upper, lower and reservoir hoses from the radiator.
   b. Disconnect and plug the automatic transmission oil cooler at the radiator. Disconnect the automatic transmission oil cooler hoses from the fan shroud clamp.
   c. Remove the radiator reservoir tank.
   d. Remove the fan shroud-to-radiator bolts and the shroud.
   e. Remove the 2 upper radiator-to-chassis nuts.
   f. Remove the middle radiator-to-chassis nut/bolts and brackets.
   g. Carefully, lift the radiator from the vehicle.
10. Remove the serpentine drive belt idler pulley bolt, cover plate and pulley.
11. Remove the right side (No. 3) timing belt cover.
12. Remove the left side (No. 3) timing belt cover by performing the following procedures:
    a. Disconnect the engine wire from both wire clamps.
    b. Disconnect the camshaft position sensor wire from the wire clamp on the left-side (No. 3) timing belt cover.
    c. Disconnect the sensor connector from the connector bracket.
    d. Disconnect the sensor connector.
    e. Remove the wire grommet from the left-side (No. 3) timing belt cover.
    f. Remove the oil cooler tube bolts and tube.
13. Remove the middle (No. 2) timing belt cover bolts and cover.
14. Remove the cooling fan bracket nuts/bolts and bracket.

➡ **If reusing the timing belt, make sure that there are 3 installation marks on the belt; if there are none, install them.**

15. Using the Crankshaft Pulley Holding tool 09213-70010, Bolt tool 90105-08076 and Companion Flange Holding tool 09330-00021, or equivalent, loosen the crankshaft pulley bolt.
16. Position the No. 1 cylinder to approximately 50 degrees After Top Dead Center (ATDC) of the compression stroke by performing the following procedures:
    a. Rotate the crankshaft pulley (CLOCKWISE) to align its groove with the timing mark "0" on the lower (No. 1) timing belt cover.
    b. Check that the camshaft sprocket timing marks are aligned with the rear timing belt plate marks; if not, rotate the crankshaft 1 revolution (360 degrees).

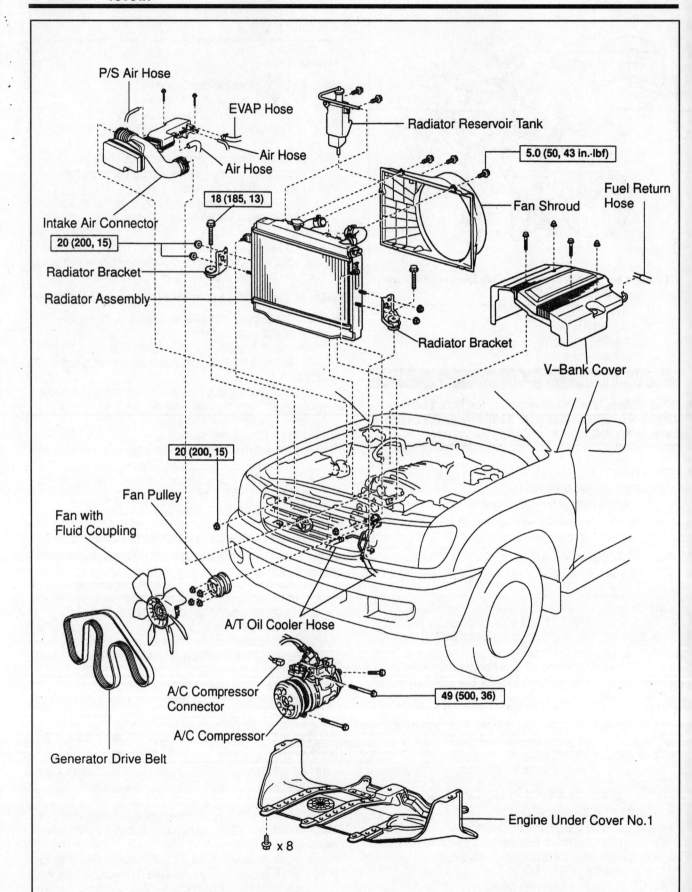

P/S Air Hose

EVAP Hose

Radiator Reservoir Tank

Air Hose

Air Hose

5.0 (50, 43 in.·lbf)

Fan Shroud

Fuel Return Hose

Intake Air Connector

18 (185, 13)

20 (200, 15)

Radiator Bracket

Radiator Assembly

Radiator Bracket

V–Bank Cover

20 (200, 15)

Fan Pulley

Fan with Fluid Coupling

A/T Oil Cooler Hose

A/C Compressor Connector

49 (500, 36)

A/C Compressor

Generator Drive Belt

Engine Under Cover No.1

x 8

Exploded view of vehicle components for timing belt replacement—Toyota 4.7L (2UZ-FE) engine—Land Cruiser

93025G24

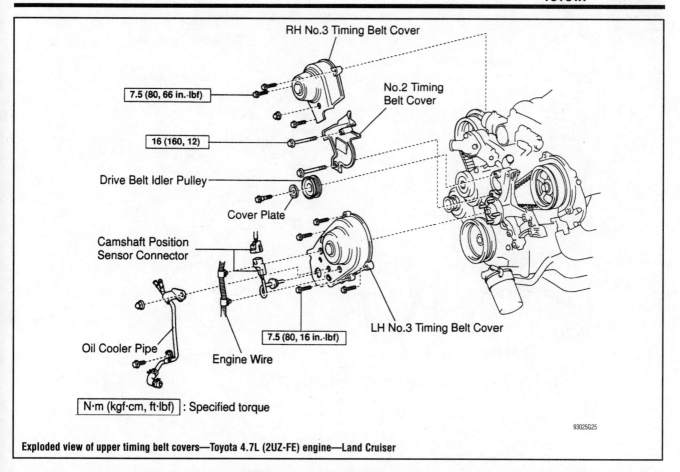

RH No.3 Timing Belt Cover

No.2 Timing Belt Cover

7.5 (80, 66 in.·lbf)

16 (160, 12)

Drive Belt Idler Pulley

Cover Plate

Camshaft Position Sensor Connector

LH No.3 Timing Belt Cover

7.5 (80, 16 in.·lbf)

Oil Cooler Pipe

Engine Wire

N·m (kgf·cm, ft·lbf) : Specified torque

93025G25

Exploded view of upper timing belt covers—Toyota 4.7L (2UZ-FE) engine—Land Cruiser

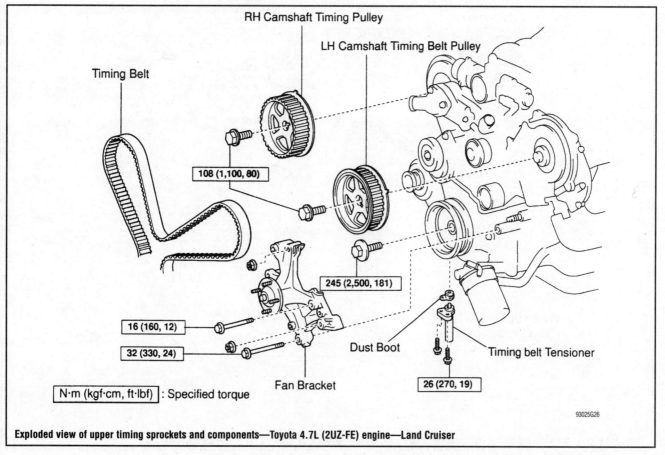

RH Camshaft Timing Pulley

LH Camshaft Timing Belt Pulley

Timing Belt

108 (1,100, 80)

245 (2,500, 181)

16 (160, 12)

32 (330, 24)

Dust Boot

Timing belt Tensioner

Fan Bracket

26 (270, 19)

N·m (kgf·cm, ft·lbf) : Specified torque

93025G26

Exploded view of upper timing sprockets and components—Toyota 4.7L (2UZ-FE) engine—Land Cruiser

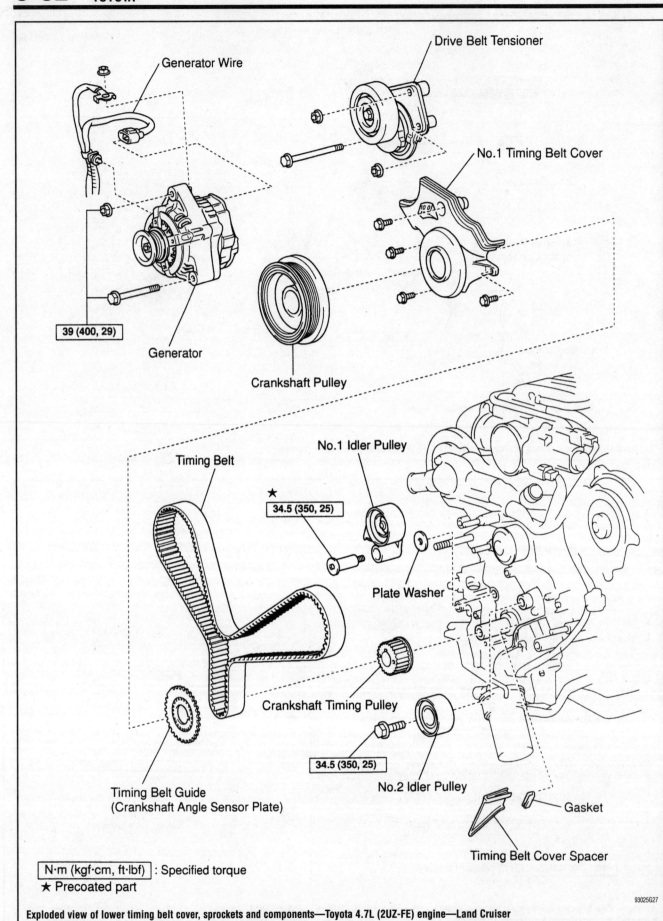

Generator Wire

Drive Belt Tensioner

No.1 Timing Belt Cover

39 (400, 29)

Generator

Crankshaft Pulley

Timing Belt

No.1 Idler Pulley

★ 34.5 (350, 25)

Plate Washer

Crankshaft Timing Pulley

Timing Belt Guide
(Crankshaft Angle Sensor Plate)

34.5 (350, 25)

No.2 Idler Pulley

Gasket

Timing Belt Cover Spacer

N·m (kgf·cm, ft·lbf) : Specified torque
★ Precoated part

Exploded view of lower timing belt cover, sprockets and components—Toyota 4.7L (2UZ-FE) engine—Land Cruiser

93025G27

c. Rotate the crankshaft pulley approximately 50 degrees (CLOCKWISE) and align the crankshaft pulley timing mark between the centers of the crankshaft pulley bolt and the idler pulley bolt.

※※ WARNING

If the timing belt is disengaged, having the crankshaft pulley in the wrong angle can cause the valve to come into contact with the piston when removing the camshaft pulley.

17. Remove the crankshaft pulley bolt.

➡️If reusing the timing belt and the installation marks have disappeared, place new installation marks on the timing belt to match the camshaft timing sprocket marks.

➡️To avoid meshing the timing sprocket and the timing belt, secure one with a string; then, place matchmarks on the timing belt and the right-side camshaft timing sprocket.

18. Remove the timing belt tensioner bolts and the tensioner.
19. Using the Camshaft Holding tool 09960-10010, or equivalent, slightly turn the left-side camshaft sprocket clockwise to loosen the tension spring. Then, disconnect the timing belt from the camshaft sprockets.
20. Remove the alternator by performing the following procedures:
    a. Disconnect the electrical connector from the alternator.
    b. Remove the rubber cap/nut and disconnect the battery wire from the alternator.
    c. Disconnect the wire clamp from the alternator cord clip.
    d. Remove the alternator-to-engine nuts/bolts and the alternator.
21. Remove the serpentine drive belt tensioner nuts/bolts and the tensioner.
22. Using the Crankshaft Puller Assembly tool 09950-50012 or equivalent, press the crankshaft pulley from the crankshaft.

※※ WARNING

DO NOT rotate the crankshaft pulley.

23. Remove the lower (No. 1) timing belt cover bolts and the cover.
24. Remove the timing belt guide, spacer and the timing belt.
To install:

➡️With the timing belt removed, this is a perfect opportunity to inspect and/or replace the water pump.

25. Inspect the timing belt tensioner by performing the following procedures:
    a. Inspect the seal for leakage; if leakage is suspected, replace the tensioner.
    b. Using both hands to hold the tensioner facing upward, strongly press the pushrod against a solid surface. If the pushrod moves, replace the tensioner.

※※ WARNING

Never hold the tensioner with the pushrod facing downward.

    c. Measure the pushrod protrusion from the housing end, it should be 0.413–0.453 in. (10.5–11.5mm). If the protrusion is not as specified, replace the tensioner.
26. Temporarily install the timing belt by performing the following procedures:
    a. Align the timing belt's installation mark with the crankshaft timing sprocket.

    b. Install the timing belt on the crankshaft timing sprocket, the No. 1 idler pulley and the No. 2 idler pulley.
27. Install the gasket to the timing belt cover spacer and install the cover spacer.
28. Install the timing belt guide with the cup side facing outward.
29. Install the lower (No. 1) timing belt cover.
30. Install the crankshaft pulley by performing the following procedures:

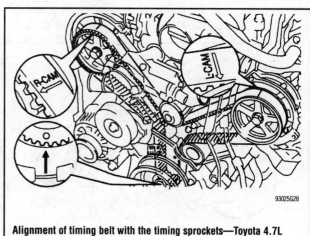

Alignment of timing belt with the timing sprockets—Toyota 4.7L (2UZ-FE) engine—Land Cruiser

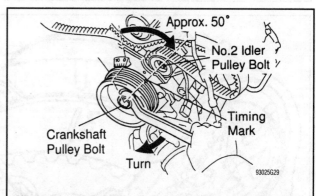

Aligning of crankshaft pulley timing mark with the center line of the crankshaft pulley bolt and the idler pulley bolt—Toyota 4.7L (2UZ-FE) engine—Land Cruiser

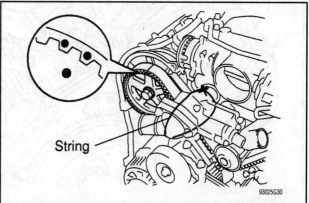

Secure the timing belt with string and matchmark the camshaft with the timing belt—Toyota 4.7L (2UZ-FE) engine—Land Cruiser

a. Align the crankshaft pulley with the crankshaft key.

b. Using the Crankshaft Installer tool 09223-46011 or equivalent, and a hammer, tap the crankshaft pulley into position.

31. Install the serpentine drive belt tensioner and torque the tensioner-to-engine bolts to 12 ft. lbs. (16 Nm).

➡**To install the serpentine drive belt tensioner, use a bolt 4.18 in. (106mm) in length.**

32. Check that the crankshaft pulley's timing mark is aligned with the centers of the idler pulley and crankshaft pulley bolts.

33. Install the alternator and torque the alternator-to-engine

nuts/bolts to 29 ft. lbs. (39 Nm). Connect the alternator's electrical connectors and clip.

34. Install the timing belt to the left-side camshaft by performing the following procedures:

a. Rotate the left-side camshaft pulley to align the timing belt installation mark with the camshaft sprocket's timing mark and slide the belt onto the camshaft timing sprocket.

b. Using the Camshaft Holding tool 09960-10010, or equivalent, slightly turn the left-side camshaft sprocket counterclock-

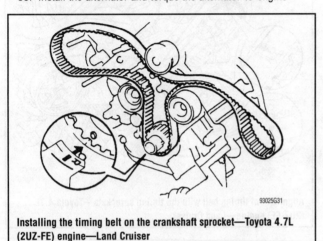

Installing the timing belt on the crankshaft sprocket—Toyota 4.7L (2UZ-FE) engine—Land Cruiser

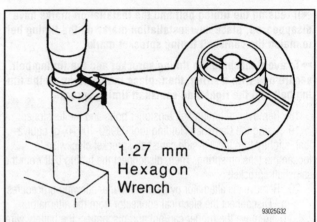

Securing the timing belt tensioner pushrod—Toyota 4.7L (2UZ-FE) engine—Land Cruiser

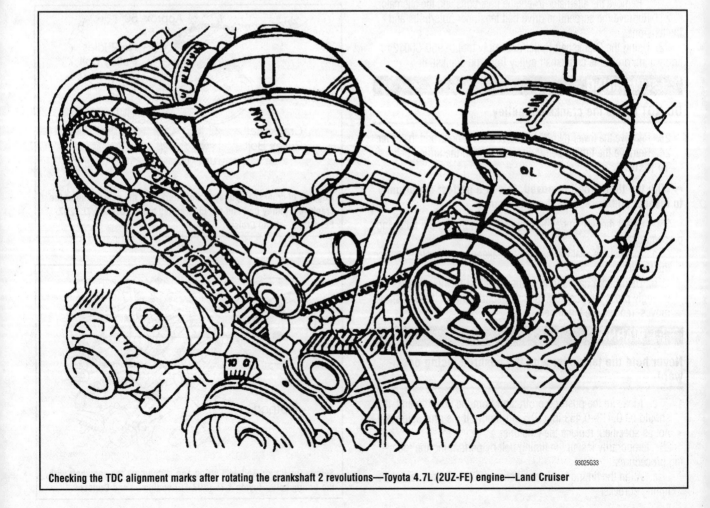

Checking the TDC alignment marks after rotating the crankshaft 2 revolutions—Toyota 4.7L (2UZ-FE) engine—Land Cruiser

wise to place tension on the timing belt between the crankshaft sprocket and the camshaft sprocket.

35. Rotate the right-side camshaft pulley to align the timing belt installation mark with the camshaft sprocket's timing mark and slide the belt onto the camshaft timing sprocket.

36. Using a vertical press, slowly press the pushrod into the housing using 200–2205 lbs. (981–9807 N) until the holes align, then, install a 1.27mm Allen® wrench to secure the pushrod and release the press. Install the dust boot on the tensioner housing.

37. Install the timing belt tensioner and torque the bolts to 19 ft. lbs. (26 Nm).

38. Using a pair of pliers, remove the Allen® wrench from the tensioner housing.

39. Check the valve timing by performing the following procedure:

a. Temporarily install the crankshaft pulley bolt.

b. Slowly, rotate the crankshaft pulley 2 revolutions (CLOCKWISE) and realign the TDC marks.

➡**If the pulley/sprocket timing marks do not realign, remove the timing belt and reinstall it.**

40. Using the Crankshaft Pulley Holding tool 09213-70010, Bolt tool 90105-08076 and Companion Flange Holding tool 09330-00021, or equivalent, torque the crankshaft pulley bolt to 181 ft. lbs. (245 Nm).

41. Install the cooling fan bracket and torque the 12mm (head size) bolt to 12 ft. lbs. (16 Nm) and the 14mm (head size) bolt to 24 ft. lbs. (32 Nm).

42. Install the air conditioning compressor.

43. Install the middle (No. 2) timing belt cover and torque the bolts to 12 ft. lbs. (16 Nm).

44. Install the upper right-side (No. 3) timing belt cover and torque the bolts to 66 inch lbs. (7.5 Nm).

45. Install the upper left-side (No. 3) timing belt cover by performing the following procedures:

a. Install the oil cooler tube and bolt.

b. Feed the Camshaft Position Sensor (CPS) through the left-side (No. 3) timing belt cover hole.

c. Install the left-side (No. 3) timing belt cover and torque the bolts to 66 inch lbs. (7.5 Nm).

d. Install the wire grommet to the left-side (No. 3) timing belt cover.

e. Install the sensor connector to the connector bracket and connect the sensor connector.

f. Install the sensor wire and the engine wire to the clamps on the left-side (No. 3) timing belt cover.

46. Install the drive belt idler pulley and cover plate; then, torque the pulley bolt to 27 ft. lbs. (37 Nm).

47. To complete the installation, reverse the removal procedures.

48. Refill the cooling system and connect the negative battery cable.

Volkswagen

1.6L DIESEL ENGINE

1982–84

Some special tools are required. A flat bar, VW tool 2065A, or equivalent, is used to secure the camshaft in position. A pin, VW tool 2064, or equivalent, is used to fix the pump position while the timing belt is removed. The camshaft and pump work against spring

pressure and will move out of position when the timing belt is removed. It is not difficult to find substitutes but do not remove the timing belt without these tools.

➡**The timing belt is designed to last for more than 60,000 miles and does not normally require tension adjustments. If the belt is removed or replaced, the basic valve timing must be checked and the belt retensioned.**

1. Disconnect the negative battery cable.
2. Loosen the alternator mounting bolts.
3. Pivot the alternator and slip the drive belt off the pulleys.
4. Unscrew the timing cover retaining nuts and remove the cover.
5. Remove the rocker arm cover.
6. Rotate the crankshaft so that the No. 1 cylinder is at Top Dead Center (TDC) of its compression stroke.
7. Using Volkswagen tool 2065A or equivalent, retain the camshaft in this position by performing the following procedure:

a. Turn the camshaft until one end of the tool touches the cylinder head.

b. Measure the gap at the other end of the tool with a feeler gauge.

c. Take ½ of the measurement and insert a feeler gauge of this thickness between the tool and the cylinder head. Turn the camshaft so that the tool rests on the feeler gauge.

d. Insert a second feeler gauge of the same thickness between the other end of the tool and the cylinder head.

8. Using Volkswagen tool 2064 or equivalent, lock the injection pump sprocket in position.

9. Check that the marks on the sprocket, bracket and pump body are in alignment (engine at TDC).

10. Loosen the timing belt tensioner.

11. Remove the timing belt.

To install:

12. Check that the TDC mark on the flywheel is aligned with the reference marks.

13. Install the timing belt and remove pin 2064 from the injection pump sprocket.

14. Tension the belt by turning the tensioner to the right.

15. Adjust the timing belt tension by performing the following procedure:

a. Using Volkswagen Tensioner Gauge tool VW210 or equivalent, install it midway on the timing belt between the longest span between 2 pulleys.

b. Tension the timing belt by turning the tensioner to the right.

c. Proper tension is when the scale reads 12–13.

16. Tighten the timing belt tensioner bolt to 33 ft. lbs. (45 Nm).

17. Torque the crankshaft damper pulley bolt to 145 ft. lbs. (196 Nm).

18. Remove the tool from the camshaft.

19. Rotate the crankshaft 2 turns in the direction of engine rotation (clockwise) and then strike the belt once with a rubber mallet between the camshaft sprocket and the injection pump sprocket.

20. Check the belt tension again. Check the injection pump timing.

21. Install the timing belt cover.

22. Reposition the spacers on the studs and install the washers and nuts.

23. Install the alternator belt and adjust its tension.

24. Connect the negative battery cable.

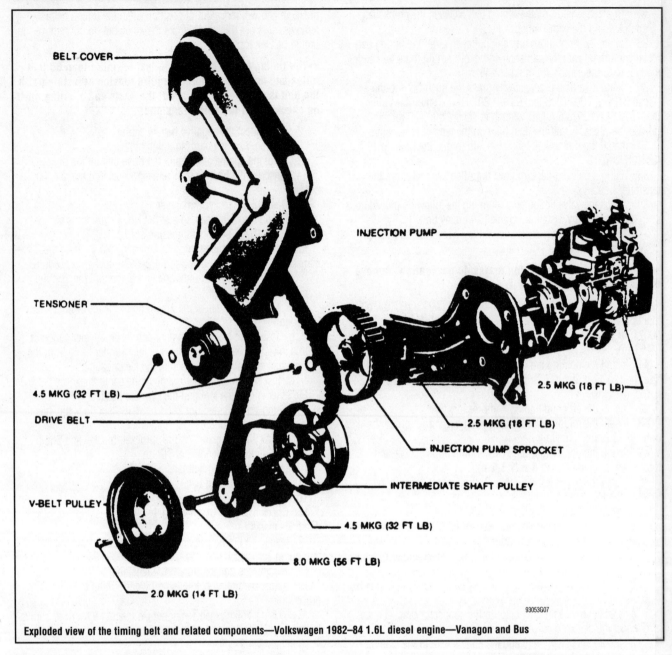

BELT COVER

INJECTION PUMP

TENSIONER

4.5 MKG (32 FT LB)

2.5 MKG (18 FT LB)

DRIVE BELT

2.5 MKG (18 FT LB)

INJECTION PUMP SPROCKET

INTERMEDIATE SHAFT PULLEY

V-BELT PULLEY

4.5 MKG (32 FT LB)

8.0 MKG (56 FT LB)

2.0 MKG (14 FT LB)

93053G07

Exploded view of the timing belt and related components—Volkswagen 1982–84 1.6L diesel engine—Vanagon and Bus

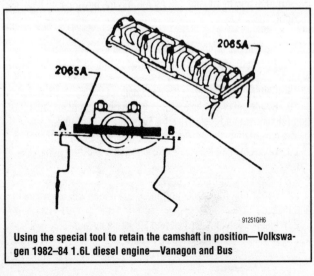

2065A

2065A

A   B

91251GH6

Using the special tool to retain the camshaft in position—Volkswagen 1982–84 1.6L diesel engine—Vanagon and Bus

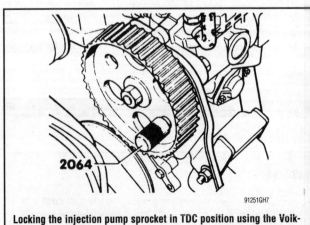

2064

91251GH7

Locking the injection pump sprocket in TDC position using the Volkswagen tool—Volkswagen 1982–84 1.6L diesel engine—Vanagon and Bus